# Percolation, Localization, and Superconductivity

# NATO ASI Series

## Advanced Science Institutes Series

*A series presenting the results of activities sponsored by the NATO Science Committee, which aims at the dissemination of advanced scientific and technological knowledge, with a view to strengthening links between scientific communities.*

The series is published by an international board of publishers in conjunction with the NATO Scientific Affairs Division

| | |
|---|---|
| A **Life Sciences**<br>B **Physics** | Plenum Publishing Corporation<br>New York and London |
| C **Mathematical and Physical Sciences** | D. Reidel Publishing Company<br>Fordrecht Boston, and Lancaster |
| D **Behavioral and Social Sciences**<br>E **Engineering and Materials Sciences** | Martinus Nijhoff Publishers<br>The Hague, Boston, and Lancaster |
| F **Computer and Systems Sciences**<br>G **Ecological Sciences** | Springer-Verlag<br>Berlin, Heidelberg, New York, and Tokyo |


***Recent Volumes in this Series***

*Volume 103*—Fundamental Processes in Energetic Atomic Collisions
edited by H. O. Lutz, J. S. Briggs, and H. Kleinpoppen

*Volume 104*—Short-Distance Phenomena in Nuclear Physics
edited by David H. Boal and Richard M. Woloshyn

*Volume 105*—Laser Applications in Chemistry
edited by K. L. Kompa and J. Wanner

*Volume 106*—Multicritical Phenomena
edited by R. Pynn and A. Skjeltorp

*Volume 107*—Positron Scattering in Gases
edited by John W. Humberston and M. R. C. McDowell

*Volume 108*—Polarons and Excitons in Polar Semiconductors and Ionic Crystals
edited by J. T. Devreese and F. Peeters

*Volume 109*—Percolation, Localization, and Superconductivity
edited by Allen M. Goldman and Stuart A. Wolf

***Series B: Physics***

# Percolation, Localization, and Superconductivity

Edited by

Allen M. Goldman

School of Physics and Astronomy
University of Minnesota
Minneapolis, Minnesota

and

Stuart A. Wolf

Metal Physics Branch
Naval Research Laboratory
Washington,D.C.

Plenum Press
New York and London
Published in cooperation with NATO Scientific Affairs Division

Proceedings of the NATO Advanced Study Institute on
Percolation, Localization, and Superconductivity,
held June 19–July 1, 1983,
at Les Arcs, Savoie, France

Percolation
Localization
Superconductivity

Library of Congress Cataloging in Publication Data

NATO Advanced Study Institute on Percolation, Localization, and Superconductivity (1983: Savoie, France)
Percolation, localization, and superconductivity.

(NATO ASI series. Series B, Physics; v. 109)
"Proceedings of the NATO Advanced Study Institute on Percolation, Localization, and Superconductivity, held June 19–July 1, 1983, at Les Arcs, Savoie, France."
Bibliography: p.
Includes index.
1. Solid state physics—Congresses. 2. Superconductivity—Congresses. 3. Percolation (Statistical physics)—Congresses. I. Goldman, Allen M. II. Wolf, Stuart A. III. Title. IV. Title: Localization, and superconductivity. V. Series: NATO advanced science institute series. Series B, Physics; v. 109.
QC176.A1N326 1983 530.4′1 84-8357

ISBN-13: 978-1-4615-9396-6 e-ISBN-13: 978-1-4615-9394-2
DOI: 10.1007/978-1-4615-9394-2


Softcover reprint of the hardcover 1st edition 1984

A Division of Plenum Publishing Corporation
233 Spring Street, New York, N.Y. 10013

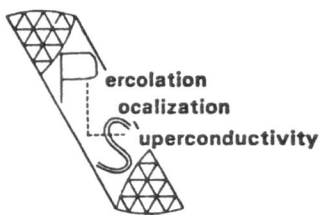

# PREFACE

The study of the effects of dimensionality and disorder on phase transitions, electronic transport, and superconductivity has become an important field of research in condensed matter physics. These effects are both classical and quantum mechanical in nature and are observed universally in "real" materials. What may at first glance seem a diverse collection of lectures which form the chapters of these proceedings is in fact, an attempt to demonstrate the commonality, inter-relationship, and general applicability of the phenomena of localization, percolation, and macroscopic quantum effects on electrical transport and superconductivity in disordered solids. The theory of these phenomena is presented in a complete, yet, self-contained fashion and the inter-relationship between the topics is emphasized. An extensive treatment of experimental results is also included, both those which have stimulated the theory as well as those that have confirmed it.

Many of the phenomena investigated in this field also have technological significance. For example, the nature of electronic localization in metals in which one or more dimensions are constrained is very important when one attempts to predict the behavior of the metallic interconnects in ultra-miniature circuits. The macroscopic quantum tunneling phenomenon which is closely associated with the concept of quantum noise may determine the ultimate sensitivity of Josephson devices to electromagnetic signals. Granular superconductors and arrays of Josephson devices which are studied here in the context of statistical mechanics and phase transitions may serve as generators or detectors of high-frequency electromagnetic radiation. A clear understanding of the nature of the metal-insulator or the superconductor-insulator transitions in random percolating systems may be important in the technological exploitation of the anomalous behavior at these critical points. Furthermore, the fabrication techniques used and in some instances developed in these contexts are currently applicable to digital superconducting electronics.

This book is based on lectures given at the NATO ASI entitled "Percolation, Localization, and Superconductivity" held from June 19th to July 1st, 1983 at Les Arcs in Savoie France. We have attempted to organize it in a generally coherent fashion, using one topic to build on the next. The topics are sufficiently well-established allowing for a pedalogical treatment, but certainly the "state-of-the-art" is rapidly evolving.

An important feature of the book are the chapters which present the current state of development of what M. Tinkham in his summary of the Institute called our "High-Tech Industrial Civilization". These review the fabrication techniques that have become available partly because of the semiconductor industry and the need for ultra-small digital circuitry. The remainder of the book is devoted to a series of experimental and theoretical topics in condensed matter physics relating to electronic transport and superconductivity in disordered and dimensionally constrained materials.

The book proceeds from macroscopic quantum tunneling (MQT) to small particles, percolation, localization, two-dimensional superconductivity, commensurate and incommensurate phases of a two-dimensional vortex lattice, and finally granular superconductivity.

The study of MQT is the newest subject in this grouping and is of fundamental significance for the quantum theory of measurement. Josephson junctions or Superconducting Quantum Interference Devices (SQUIDs) are ideal model systems for work in this field. Low temperatures are needed to suppress thermal activated processes. A complete understanding of the role of damping in these systems appears to require more experimental and theoretical work.

In the study of percolation in real superconducting systems experiments show a remarkable semiquantitative agreement with simple model predictions. Advances in the characterization of sample geometries should result in even more quantitative comparison of experiment and theory in the future.

The theory of weak localization bears a remarkable resemblance to the theory of superconductivity. It seems to describe experiments extremely well. The strong localization regime is not yet served by an elegant formal theory combining all effects.

Two-dimensional superconducting systems such as amorphous and granular thin films and weakly-coupled arrays appear to provide excellent model systems for the investigation of both the continuum and discrete realizations of the Kosterlitz-Thouless transition. Measurements of electrical resistance vs temperature above the transition yield the temperature-dependent free vortex density. Studies of the nonlinear voltage-current characteristic and the kinetic inductance yield the temperature-dependent superelectron density. Additional experiments, in which various parameters of the theory are determined independently, are needed to further clarify the issues in this area.

The study of granular superconductors and composites of small particles is a relatively old subject. It involves the understanding of systems of isolated grains as well as coupled systems. It is important in the context of the other subjects of this Institute because randomness, percolation, localization, and quantum tunneling are theoretical concepts recognized as being relevant to the understanding of granular materials.

Our intent in organizing an Advanced Study Institute was to make evident the opportunities in this field. We hope that the reader of these proceedings will agree with us that that field is a fertile area of research.

Allen M. Goldman and Stuart A. Wolf

# ACKNOWLEDGEMENTS

This ASI was organized on a rather short time scale and many individuals made important contributions to its ultimate success. Dr. R. Sinclair of the NATO Advanced Study Institutes Programme deserves special thanks for his patience in dealing with the neophyte organizers. A significant financial contribution to the travel of participants from the USA was made by the NSF-Low Temperature Physics Program under Grant NSF/DMR-8302327. S. Foner, K. Gray, and J. Ruvalds were very helpful in sharing with us their previous experience in organizing NATO ASI's. Bernard Croise and Philipe Fabing of the Les Arcs resort deserve special thanks for their efforts on behalf of the Institute. John Claassen, Tom Francavilla, Wendy Fuller, Don Gubser, Deborah Van Vechten, and Beverly Wood made important contributions to a number of activities including the preparation of these proceedings. Kay Reilly and Ina Rubenstein helped with financial and travel arrangements, respectively. The Naval Research Laboratry and the School of Physics and Astronomy of the University of Minnesota contributed materially to the preparation of these proceedings and to the organizational effort for the ASI. The organizers especially acknowledge the rather substantial efforts of their secretaries, Evelyn Backman of the University of Minnesota and Lahni Blohm of the Naval Research Laboratory, in making the Conference and these proceedings a success.

CONTENTS

# MACROSCOPIC QUANTUM TUNNELLING AND RELATED EFFECTS IN JOSEPHSON SYSTEMS

A.J. Leggett*

Laboratory of Atomic and Solid State Physics and Materials Science Center, Cornell University
Ithaca, NY 14853

These lectures discuss the possibility of using Josephson systems (current-biased junctions and SQUID rings) to examine the validity of the concept of superposition of states with macroscopically different properties. In particular, I examine the phenomena of "macroscopic quantum tunnelling" (the macroscopic analogue of the decay of a heavy nucleus by alpha-emission) and "macroscopic quantum coherence" (the analogue of the inversion resonance of the $NH_3$ molecule), paying special attention to the question of the influence of dissipation on these phenomena. The experimental situation with regard to tunnelling is briefly reviewed.

## LECTURE 1. The Quantum Mechanics of a Macroscopic Variable

### INTRODUCTION

In discussions of the quantum theory of measurement, a crucial question is whether the usual laws of quantum mechanics can be applied to macroscopic bodies, and in particular, whether it is legitimate to assume the occurrence in nature of linear superpositions of states with macroscopically different properties. In recent years, with the development of the technology of Josephson devices and of submillidegree cryogenics, it has become clear that this is not entirely a matter of "quantum theology" but can be tested, at least indirectly, by experiment. The execution and

---

*On leave of absence from University of Sussex, UK and University of Illinois

interpretation of such tests presents a stimulating challenge both to the experimentalist, who needs to use the most advanced microfabrication, cryogenic and noise isolation techniques, and to the theorist, who is faced with the necessity of describing the quantum mechanics of systems where dissipation plays a crucial role in circumstances far from the correspondence limit. Although other systems besides superconductors can in principle be used for such tests, in practice most experiments explicitly designed with this aim in mind have exploited the Josephson effect, and I shall confine myself in these lectures to this case. For general background I refer to Refs. 1-3 (Ref. 1 discusses the relevance of this topic to the quantum theory of measurement, while Ref. 3 concentrates specifically on the Josephson case); for further details of some of the calculations referred to here, see Ref. 2 and also the lectures[4] of V. Ambegaokar at this Institute. A review of some recent work in this area is given in Ref. 5. My general approach will be to assume that the linear laws of quantum mechanics do apply without modification to macroscopic bodies and to explore the consequences of this assumption. Naturally, if the experiments were to fail to show the predicted results, the assumption might have to be re-examined.

## BRIEF REVIEW OF SOME PROPERTIES OF SUPERCONDUCTING SYSTEMS

Consider the electrons in a superconducting system at zero temperature. The system may be of a quite general geometry - e.g., a single bulk superconducting block, a thin superconducting ring or two bulk superconductors connected by a Josephson junction - provided only that there is some mechanism of contact by which an electron in one part of the system can reach any other part. Then, quite generally, the wave function of the N-particle system which corresponds to the (particle non-conserving) BCS ansatz is of the general form[6]

$$\Psi(r_1 r_2 \ldots r_N : t) = A\,\chi(r_1 r_2 : t)\,\chi(r_3 r_4 : t) \ldots \chi(r_{N-1}\, r_N : t) \qquad (1)$$

where A is the anti-symmetrizing operator and where for notational simplicity we have suppressed the spin indices and assumed N to be even. Note that the "pseudo-molecular wave function" $\chi(r_1 r_2)$, which for spin singlet pairing is symmetric under interchange of its indices, is <u>the same</u> for all pairs of electrons: in some sense we have something like a Bose condensate of "diatomic molecules" (Cooper pairs). An alternative, and usually more convenient, description of a system possessing a wave function of this form is in terms of the so-called "anomalous average"

$$F(\underset{\sim}{r}_1 \underset{\sim}{r}_2 : t) \equiv \langle \psi_\uparrow^+ (r_1) \psi_\downarrow^+ (r_2) \rangle (t) \qquad (2)$$

which, although not identical to $\chi(r_1 r_2)$, is related to it by a simple integral equation.[6,7] It is convenient to introduce the relative and centre-of-mass coordinates $\underset{\sim}{R} \equiv (r_1+r_2)/2$, $\underset{\sim}{r} \equiv r_1-r_2$. The relative coordinate plays no role in our considerations and we shall not write it out explicitly in future; it is the dependence of the wave function on the centre-of-mass coordinate $\underline{R}$ of the Cooper pair which is essential. We denote a wave function of the form (1) schematically by the notation

$$\Psi(r_1 r_2 \ldots . r_N : t) = [\chi(\underset{\sim}{R}:t)]^{N/2} \tag{3}$$

thereby emphasizing the fact that all N/2 Cooper pairs are behaving identically.

In a homogeneous bulk, simply connected, superconductor energy considerations usually ensure that in most physically interesting situations the centre-of-mass wave function $\chi(\underline{R}) \equiv |\chi(\underline{R})|\, e^{i\phi(\underset{\sim}{R})}$ is nearly constant both in magnitude and in phase. In a superconducting ring more interesting behaviour is possible: e.g., we can have

$$\chi(\underset{\sim}{R}) \stackrel{\sim}{=} |\chi(\underset{\sim}{R})| \exp in\Theta \equiv \chi_n(\underset{\sim}{R}) \qquad (n = \text{integer}) \tag{4}$$

where, in most interesting cases, the modulus $|\chi(\underline{R})|$ is nearly constant. As is well known (see e.g., Ref. 8) the superfluid velocity $\underline{v}_s$, which for a given $|\chi(R)|$ is proportional to the electrical current $\underline{j}(\underline{r})$ carried by the Cooper pairs, is given by the expression

$$\underset{\sim}{v}_s(\underset{\sim}{R}) = \frac{\hbar}{2m}[\underset{\sim}{\nabla}\phi(\underset{\sim}{R}) - 2e\underset{\sim}{A}(\underset{\sim}{R})/\hbar] \tag{5}$$

where $\underline{A}(\underline{R})$ is the electromagnetic vector potential; thus, in state (4) it satisfies the condition of fluxoid quantization

$$\oint 2(m\underset{\sim}{v}_s(R) + e\underset{\sim}{A}(R))\cdot d\ell = 2\pi n\hbar \tag{6}$$

In particular if the thickness of the ring is large compared to the London penetration depth $\lambda_L$ and we take a circuit deep inside it, then the vanishing of the electrical current required by the Meissner effect, when combined with Eq. (6), yields the well-known result that the flux $\phi$ through the circuit is quantized:

$$\Phi = n\Phi_o\,, \qquad \Phi_o \equiv h/2e. \tag{7}$$

Consider next two isolated bulk superconductors connected by a Josephson junction. In this case the Cooper pair wave function can be a linear super position of an amplitude for being on the left of the junction and an amplitude for being on the right (we assume for simplicity that within each bulk superconductor the wave function is uniform in space):

$$\chi(\underset{\sim}{R},t) = a(t)\chi_L(\underline{R}) + b(t)\chi_R(\underline{R}) \tag{8}$$

and thus the N-particle wave function of the entire system is, schematically,

$$\Psi(r_1 r_2 \ldots r_N : t) = (a(t)\chi_L(\underline{R}) + b(t)\chi_R(\underline{R}))^{N/2} \tag{9}$$

This is the general type of wave function which is implicit in the classic calculations of the Josephson effect (cf., e.g., Ref. 9); note carefully that, in distinction to the cases to be discussed below, it does not correspond to a super position of states with macroscopically different properties.*

It is often helpful to introduce explicitly the "phase difference of the pairs on opposite sides of the junction" $\Delta\phi(t)$ by writing (8) in the form (apart from an irrelevant overall phase factor)

$$\chi(R,t) = |a(t)|\chi_L(R) + |b(t)|e^{i\Delta\phi(t)}\chi_R(R) \tag{8'}$$

We recall that in the standard theory of the Josephson effect[10] the energy of the system depends periodically on the quantity $\Delta\phi$, the simplest and most widely used form of the dependence being simply

$$U(\Delta\phi) = -\frac{I_c\phi_o}{2\pi}\cos\Delta\phi \tag{10}$$

where $I_c$ is the critical current of the junction. The dynamics of $\Delta\phi$ is given by the Josephson equation

$$\frac{d}{dt}(\Delta\phi) = 2eV/\hbar \equiv (2\pi/\phi_o)V \tag{11}$$

where V is the voltage (or more strictly the electrochemical potential) applied across the junction. By equating the change of the energy (10) to the reversible work done by the voltage, we find the familiar result[10] for the non-dissipative current (super-current through the junction when part of a circuit to be:

---

* This statement is perhaps somewhat ambiguous. In fact, if $\chi_L$ and $\chi_R$ are normalized and a, b, $\sim$ 1, then the quantum fluctuations in the number of electrons in either one of the bulk superconductors are of order $N^{1/2}$. (However, as we will see below, the actual fluctuations in a realistic junction are much smaller than this.) A more precise statement about the system described by the wave function (9) is that its "disconnectivity"[1] is not macroscopic (in fact it is 2, see Ref. 1).

$$I_s = I_c \sin\Delta\phi. \tag{12}$$

We know that in equilibrium (i.e., when the state of the system is time-independent and no dissipation is present) the supercurrent (12) is the only current across the junction (the normal current by definition involves dissipation), so under these circumstances it must be equal to the external current $I_{ext}$. For many purposes it is convenient to derive this result by adding an extra term $-I_{ext}$ to the potential energy (10) of the junction, that is, by writing

$$U(\Delta\phi) = -\frac{I_c\phi_o}{2\pi}\cos\Delta\phi - I_{ext}\frac{\phi_o}{2\pi}\Delta\phi \tag{13}$$

so that minimization with respect to $\Delta\phi$ leads to $I_{ext} = I_c \sin\Delta\phi$ as required. However, it should be carefully noted that this is a formal trick which should not be taken too seriously: in particular, it must be carefully borne in mind that $\Delta\phi$ is only defined modulo $2\pi$, so that the "washboard potential" defined by Eq. (13) really has no direct physical meaning. (Compare the case of a simple pendulum driven by a constant external torque N, where the potential energy as a function of angle $\theta$ is identical to (13) if we make the replacements $\Delta\phi \longrightarrow \theta$, $I_c\,\phi_o/2\pi \longrightarrow mg\ell$, $I_{ext}\,\phi_o/2\pi \longrightarrow N$).

If the external current $I_{ext}$ exceeds $I_c$, then no value of $\Delta\phi$ will permit all the current across the junction to be carried by the supercurrent: we then find a normal current $I_n$ generated in parallel with the supercurrent. Associated with this current is some energy dissipation which in general may have a very complicated form.[11] In the case of an ideal oxide tunnel junction, the current is associated with the thermally excited normal quasiparticles and hence, vanishes in the limit $T \longrightarrow 0$ (for frequencies less than the "gap frequency" $2\Delta/\hbar$). However, in real life most junctions appear to have some parasitic shunting conductance even at zero temperature: this is often described by the "resistively shunted junction" model which assumes that there is an effective resistance $R_n$ in parallel with the junction, i.e.

$$I_{ext} = I_s + I_n, \qquad I_n = V/R_n \tag{14}$$

A slight generalization of this model would allow $R_n$ to depend both on frequency and (periodically) on $\Delta\phi$.

To complete this brief review of familiar results, we consider the case in which we insert our Josephson junction in a closed thick superconducting ring. (This system is the active element of an rf SQUID (Ref. 12, Ch. 7) and will henceforth be referred to for brevity simply as a SQUID. In this case, for all phenomena taking place at frequencies well below the bulk plasma frequency ( $\sim 10^{16}\text{secs}^{-1}$), the vanishing of the current deep in the bulk of the ring implies the well-known relation between the trapped flux $\phi$ and the phase difference across the junction $\Delta\phi$:

$$\Delta\phi = 2\pi(\phi/\phi_0) + 2\pi n \tag{15}$$

In addition to the Josephson coupling energy (10) there is now a purely electromagnetic energy $(\Phi - \Phi_x)^2/2L$ due to the finite self-inductance L of the ring, where $\Phi_x$ is the externally applied flux. Thus, as a function of the trapped flux $\Phi$ the "potential energy" has the form

$$U(\Phi) = \left(\frac{\Phi-\Phi_x}{2L}\right)^2 - \frac{I_c\phi_0}{2\pi}\cos 2\pi\Phi/\phi_0 \tag{16}$$

For $\beta_L \equiv 2\pi LI_c/\phi_0 > 1$ this expression has at least one metastable minimum at least for $\Phi_x = \phi_0/2$. Note that, in distinction to $\Delta\phi$ which is defined only modulo $2\pi$, values of $\Phi$ differing by integral numbers of flux quanta correspond to physically distinct states. This makes the SQUID conceptually simpler to discuss than an isolated current-biased junction (hereafter referred to as a CBJ). The considerations concerning dissipation in the junction are precisely the same in the two systems, but the SQUID may also have additional mechanisms of dissipation associated with the bulk ring (see below).

## MACROSCOPIC SUPERPOSITION IN JOSEPHSON SYSTEMS

What we would like to do is to seek experimental evidence for (or against) the superposition of quantum-mechanical states of our system corresponding to macroscopically different properties. Consider for example the simple superconducting ring described above. Then a state of the "interesting" kind might have the form

$$\Psi(r_1, r_2 \dots r_N) = \sum_n c_n\, [\chi_n(R)]^{N/2} \equiv \sum_n c_n \Psi_n \tag{17}$$

with $\chi_n(R)$ given by Eq. (4). Since the states $\Psi_n$ correspond to different quantized values of the trapped flux (Eq. (7)), this clearly does represent a superposition of macroscopically different states. Unfortunately it is not at all clear how one could ever obtain any evidence for a wave function of the type (17). The problem is that since all N particles (or N/2 pairs) are behaving differently in the different components, a direct measurement which would detect the difference between the pure state (17) and the (uninteresting) classical probabilistic mixture of the states $\Psi_n$ with probability $|c_n|^2$ would require us to measure, as a minimum, N-particle correlations -- a task beyond the powers of even the most dedicated experimentalist.* Indeed, direct detection of superpositions of the type (17) are always out of the question, and the only hope is to let Nature do the "measurement" for us. The reason that Nature, unlike us, can in principle detect superpositions of the type (17) is that she has at her disposal one operator which we do not, namely the time evolution operator $\hat{U}(t) \equiv \exp -i\hat{H}t$. In the general case such an operator will give different results when acting on the pure states (17) and the associated mixture. Thus we can, in principle, infer which of the two is occurring from the subsequent dynamics. Unfortunately, there is a special feature about the specified example just discussed which makes it impossible to do so in this case: namely, the states $\Psi_n$ are eigenfunctions of the Hamiltonian (for a uniform ring, the angular momentum operator is a constant of the motion) and hence $\langle n| \hat{U}(t) | n'\rangle \sim \delta_{nn'}$. Thus, the time evolution in this case is trivial ($\Psi_n(t) = \sum_n c_n e^{-ie_n t/\hbar} \Psi_n$) and simply preserves the form of the original superposition, with only the (not directly measurable) relative phases of the different components changed; we are therefore no further forward.

We need, therefore, a case in which the macroscopic property whose different eigenstates are to be superimposed is <u>not a conserved quantity</u>. Consider for example a SQUID, and take the relevant macroscopic variable to be the trapped flux $\Phi$. In the analogous mechanical problem the coordinate x is not conserved because of the presence of the kinetic energy $\frac{1}{2}m\dot{x}^2$, and by analogy we would expect that $\Phi$ would not be conserved if the Hamiltonian contains a term proportion to $\dot{\Phi}^2$. This is indeed the case since $\dot{\Phi}$ is simply the voltage developed across the junction, the energy associated with the finite junction capacitance C is just $\frac{1}{2}C\dot{\Phi}^2$ and plays precisely the role of a "kinetic energy" in this problem. At the classical level this is very well known: indeed the phenomenological equation of motion used, in the RSJ model, to descrilbe the behavior of the flux is[13]

$$C\ddot{\Phi} + \dot{\Phi}/R_n + \left\{ \frac{(\Phi - \Phi_x(t)}{L} + I_c \sin 2\pi\Phi/\Phi_o \right\} = 0 \qquad (18)$$

(where the term in brackets is the derivative of the potential energy (16) with respect to $\Phi$). We see that C plays precisely the role of the mass in the analogous mechanical problem. The classical equation for the motion of the phase difference $\Delta\phi$ in a CBJ has a structure very similar to 18) and can in fact be obtained directly from it by making the substitutions

---

* See Ref. 1, section 4, for a discussion of this point.

$$\Phi \to (\Phi_0/2\pi)\Delta\phi, \qquad \Phi_x(t) \to LI_{ext}(t) \tag{19}$$

and letting L tend to infinity. Provided we avoid questions where the periodic nature of the variable $\Delta\phi$ plays an essential role (cf. below), all the subsequent results, both classical and quantum-mechanical, for a CBJ can be obtained from those for a SQUID by the above prescription: I shall therefore from now on usually discuss explicitly only the SQUID case. Note that the "mass" for the CBJ case is $C(\Phi_0/2\pi)^2$.

It should be noted at this point that the ascription to a Josephson junction (whether or not contained in a SQUID ring) of a single unique capacitance C may be somewhat problematical, particularly if the inferred value of this parameter is much smaller than the typical geometrical capacitances in the problem (which for a typical SQUID geometry may be of the order of 1 pF). In such cases it may be necessary to represent the system by a circuit more complicated than that corresponding to Eq. (18) and to introduce more than one flux-type variable to describe the motion: cf. Refs. 14 and 15. For present purposes we shall ignore this possible complication, although it needs to be borne in mind when designing concrete experiments. See also the remarks in the second lecture.

What we are going to do now is to <u>quantize</u> the motion of the flux. Let us for the moment neglect the presence of the dissipation term (i.e., set $R_n = \infty$). Then the classical equation of motion can be obtained by treating $\Phi$ as a generalized coordinate in the usual sense of mechanics and constructing a Lagrangian

$$L(\Phi,\dot{\Phi}) = \tfrac{1}{2} C\dot{\Phi}^2 - U(\Phi) \tag{20}$$

with $U(\Phi)$ given by Eq. (16). We then define in the usual way a generalized momentum

$$p_\Phi \equiv \partial L/\partial\dot{\Phi} = C\dot{\Phi} = -Q \tag{21}$$

where $Q$ is the charge stored in the capacitance, and obtain the quantum-mechanical formulation by the usual quantization postulate

$$[p_\Phi,\Phi] = -i\hbar \qquad \text{or } p_\Phi \to -i\hbar\frac{\partial}{\partial\Phi} \tag{22}$$

In this way we obtain a familiar-looking Schrödinger equation:

$$\left\{-\frac{\hbar^2}{2C}\frac{\partial^2}{\partial\Phi^2} + U(\Phi)\right\}\Psi(\Phi) = i\hbar\frac{\partial\Psi}{\partial t}(\Phi) \tag{23}$$

whose solution gives the amplitude $\Psi(\Phi,t)$ for the trapped flux in the system at time t to have the value $\Phi$. In so far as this procedure is correct, it is clear that in principle we should be able to see the effects of interference between the amplitudes for different $\Phi$ in precisely the same way as in microscopic systems we see the interference between amplitudes for different values of the coordinate (e.g., in a two-slit experiment). But now, since $\Phi$ is a macroscopic variable, the states which interfere will in general be macroscopically different Thus we attain our goal of constructing superpositions of macroscopically different states which may be in principle observable.

Is it, in fact, legitimate to quantize the motion of $\Phi$ in this way? Suppose we start from a Lagrangian formulated in basic quantum electro-dynamic terms as a function of the electromagnetic potentials $A_\mu(\underset{\sim}{r},t)$, the electron coordinates $\underset{\sim}{r}_j(t)$ (j = 1, 2,.... N) and any other variables (e.g., nuclear coordinates) which are relevant. At this level we are sure of the commutation relations[16]

$$[A_\mu(\underset{\sim}{r}), \partial L/\partial \dot{A}_\nu(\underset{\sim}{r}')] = -i\hbar\delta_{\mu\nu}\delta(\underline{r} - \underline{r}') \qquad (\mu,\nu = 1,2,3) \tag{24}$$

But since the trapped flux $\Phi$ is simply a line integral of $A(\underset{\sim}{r})$, we can just perform a transformation to new coordinates one of which is $\Phi$ and the invariance of the commutation relations under such a coordinate transformation will immediately imply the result* $[\Phi, \partial L/\partial\dot{\Phi}] = i\hbar$. Naturally, the question of whether the dependence of the Lagrangian on $\dot{\Phi}$ is one of the form $C\dot{\Phi}^2$ is a separate issue: however, we can settle this by requiring the formalism to produce the correct classical equation for $\Phi$ in the correspondence limit. This agrument gives clear-cut results in the case presently considered, where we neglect dissipation in the classical equation of motion: the question of the effect of dissipation is taken up in the second lecture.+

---

*Indeed, if we follow this line of argument (or the similar argument concerning the quantization of the centre-of-mass motion of a macroscopic body, which is even more clear-cut since in this case there is no possible ambiguity in the kinetic-energy term) it seems at first sight totally impossible to avoid the conclusion that if quantum mechanics applies to microscopic coordinates it must inescapably apply also to macroscopic ones. On closer examination, however, we see that we are making an implicit (albeit very natural) assumption about the validity of the quantum-mechanical formalism for states of high "disconnectivity" for which we have at present no direct evidence. Compare Ref. 1, section 3.

+On the objection that $\Phi$ is already in some sense quantized since the potential energy $U(\Phi)$ explicitly involves h through its dependence on the flux quantum $\hbar/2e$, see Ref. 2, section 1. This reference also discusses briefly the extra conceptual difficulties associated with the quantization of $\Delta\Phi$ for a CBJ.

ENERGETICS AND CLASSICAL DYNAMICS OF A SQUID

Before proceeding to detailed quantum-mechanical considerations, let us note a few results concerning the classical motion described by Eq. (18). For the moment we set $R_N = \infty$ (no damping). First, suppose the external flux $\Phi_x$ is set equal to zero and we consider the motion of the system near the bottom of the deepest potential well ($\Phi = 0$). The small oscillation is the junction plasma resonance[17] and its frequency given by

$$\omega_J = \left\{ \frac{1}{C} \left( \frac{1}{L} + \frac{2\pi I_c}{\phi_0} \right) \right\}^{\frac{1}{2}} \equiv \omega_{LC}(1 + \beta_L)^{\frac{1}{2}} \qquad (25)$$

where $\omega_{LC}$ is the classical resonance frequency $(LC)^{-1/2}$ in the normal phase and $\beta_L \equiv 2\pi LI_c/\phi_0$. Next, suppose that the external flux is increased from zero and the system follows the changing potential $U(\Phi,\Phi_x)$ adiabatically (as in the typical operation of a practical rf SQUID[12]). The originally stable minimum in which the system is trapped becomes metastable (for $\beta_L > 1$) and as we increase $\Phi_x$ further we eventually approach the classical "break point" at which it becomes unstable (for $\beta_L >> 1$ this happens when $\Phi_x \sim LI_c$). For $\Phi_x$ just below the break point value $\Phi_{x0}$ the curve $U(\Phi:\Phi_x)$ near the metastable minimum is approximately cubic in $\Phi$ and in the limit $\beta_L >> 1$ we find that the frequency of the small oscillations around the metastable minimum is[13]

$$\omega_0 = \omega_{LC} \,(2\pi\sqrt{2})^{\frac{1}{2}}\, (LI_c/\phi_0)^{\frac{1}{4}}\, (\delta\Phi_x/\phi_0)^{\frac{1}{4}} \qquad (26)$$

where $\delta\Phi_x$ is $\Phi_{xo} - \Phi_o$. Moreover the height of the energy barrier confining the metastable minimum is

$$U_0 = \frac{2\sqrt{2}}{3\pi}\, \frac{\phi_0^2}{L} \left( \frac{\phi_0}{LI_c} \right)^{\frac{1}{2}} (\delta\Phi_x/\phi_0)^{3/2} \qquad (27)$$

For reference we note that the results for a current-biased junction, obtained from (26) and (27) as indicated above,* are

$$\omega_0 = \tilde{\omega}_J\, (2\delta I/I_c)^{\frac{1}{4}}\,, \qquad U_0 = \frac{2\sqrt{2}}{3\pi}\, \phi_0 I_c (\delta I/I_c)^{3/2} \qquad (28)$$

where $\tilde{\omega}_j \equiv (2\pi I_c/C\phi_0)^{1/2}$ is the frequency of the isolated-junction plasma resonance (cf. Eq. (21)) and $\delta I \equiv I_{ext} - I_c$. One further case of interest, which applies only to the SQUID, is where $\beta_L$ is only just greater than 1, so that the metastable well exists only for $\Phi_x$ close to the value $\phi_0/2$. The form of the potential for $\Phi$ and $\Phi_x$ both close to $\phi_0/2$ is

*Note that the CBJ case corresponds to $\beta_L \longrightarrow \infty$, so the use of Eqs. (26) and (27) is legitimate.

$$U(\Phi) = \frac{\Phi_o^2}{2L} \{(\Delta\Phi - \Delta\Phi_x)^2 + \beta_L(-(\Delta\Phi)^2) + \pi^{2/3}(\Delta\Phi)^4\} \tag{29}$$

where $\Delta\Phi \equiv \Phi/\Phi_o - 1/2$, $\Delta\Phi_x \equiv \Phi_x/\Phi_o - 1/2$. The stable and metastable equilibrium positions, barrier height $U_o$ and small oscillation frequency $\omega_o$ can be calculated from (29) as a function of $\Delta\Phi_x$: note that for the special case $\Delta\Phi_x = 0$ we have a symmetric double-well potential with a separation between the minima of $([6(\beta_l - 1)]^{1/2}/\pi)\phi_o$, and the quantities $U_o$ and $\omega_o$ are given by

$$U_o = \frac{3}{8\pi^2} \frac{\phi_o^2}{L} (\beta_L - 1)^2, \quad \omega_o = [2(\beta_L - 1)]^{\frac{1}{2}} \; \omega_{LC} \tag{30}$$

Finally, it is convenient to introduce the quantity $\alpha \equiv (2CR\omega_o)^{-1}$ which gives a measure of the degree of damping of the small oscillation in the various cases. ($\alpha = 1$ corresponds to critical damping.) For oscillations around the stable minimum at zero external flux bias (or zero current bias for a CBJ) we have $\alpha = \frac{1}{2}\beta_c^{-1/2}$ where $\beta_c \equiv 2\pi C I_c R^2/\phi_0$ is the conventionally defined parameter: for a highly hysteretic SQUID ($\beta_L \gg 1$) near its classical break point we get from Eq. (26)

$$\alpha = \tfrac{1}{2}\beta_c^{-\frac{1}{2}}(\beta_L/4\pi)^{\frac{1}{4}} (\Delta\Phi_x/\phi_o)^{-\frac{1}{4}} \tag{31}$$

(or, for a CBJ, $\alpha = \frac{1}{2}\beta_c^{-1/2}(2\delta I/I_c)^{-1/4}$). We see, therefore, that in the limit as the classical break point is approached the small oscillations of the flux are always overdamped, however small the shunting conductance.

## QUANTUM FLUCTUATIONS AROUND EQUILIBRIUM

In the case of this lecture we consider the effects of naively applying the (conservative) Schrödinger Eq. (23) to the flux in a SQUID, that is, of applying quantum mechanics without taking account of the dissipation which in real life characterizes the classical motion. First we briefly consider the motion near the stable minimum ($\phi = 0$) for zero external flux (cf. Ref. 17). Provided the capacitance C is not too small (see below) we expect the system to behave as a simple quantum harmonic oscillator, with energy levels $(n + \frac{1}{2})\hbar\omega_J$ where $\omega_J$ is given by Eq. (25). The ground state mean-square fluctuations of the flux $\Phi$ and the imbalance $\Delta N$ of the electrons on the two sides of the junction (which by Eq. (21) is $p_\Phi/e$) are given by

$$\langle\phi^2\rangle = \hbar/2C\omega_J \cong \pi^{-2}\lambda^{-1}\phi_o^2 \tag{32}$$

$$<(\Delta N)^2> = \tfrac{1}{2}C\hbar\omega_J/e^2 \cong \lambda \tag{33}$$

where the dimensionless parameter $\lambda$ is defined by

$$\lambda \equiv (8CI_c\,\phi_o^3/\pi^3\hbar^2)^{\frac{1}{2}} \tag{34}$$

$\lambda$ (or rather $\lambda^{-1}$) is the characteristic parameter which measures the importance of quantum effects (see below). We see that for $\lambda$ large the fluctuations of the particle (or charge) imbalance across the junction are large (though not in general as large as estimated earlier) and the flux is localized on a scale small compared to the typical "length" scale of the potential ($\sim \phi_o$). On the other hand, if $\lambda$ were to become comparable to 1 then the flux would become effectively delocalized and the number of electrons on the two sides of the junction would become nearly well-defined. Thus the parameter $\lambda$ gives a measure of how semi-classical the motion of $\Phi$ in the "deep" wells is: if $\lambda >> 1$ we are in the correspondence limit (cf. also below). Now although in principle the critical current $I_c$ can be made as small as we wish, the whole quantum behavior will certainly be upset by thermal fluctuations unless we satisfy the condition that the Josephson barrier height $I_c\phi_o/\pi$ which separates (approximately) neighboring minima of the potential is large compared to kT; however the parameter $\lambda$ can be comparable to 1 only if $C \lesssim e^2/8kT$ (i.e., roughly, the capacitative energy $e^2/2C$ of a single electron is larger than kT). At the lowest temperatures available ($\sim$ 1 mK) this would read $C \sim 0.1$ pF, which is about the minimum genuine capacitance (see above) of a typical SQUID. Thus it would be just about possible to see quantum effects in the motion of the flux around the stable minimum, but it would require extreme values of the parameters. Moreover, such effects would show up only as corrections of detail to the classically predicted motion: they would not change its overall nature.

## MACROSCOPIC QUANTUM TUNNELLING

A much more promising way of looking for evidence of quantum-mechanical behaviour is to exploit the phenomenon of tunnelling through a potential barrier, which has no classical analogue. The simplest experiment one could do is to look for the decay of a metastable state due to tunnelling. Let us briefly review the theory of such decay; for the sake of familiarity we shall talk explicitly in terms of a mechanical system possessing a geometrical coordiante q and a potential energy V(q). As we shall see below, the attainment of conditions for a reasonable tunnelling rate is very much easier if we bias the system to near its classical instability; for this reason we pay special attention to the case of a cubic potential ($V(q) = \frac{1}{2} m\omega_o^2 q^2 - bq^3$) which describes the relevant region of the potential in this case. The height of the barrier is denoted by $U_o$ as above and the "exit point" is written as $q_o$.

As is well-known, in the WKB limit ($\hbar\omega_o << U_o$) the rate of decay of the metastable state at zero temperature is given by

$$P_{QM} = \text{const.}\ \omega_o \left(\frac{U_o}{\hbar\omega_o}\right)^{\frac{1}{2}} \exp - 2\int_0^{q_o} (2mV(q))^{\frac{1}{2}}\, dq/\hbar \qquad (35)$$

For any potential which is not too pathological in shape it is possible to rewrite the argument of the exponent in the form $a_o U_o/\hbar\omega_o$, where $a_o$ is a dimensionless constant, generally of order $2\pi$, which is a function of the shape of the barrier but not of its scale. In particular, for a cubic potential we have[18] $a_o$ = 36/5. Since the constant in the prefactor is in this case $(60)^{1/2}$ $(18/5\pi)^{1/2}$, we get

$$P_{QM}(\text{cubic}) = \omega_o(60U_o/\hbar\omega_o)^{1/2}(18/5\pi)^{1/2}\exp - \left(\frac{36}{5}\frac{U_o}{\hbar\omega_o}\right) \qquad (36)$$

It is evident from Eq. (36) that to obtain a reasonable tunnelling rate $U_o$ must be not too large compared to $\hbar\omega_o$. Now we can, of course, make $U_o$ in a SQUID as small as we like by letting the external biasing flux $\Phi_x$ approach arbitrarily close to the value $\Phi_{xo}$ at which the classical instability occurs. However, this is unsatisfactory, for the following reason. The purely thermal (classical) decay rate at a temperature T is given by the Kramers[19] formula*

$$P_{CL} = \text{const.}\ \omega_o \exp - U_o/k_BT \qquad (37)$$

and therefore to be sure that any decays we are seeing are genuinely due to quantum tunnelling we need $kT << \hbar\omega_o/a_o$ (so that $P_{QM} >> P_{CL}$). Unfortunately, in the limit $\Phi_x \longrightarrow \Phi_{xo}$ this condition is automatically violated, since according to Eq. (26) $\omega_o$ tends to zero in this limit. Thus it is necessary that quantum tunnelling becomes reasonably probable while we are still far enough from the classical break point to neglect thermal decay.

A more quantitative formulation of this condition is obtained if we consider how the experiment is likely to be done in practice. (We revert now explicitly to the SQUID case.) Typically, one

---

*The constant in the prefactor is of order 1 only for reasonable damping: it tends to zero in the limit of very strong or very weak damping.[19]

would sweep the external flux up towards the classical break point at some sweep frequency $\omega_s$ which in practice would be at most of the order of a few hundred MHz (and hence, as we shall verify, small compared to the characteristic frequency $\omega_0$ in the region of interest); and one would monitor the distribution of the decay occurrences as a function of $\delta\Phi_x \equiv \Phi_x - \Phi_{xo}$ (or, in the CBJ case, of the variable $\delta I \equiv I_{ext} - I_c$). In principle the shape of this curve should distinguish between classical and thermal decay even without consideration of its scale,[14] since the dependence of the argument of the exponent on $\delta\Phi_x$ is different for the two cases (see Eqs. (36)-(37) and (26)-(27)); however, since the difference is only between the 3/2 and 5/4 powers of $\delta\Phi_x$, this may or may not be an accurate diagnostic technique in practice.) A convenient measure of this distribution is its root-mean-square width, and the condition that we expect, theoretically, to see quantum tunnelling rather than thermal decay is that the width $\sigma_{CL}$ calculated from the classical formula (37) is small compared to the width $\sigma_{QM}$ calculated from Eq. (36). We can estimate the width, to within a small correction factor, in each case by equating it* to the value of $\delta\Phi_x$ at which the argument of the exponent becomes equal to some constant of the order of the logarithm of the prefactor in the decay rate divided by the sweep frequency (so that the probability of decay in a sweep cycle is of order 1); to within the accuracy of the calculation the constant may be taken to be the same for the two cases (cf. Ref. 14). In this way we find from Eqs. (36), (37), (26), and (27)

$$\sigma_{CL} = \text{const.}\left[\frac{3\pi}{2\sqrt{2}}\;\frac{k_BT}{\phi_o^2/L}\right]^{2/3}\left(\frac{LI_c}{\phi_o}\right)^{1/3}\phi_o \tag{38}$$

$$\sigma_{QM} = \text{const.}\ \tfrac{1}{2}\left(\frac{5\pi}{12}\;\frac{\hbar\omega_J}{\phi_o^2/L}\right)^{4/5}\left(\frac{LI_c}{\phi_o}\right)^{1/5}\phi_o \tag{39}$$

and the condition $\sigma_{QM} \gg \sigma_{CL}$ therefore requires

$$k_BT \ll 0.15(\hbar\omega_J/I_c\phi_o)^{0.2}\,\hbar\omega_J \equiv 0.15\,c\hbar\omega_J \tag{40}$$

Since the factor c is typically of order 0.2, we see that the temperature should be low compared to 0.03 $\hbar\omega_J$; for reasonable parameters this means it should be well below 100 mK, which is not too difficult a condition to satisfy. Since the criterion (41) does not involve L, it is equally valid for a current-biased junction.

*For a more rigorous calculation for the thermal case, see Ref. 13.

It is interesting to determine the order of magnitude of various quantities associated with the tunnelling in the "interesting" region (i.e., for $\delta\Phi_x \sim \sigma_{QM}$). For this purpose we assume that the constant in Eq. (39) is not too different from unity. From Eqs. (26) and (27) we see that the barrier height and small-oscillation frequency are given by the order-of-magnitude estimates

$$U_o \sim \left(\frac{\hbar\omega_J}{I_c\phi_o}\right)^{1/5} \hbar\omega_J \sim \lambda^{-0.2}\,\hbar\omega_J \tag{41}$$

$$\omega_o \sim \left(\frac{\hbar\omega_J}{I_c\phi_o}\right)^{1/5} \omega_J \sim \lambda^{-0.2}\,\omega_J \tag{42}$$

(where $\lambda$ is the characteristic parameter defined by Eq. (34)). Also the width of the barrier through which the tunnelling takes place is given by the order-of-magnitude estimate $(U_o/C\omega_o^2)^{1/2}$ $\sim\lambda^{0.1}(\hbar/C\omega_J)^{1/2}\sim\lambda^{-0.4}\,\phi_o$ (where we used Eq. (25)). All these estimates are equally valid for a current-biased junction (in the last case, the analogous statement is that the distance through which the relative phase $\Delta$ has to tunnel is of order $\lambda^{-0.4}$ radians). All the above formulae are valid only for $\lambda >> 1$ (for $\lambda \lesssim 1$ tunnelling takes place already for "deep" wells). (See below.) We note finally, from (36), that the quantity $U_o/\hbar\omega_o$ cannot in practice be much greater than 4, since $\omega_o$ cannot be much greater than $\sim 10^{12}$ without totally destroying the self-consistency of the model (even for a CBJ)*, and we do not wish to wait more than say 10 secs. for a tunnelling event to occur.

## MACROSCOPIC QUANTUM COHERENCE

A phenomenon which is based on tunnelling but which is considerably more subtle is what I have elsewhere christened "macroscopic quantum coherence", in which the system tunnels periodically backwards and forwards between two (exactly or nearly) degenerate potential wells, in analogy with the behavior of the nitrogen atom in the inversion resonance of the $NH_3$ molecule. To see this (if it can be seen: cf. next lecture) we need to bias the external flux close to $\phi_o/2$, so that the two lowest wells are nearly degenerate. Suppose first that $\Phi_x$ is exactly $\phi_o/2$, so that we have a perfectly symmetric double well. Then, as is well-known,[20] the groundstate splits into an even and an odd state which are approximately symmetric and antisymmetric combinations, respec-

*Since, inter alia, $\omega_o$ is then large compared to the gap freqency $2\Delta/\hbar$.

tively, of the ground states in the two wells separately. The energy splitting (or tunnelling amplitude) is $\hbar\Gamma$, where the oscillation rate $\Gamma$ (see below) is given by the expression

$$\Gamma = \text{const. } \omega_o \exp - \int (2mV(q)dq)^{\frac{1}{2}}/\hbar \tag{43}$$

where for the sake of familiarity we have written q instead of the flux variable $\Phi$ (and m for C), the integral is taken between the two degenerate minima, and $\omega_o$ is the small-oscillation frequency in either of the wells. (Note that there is no factor of 2 in the exponent, unlike in Eq. (35).)

Consider first the case $\beta_L >> 1$. Then we can neglect the self-inductance term in Eq. (16) in the region of the two wells, and the potential in this region is a pure cosine one. In this case we can easily evaluate the integral in the WKB expression and find

$$\Gamma = \text{const. } \omega_o \exp - \lambda \tag{44}$$

with $\lambda$ given by Eq. (34). Since in all but the most extreme cases $\lambda$ is large compared to unity (cf. above), it follows that it will be extremely difficult, if not impossible, to see macroscopic quantum coherence if $\beta_L >> 1$.

Let us therefore consider the case in which the SQUID is only weakly hysteretic ($\beta_L$ just greater than 1), and define $K \equiv \beta_L - 1$. In that case the potential has approximately the quadratic-plus-quartic form, and for this case the WKB amplitude can be written in the form $a_o'(U_o/\hbar\omega_o)$, where as above $U_o$ is the barrier height and the constant $a_o'$ is 16/3. Thus, using Eq. (30), we see that the oscillation rate $\Gamma$ is given, up to a numerical constant in the prefactor, by*

$$\Gamma \sim \omega_o \exp - (2^{-\frac{1}{2}} K^{3/2} \lambda) \tag{45}$$

with K defined as above. We must, of course, satisfy simultaneously the condition $U_o >> k_B T$, otherwise incoherent thermal transitions will occur and spoil the simple quantum behaviour. To satisfy this condition while making $\Gamma$ not negligibly small requires, crudely speaking, the inequality

$$k_B T/I_c \phi_o << \lambda^{-4/3} \tag{46}$$

which may be just satisfied in a narrow window of the parameter space provided C can be made as small as $\sim 10^{-14}$F. (Which places, of course, very stringent requirements on the SQUID geometry as well as on the junction fabrication.) A more detailed quantita-

---

*We neglect terms of higher order in K relative to those kept.

tive study of this problem has recently been carried out by Chakravarty.[21]

What shall we look for? In principle we could do the analogue of the spectroscopic experiment for the $NH_3$ molecule, that is, measure the power absorbed in the induced transition between the even and odd levels. However, since we have only one SQUID instead of $\sim 10^{23}$ $NH_3$ molecules, the power absorption would be extremely small and it is doubtful whether it would be detectable. A more spectacular experiment, prima facie, would be to start the system in one well (so that it is in a linear combination of the even and odd energy levels) and actually "watch" it oscillating backwards and forwards: the prediction is[20] that the probability amplitude for being in the original well a time t after starting out, P(T), is given by

$$P(t) = \cos^2 \Gamma t/2 \tag{47}$$

There is, however, a rather devasting objection to the idea that we should monitor the flux continuously: by the usual arguments of the quantum theory of measurement, any such continuous monitoring would "collapse the wave packet" and thus totally destroy the coherent oscillating behavior. Indeed, the whole point of the phenomenon in the context of tests of quantum mechanics at the macroscopic level is that the time evolution operator U(t' - t), when acting on the linear combination

$$\psi(t) = a(t)\psi_L + b(t)\psi_R \tag{48}$$

of states $\psi_L$, $\psi_R$ localized in the left and right wells, gives a result at time t' different from what it would give acting on the incoherent mixture of $\psi_L$ and $\psi_R$ (compare the discussion given earlier); thus, even if we were to choose voluntarily not to inspect or record the results of our monitoring at intermediate times, the mere fact that it is carried out totally destroys the interference phenomenon we are after (just as continuous monitoring of the slits in the classic two-slit experiment destroys the interference pattern). However, in principle there is nothing to stop us initially switching on our measuring apparatus and verifying that (e.g.) the system is in the left-hand well, then turning the apparatus off and letting the system propagate undisturbed, and finally turning the apparatus back on after some time t and observing which well the system is now in. This is almost an exact analogue of the way in which particle physicists detect the strangeness oscillations in the $K_0 - \bar{K}_0$ meson system,[22] and we should expect the statistics obtained in such an experiment to conform to Eq. (47).

Finally, I should mention that there is a third phenomenon, in a sense intermediate between "tunnelling" and "coherence", in which we constrain the particle to move in a potential with two finite but unequal wells and observe it "hopping" from the ground-state of the metastable one to that of the stable one with emission of an excitation. This is similar to tunnelling in being an irreversible prcess, but differs from it in that the spectrum of the isolated system outside the barrier cannot be treated as continuous.* Some detailed predictions about this process and its inverse (photo-assisted tunnelling) have been made in a recent paper by Chakravarty and Kivelson.[23]

All the above considerations completely neglected the fact that the classical motion of the flux in a SQUID involves considerable dissipation. This is a serious defect and will be partially remedied in the second lecture.

## LECTURE 2. The Effects of Dissipation

### THE IMPORTANCE OF DISSIPATION IN MACROSCOPIC SYSTEMS

In discussing the quantum mechanics of a macroscopic variable, and in particular the phenomena of macroscopic quantum tunnelling and coherence, it is impossible to overestimate the importance of taking into account the irreversible interactions of the system with its environment which show up, in the classical equations of motion, as dissipative terms. Indeed, in discussions of the quantum theory of measurement over the last two or three decades one recurring theme which has been at least implicitly present is that under most normal circumstances these interactions will be adequate to destroy all quantum-mechanical coherence effects and hence, in the opinion of some authors (cf., e.g., Ref. 24), make it impossible to solve the quantum measurement problem.

Consider a macroscopic system characterized by some macroscopic variable X (for example a SQUID characterized by the flux $\Phi$ , as discussed in the last lecture). Since such a body automatically has very many degrees of freedom, it must in addition to X be characterized by a large number of microscopic coordinates $\xi_i$, (i = 1, 2,....N). (A familiar example is the mechanical motion of a macroscopic body, where X is the centre-of-mass coordinate and the $\xi_i$ are atomic coordinates relative to the centre-of-mass.) In certain very special cases, such as the motion of a body falling freely in a spatially constant gravitational field, the Hamiltonian may separate into a term referring only to X and its conjugate momentum and a term referring only to the $\xi_i$ and their conjugate momenta; in such cases it is possible (though not,

*Note that neither coherence nor (prima facie at least) hopping is defined for a current-biased junction, since in this system the various minima are not physically distinct.

of course, necessary) that the total wave function of the system is a product:

$$\psi\ (X,\{\xi_i\}: t) = \psi(X,t)\,\chi(\{\xi_i\}, t) \tag{2.1}$$

in which case the motion of the macroscopic coordinate satisfies its own Schrödinger's equation (such as* Eq. (23) and the considerations of the last lecture apply without modification. However, in general, the Hamiltonian will couple X and the $\xi_i$, and the wave function, even it it starts out in the form (2.1), will not remain so for arbitrary times.

The presence of this coupling between macroscopic and microscopic coordinates does not, of course, inevitably imply the existence of dissipation, i.e., of the irreversible transfer of energy between the macroscopic and microscopic degrees of freedom. For example, an electrically insulating macroscopic body moving in an inhomogeneous magnetic field will generally suffer negligible disdipation, even though there is no decoupling, in this case, between the macroscopic and microscopic coordinates. The reason lies in the possibility of making an adiabatic (Born-Oppenheimer) approximation[25] for the wave function of the body: if the characteristic frequencies associated with the microscopic variables are very high compared to those characterizing the macroscopic motion, then the energy eigenfunctions of the body are well represented in the form

$$\psi_n(X,\{\xi_i\}) = \psi_{j\alpha}(X)\,\chi_\alpha(\xi_i:X) \tag{2.2}$$

Here the "microscopic" wave functions obey an equation of the general form (assuming for definiteness that X and the $\xi_i$ have the significance of geometrical coordinates)

$$[\hat{T}(\{\xi_i\}) + V(\{\xi_i\}, X)]\,\psi_\alpha(\xi_i:X) = U_\alpha(X)\psi_\alpha(\{\xi_i\}:X) \tag{2.3}$$

where $\hat{T}(\{\xi_i\})$ is a kinetic-energy operator which depends only on the $\xi_i$ (not on X). The X-dependent eigenvalue $U_\alpha(X)$ of this equation now serves as an effective potential for the motion of the macroscopic variable X, whose wave function $\psi_{j\alpha}(X)$ satisfies the Schrödinger equation

$$\left\{-\frac{\hbar^2}{2M}\frac{\partial^2}{\partial X^2} + U_\alpha(X)\right\}\psi_{j\alpha}(X) = E_{j\alpha}\,\psi_{j\alpha}\,(X) \tag{2.4}$$

To the extent that the adiabatic approximation is valid, there is

*In this lecture equation numbers such as (23) refer to equations of Lecture 1. Numbers such as (2.23) refer to the equations of this Lecture.

therefore no mechanism of irreversible energy transfer between the macroscopic and microscopic degrees of freedom: the effect of the microscopic variables on the macroscopic one is wholly contained in the elastically stored potential energy $U_\alpha(X)$. Moreover, for most systems which are genuinely macroscopic we can invoke a simplifying assumption, namely that the effective potential $U(X)$ is nearly independent of the microscopic state $\alpha$ provided this corresponds to weak excitation, and therefore the macroscopic wave functions $\psi_{j\alpha}(X)$ and energies $E_{j\alpha}$ are also independent* of $\alpha$. We can therefore write the eigenfunctions (2.2) as

$$\Psi_{j\alpha}(X,\{\xi_i\}) = \psi_j(X)\chi_\alpha(\{\xi_i\} : X) \tag{2.5}$$

Thus the net effect is that while the microscopic variables follow the macroscopic one adiabatically, the motion of the latter is quite independent of what is going on at the microscopic level except in so far as is reflected in the effective potential $U(X)$ (which can be determined once and for all).** Effectively, therefore, we recover Eq. (2.1) with however the microscopic wave function $\chi$ now parametrically dependent on $X$.

If this were the whole story, we could use the results obtained in the last lecture with no qualms of conscience. However, we know that the adiabatic approximation must break down whenever the microscopic degrees of freedom have energy level spacings which are comparable to those of the macroscopic variable; this immediately allows the irreversible transfer of energy between them. It is not, of course, a priori obvious that such a situation need always exist. For example, if we take seriously the standard tunnelling-Hamiltonian model of an oxide tunnel junction,[4,11] then at zero temperature the gap in the "microscopic" (quasiparticle) excitation spectrum is $2\Delta$, and if the relevant frequencies of the macroscopic variable $\Delta\phi$ are small compared to this we should predict no dissipation at all. However, in real life all Josephson junctions, even the most "ideally" fabricated, show some finite dissipation even at temperatures very small compared to $T_c$. More generally, one can probably safely say that there is no case known in which the motion of a macroscopc coordinate does not dissipate some energy, even if very little. Moreover, it is a very common situation that we

*Contrast the case of a diatomic molecule, where the interatomic potential, and hence the vibrational energy levels and wave functions, does in general, depend strongly on the electronic state.

**In certain cases the use of the adiabatic approximation can lead to renormalization of the effective mass associated with the macroscopic variable instead of (or as well as) an effective potential: cf. Ref. 4).

have a perfectly good phenomenological description of the dissipation without having any very clear idea of its microscopic origins. For example, in many kinds of Josephson junctions the resistively shunted junction (RSJ) model represented by Eq. (18) seem to work quite well, yet the origin of the shunting conductance $R_n^{-1}$ can only be guessed at.

Thus, it becomes an urgent task to construct a theoretical framework to take into account the effect of dissipation on the quantum mechanics of a macroscopic variable, in particular on macroscopic quantum tunnelling and coherence. Ideally, one would like such a framework to be independent of any knowledge, or lack of it, about the underlying mechanism; in other words, one would like to be able to infer directly from the experimentally observed dissipation in the classical motion to the effect on quantum-mechanical behavior. As we shall see, under certain conditions this goal can be achieved.

## EFFECT OF DISSIPATION ON MACROSCOPIC QUANTUM TUNNELLING: THE SIMPLEST CASE

Of course, dissipation may be of many kinds, and it is not a priori obvious that they can all be treated in the same way. Let us therefore start by considering the simpest possible case: we have a system characterized by some (macroscopic) variable q whose motion in the classically accessible regime satisfies the equation

$$M\ddot{q} + \eta\dot{q} + \partial V/\partial q = F_{ext}(t) \tag{2.6}$$

where $F_{ext}(t)$ is a c-number external force and the conservative potential $V(q)$ has a metastable minimum. The friction coefficient $\eta$ is just a constant, so the rate of dissipation of energy by the system into its environment* is simply $\eta\dot{q}^2$. Evidently, the RSJ Eq. (18) for a SQUID is a special case of (2.6), with the replacements $q \longrightarrow \Phi$, $M \longrightarrow C$, $\eta \longrightarrow R_n^{-1}$, $F_{ext}(t) \longrightarrow \Phi_{ext}(t)/L$. We now ask: what is the rate of tunnelling, at T = 0, out of the metastable groundstate ?

It is my no means obvious that this question has a unique answer: it could well be that we cannot calculate the tunnelling rate without knowing many more specific details about the interaction responsible for the dissipation than are encapsulated in

*From now on I shall refer, for brevity, to the macroscopic variable or the "system" and the microscopic ones as the "environment" even though the two are not, of course, necessarily geometrically separated.

the single constant $\eta$. Perhaps the most striking feature of the results to be obtained below is that, given one very general condition which is often satisfied in practice, this is not so: the parameters entering Eq. (2.6) are sufficient, alone, to determine the tunnelling rate uniquely.

It is possible that there is a clever trick by which one can go directly from the classical equation of motion (2.6) to an expression for the tunnelling rate: if so, I do not know it. What we must do, rather, is "reculer pour mieux sauter": that is, we must first go back a step and ask: Where did the Eq. (2.6) actually come from? In other words we must try to construct a microscopic Hamiltonian (or Lagrangian) for the system interacting with its environment which will give Eq. (2.6) in the correct circumstances, and then use that Hamiltonian to calculate the tunnelling behaviour (and, if we wish, also other quantum phenomena such as coherence).

In the following we make essential use of the fact that in a macroscopic body and one degree of freedom will be only weakly perturbed by the motion of the system; the dissipation arises from the fact that the system interacts simultaneously with a macroscopic number of environmental degrees of freedom. This statement needs some explanation. Suppose that we take as the basis of description the adiabatic-approximation energy eigenfunctions (2.5). These are eigenfunctions of the Hamiltonian apart from a term whose matrix elements are, in this basis, (we write X⟶q for clarity)

$$H'_{j\alpha,j'\alpha'} = -\frac{\hbar^2}{2M}\int dq \prod_i \int d\xi_i \left\{ 2\left(\psi_j^*(q)\frac{\partial}{\partial q}\psi_{j'}(q)\right)\left(\chi_\alpha^*(\{\xi_i\},q)\frac{\partial}{\partial q}\chi_{\alpha'}(\{\xi_i\},q)\right) + \left(\psi_j^*(q)\psi_{j'}(q)\right)\left(\chi_\alpha(\{\xi_i\},q)\frac{\partial^2}{\partial q^2}\chi_{\alpha'}(\{\xi_i\},q)\right)\right\} \quad (2.7)$$

We now argue that in a macroscopic body the dependence of the environment wave functions $\chi_\alpha(\{\xi_i\},q)$ on q will always be proportional to some inverse power of the number of particles involved, and hence the individual matrix elements $H'_{j\alpha,j'\alpha'}$ will tend to zero in this limit. Although it is difficult to formulate a rigorous argument for this conclusion in general terms, consideration of specific cases always seems to lead to this result. Consider for example a SQUID, for which the "system" is the trapped flux $\Phi$ and the "environment" is given by the normal quasiparticles present. In this case it is easy to determine the general nature of the dependence of the single-electron wave functions on $\Phi$ (see Ref. 26) and to verify the above statement. Again, for a macroscopic object such as a ball-bearing moving in a gas or fluid, consideration of the dependence of the one-particle atomic wave functions of the gas atoms leads to a similar conclusion. A more general, if less rigorous argument, is that the occurrence of irreversibility implies that the density of environmental levels tends to infinity with increasing N, so that except in very special circumstances the coupling to each of them must tend to zero.

We therefore conclude that any individual level of the environment other than the groundstate will be only very weakly excited by the motion of the systems. This situation is very reminiscent of what happens when a not too powerful pulse of radiation passes through a gas: each individual atom (and each of its levels) is only weakly excited even though the collective effect may be to absorb the beam entirely. Now it is a striking fact that although we now know, of course, that a correct description of the absorption process requires that the atoms be described by quantum mechanics, for several decades in the nineteenth century physicists got along perfectly well with this problem by using a classical harmonic oscillator model. Just so, in our case we can get correct results by replacing the exact, very complicated model of the environment by a bath of harmonic oscillators with appropriately chosen parameters. Moreover, in the interaction coupling the system to the environment we need to keep only terms of zeroth and first order in the oscillator variables. The detailed justification for this claim is given in Ref. 2, Section 3.

Thus, the most general Hamiltonian we need to write down for any system* interacting dissipatively with its environment under the above restriction is

$$\hat{H}(q:\{x_i\}) = \frac{p^2}{2M} + V(q) + \sum_i \left(\frac{p_i^2}{2m_i} + \frac{1}{2}m_i\omega_i^2 x_i^2\right)$$
$$- \sum_i \{F_i(p,q)x_i + G_i(p,q)p_i\} + \Delta H(p,q) \tag{2.8}$$

where $x_i$, $p_i$ represent the cordinate** and momentum of the i-th oscillator which has "mass" $m_i$ and frequency $\omega_i$. We will always choose the origin of q to lie at the metastable minimum of V(q). Now, we are specifically interested in the case of linear frequency-independent dissipation described by Eq. (2.6). It is very

*Subject, of course, to certain conditions about the conservative terms (no relativistic or explicitly velocity-dependent terms, etc.).

**Naturally, $x_i$ need not have the dimensions of a true geometrical coordinate.

easy to verify, by writing down the classical equations of motion which follow from the Hamiltonian (2.8) and eliminating the environment variables, that Eq. (2.6) results if we set*

$$G_i(p,q) = 0, \quad F_i(p,q) = qC_i, \quad \Delta H(p,q) = \sum_i C_i^2/2m_i\,\omega_i^2 \qquad (2.9)$$

and impose a constraint on the spectral density $J(\omega)$ defined by

$$J(\omega) \equiv \frac{\pi}{2}\sum_i (C_i^2/m_i\,\omega_i)\,\delta\,(\omega - \omega_i) \qquad (2.10)$$

namely**

$$J(\omega) = \eta\omega \qquad (2.11)$$

Is the choice embodied in Eq. (2.9) unique (up to canonical transformations)? This is a somewhat delicate question, which is discussed in detail in Ref. 2, Appendix C. One can argue that apart, possibly, from very pathological coincidences the need to obtain Eq. (2.6) requires $G_j$ to be zero and $F_j$ to be a function only of q. However, it is much more difficult to exclude the possibility that $F_j(q)$ has a non-factorizable form ($\neq qC_j$), while Eq. (2.14) holds with $C_j$ replaced by $\partial F_j(q)/\ q$ in the definition of $J(\omega)$. Such a form of Hamiltonian will in fact lead to extra terms in the equation of motion (2.6), but these will not necessarily show up at velocities of the order of "typical" velocities of the problem (e.g., the classical "escape velocity" for the metastable well). Nevertheless, such a coupling term gives quantitatively quite different results for the quantum tunnelling rate. This somewhat counter-intuitive result --that we can produce on infinity of different microscopic Hamiltonians which give, to all practical intents and purposes, the same classical behaviour but quite different quantum behaviour

*The last term in Eq. (2.8) ("counterterm") is included so that the potential V(q) occurring in the phenomenological Eq. (2.6) is the same as that in Eq. (2.8). This point is discussed in detail in Ref. (2), Section 4 and Appendix A.

**Evidently we not expect the form Eq. (2.11) to be valid to arbitrarily high frequencies (any more than we expect the phenomenological equation of motion Eq. (2.6) to be universally valid). There will be some cutoff frequency $\omega_c$ beyond which Eq. (2.1) needs to be modified: e.g., if $J(\omega)$ has the Drude form $\eta\omega/(1+\omega^2\tau^2)$ then $\omega_c$ is of order $t^{-1}$. We assume only that $\omega_c$ is large compared to all characteristic frequencies of the classical motion.

even in the WKB (semi-classical) limit--means that the phenomenological friction coefficient $\eta$ certainly does not uniquely determine the tunnelling behaviour, unless we can on some experimental or a priori ground determine $F_i(q)$ to have the factorizable form $qC_i$. Fortunately, in many cases of practical interest we can do just this. In particular, in most cases involving Josephson junctions or SQUID's the nature of the basic electromagnetic coupling plus the fact that the barrier width is small compared to $\phi_0$ implies that we can expand the coupling up to lowest (first) order in $\phi$, (or $\Delta\phi$), from which the factorizability immediately follows.

Thus, given that this condition is fulfilled, the most general Hamiltonian we shall need to deal with is of the form

$$H = \frac{p^2}{2M} + V(q) + \frac{1}{2}\sum_i (p_i^2/m_i + m_i \omega_i^2 x_i^2) - q\sum_i C_i x_i + q^2 \sum_i C_i^2/2m_i\omega_i^2 \tag{2.12}$$

with the parameters $C_j$, $m_j$, and $\omega_j$ constrained by Eq. (2.11). This concludes the "backwards" step of the calculation; we now have to find a way of using Eq. (2.12) to calculate the tunnelling rate.

If one needs only qualitative results it is quite helpful to think of the problem as a tunnelling problem in the many-dimensional space spanned by q and the $\{x_i\}$ . The potential contours are shown schematically in Fig. (1a) for the case $C_i$=0, and in Fig. (1b) for $C_j \neq 0$. In each case the metastable groundstate is at the origin, and to escape the system will take, as it were, the easiest path out to the "lower ground" (region S in the figure), that is, a path going over (or at least near) the saddlepoint. In the non-dissipative case ($C_i$=0) the saddlepoint lies on the q-axis at some point $q_0$, and the "easiest" path is straight along the q-axis as shown; the environmental degrees of freedom are clearly irrelevant in this case. In the dissipative case ($C_j \neq 0$) we easily verify that the saddlepoint occurs at the same value of q, namely $q_0$, and is of the same height as before, but is displaced out along the $x_i$-axis as shown. We therefore immediately conclude that thermal decay, for which the Gibbs exponent in Eq. (37) depends only on the height of the saddlepoint, will be almost insensitive to the presence of dissipation;* on the other hand the quantum tunnelling rate, for

*The prefactor does depend on the dissipation.

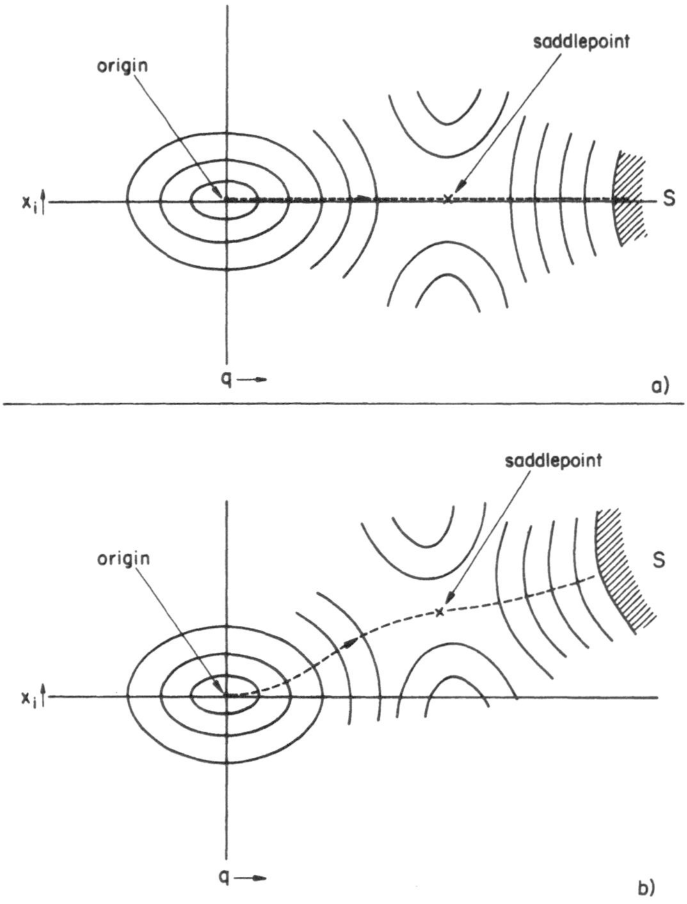


Fig. 1 Potential contours in the many dimensional space spanned by q and $x_i$. a) for the case $C_j = 0$ and b) $C_j \neq 0$.

which the exponent is generally proportional to $\int [V(q,x_i)]^{1/2} ds$, will be reduced in the dissipative case since the distance the system has to travel to and away from the saddlepoint is greater in this case (see Fig. (1b)).

It is possible, but rather combersome, to obtain quantitative formulae for the decay rate by applying the WKB method in the many-dimensional space of q and the $\{x_i\}$. A more elegant and powerful technique is to use some variant of the idea pioneered by Feynman and Vernon[27] for the study of dissipative quantum systems: one writes down an expression for some quantity characterizing the system in terms of a functional integral over both the system and environmental coordinates and then integrates out the latter explicitly. What results is a functional integral over the trajectories of the system variable q alone, which in general will contain non-local terms (cf. below), and from which one may hope to extract the required information. A number of variants of this approach can be found in the literature; here I shall describe briefly the one developed by A.O. Caldeira and myself[28,2] (and independently by Sethna[29]), which is based on calculation of the system density matrix at temperature $T \equiv 1/k_B\beta$:

$$p(q_i, q_f : \beta) \equiv \sum_n e^{-\beta E_n} \prod_\alpha \int dx_\alpha \, \Psi_n^*(q_i : \{x_\alpha\}) \Psi_n(q_f : \{x_\alpha\}) \tag{2.13}$$

By the Feynman-Kac formula this quantity can be represented as a functional integral:

$$p(q_i, q_f : \beta) = \prod_i d\bar{x}_i \int_{q(o)=q_i}^{q(\beta)=q_f} \mathcal{D}q(\tau) \int_{\dot{x}_i(o)=\bar{x}_i}^{x_i(\beta)=\bar{x}_i} \mathcal{D}x_i(\tau) \exp\left(-\int_o^\beta L(q, \dot{q} : \{x_i, x_i\})\, d\tau/\hbar\right) \tag{2.14}$$

where L is the "Euclidean" Lagrangian corresponding to the Hamiltonian Eq. (2.12), which is in fact nothing but the Hamiltonian itself expressed in terms of $\dot{q}$ and $\dot{x}_i$ rather than p and $p_i$. Since the $x_i$ and $\dot{x}_i$ occur only up to second order in this expression, we can perform the functional integrals over $\mathcal{D}x_i(t)$ and then the integrals over $\bar{x}_i$ explicitly (see Ref. 20, pp. 82-3). The result is

$$p(q_i, q_f : \beta) = \int_{q(o)=q_i}^{q(\beta)=q_f} \mathcal{D}q(\tau) \exp(-S_{eff}[q(\tau)]/\hbar) \tag{2.15}$$

where the "effective action" $S_{eff}[q(t)]$ is given by the expression

$$S_{eff}[q(\tau)] = \int_o^\beta d\tau \{\tfrac{1}{2}M\dot{q}^2 + V(q)\} + \tfrac{1}{2}\int_o^\beta d\tau \int_o^\beta d\tau' \alpha(\tau-\tau')\{q(\tau)-q(\tau')\}^2 \tag{2.16}$$

where the quantity $\alpha(\tau - \tau')$ is given by*

$$\alpha(\tau-\tau') \equiv \int_0^\infty J(\omega) e^{-\omega|\tau-\tau'|} d\omega \tag{2.17}$$

*Strictly speaking this expression can only be used if the trajectory $q(\tau)$ is continued periodically so that $q(\tau+\beta) \equiv q(\tau)$, and the integral over $\tau'$ in the second term in Eq. (2.16) extended over the whole real axis. However, in the limit $\beta \to \infty$ which is of interest here this complication can be neglected. See Ref. 2, Section 4.

with the spectral density $J(\omega)$ defined by Eq. (2.10). The validity of Eqs. (2.15)-(2.17) is, of course, completely independent of whether or not $J(\omega)$ has the specific form given in Eq. (2.11). If it does, then $\alpha(\tau - \tau')$ is given by the simple expression

$$\alpha(\tau-\tau') = \frac{\eta}{2\pi} \frac{1}{(\tau-\tau')^2} \tag{2.18}$$

down to values of $|\tau - \tau'|$ of the order of the inverse cutoff frequency $\omega_c^{-1}$. Since small values of $|\tau - \tau'|$ do not contribute appreciably to the functional integral (2.16), we can take Eq. (2.18) to be generally valid without appreciable error.

In principle the expression (2.15) will allow us to obtain all the equilibrium properties of the system. We are specifically interested in obtaining the tunnelling rate out of the groundstate. Although this is not, strictly speaking, an equilibrium property, it may plausibly be obtained by taking over an argument due to Langer,[31] Stone,[32] and Callan and Coleman[33] for the non-interacting case ($\alpha(\tau - \tau') \equiv 0$). This argument is based on the observation that the functional integral (2.15) will be dominated by paths which are extrema of the action $S[q(\tau)]$, that is, those trajectories for which $q(t)$ is a solution of the classical equations of motion in the inverted potential $-V(q)$. Now, if we consider for definiteness the case $q_i = q_f = 0$, there are only two such paths in the limit $\beta \longrightarrow \infty$: one is the trivial trajectory $q(\tau) \equiv 0$ (which is obviously a local minimum of S), and the other is the so-called "bounce", in which the system "rolls off" the maximum of the inverted potential at the origin, moves across to the exit point and then rolls back to the origin (see Ref. 33). This trajectory is not a minimum but a saddlepoint of the action, and an appropriate analytic continuation argument then indicates that it should produce, in effect, an imaginary part in the groundstate energy (which can be identified from the consideration that in the limit $\beta \longrightarrow \infty$ the $\beta$-dependence of $p(q_i, q_f:\beta)$ is, from Eq. (2.13), $\exp -\beta E_0$). This is then identified with the tunnelling rate (times h). The upshot of the argument is that the tunnelling rate $\Gamma$ has the form

$$\Gamma = \bar{\omega} \exp -B_0/\hbar \tag{2.19}$$

where $B_0$ is the value of the action for the "bounce" (saddlepoint) trajectory, and $\bar{\omega}$ is a prefactor which can be calculated from the small fluctuations around this trajectory. A detailed calculation shows that (2.19) agrees precisely with the result (35) obtained by the familiar WKB method.

We now take this argument over directly to the dissipative

case. All stages proceed exactly in parallel, and we finally obtain the result

$$\Gamma = \tilde{\omega} \exp - B/\hbar \tag{2.20}$$

where, however, B is now the action (2.16) evaluated along the trajectory which is a saddlepoint of this functional, and $\tilde{\omega}$ can be calculated from the small fluctuations around this path.

Thus, the main part of the problem is reduced to the purely mathematical problem of finding the saddlepoint value of the action functional (2.16). It is possible to analyze this question in some detail for the physically important case of a quadratic-plus-cubic potential (see Ref. 2, Section 5). The result is that we can always write B in the form $B_0 + \Delta B$, where $B_0$ is the action in the absence of the last term in (2.16) and $\Delta B$ has the form

$$\Delta B = A_0 \eta (\Delta q)^2/\hbar \tag{2.21}$$

Here $\eta$ is the phenomenological friction coefficient, $\Delta q$ is the distance which the undamped system would have to travel under the barrier to escape, and $A_0$ is a constant of order 1 which is a weak function of the degree of damping, varying smoothly from the value $12\zeta(3)/\pi^3 \cong 0.47$ for light damping to $2\pi/9 \cong 0.70$ for strong damping. It is possible to analyze the behavior of the prefactor $\tilde{\omega}$ and in particular to show that in limit of strong damping it is proportional to the 9/2 power of the damping. Since this dependence, though unexpectedly strong, is still negligible under most realistic circumstances compared to the dependence through $\Delta B$, the upshot of all this is that we can say that to a good approximation the effect of the damping term $\eta\dot{q}$ in the phenomenological equation of motion (2.6) is to multiply the tunnelling rate calculated in its absence by $e^{-\Delta B/\hbar}$. Thus, if $P_0$ denotes the tunneling rate calculated for a system obeying Eq. (2.6) with $\eta = 0$ and $P(\eta)$ is the rate for a system obeying the actual Eq. (2.6), then our fundamental result is

$$P(\eta)/P_0 \sim \exp(-A_0\eta (\Delta q)^2/\hbar) \tag{2.22}$$

Since $\Delta q$ for a quandratic-plus-cubic potential can be expressed in terms of $V_0$ and $\omega_0$, we can equally well write the result for the actual tunnelling rate in terms of the dimensionless damping parameter $\alpha \equiv \eta/2M\omega_0$ defined in Lecture 1:

$$P(\alpha) = f(\alpha)\omega_0(60)^{1/2}(18\, U_0/5\pi\hbar\omega_0)^{1/2} \exp - \frac{36}{5}\frac{V_0}{\hbar\omega_0}\{1 + \frac{15}{4}A_0\alpha\} \tag{2.23}$$

where $f(\alpha)$ is of order 1 provided we have $\alpha < 1$. These results may be applied directly to the physically interesting problem of macroscopic quantum tunnelling in SQUID's or CBJ's, provided they are assumed to be adequately described by the RSJ model: we need only take the quantities $V_o/\hbar\omega_o$ and $\alpha$ to be given by the appropriate expressions, that is (see. Eqs. (26), (27) and (32)).

$$(36/5)V_o/\hbar\omega_o = (3.2^{1/4}/5)\lambda^{1/2}\ (\delta I/I_c)^{5/4} \tag{2.24}$$

$$\alpha = \frac{1}{2}\beta_c^{-1/2}(2\delta I/I_c)^{-1/4} \tag{2.25}$$

COMPARISON WITH EXPERIMENT

We may now compare the prediction with the existing experimental results. Apart from some early experiments[34] which probably reached only the borders of the tunnelling regime, the only experiments done in a region of parameters where the above formulae should be quantitatively valid* are those of Jackel and co-workers[35] at Bell and of Voss and Webb[36] at IBM. Both of these were performed on current-biased junctions, and consisted essentially of sweeping the external current up towards $I_c$ and measuring the distribution of values of $\delta I$ at which the onset of a finite-voltage state occurred (indicating that the system has tunnelled through, or fluctuated over, the barrier in the washboard potential and is "running downhill"). From these data it is possible to deduce the tunnelling rate as a function of $\delta I$. Both experiments suffer from some uncertainty concerning the exact value of the relevant capacitance, which could only be estimated. Jackel et al. used Pb - Pb(In) tunnel junctions with critical currents of a few hundred μA and capacitances estimated at 5 10 x $10^{-15}$ F. With these parameters the crossover between thermal decay and quantum tunnelling (which should show up as a flattening-off of the graph of $\ln P$ versus temperature for fixed $\delta I$, or of the root-mean-square $\sigma$ of $\delta I$ (see Lecture 1) as a function of T) is predicted by the simple WKB theory of Lecture 1 to occur at about 8K. In fact, the data of Jackel et al. do show a flattening-off, but only below about 2K. (Their data go down to ∿ 1.5K.) Thus their data are qualitatively compatible with the above theoretical predictions about the effect of damping;

*The SQUID experiments of the Leiden group[37-8] were done with nearly degenerate potential wells and are more naturally discussed in the context of quantum coherence, while the SQUID experiments done at Sussex[39] and Kharkov[40] are too indirect to be easily compared with theory.

however, since they could not measure the shunting resistance of their junction directly it is impossible to draw quantitative conclusions. Voss and Webb[36] used all Nb junctions with critical currents of the order of 100 nA - 1 A and capacitances estimated[40a] at 0.1 pF with an upper limit of 0.15 pF; they were able to take data down to 3 mK. For a junction with very weak damping ($\beta_c \equiv 2\pi I_c R^2 C/\phi_0 \sim 5000$) they found that their data were in fairly good agreement with the above theoretical predictions. For a second junction with moderate damping ($\beta_c \sim 50$) they found reasonable agreement with Kramers prediction (37) at high temperature and a flattening-off at low temperature; however, the asymptotic value of the rate in the low-temperature limit suggested a value of the constant $A_0$ in (2.23) of about 2.8*, whereas the value calculated in Ref. 2 is 0.47 in the weak-damping limit and never rises above $\sim 0.7$. If one believes that the general lines of the theoretical predictions are correct, the most likely resolution of the discrepancy is that the actual shunting conductance of the junction has an appreciable frequency-dependence, so that the simple RSJ model with constant $R_n$ is not strictly applicable. We would then expect formula (2.22) still to be qualitatively right, but with $\eta$ replaced by the inverse value of resistance at some frequency of the order of the attempt frequency $\omega_0$.

MACROSCOPIC QUANTUM TUNNELLING: MORE GENERAL CASE

A more detailed consideration of the experimental setup in experiments of this type raises the possibility that the simple RSJ Eq. (18) may be an oversimplification, not only because the shunting conductance may be frequency-dependent but also because the circuit may be more complicated than is describable by a single capacitance or resistance. Nevertheless, we should expect in principle that all elements of the circuit other than the (obviously nonlinear) Josephson element which provides the metastability should be describable by appropriate linear impedances and hence that the classical equation of motion should be linear apart from the term generated by this element. This prompts the following question: Assume that we have a tunnelling variable q(t) such that its Fourier transfer $q(\omega)$ obeys, in the classical regime, an equation of the form

$$K(\omega)q(\omega) = -(\partial V/\partial q)(\omega) \tag{2.26}$$

where $K(\omega)$ is an arbitrary function of $\omega$ subject only to the conditions imposed by causality, etc. Can we then predict the tun-

*Ref. 36 underestimates the quantity $(\Delta s)^2$ for a quadratic-plus-cubic potential for the factor 16/27, and hence overestimates $\alpha$ (or equivalently A) by the inverse of this.

nelling rate of the variable q out of the metastable minimum. Evidently the problem we considered above is a special case of this with $K(\omega) = -M\omega^2 + i\omega\eta$.

Rather remarkably, the answer to this question is yes, provided that the actual mechanism giving rise to $K(\omega)$ is itself linear in q (i.e., the Hamiltonian of the system interacting with its environment contains, other than V(q) only quadratic terms referring to the system variables alone and linear terms involving coupling to the environment).* In many cases, especially those involving electromagnetic effects, we can be reasonably sure on a priori grounds that this condition is fulfilled. Then the prescription for calculating the tunnelling rate is remarkably simple: The rate is calculated, as indicated above, by finding the saddlepoint value (and the fluctuations around it) of an effective action $S_{eff}$ given in terms of the Fourier transform $\tilde{q}(\omega)$ of $q(\tau)$ with respect to $\tau$ as follows:

$$S_{eff}[\tilde{q}(\omega)] = \int d\omega \{\tfrac{1}{2}\tilde{K}(\omega)|\tilde{q}(\omega)|^2\} + S_V[\tilde{q}(\omega)] \qquad (2.27)$$

where $\tilde{K}(\omega)$ is obtained from $K(\omega)$ by the simple replacement $i\omega \longrightarrow |\omega|$, i.e., $\tilde{K}(-\omega) = \tilde{K}(\omega) = K(-i\omega)(\omega \geq 0)$ and $S_V$ corresponds to the term V(q) in Eq. (2.16. It is easily verified that the result Eq. (2.16), with Eq. (2.18) corresponds to $\tilde{K}(\omega) = M\omega^2 + \eta|\omega|$ , and is therefore a special case of this general prescription. Although the above result, once stated, looks obvious, the proof is not entirely trivial; it will be given in detail elsewhere.**

As an example, consider the circuit of Fig. 2, which may well be a reasonable description of SQUID's with point-contact junctions (see above). In this case the classical equation of motion of the flux $\Phi$ through the bulk inductance corresponds to a $K(\omega)$ of the form

$$K(\omega) = i\omega\{i\omega c + R_n^{-1} + \frac{i\omega C_G + 1/i\omega L}{(1+\frac{L'}{L}) + (i\omega L')(i\omega C_G)}\} \qquad (2.28)$$

where it should be carefully noted that the original self-inductance term $\Phi^2/2L$ has been incorporated into $K(\omega)$ (so that V(q) is now simply the Josephson term in Eq. (16)). From the form of $\tilde{K}(\omega)$ it is then obvious that, as we might intuitively expect, if the frequencies important in the integral (2.27) are small

*This proviso is obviously a generalization of the one we found necessary for the case of simple linear friction.

**A paper on this subject is in preparation for submission to Phys. Rev. B.

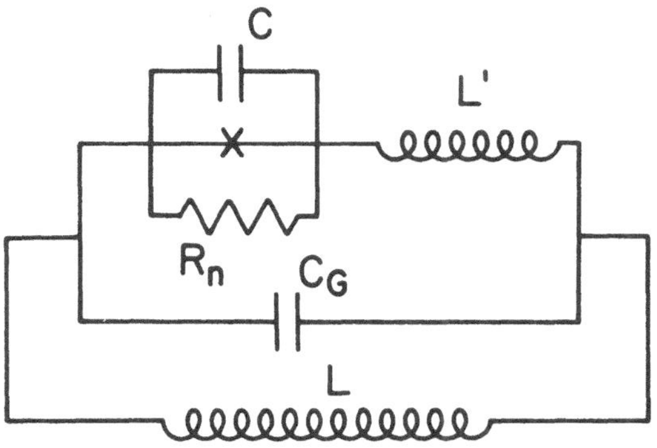


Fig. 2 Equivalent circuit of a SQUID with a point contact junction.

compared to the characteristic frequency* $\bar{\omega} = (LC_g)^{-1/2}$ then the correct tunnelling formula can be obtained from a simple RSJ model using an effective capacitance and inductance given by

$$C_{eff} = C_G/(1 + L'/L), \quad L_{eff} = L + L' \tag{2.29}$$

while if the important frequencies are large compared to $\bar{\omega}$ we can again use a simple RSJ model with**

$$C_{eff} = C, \quad L_{eff} = L' \tag{2.30}$$

Since, generally speaking, the tunnelling rates given by the latter choice will be the faster, we can conclude that is is self-consistent to use the values (2.30) provided that the characteristic frequency of the "bounce" motion which results from their use is indeed large compared to $\bar{\omega}$. Whether the resultant tunnelling is correctly described as "macroscopic" in this case is however, a delicate question: as is intuitively obvious from the circuit diagram (and confirmed by an explicit introduction of extra flux variables) the actual tunnelling event takes place in this case exclusively in the region very close to the junction itself, and the new value of flux than spreads around the bulk

*In this order-of-magnitude argument we assume that $C_G >> C$ and $L \sim L'$.

**However, note that L' is an effective inductance only for the small deviations of $\Phi$ from the metastable equilibrium value. The static potential curve is of course, determined by the effective low-frequency inductance (L + L').

SQUID ring by a purely classical process.

As the results quoted above are valid only at zero temperature, it is clearly highly desirable to generalize them to the finite-temperature case. In particular, it would be satisfying to confirm the intuitively plausible conjecture[41,2] that for a heavily damped system the crossover form classical to quantum behaviour occurs not for $k_bT \sim \hbar\omega_o$, as in the undamped case, but for $k_BT/\hbar$ of the order of the characteristic frequency of overdamped motion, $2M\omega_o^2/\eta$. However, the finite-temperature calculation is not an entirely trivial generalization of the zero-temperature one, and at the moment only limited results[42] have been obtained in this area.

DISSIPATION AND MACROSCOPIC QUANTUM COHERENCE

I want finally, to say a few words about the question of the effect of dissipation on macroscopic quantum coherence. The reason it is only a few words is that, although a number of significant results have been obtained in this area, our overall understanding of the effect of dissipation on coherence is (in my opinion) much less complete than our knowledge of its effects (at least at zero temperature) on tunnelling. I will consider here only the case where in the absence of dissipation the system tunnels between two exactly degenerate wells, as in Eq. (47): the distance between the wells along the coordinate (q) axis will be denoted $q_o$. I will assume that the coupling of the system to its environment is adequately described by the Hamiltonian Eq. (2.12), where the spectral density $J(\omega)$ (Eq. (2.10)) is not necessarily constrained to satisfy Eq. (2.11) but may have a more general form; also, I assume that $V(q)$ is a symmetric function of q (e.g., $V(q) = -\alpha q^2 + \beta q^4$). Note that with these assumptions parity is a good quantum number even in the presence of the system-environment coupling and in particular the two degenerate minima in the many-dimensional space of q and the $x_i$ are symmetrically placed with respect to the origin (Fig. 3).

The main reason why the coherence problem is more subtle than the tunnelling one is that a much greater range of frequencies plays a role. We recall that typically in the WKB regime the "bare" tunnelling frequency $\Gamma_o$ defined by Eq. (43) is smaller by an exponential factor than the "attempt" frequency $\omega_o$. Thus, except under very special circumstances, $\Gamma_o$ is unlikely to be greater than, say a few MHz (and may well be much smaller than this)* while $\omega_o$ is typically of the order of several $\Gamma_o$ by a

*Even if it is possible to set the parameters so as to obtain higher values, would presumably be extremely difficult to do an "oscillation" experiment at these frequencies, in view of the need to switch the electronics on and off (see Lecture 1).

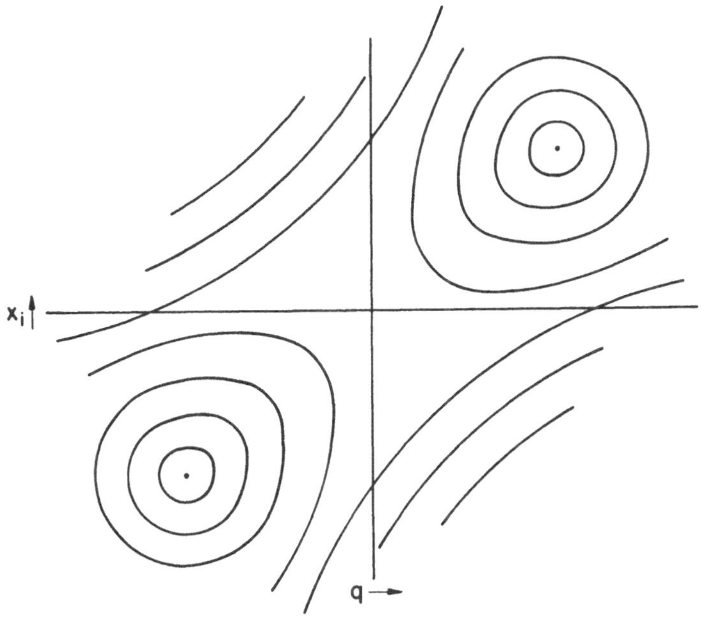


Fig. 3 Symmetric contours in the many dimensional space of q and $x_i$.

factor similar to that in the tunnelling problem. Now let us consider the interaction with oscillators in the frequency range $\Gamma \ll \omega \ll \omega_0$. Since the nature of the (damped-harmonic-oscillator) density matrix of the system in the region near the two wells is determined mainly by the oscillators with $\omega \gtrsim \omega_0$ and fairly insensitive to what goes on in the range $\Gamma \ll \omega \ll \omega_0$, we can now preseumably approximate the problem by a "two-state" problem of a form very familar in, for example, the theory of defect tunnelling in solids. If we represent this two-state problem in the familiar way as the problem of a particle of spin 1/2 such that $\sigma_z = +1(-1)$ corresponds to being in the left-hand (right-hand) well with $q = \pm q_0$, then the appropriate Hamiltonian is

$$\hat{H} = -\Gamma\sigma_x - q_o\sigma_z\Sigma_j C_j x_j + \tfrac{1}{2}\Sigma_i(p_i^2/m_i + m_i\omega_i^2 x_i^2) \tag{2.32}$$

where the sum is cut off at some frequency $\bar{\omega}$ such that $\Gamma \ll \bar{\omega} \ll \omega_0$. Let us consider the effect of the oscillators with $\Gamma \ll \omega \ll \bar{\omega}$. These will follow the motion of the variable $\sigma_z$ nearly adiabatically, so that the wave function of the whole system is to a good approximation of the form

$$\Psi(\sigma_z:\{x_i\}:t) \cong \sum_{\alpha=\pm 1} C_\alpha(t)\chi_\alpha(\sigma_z)\prod_i \psi_\alpha^0(x_i) \tag{2.33}$$

where $\chi_\alpha(\sigma_2)$ is the spin function corresponding to $\sigma_2 = \alpha = \pm 1$, and the wave functions $\psi^0_\alpha(x_i)$ are the groundstate of the last three terms in (2.32) for fixed $\sigma_2 = \alpha$, i.e.,.

$$\psi^0_\alpha(x_i) = \psi_0(x_i - \alpha C_i q_0 / m_i \omega_i^2) \tag{2.34}$$

where $\psi_0(x)$ is the usual harmonic-oscillator groundstate wave function. The matrix element of the term $\Gamma\sigma_x$ in the Hamiltonian is now reduced by the incomplete overlap of the wave functions $\psi^0_\alpha(x_i)$ corresponding to $\alpha = \pm 1$, so that it becomes $\tilde{\Gamma}\sigma_x$, where

$$\begin{aligned}\tilde{\Gamma} &= \Gamma \prod_i (\psi^0_{+1}(x_i), \psi^0_{-1}(x_i)) \approx \Gamma \prod_i \left(1 - \frac{q_0^2 h C_i^2}{m_i \omega_i^2}\right) \\ &\approx \Gamma \exp - \sum_i \frac{q_0^2 C_i^2}{\hbar m_i \omega_i^3} \equiv \Gamma \exp\left(- \int_{\omega_\Gamma}^{\bar{\omega}} \frac{2}{\pi} \frac{q_0^2}{\hbar} \frac{J(\omega)}{\omega^2}\, d\omega\right)\end{aligned} \tag{2.35}$$

where $\omega_\Gamma$ is a lower cutoff on the frequency of the oscillators taken into account. In deriving this result we used the fact that the $C_i$ individually are very saml1 compared to 1 ($\sim N^{-1/2}$).

Suppose, crudely, that the adiabatic approximation is applicable for oscillator frequencies $\omega$ greater than n times the tunnelling frequency, where n is some fairly large number. Thus, originally we shall have to take $\omega_\Gamma = n\Gamma$. But, as a result of the above argument, $\Gamma$ has been reduced to $\tilde{\Gamma}$. Thus, we can now take into account in the adiabatic approximation a new set of oscillators, namely those with frequencies between $n\Gamma$ and $n\tilde{\Gamma}$, as a result of which the effective matrix element is further reduced: and so on ad infinitum. It is clear that the condition for the process to converge to a finite value for the renormalized tunnelling matrix element $\tilde{\Gamma}$ is that for some value of $\tilde{\Gamma}$ we reach the condition $d(\ln\tilde{\Gamma})/d(\ln\omega_\Gamma) < 1$, i.e.,

$$\frac{2q_0^2 J(\omega_\Gamma)}{\pi \hbar \omega_\Gamma} < 1 \tag{2.36}$$

This condition is clearly always met whenever $J(\omega)$ is proportional to $\omega^n$, $n > 1$, and hence in all such cases the renormalized tunnelling element remains finite, even though it may be much reduced from its original value.*

The case $n = 1$ ($J(\omega) = \eta\omega$) corresponding to the RSJ model is an interesting one. We see that there is a critical value $\eta_c \equiv \hbar/4q_0^2$ such that for $\bar{\alpha} \equiv \eta/\eta_c < 1$ the tunnelling matrix element

*In particular, this is the case for defect tunnelling in solids, where n is 3: cf. Ref. 29.

$\tilde{\Gamma}$ renormalized in this way takes a finite value of order

$$\tilde{\Gamma} \sim \bar{\omega}\,(\Gamma/\bar{\omega})^{1/(1-\bar{\alpha})} \tag{2.37}$$

(this follows from Eq. (2.35) if we set $\omega_\Gamma \sim \tilde{\Gamma}$), while for $\bar{\alpha} > 1$ $\tilde{\Gamma}$ iterates to zero. Thus for $\bar{\alpha} > 1$ there is no tunnelling and an infinitesimal bias will suffice to localize the system in one well or the other. The above argument is essentially a "poor man's version" of the more sophicticated renormalization-group arguments given by Chakravarty[44] and Bray and Moore[45] which lead to the same conclusion.

Thus, for linear ohmic dissipation given by a friction coefficient $\eta$ greater than the critical value $\eta_c$ the problem of macroscopic quantum coherence does not not arise. For $\eta < \eta_c$ (or for dissipation other than linear) we still have the problem of taking into account the effect of the oscillators with frequencies $\omega \lesssim \tilde{\Gamma}$. To the best of my knowledge there is at present no completely quantitative solution of this problem outside the weak-coupling limit, but one may get some feeling for the likely qualitative effects by the following argument (cf. Ref. 1). Imagine that we take our two-level system, uncoupled to its environment but with effective tunnelling frequency $\tilde{\Gamma}$, and "detune" the two wells by applying a biasing potential $\Delta\varepsilon$ to one of them (equivalent to a "magnetic field" 1/2 $\Delta E$ in the z-direction). It is very easy to see that for $\Delta E \ll \hbar\tilde{\Gamma}$ the tunneling process occurs much as before, while $\Delta E > \hbar\tilde{\Gamma}$ the energy eigenfunctions are effectively localized in one well or the other and tunnelling is suppressed. Now we imagine the environment as providing, through the coupling terms in (2.32), a fluctuating biasing potential $\Delta E(t) = 2q_o \Sigma C_i x_i$: if we assume that the dynamics of the "field" $q_o \sum_i C_i x_i$ are independent of the coupling to the system and determined only by the intrinsic environment Hamiltonian, (which is, of course, an approximation) then we can calculate the correlations $\langle \Delta E(t) \Delta E(t') \rangle$ and hence the root-mean-square detuning $\langle \overline{\Delta E})^2 \rangle^{1/2}$, where $\overline{\Delta E}$ is the time average of the operator $\Delta E(t)$ over a period $\sim \tilde{\Gamma}^{-1}$ (the fluctuations of frequency $\omega \gg \tilde{\Gamma}$ have already been taken into account in the replacement of $\Gamma$ by $\tilde{\Gamma}$). For linear ohmic dissipation ($J(\omega) = \eta\omega$) we find

$$\langle (\overline{\Delta E})^2 \rangle^{\frac{1}{2}} \sim (q_o^2 \eta \tilde{\Gamma}\, k_B T)^{\frac{1}{2}}, \quad k_B T \gtrsim \hbar\tilde{\Gamma} \tag{2.38a}$$

$$\langle (\overline{\Delta E})^2 \rangle^{\frac{1}{2}} \sim (q_o^2 \eta \hbar \tilde{\Gamma}^2)^{\frac{1}{2}}, \quad k_B T \ll \hbar\tilde{\Gamma} \tag{2.38b}$$

If we now assume that the criterion for appreciable coherence effects to be observable is $\langle(\Delta\tilde{E})^2\rangle^{1/2} \lesssim \hbar\tilde{\Gamma}$, then this criterion becomes, in terms of the dimensionless parameter $\alpha \equiv \eta/\eta_c$

$$\bar{\alpha} \lesssim 1 \text{ for } k_B T \ll \hbar\tilde{\Gamma} \qquad (2.39a)$$

$$\bar{\alpha}\left(\frac{k_B T}{\hbar\tilde{\Gamma}}\right) \lesssim 1 \quad \text{for } k_B T \gtrsim \hbar\tilde{\Gamma} \qquad (2.39b)$$

As pointed out above, for a realistic experiment it is probably the second criterion which is more relevant. Going back to Eq. (2.37) and expanding it for small $\bar{\alpha}$, we see that (2.39a) implies (assuming $\bar{\omega} \sim \omega_0$ and $\Gamma \sim \Gamma_0$)

$$k_B T/\hbar\Gamma_0 \lesssim \bar{\alpha}^{-1}\exp(-\bar{\alpha}\ln(\omega_o/\Gamma_o)) = \bar{\alpha}^{-1}\exp -\bar{\alpha}\, B_0/\hbar$$

Thus we need, as a minimum, $\bar{\alpha}$ small compared with the inverse of the (undamped) bounce exponent $B_0/\hbar$. For a SQUID, in view of the results of Lecture 1 and the definition of $\bar{\alpha}$, this means the resistance should satisfy the inequality

$$R_n \gtrsim K^{5/2}\lambda \quad \phi_0^2/\hbar \qquad (2.40)$$

which is a difficult condition, but not a priori obviously impossible. Clearly further work along these lines is necessary to take account of other possible dissipative mechanisms not included in the RSJ model, and to obtain quantitative results for intermediate values of $\bar{\alpha}$.

## CONCLUSION

Generally speaking, it is probably fair to say that the theory of macroscopic quantum tunnelling in SQUIDs and Josphson junctions is reasonably well understood, at least at zero temperature, although there are many points of detail which require further investigation. Moreover, existing experiments appear to be in at least qualitative agreement with the theoretical predictions, although some discrepancies remain to be explained. With regard to macroscopic quantum coherence, the theory is at present less well-developed, and it appears probable that the few existing experiments in this area[38] have operated in a region where we would not expect to observe coherence. It is clear that both theorists and experimentalists have a lot of work ahead of them.

I am very grateful the members of the Cornell Quantum Tunnelling Seminar, and in particular to Vinay Ambegaokar, Sudip

Chakravarty, Jim Sethna, and David Waxman for very helpful discussions of this topic.

REFERENCES

1. A.J. Leggett, Prog. Theor. Phys. Supp. 69, 80 (1980).

2. A.O. Caldeira and A.J. Leggett, Ann. Phys. (NY), in press.

3. K.K. Likharev, Usp. Fiz. Nauk 139, 170 (1983) (translation to appear in Soviet Physics Uspekhi).

4. V. Ambegaokar, this volume, p. 43.

5. A.J. Leggett, to be published

6. V. Ambegaokar, in Superconductivity, ed. R.D. Parks (Marcel Dekker, NY 1969) Vol. 1, p. 259 (see pp. 289-292).

7. M. Ishikawa, Prog. Theor. Phys. 57, 1836 (1977).

8. P.G. de Gennes, Superconductivity in Metals and Alloys, (Benjamin, NY 1966).

9. R.P. Feynman, R.B. Leighton, and M. Sands, The Feynman Lectures on Physics, Vol. 3 (Addison-Wesley, Reading, MA 1965).

10. B.D. Josephson, Adv. Phys. 14, 419 (1965).

11. A. Barone and S. Paterno, Physics and Applications of the Josephson Effect (Wiley, NY 1982), Ch. 2.

12. O.V. Lounasmaa, Experimental Principles and Methods Below IK (Academic Press, NY 1974).

13. J. Kurkijärvi, Phys. Rev. B6, 832 (1972).

14. A.J. Leggett, in Proc. 6th International Conf. on Noise in Physical Systems, ed. R.J. Soulen, Gaithersburg, MD 1981 (NBS Special Publ. No. 614), p. 355.

15. R. de Bruyn Ouboter and D. Bol, Physica 112B, 15 (1982).

16. S. Gasiorowicz, Elementary Particle Physics (Wiley, NY 1966), pp. 50-54.

17. P.W. Anderson, in Lectures on the Many Body Problem, ed. E. Caianello, (Academic Press, NY 1964), p. 113.

18. A.O. Caldeira, D. Phil, Thesis, Univ. of Sussex 1980 (unpublished).

19. H.A. Kramers, Physica 7, 284 (1940).

20. D. Park, Quantum Mechanics (McGraw-Hill, NY 1964) p. 100.

21. S. Chakravarty, to be published.

22. D.H. Perkins, An Introduction to High Energy Physics, (Addison-Wesley, Reading, MA, 1972) Ch. 4.

23. S. Chakravarty and S. Kivelson, Phys. Rev. Lett. 50, 1811 (1983).

24. M. Cini, Nuovo Cimento 73B, 27 (1983).

25. A.B. Migdal, Qualitative Methods in Quantum Mechanics, (Benjamin, Reading, MA 1977), Ch. 2.

26. F. Bloch, Phys. Rev. B2, 109 (1970).

27. R.P. Feynman and F.L. Vernon, Jr., Ann. Phys. (NY) 24, 118 (1963).

28. A.O. Caldeira and A.J. Leggett, Phys. Rev. Lett. 46, 211 (1981).

29. J.P. Sethna, Phys. Rev. B24, 698 (1981).

30. R.P. Feynman, Statistical Mechanics (Benjamin, NY 1972).

31. J.S. Langer, Ann. Phys. (NY) 41, 108 (1967).

32. M. Stone, Phys. Lett. 67B, 186 (1977).

33. C.G. Callan and S. Coleman, Phys. Rev. D16, 1762 (1977).

34. T.A. Fulton and G. Dunkleberger, Bull. Am. Phys. Soc. 19, 205 (1975).

35. L.D. Jackel et al., Phys. Rev. Lett. 47, 697 (1981).

36. R.F. Voss and R.A. Webb, Phys. Rev. Lett. 47, 265 (1981).

37. W. den Boer and R. de Bruyn Ouboter, Physics B + C 98, 185 (1980).

38. D. Bol, R. van Weelderen, and R. de Bruyn Ouboter, to be published.

39. R.J. Prance et al., Nature 289, 542 (1981).

40. I.M. Dmitrenko, G.M. Tsoi, and V.I. Shnyrkov, Fiz. Nizk. Temp. 8, 330 (1982) (translation: Soviet J. Low Temp. Phys. 8, 330 (1982).

40a. R.F. Voss and R.A. Webb, Physics 108B, 1307 (1981).

41. K.K. Likharev, Proc. 16th Inter. Conf. on Low Temperature Physics, Los Angeles, CA 1981 (Physica 108B, 1079 (1981).

42. G. Zwerger, to be published.

43. G. Zwerger, Z. Phys. B47, 129 (1982).

44. S. Chakravarty, Phys. Rev. Lett. 49, 681 (1982).

45. A.J. Bray and M.A. Moore, Phys. Rev. Lett. 49, 1545 (1982).

# QUANTUM DYNAMICS OF SUPERCONDUCTORS AND TUNNELING BETWEEN SUPERCONDUCTORS

Vinay Ambegaokar

Laboratory of Atomic and Solid State Physics
Cornell University
Ithaca, New York 14853

## 1.1 INTRODUCTION

In weakly coupled superconducting systems the phase difference of the complex order parameters on the two sides is the essential dynamical variable. The switching of such systems from superconducting to resistive states involves a transition from a phase value that is on the average constant to one that grows with time. At sufficiently low temperatures it is natural to think of this transition as taking place by quantum mechanical tunneling through a potential barrier. However, the phase is itself an attribute of an intrinsically quantum mechanical quantity. Thus, the sense in which it can be thought of as a quantum mechanical degree of freedom is not completely obvious. Furthermore, the dynamics of the phase cannot in general be separated from the microscopic degrees of freedom--quasiparticles--so that the correct equations of motion must describe a kind of quantum Brownian motion.

My aim in these lectures is to provide a framework which addresses these conceptual and technical difficulties head on. The approach is tied to a microscopic model of a tunnel junction. This has advantages and drawbacks. On the positive side, the development is explicit and self contained. The explicitness, however, ties one to a simple tractable model, whereas we all know that phenomenology, when used with good judgment, can get at essential physics while avoiding irrelevant details. Since A. J. Leggett's lectures at this institute take very much a phenomenological point of view, mine should complement his rather well.

One matter I leave entirely to Leggett. That is the general question of whether ordinary quantum mechanics describes transitions between macroscopically distinct quantum states in superconducting devices. (See his lectures, and references therein.) I would be most surprised if it does not, and it would never occur to me to doubt that it does. What follows is a technical but straightforward application of the quantum mechanical machinery which--basically mysterious though it may be--we have all learned to operate with instructions from Copenhagen. As for Schrödinger's cat, my way out of that conundrum is to remark that, as a reluctant co-owner of one, I know that cats are more devious--for which read complex--than superconductors.

A natural way to introduce a complex order parameter into a discussion of superconductivity--without at the same time limiting one's self to self-consistent field theory--is to introduce the order parameter as a variable in a functional integral. This is the method I shall be using. It has certain artificialities which will not be concealed, but it has merits as well. In addition to the one mentioned above, it has the advantage of allowing a nice treatment of dissipation and fluctuations.

Since I have been asked to give two lectures and to give a somewhat tutorial survey, this first lecture will concentrate on a single bulk superconductor. We shall see how the method allows one to recover the mean field theory and to deal easily with small fluctuations of the amplitude and phase of the order parameter. In the second lecture, the same method will be applied to a model of an oxide layer tunnel junction. The first part of this lecture will focus on the imaginary time version of the effective action, which is appropriate for calculating the rate of escape from a metastable potential minimum at zero temperature. Here there occurs an interesting physical consequence of the energy gap characteristic of superconductors: the adiabatic coupling of quasiparticle fluctuations to slow motions of the phase leads to an increase in the capacitance of the junction. Until I started to write these lectures, I guessed that this would be an extremely small effect. However, a numerical estimate for a feasible junction shows that the correction can be appreciable; I would be interested in finding out here if this effect has been observed. In the rest of lecture II I go on to discuss the real time finite temperature formalism. Here again one finds that the self-consistent theory reappears as an extremum or least action path. At this point I shall reach current research, stop talking, and be prepared to listen.

These are rough lecture notes largely based on private working papers. I apologize in advance to other authors: references are only given to clarify or complement the treatment given here.

Let me begin, then, with a very few words about functional integrals in the quantum mechanics of one particle. The Schrödinger equation for a free particle of mass m is a diffusion equation in imaginary time, with diffusion constant $(\hbar/m)$. The probability amplitude that a free particle at point x = 0 is at point x a time t later is--taking one dimensional motion for simplicity, and writing $\tau = it$:

$$<x|e^{-(\tau H/\hbar)}|0> = \int \frac{dk}{2\pi} e^{ikx} \exp(-\tau\hbar k^2/2m)$$

$$= (m/2\pi\tau\hbar)^{\frac{1}{2}} \exp-(mx^2/2\hbar\tau) \qquad (1)$$

One can view this transition as taking place in N steps, corresponding to time intervals $\delta\tau = \tau/N$, with no attempt being made to determine the position of the particle at intermediate times. The previous answer (1) must, of course, reappear. A calculation, the details of which are left as an exercise, proceeds as follows:

$$<0|e^{-\tau H}|0> = (2\pi\delta\tau)^{-\frac{1}{2}} \int \prod_{j=1}^{N-1} \frac{dx_j}{(2\pi\delta\tau)^{\frac{1}{2}}} \exp-\sum_{j=0}^{N-1} (x_j-x_{j+1})^2/2\delta\tau \Big|_{x_0=x_N=0}$$

$$= (2\pi N\delta\tau)^{-\frac{1}{2}} = (2\pi\tau)^{-\frac{1}{2}} \qquad (2)$$

Here, the diffusion constant $\hbar/m$ has been set equal to one to simplify writing. The second expression in (2), in the limit of very large N, is an extremely explicit example of a path integral. Indeed, that multiple integral is what is meant by the notation

$$<0|e^{-\tau H}|0> = \int \mathcal{D}x \ e^{-S[x]} \qquad (3)$$

with $S[x] = \int_0^\tau du\frac{1}{2}(dx/du)^2$ and $x(0)=x(\tau)=0$. This is Feynman's form of quantum mechanics:[1] the transition amplitude is a weighted sum over paths. For a particle in a potential V(x), a similar decomposition into small time steps leads to Eq. (3) with $S[x] = \int_0^\tau du[\frac{1}{2}(dx/du)^2 + V(x)]$. The "action" S is a c-number functional of x; the non-commuting operators of the Heisenberg-Schrödinger quantum mechanics are replaced by the coherent sum over paths. The transition amplitude Eq. (1) is easily seen to be given by (3) with $x(0)=0$ and $x(\tau)=x$.

If one is interested in the escape of a particle from a metastable potential minimum at x = 0, Eq. (3) is the interesting thing to calculate. One would expect an exponential decay in real time $<0|e^{-\tau H}|0> = A\exp-(2\Gamma t)$. The formalism of (3) is then specially convenient in the semi-classical limit where there are only small deviations from the extremum of S corresponding to the classical path.[2]

The path integral formalism may seem unfamiliar and unnecessary in dealing with single particle problems, since it is completely equivalent to the Schrödinger equation. It comes into its own in a many particle context, and especially in superconductivity where the interesting degree of freedom is a complex order parameter, associated with off-diagonal order, for which the analog of the operator formalism is not at all obvious. Rather than discussing these questions in the abstract, I turn to a specific model of a superconductor.

I.2 Bulk Superconductor

A superconductor may be modelled by the Hamiltonian

$$H = -\int d^3x\psi_\sigma^+(x)\,\frac{\hbar^2\nabla^2}{2m}\,\psi_\sigma(x) - \frac{g}{2}\int d^3x\psi_\sigma^+(x)\psi_{\underline{\sigma}}^+(x)\psi_{\underline{\sigma}}(x)\psi_\sigma(x) \tag{4}$$

where $\psi_\sigma(x)$ is the electron field operator for spin $\sigma$, and a sum over spins is implied. The first term in (4) is the electronic kinetic energy and the second describes the effective attraction, in a shell around the Fermi surface, due to the exchange of phonons.[3]

The order in superconductors is associated with Cooper pairing. We would like to write an expression of the form of Eq. (3) for the transition amplitude between states characterized by different order parameters. There is a mathematical trick that allows one to do something like this, but it is important to emphasize that the method is less fundamental than that of the last section. This fact will emerge more clearly as we proceed. On the positive side, we can express an unrestricted trace as a functional integral over the order parameter. Consider the grand partition function for the system described by (4), i.e., $Z_G \equiv \mathrm{Tr}\{\exp[-\tau(H-\mu N)/\hbar]\}$ where $N$ is the number operator for electrons. ($H-\mu N$ is obtained from (4) by the replacement $\hbar^2\nabla^2/2m \to (\hbar^2\nabla^2/2m) + \mu$.) As in the last section, we break the interval $\tau$ into N subintervals. In each subinterval, the fact that the two parts of (4) do not commute with each other may be ignored to order $N^{-1}$. Now we use a Gaussian integral representation for the interaction term--quartic in $\psi$:

$$\exp|a|^2 = \int \frac{d^2z}{\pi}\exp\{-[|z|^2 - za^* - z^*a]\} \tag{5}$$

to obtain the expression

$$Z_G = \mathrm{Tr}\,\{\int \mathcal{D}^2\Delta(xu)\ T\exp[-\int_0^\tau du\ H_{eff}(u)/\hbar]\} \tag{6}$$

In Eq. (6) the integral is over a complex parameter $\Delta$ for each cell in space and (imaginary) time. From the operations described above one sees that the meaning of the functional integral is $\mathcal{D}^2\Delta(xu) = \prod_i d^2\Delta_i[(\delta\tau)(\delta\mathrm{Vol})/\pi g\hbar]$, where i labels the cell of volume $(\delta\tau)(\delta\mathrm{Vol})$. Eq. (6) is similar to Eq. (3) with the set of $\Delta_i$'s corresponding to a given space cell specifying a "path." One difference is that in (3) we have evaluated one diagonal matrix element, indicated by the fact that the paths start and end at x = 0, whereas in (6) the cyclic invariance of the trace only requires $\Delta(x\tau) = \Delta(x0)$, and not that some particular value be taken at these end points. The parallel suggests, but does not prove, that fixing the function

$\Delta(x)$ at $u = 0$ and $u = \tau$ gives a transition amplitude between states of the system. I am not sure, and indeed somewhat doubt, that this argument by analogy, which I will use in the next lecture, is strictly correct, but I do not know how to prove or disprove it. There is, however, no difficulty with calculating an unrestricted trace, as in the partition function. The symbols in Eq. (6) not yet explained are T--a shorthand to remind one that the exponential is actually an ordered product of exponentials each corresponding to an infinitesimal range of u--and $H_{eff}(u)$ which by the construction (5) is given by

$$H_{eff}(u) = \int d^3x\ \psi_\sigma^+(x)[-\frac{\hbar^2\nabla^2}{2m} - \mu]\psi_\sigma(x) + \int d^3x\Delta^*(xu)\psi_\downarrow(x)\ \psi_\uparrow(x)$$
$$+ \int d^3x\Delta(xu)\ \psi_\uparrow^+(x)\ \psi_\downarrow^+(x) + g^{-1}\int d^3x\Delta^*(xu)\Delta(xu) \quad (7)$$

The symbol Tr in (6) refers to a trace over electron states. Because $H_{eff}$ is bilinear in electron field operators, this trace can be done exactly. The key for this step is the theorem that for a u-dependent linear hermitean operator h(xu):

$$Z_G^0 \equiv \mathrm{TrT}\ \exp-\int_0^\tau (du/\hbar)\int d^3x\ \psi^+(x)h(xu)\psi(x)$$
$$= \exp\ \mathrm{Tr}\ell n G^{-1}(xu,x'u') \quad (8)$$

where $G^{-1}(xu,x'u') = [(-\hbar\partial/\partial u) - h(xu)]\ \delta^3(xx')\delta(uu')$, with anti-periodic boundary conditions, see below, at the ends of the interval $0 < u,\ u' < \tau$. A proof of (8) can be given with the use of fermion coherent states.[4] This, and the associated Grassmann variables, would take me far afield with little physical reward, so I shall content myself with showing that (8) is true, and also with what precisely it means, for the special case in which h is independent of u.[5] Then one is simply evaluating the grand partition function for a system of non-interacting fermions. If $\varepsilon_r$ are the eigenvalues of h(x), direct evaluation of the trace in (8) gives

$$Z_G^0 = \prod_r \{1 + \exp[-(\tau\varepsilon_r/\hbar)]\} \quad (9)$$

A single term in the product (9), with $\beta \equiv (\tau/\hbar)$, can be manipulated as follows:

$$(1+e^{-\beta\varepsilon}) = \exp[\ell n(1 + e^{-\beta\varepsilon})] = \exp[\int_{-\infty}^{\varepsilon} dx\ e^{x0^+}\frac{\partial}{\partial x}\ell n(1+e^{-\beta\varepsilon})]$$
$$= \exp\{\frac{\beta}{2\pi i}\int_{\Gamma_0} dz\ e^{z0^+}[\ell n(z-\varepsilon)](e^{\beta z}+1)^{-1}\} \quad (10)$$

where $\Gamma_0$ is a contour in the complex z plane going from $-\infty$ to $+\infty$ just below the real axis and returning from $+\infty$ to $-\infty$ just above. Deforming this contour to surround the simple poles, at

$\zeta_\ell = (2\ell + 1)\pi i/\beta$, $\ell=0, \pm 1, \ldots$, of $(e^{\beta z}+1)^{-1}$ and using the resulting form in (9) one obtains

$$Z_G^0 = \exp \sum_{r\ell} e^{\zeta_\ell 0^+} \ell n(\zeta_\ell - \varepsilon_r) \tag{11}$$

Now $(\zeta_\ell - \varepsilon_r)$ are the eigenvalues of the matrix $G^{-1}$, subject to the boundary conditions $G^{-1}(\tau,u') = -G^{-1}(0,u')$ and $G^{-1}(u,\tau) = -G^{-1}(u,0)$, for $h \neq h(u)$. Thus (11) is the explicit form of (8) in this special case. The factor $e^{\zeta_\ell 0^+}$ comes about because of the order of operators [$\psi^+$ to the left of $\psi$] in (8), and is needed to define the sum.

To use Eq. (8) in the context of Eq. (6), we view (7) as a matrix generalization of the bilinear form in (8). Following Nambu[6] we introduce the object

$$\psi_i(x) = \begin{pmatrix} \psi_\uparrow(x) \\ \psi_\downarrow^+(x) \end{pmatrix} \tag{12}$$

The contribution to the kinetic energy from states within the shell around the Fermi surface mentioned below (4) can be expressed in terms of $\psi_i$ and its adjoint vector as

$$\int d^3x \psi_\sigma^+(x) \left( - \frac{h^2\nabla^2}{2m} - \mu \right) \psi_\sigma(x) = \int d^3x \; \psi_i^+(x) \left( - \frac{\hbar^2\nabla^2}{2m} - \mu \right) (\tau_3)_{ij} \; \psi_j(x) + {\sum_k}' \varepsilon_k \tag{13}$$

where $\tau_3$ is the third Pauli matrix, $\varepsilon_k = (\hbar^2k^2/2m) - \mu$, and k is summed over the shell. The second and third terms in (7) can be written as $\int d^3x \; \psi_i^+(x) \; \hat{\Delta}_{ij}(xu) \; \psi_j(x)$ with

$$\hat{\Delta}_{ij} = \begin{pmatrix} 0 & \Delta(xu) \\ \Delta^*(xu) & 0 \end{pmatrix} \tag{14}$$

Proceeding by analogy with (8), one expresses (6) in the form

$$Z_G = \int \mathcal{D}^2\Delta(xu) e^{-S[\Delta]} \tag{15}$$

where

$$-S[\Delta] = \mathrm{Tr} \; \ell n \; \hat{G}^{-1} - g^{-1} \int d^3x du |\Delta(xu)|^2 - (\tau/h) {\sum_k}' \varepsilon_k \tag{16}$$

and

$$\hat{G}^{-1}(xu,x'u') = [-\hbar \frac{\partial}{\partial u} + \left(\frac{\hbar^2\nabla^2}{2m} + \mu\right)\tau_3 - \hat{\Delta}(xu)]\ \delta^3(xx')\delta(uu') \tag{17}$$

Eq. (15) is a quite remarkable formula. A sum over electronic states has been transformed into an integral over paths in order parameter space. In some sense, the progress is, however, purely formal: nothing has been explicitly calculated. The important physics in (15), alluded to by the use of the words "order parameter" in describing $\Delta$, is that the functional variables of integration correspond to the characteristic order underlying superconductivity, i.e., pairing. To emphasize this point, I pause to make connection with the Bardeen, Cooper, Schrieffer (BCS) theory of superconductivity.

## I.3 Mean Field Theory

The BCS theory emerges from (15) as the most probable or least action value for constant $\Delta$. One nice feature of the present formalism is that the broken symmetry is explicit: any constant value of the phase of $\Delta$ is, as we shall see, allowed. This is analogous to the fact that in an isotropic ferromagnet, the macroscopic magnetization is free to point in any direction. Quite generally it is useful to display the phase dependence of $\hat{\Delta}(xu)$ in (17). Let $-\phi(xu)$ be the argument of the upper right hand element in (14). Then one verifies that the matrix in (17) may be written as

$$\hat{G}^{-1}(xu,x'u') = e^{-\frac{i}{2}\phi(xu)\tau_3}\ G^{-1}(xu,x'u')\ e^{+\frac{i}{2}\phi(x'u')\tau_3} \tag{18}$$

where $G^{-1}$ is obtained from (17) by the replacements $(\partial/\partial u) \to (\partial/\partial u) - (i/2)(\partial\phi/\partial u)\tau_3$, $\vec{\nabla} \to \vec{\nabla} - (i/2)(\vec{\nabla}\phi)$, and $\Delta(xu) \to |\Delta(xu)|$ . This gauge transformation shows that a constant $\phi$ contributes nothing to the action (17). It therefore suffices for the mean field theory to evaluate (16) for constant real $\Delta$. One finds

$$-S[\Delta] = \sum_{k,\zeta_\ell} e^{\zeta_\ell 0^+}\ \ell n(\zeta_\ell^2 - E_k^2) - \frac{(Vol)\beta}{g}\Delta^2 - \beta\sum_k{}' \varepsilon_k \tag{19}$$

where again $\beta = \tau/\hbar$, and $E_k^2 \equiv \varepsilon_k^2 + \Delta^2$. The least action occurs for $\Delta = \bar{\Delta}$ where $\bar{\Delta}$ obeys the equation

$$[-\delta S/\delta\Delta]_{\Delta=\bar{\Delta}} = -\bar{\Delta}\sum_{k\zeta_\ell}(\zeta_\ell^2 - E_k^2)^{-1} - \frac{Vol\beta}{g}\bar{\Delta} = 0 \tag{20}$$

This is the BCS gap equation, as is seen by doing the sume over $\zeta_\ell$ -- whereupon one finds the well-known condition

$$1 = N(0)g \int_{-\omega_D}^{\omega_D} \frac{d\varepsilon}{2E(\bar{\Delta})} \tanh\tfrac{1}{2}\beta E(\bar{\Delta}) \tag{21}$$

where N(0) is the electron density of states at the Fermi surface, and $\omega_D$ specifies the region of phonon induced attraction. If $\bar{\Delta}$ is substituted into (19), the BCS free energy is obtained.

Eq. (19) also allows the estimation of fluctuations of $\Delta$ about $\bar{\Delta}$. One has

$$[\delta^2 S/\delta^2\Delta]_{\Delta=\bar{\Delta}} = -\bar{\Delta}\frac{\partial}{\partial\bar{\Delta}}\left[N(0)\beta(\mathrm{Vol})\int_{-\omega_D}^{\omega_D}\frac{d\varepsilon}{2E}\tanh\tfrac{1}{2}\beta E\right] \tag{22}$$

For $\beta \to \infty$ (zero temperature) the right side of (22) is $N(0)\beta(\mathrm{Vol})$, whereas for $\beta = \beta_c + \delta\beta$, where $\beta_c$ corresponds to the critical temperature, it is of order $N(0)(\delta\beta)(\mathrm{Vol})$. The correct volume to put into these expressions is the cube of the coherence length $\xi(T) \sim \hbar v_F/k_B(T_c\Delta T)^{\frac{1}{2}}$. ($k_B$ is Boltzmann's constant.) It follows that fluctuations in the magnitude of $\Delta$ are controlled at low temperatures, where $\bar{\Delta} \sim k_B T_c$, by the quantity $(\overline{\Delta^2}/\bar{\Delta}^2) \sim (T/N(0)\xi_0^3\, k_B T_c) \sim (TT_c/T_F)^2$. ($T_F$ is the Fermi degeneracy temperature.) Correspondingly, near $T_c$, where $\bar{\Delta} \sim k_B(T_c\Delta T)^{\frac{1}{2}}$, one finds the estimate $(\overline{\Delta^2}/\bar{\Delta}^2) \sim (T_c/T_F)^2(T_c/\Delta T)^{\frac{1}{2}}$. Because $(T_c/T_F)$ is typically $\sim 10^{-3}$, fluctuations in $\Delta$ are miniscule except in the tiny critical region $(\Delta T/T_c) \sim (T_c/T_F)^4$.

I have gone through these simple things to emphasize that the expression (15) contains BCS theory for a single bulk superconductor, and that $S[\Delta]$ has a deep minimum at $\Delta = \bar{\Delta}$ which constrains the bulk order parameter to the BCS value.

Eq. (15) also contains information about oscillations of spatially varying phases. These collective modes are the subject of the next section. Before closing this section, however, let me mention in passing that several other interesting consequences follow from the formalism developed above. The Ginsburg-Landau theory for spatial variations of the order parameter is obtained by working out the extremum of S for general $\Delta(x)$.[7] More interestingly, it is possible to work out in a quite straightforward way the effects of (non-critical) fluctuations on thermodynamic properties, such as the magnetic susceptibility.[8] For the examples mentioned, the results are sensitive to impurity scattering mechanisms, and the simple Hamiltonian (4), suitable for a pure system, must be modified.[9]

## I.4 Collective Modes

In a discussion of weakly coupled superconductors, the phase difference of the order parameters across the junction is the essential dynamical variable. So far, in my description of a single bulk superconductor, I have considered only a constant (and irrelevant) phase. Space and time variations of the phase couple to charge and current fluctuations. This coupling is central to the way in which the difference in the electrochemical potential between two weakly connected superconductors determines the supercurrent flowing between them. Let us, therefore, examine this question.

When considering charge and current fluctuations, we cannot ignore the Coulomb interaction between electrons. Part of the effect of the Coulomb repulsion has been included in the effective interaction specified by the constant g in Eq. (4). Now, however, we are interested in the effect of the Coulomb repulsion in the particle-hole rather than the pair channel. It is a weakness of the method being used here that it is difficult to treat these two channels consistently. I believe that for the purpose at hand no physical error is made by ignoring this difficulty. Thus we add to (4) the Coulomb energy

$$\mathcal{U}_C = \frac{e^2}{2}\int d^3x\, d^3x' \,\frac{:\hat{n}(x)\hat{n}(x'):}{|x-x'|} \tag{23}$$

where $\hat{n}(x)$ is the number density operator $\psi^+_\sigma(x)\psi_\sigma(x)$, and the colons mean that the self energy has been subtracted, i.e., that the operators are in normal form.[10] Now we can again use a Gaussian integral representation analogous to (5) but with a slight difference because we are dealing with a repulsive interaction. The formula now needed is

$$\exp[-b^2] = \int_{-\infty}^{\infty} \frac{dx}{\sqrt{\pi}} \exp[-x^2 - 2ibx] \tag{24}$$

To put (23) into the form of the left side of (24) we introduce a potential $\hat{V}(x)$ such that $\nabla^2\hat{V}(x) = -4\pi e\hat{n}(x)$. Then (23) becomes $\mathcal{U}_C = (8\pi)^{-1}\int d^3x{:}\vec{\nabla}\hat{V}(x)\cdot\vec{\nabla}\hat{V}(x){:}$, and, following the steps that led to (6) we find an expression for $Z_G$ which has two functional integrals. One is as before over complex functions $\Delta(xu)$; the other via (24) is over a real function $\nabla V(xu)$. The effective Hamiltonian in the consequent generalization of (6) is given by (7) with the addition of two new terms, so that

$$\mathcal{H}_{eff} = \mathcal{H}_{eff}(\text{Eq.}(7)) + ie\int d^3x\, \hat{n}(x)V(xu) + (8\pi)^{-1}\int d^3x|\nabla V(xu)|^2 \tag{25}$$

This can now be written in the Nambu representation, and steps similar to those that led from (7) to (15) performed. The result is of the form

$$Z_G = \int \mathcal{D}^2\Delta(xu)\; \mathcal{D}\vec{\nabla}V(xu)\; e^{-S[\Delta,V]} \tag{26}$$

It is convenient to transform to real $\Delta$ via (18). One then finds that the V of (26) and $(\partial\phi/\partial u)$ enter in the combination $\tilde{V} \equiv [V-(\hbar/2e)(\partial\phi/\partial u)]$. In addition, $\nabla\phi$ enters explicitly as described below (18). After these manipulations, the action S in (26) has the form given in (16) with V entering $G$ as described above and the additional explicit term, from the Gaussian transformation, $(8\pi)^{-1}\int d^3x(\nabla V)^2$. In order to see how the phase and electric potential couple we expand in the quantities $\tilde{V}$ and $\nabla\phi$. These quantities occur in $G$. The expansion needed--valid for non-commuting matrices --is

$$\begin{aligned} \mathrm{Tr}\ln G^{-1} &\equiv \mathrm{Tr}\ln(G_0^{-1} - \delta G^{-1}) \\ &= \mathrm{Tr}\ln G_0^{-1} - \mathrm{Tr}G_0\delta G^{-1} - \tfrac{1}{2}\mathrm{Tr}G_0\delta G^{-1}G_0\delta G^{-1} + \ldots \end{aligned} \tag{27}$$

One can thus calculate the effective susceptibilities controlling fluctuations in $\tilde{V}$ and $\nabla\phi$. For the first of these one finds

$$\delta S_{\tilde{V}} = +\frac{e^2}{2}\sum_{q\nu_m} |\tilde{V}(q,\nu_m)|^2\; \chi_{\tilde{V}}(q,\nu_m) \tag{28}$$

with

$$\chi_{\tilde{V}}(q,\nu_m) = -\tfrac{1}{2}\sum_{k,\zeta_\ell} \mathrm{Tr}\{G_0(k-q,\zeta_\ell-\nu_m)\tau_3 G_0(k,\zeta_\ell)\tau_3\} \tag{29}$$

where $\nu_m = (2\pi m i/\beta)$. One can similarly evaluate the other second order term, and calculate both in the long wavelength, low frequency limit. One finds the result

$$\begin{aligned} \delta S &= N(0)e^2\int d^3x\int du[\tilde{V}(xu)]^2 \\ &\quad + \int d^3x\int du\; \tfrac{1}{2}\rho_s\, \bar{v}_s^2 + (8\pi)^{-1}\int d^3x\int du(\nabla V)^2 \end{aligned} \tag{30}$$

Here $v_s = (\hbar/2m)\nabla\phi$ and $\rho_s$ is a temperature dependent quantity. Near T = 0 one calculates $\tfrac{1}{2}\rho_s v_s^2 \to (N(0)v_F^2/3)(\hbar\nabla\phi/2)^2$. It is straightforward to see that for a neutral system $[V \to 0]$ (30) is the (imaginary time) action for oscillations obeying the frequency wave number relation $\omega_q = qv_F/\sqrt{3}$.[11] For a charged system, the least action obtained from (30) in the limit $q \to 0$ describes plasma oscillations. The interesting physics obtained from the first term of

(30) is that the density of states and an effective fluctuation volume determine how far $\tilde{V}$ deviates from 0. The first term of (30) shows that the power spectrum of $\tilde{V}$ in the long wavelength low frequency limit has the form[12]

$$P_{\tilde{V}}(\omega) = \frac{\hbar}{e^2} \frac{1}{(\mathrm{Vol})N(0)} \sim \left(\frac{\hbar}{e^2}\right) \frac{k_B T_F}{N} \tag{31}$$

where N is the number of electrons in the fluctuation volume. Eq. (31) is to be compared with the Johnson noise formula for a resistance R $P_V(\omega) = k_B TR$. For a junction, N would be the number of electrons in a volume of the junction area times the smaller of the film thickness or the coherence length. Estimates by Arai[12] show that these fluctuations are negligible in junctions used to determine $(2e/\hbar)$ by the Josephson effect. An important result of this section is thus that the cost of charge fluctuations in a single superconductor pins $(\partial\varphi/\partial u)$ to a local electric potential.

This completes my first lecture. In the second we shall use these results to examine the quantum mechanical behavior of the phase difference across an oxide layer separating two superconductors.

## II.1 Functional Integral Description of Tunnel Junctions

I shall start by briefly sketching how the methods described in the last lecture can be used to derive an effective action for a tunnel junction. Most but not all of what is presented in this section is contained in a brief communication published with U. Eckern and G. Schön.[13] Everything that may make that paper difficult to read has been explained in lecture I, so I shall basically just describe the procedure we follow. We start with the Hamiltonian

$$H = H_L + H_R + H_T + H_Q \tag{32}$$

where $H_L$ and $H_R$ are of the form of Eq. (4) and describe the superconductors on the left and right of the junction. $H_T$ describes the coupling of the two sides by the junction:

$$H_T = \int_{\substack{\vec{x} \subset L \\ \vec{x} \subset R}} d^3x d^3x' \, \{T(\vec{x},\vec{x}')\psi^+_{L\sigma}(\vec{x})\, \psi_{R\sigma}(\vec{x}') + \text{h.c.}\} \tag{33}$$

and $H_Q$ is that part of the Coulomb energy which couples the two sides, namely

$$H_Q = \frac{1}{2C}(Q_L - Q_R)^2 \tag{34}$$

with

$$Q_L = e\int d^3x\ \psi^+_{L\sigma}(\vec{x})\ \psi_{L\sigma}(\vec{x}) \tag{35}$$

and a similar expression for $Q_R$. We have thus ignored in (32) the Coulomb energies in the bulk, whose role for the present purposes was discussed in Section I.4. It pins the phase of the order parameter to a local fluctuating potential. The part of the Coulomb energy we have kept can be obtained from (25) by applying appropriate boundary conditions between the two sides. The usual capacitance--determined by geometry and the dielectric constant of the insulator--then enters (34).

We now proceed, exactly as in the last lecture to express the partition function corresponding to (32) as a functional integral--by introducing three auxilliary fields to eliminate the terms quartic in field operators in $\mathcal{H}_L$, $\mathcal{H}_R$ and $\mathcal{H}_Q$. The trace over electron coordinates can then be done by using an appropriate generalization of Eq. (8). The coupling between the two sides is then treated in perturbation theory by an expansion of the form (27). We next make the gauge transformation (18) on the left and right to make the phases of the order parameters explicit. The result of these steps is to put the partition function into the form

$$Z_G = \int \mathcal{D}^2\Delta_L \mathcal{D}^2\Delta_R \mathcal{D}V\ \exp\{-S[\Delta,\Delta^*,V]\} \tag{36}$$

where

$$S = S_L[\Delta_L] + S_R[\Delta_R] + S_\mathcal{T} + \int_0^\tau \frac{dt}{h}\,\frac{1}{2}\,CV^2 \tag{37}$$

Here $S_L$ and $S_R$ are given by (16) with appropriate subscripts, the last term in (37) comes from a Gaussian representation of the form (24) used to eliminate $\mathcal{H}_Q$, and $S_\mathcal{T}$ is a new and interesting contribution. It describes the way in which supercurrent and normal current fluctuations are related to differences in the order parameter phase across the junction.

Explicit evaluation[13] yields

$$S_\mathcal{T} = |\mathcal{T}|^2 \int \frac{d^3p_L}{(2\pi\hbar)^3} \int \frac{d^3p_R}{(2\pi\hbar)^3} \int_0^\tau \frac{du}{\hbar} \int_0^\tau \frac{du'}{\hbar}$$

$$\Big(\exp\{\tfrac{-i}{2}[\phi(u)-\phi(u')]\}\ G_L(u-u',p_L)\ G_R(u',u,p_R) \tag{38}$$

$$-\exp\{\tfrac{-i}{2}[\phi(u)+\phi(u')]\}F_L(u-u',p_L)F_R(u'-u,p_R)+(R\leftrightarrow L)\Big)$$

Above $\phi$ is the phase difference $\phi \equiv \phi_L-\phi_R$ and $|\mathcal{T}|^2$ is the appropriate average [14] of tunneling matrix elements. G and F are the 11 and 12 components of the Nambu Green's function matrix (for constant real

$\Delta$) whose inverse is given in Eq.(17). Explicitly, the imaginary time Fourier transforms of G and F are given by $G(p,\zeta_\ell)=(\zeta_\ell+\varepsilon_p)/(\zeta_\ell^2-E_p^2)$ and $F(p,\zeta_\ell)=\Delta/(\zeta_\ell^2-E_p^2)$, where all symbols have been previously defined in Sec. I.3. There is a great deal of information in Eq.(37) which we shall try to extract in the remainder of this lecture.

First, note that $S_L[\Delta_L]$ and $S_R[\Delta_R]$ fix the magnitude of $\Delta_L,\Delta_R$ at their equilibrium values, and also pin $\hbar(\partial\phi/\partial u)_{L,R}$ to $\pm eV$, by the mechanisms explained in Sec. I.3 and I.4. When these values are fixed, the multiple functional integral in (36) is reduced to a single one over $\phi$. The effective action for $\phi$ contains the last two terms in (37) and contributions due to supercurrents in the left and right of the form of the second term in (30). In a mechanical analogy the last term of (37), with $V=(\hbar/2e)(\partial\phi/\partial u)$, plays the role of kinetic energy, whereas the potential energy and the effects of fluctuations are contained in the remaining pieces. Let us see how this comes about. By absorbing the momentum integrals into the definitions of $\alpha$ and $\beta$ one may write (38) as

$$S_T = -\int_0^\tau du \int_0^\tau du' [\alpha(u-u') \cos \frac{\phi(u)-\phi(u')}{2} + \beta(u-u') \cos \frac{\phi(u)+\phi(u')}{2} ] \qquad (39)$$

The functions $\alpha$ and $\beta$ are rather easy to evaluate at zero temperature ($\tau \to \infty$). One finds

$$\alpha(u) = (2/\hbar^2)|T|^2[N(0)]^2 \{\int_\Delta^\infty dE \frac{E}{\sqrt{E^2-\Delta^2}} e^{-E|u|/\hbar}\}^2$$

and

$$\beta(u) = (2/\hbar^2)|T|^2[N(0)]^2 \{\int_\Delta^\infty dE \frac{\Delta}{\sqrt{E^2-\Delta^2}} e^{-E|u|/\hbar}\}^2 \qquad (40)$$

The physics of these two terms is quite different. Suppose $\phi(u)$ varies slowly on the time scale determining the extent of $\alpha$ and $\beta$, namely ($\hbar/\Delta$). This can be arranged by appropriate choice of junction parameters.* Then both terms in (39) become local--depend on only one u. However the first has a leading $\phi$ dependence of the form $(\partial\phi/\partial u)^2$, whereas the second behaves like $\cos\phi$.† The term containing $\alpha$, which reflects the coupling of single particle degrees of freedom, thus at T = 0 simply changes (in fact increases--see below) the capacitance of the junction. What is happening here is that the energy gap prevents tunneling of quasiparticles across the oxide

*Increasing the impressed current, I, also decreases the characteristic value of $(\partial\phi/\partial u)$. †See Added Note.

layer but allows for an additional accumulation of charge at a given voltage difference. To calculate the size of this effect note that the normal state resistance of the junction is given[14] by $R_N^{-1} = 4\pi e^2|T|^2[N(0)]^2/\hbar$. The integral expression for $\alpha$ involves a representation of the Bessel function $K_1$, of imaginary argument. Explicitly;

$$\alpha(u) = \frac{\hbar}{2\pi e^2 R_N}\left(\frac{\Delta}{\hbar}\right)^2 \{K_1\left(\frac{\Delta|u|}{\hbar}\right)\}^2$$

$$= \frac{\hbar}{2\pi e^2 R_N}\begin{cases} u^{-2} & \text{for } \Delta|u| \ll \hbar \\ \frac{\pi}{2}\left(\frac{\Delta}{\hbar}\right)^2 \frac{\hbar}{\Delta|u|} e^{-(2\Delta|u|/\hbar)} & \text{for } \Delta|u| \gg \hbar \end{cases} \tag{41}$$

To estimate the capacitance correction we approximate the first term of (39) as

$$\delta S_T(\alpha) \approx \frac{1}{4}\int_0^{\tau} du(\partial\phi/\partial u)^2 \int_0^{\infty} du'\ \alpha(u')u'^2 \tag{42}$$

Performing the last integral using the first expression in (41), one finds a correction

$$(\delta C/C) = (3\pi\hbar/32\Delta R_N C) \tag{43}$$

For a junction of capacitance $10^{-14}$F and critical current $(I_J)10^{-4}$A, the estimate (43) yields--assuming the relationship[14] $I_J R_N = (\pi\Delta/2e)$ --a correction of about 70%.* It would be interesting to observe this effect, or to check if it has already been indirectly observed. It will decrease tunneling rates at low temperature compared to those calculated using the ordinary capacitance described in the paragraph following Eq. (35).

The second term in (39) in the local limit is the potential in mechanical analogies of the Josephson effect. (See, e.g., A. J. Leggett lectures at this Institute.) It reflects the coupling of the phases due to pair fluctuations. From the expression (39) one finds that $\int_{-\infty}^{\infty} du\beta(u) = (2e)^{-1} I_J$. Thus for slow variations of $\phi$ this is the "washboard potential" $-(2e)^{-1} I_J \int_0^{\tau} du \cos\phi(u)$. In the mechanical analogy there is another term--which tilts the "washboard"--due to the impressed current. This effect is contained in the superflow

*A gap $\Delta$ of 15°K (lead) has been assumed.

energy $\frac{1}{2}\rho_s v_s^2$ [cf. Eq. (30)] on the left and right. If one assumes a constant supercurrent I on the two sides these superflow energies contribute $-\int du(I/2e)\phi$ to the action.

Finally, then, at T = 0 we have a very simple effective action for $\phi$ variations slow on the scale $\hbar/\Delta$. It is

$$S[\phi] = \int_0^{\tau} du[\frac{\hbar}{8e^2}(C+\delta C)(\partial\phi/\partial u)^2 - (I_J/2e)\cos\phi - (I/2e)\phi] \tag{44}$$

Note that this local action, with caveats made below, corresponds to the non-dissipative quantum mechanics of a particle moving in a potential. The caveats are as follows. (i) In using (43) to obtain a Schrödinger equation, or to calculate the rate of switching from a metastable minimum, one would interpret the functional integral between given end states as the transition amplitude. We have not shown that this is a correct or consistent procedure. (ii) Our development is for an oxide layer junction with the ideal BCS non-linear resistive behavior. We can mimic the effect of a linear resistance by assuming it to be described by a normal metal--oxide layer--normal metal junction. In that case $\alpha(u)$ behaves like $u^{-2}$ for all u: the dissipative form used by Leggett and others.[15]

## II.2 Dissipation and Noise at Finite Temperature

The theory presented in the last section also contains information about the time development of the system at finite temperatures. In what follows, I shall try to display this information in a physically transparent form. Calculations of physical effects using the formalism to be developed here await the future.

All of the work reported here has been done in collaboration--and will shortly be prepared for publication--with U. Eckern and G. Schön.

The time development of a density matrix $\rho(t_1)$ in quantum mechanics is governed by the equation

$$\rho(t_1) = U(t_1,t_o)\ \rho(t_o)\ U(t_o,t_1) \tag{45}$$

where $U(t_1,t_o)$--the time evolution operator--is the time ordered exponential $T\exp[-i\int_{t_o}^{t_1} dt'H(t')]$. For a system of one particle it is straightforward to take matrix elements of (45). A system of many degrees of freedom is also governed by (45), but one might be interested in the development of some degrees of freedom and not others, in which case one would trace (45) over the unobserved variables and calculate matrix elements between states of the observed variables. In our case of a tunnel junction the interesting variable is the phase, but we have not identified a set of simulteneously measurable variables that together with the phase completely specify the states of the system.

If we boldly assume that such a complete set exists, we obtain from (45) for the density matrix traced over all degrees of freedom except $\phi$:

$$<\phi'|\rho(t_1)|\phi''> = \mathrm{Tr}\{\rho(t_0)U(t_0 t_1)P_{\phi''\phi'}(t_1)U(t_1 t_0)\} \tag{46}$$

where $P_{\phi''\phi'}$ is a projection operator, and Tr indicates a trace over all variables. In the functional integral formalism, the natural phenomenological byt quite unproved interpretation of $P_{\phi''\phi'}$ is that it is a filter which only permits paths in $\phi$-space with specified end points. If this is a correct procedure, then (45) may be evaluated as in the last section. Let the system be in thermal equilibrium at $t_0$ with a temperature $T = (k_B\beta)^{-1}$. Then the formal analogy between the statistical equilibrium Boltzmann factor and the time evolution operator U gives $\rho(t_0) = U(t_0 - i\beta, t_0)/Z_G$. Thus one obtains

$$<\phi'|\rho(t_1)|\phi''> = \frac{1}{Z_G}\int' \mathcal{D}\phi \; e^{-S[\phi]} \tag{47}$$

where the paths are along the Baym-Kadanoff[16] contour given in Figure 1.

All of the steps of the last section have been done for each time cell along this contour. The prime is to remind one that the value of $\phi$ is $\phi'$ at the end of the path from $t_0$ to $t_1$, and $\phi''$ at the start of the path from $t_1$ to $t_0 - i\beta$.

With the equilibrium boundary condition at $t_0$, the effective action $S[\phi]$ is given by the analytic continuation $u \to it$ of the

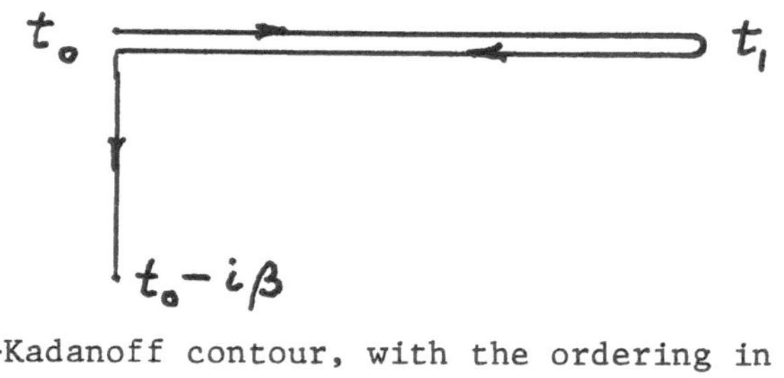


Fig. 1. Baym-Kadanoff contour, with the ordering in time indicated by the arrows.

results of the last section. [Note that we assume that there are no other perturbations in real time to drive $|\Delta|$ and the other internal degrees of freedom.] The quantities $\alpha$ and $\beta$ are discontinuous functions of their time argument. They can be expressed in terms of interesting physical response functions. Comparing the terms entering (38) and (39) with the usual theory of the Josephson effect[14] one finds:

$$\alpha^{>}_{<}(t) = \frac{\pm 1}{2\pi e} \int_{-\infty}^{\infty} d\omega e^{-i\omega t} \frac{1}{1-e^{\pm\beta\omega}} I_N(\omega)$$

$$\beta^{>}_{<}(t) = \frac{\pm 1}{2\pi e} \int_{-\infty}^{\infty} d\omega e^{-i\omega t} \frac{1}{1-e^{\pm\beta\omega}} I_c(\omega) \tag{48}$$

Here $I_N(\omega)$ is the quasiparticle current at voltage ($\hbar\omega/e$) and $I_c(\omega)$ is related to the zero-voltage critical current $I_J$ by*

$$I_J = P\int_{-\infty}^{\infty} \frac{d\omega}{\pi} \frac{I_c(\omega)}{\omega} \tag{49}$$

For real t, $\alpha^{>}$ and $\beta^{>}$ have real and imaginary parts which we shall call $\alpha_R$, $\alpha_I$, $\beta_R$, $\beta_I$. To put (47) into a more transparent form it is convenient to distinguish between the forward and backward paths connecting $t_o$ and $t_1$. Calling the path function $\phi_1(t)$ on the forward segment from $t_o$ to $t_1$ and $\phi_2(t)$ on the backward segment from $t_1$ to $t_o$, and being careful to note that the time on the second path is always "later" in the ordering indicated in Figure 1, one can write the right side of (47) as a double functional integral

$$\int \mathcal{D}\phi_1 \mathcal{D}\phi_2 \; e^{-S[\phi_1,\phi_2]} \tag{50}$$

where both $\phi_1$ and $\phi_2$ are functions over the domain $t_o \le t \le t_1$. Somewhat tedious algebra of the kind usually called straightforward allows one to express S in terms of the parameters and response functions of the system. Equation (50) is of the Feynman-Vernon[17] form. In the present context Schmid[18] has discussed a simpler version of the problem in a similar form. In Schmid's notation $\frac{1}{2}(\phi_1 + \phi_2) \equiv x$ and $(\phi_1 - \phi_2) \equiv y$. In terms of these variables our result reads:

*$I_c$ is the $\cos\phi$ or "quasiparticle-pair interference" current. Insofar as this requires additional bulk scattering terms to be given correctly,[19] our model would also require modification.

$$-S[xy] = \frac{i}{\hbar}\int_{t_o}^{t_1} dt\left[\frac{\hbar^2}{4e^2}\, C\dot{x}\dot{y} + \frac{I\hbar}{2e}\, y\right]$$

$$+ 8i\int_{t_o}^{t_1} dt \int_{t_o}^{t_1} dt'\; \Theta(t-t')[\alpha_I(t-t')\sin\tfrac{1}{2}(x-x')$$

$$+ \beta_I(t-t')\sin\tfrac{1}{2}(x+x')]\sin\tfrac{1}{4}y\; \cos\tfrac{1}{4}\, y'$$

$$-4\int_{t_o}^{t_1} dt \int_{t_o}^{t} dt'[\alpha_R(t-t')\cos\tfrac{1}{2}(x-x')$$

$$+\beta_R(t-t')\cos\tfrac{1}{2}(x+x')]\sin\tfrac{1}{4}y\; \sin\tfrac{1}{4}\, y' \qquad (51)$$

There are limiting cases in which one can see that Eq. (51) contains the correct physics, but the general form remains unexploited. The difficulty is that such real time functional integrals are not well understood even for linear dissipation, and we have included the complete non-linear dependence of the BCS model.

I shall discuss the physics of (51)in two approximations: the "Feynman-Vernon" limit describing the quantum Brownian motion of the phase in the washboard potential with linear damping and the associated noise; and the "least action" limit for the non-linear case.

To get the first result, one must calculate the superconducting fluctuations (described by $\beta_R$ and $\beta_I$) for phase variations slow compared to $\hbar/\Delta$, and the quasiparticle fluctuations (described by $\alpha_R$ and $\alpha_I$) for zero gap, thus using the junction to model a normal resistance. One finds then that $\beta_I$ dominates $\beta_R$ and that the former can be well approximated by $\beta_I = -(I_J/2e)\delta(t-t')$. Substituting this form into (51) one obtains the contribution $-i\int dt(I_J/2e)[\cos(x-\tfrac{1}{2}y) -\cos(x+\tfrac{1}{2}y)]$, i.e., exactly the effect on the forward and backward time evolution of the potential $-(I\hbar/2e)\cos\phi$. For a normal junction $I_N(\omega) = \hbar\omega/eR_N$; it follows from (48) that

$$\alpha_I^N(t) = -(\hbar/2e^2R_N)\int_{-\infty}^{\infty}\frac{d\omega}{2\pi}\,\omega\,\sin\omega t = (\hbar/2e^2R_N)\,\frac{\partial}{\partial t}\,\delta(t) \qquad (52)$$

and $\alpha_R^N(t) = (e^2 R_N \beta)^{-1} K(t)$, where

$$K(t) \equiv \int_{-\infty}^{\infty} \frac{d\omega}{2\pi} \frac{\beta\hbar\omega}{2} \coth \frac{\beta\hbar\omega}{2} \cos\omega t \tag{53}$$

When these forms are substituted into (51) the total action is $S = S_1 + S_2$ with

$$S_1 = -(i/\hbar) \int_{t_o}^{t_1} dt [y(m\ddot{x} + m\gamma\dot{x}) + V(x + \frac{y}{2}) - V(x - \frac{y}{2})] \tag{54}$$

and

$$S_2 = -(16m\gamma/\hbar^2\beta) \int_{t_o}^{t_1} dt\, dt' \sin \frac{1}{4}y(t)K(t-t')\sin \frac{1}{4}y(t') \times \cos \frac{1}{2}[x(t) - x(t')] \tag{55}$$

To bring out the mechanical analogy we have defined $m \equiv (\hbar^2/4e^2)C$, $\gamma = (R_N C)^{-1}$, and $V(x) = -(\hbar I_J/2e) \cos x - (\hbar I/2e)x$. Eqs. (54) and (55) are as close as one can come to the linear damping model of other authors (cf. Schmid,[18] Eq. (5)) starting from a tunnel junction model of the Josephson effect and of the resistive mechanism. A natural way to deal with (55) is to interpret it as the average effect of a stochastic force. Evidently

$$e^{-S_2} = \langle e^{+i(4/\hbar) \int dt\, f[xt] \sin\frac{1}{4}y(t)} \rangle \tag{56}$$

where f is a Gaussian random quantity with an autocorrelation function

$$\langle f[xt]\, f[x't'] \rangle = (2m\gamma/\beta)\, K(t-t') \cos\frac{1}{2} [x(t)-x(t')] \tag{57}$$

The motion of a classical wave packet under the influence of the action described by (54) and (55) is governed by

$$\left.\frac{\delta S}{\delta y}\right|_{y=0} = 0 \tag{58}$$

Within the above approximations (58) yields

$$m\ddot{x} + m\gamma\dot{x} + \frac{\partial}{\partial x} V(x) = f[x,t] \tag{59}$$

Discussions of the limits of validity of this Langevin-like equation are to be found in the literature.[18,20] I am inclined to doubt that it can be used to describe true quantum mechanical tunneling. It certainly does apply when $\beta h\omega \ll 1$ in (53); in this limit it is the RSJ model with thermal noise.

In the limit in which Eq. (58) is valid, (51) can be seen to make physical sense including all the non linearities. The least action value of all the imaginary terms gives a correct statement of conservation of charge. The impressed current I is seen to be equal to the current stored in the capacitor plus the currents carried by quasiparticles and Cooper pairs, these last being given by the correct causal linear response functions.[21] The real terms in (51) then presumably describe noise currents with $\alpha_R$ and $\beta_R$ entering an autocorrelation function similar to (57). It remains an interesting challenge to learn how to extract the considerable physics contained in Eq. (51), and to use it to calculate quantum effects in superconducting tunnel junctions.

Acknowledgements

I have described work done and being done in collaboration with Ulrich Eckern and Gerd Schön. They should not, however, be blamed for these lecture notes which have been written under pressure of a deadline, and without the benefit of their comments.

It is a pleasure to thank Max Arai, Sudip Chakravarty, Tony Leggett, and Albert Schmid for informative discussions.

This work was supported in part by the US Office of Naval Research.

References

1. See, e.g., R. P. Feynman and A. R. Hibbs, Quantum Mechanics and Path Integrals, (McGraw-Hill, New York 1965).
2. J. S. Langer, Ann. Phys. (NY) 41, 108 (1967); C. G. Callan and S. Coleman, Phys. Rev. D16, 1762 (1977).
3. See, e.g., V. Ambegaokar, in Superconductivity, ed. R. D. Parks (M. Dekker, New York 1969) Vol. I, p. 259.
4. D. E. Soper, Phys. Rev. D18, 4590 (1978); see also G. Baym, Phys. Rev. 127, 1391 (1962).
5. J. M. Luttinger and J. C. Ward, Phys. Rev. 118, 1417 (1960).
6. Y. Nambu, Phys. Rev. 117, 648 (1960).
7. J. S. Langer, Phys. Rev. 134, A553 (1964).
8. P. A. Lee and M. G. Payne, Phys. Rev. Lett. 26, 1537 (1971); J. Kurkijärvi, V. Ambegaokar and G. Eilenberger, Phys. Rev. B5, 868 (1972).
9. See, e.g., N. R. Werthamer in ref. (3), Vol. I, p. 321.
10. See, e.g., V. Ambegaokar in ref. (3), Vol. II, p. 1359.
11. P. C. Martin in ref. (3), Vol. I, p. 371, and references therein.
12. M. R. Arai, Ph.D. dissertation, Cornell University, Ithaca, NY (1983).
13. V. Ambegaokar, U. Eckern and G. Schön, Phys. Rev. Lett. 48, 1745 (1982).

14. B. D. Josephson, Phys. Lett. 1, 251 (1962); V. Ambegaokar and A. Baratoff, Phys. Rev. Lett. 10, 486 (1963), and 11, 104(E) (1963).
15. A. O. Caldeira and A. J. Leggett, Phys. Rev. Lett. 46, 211 (1981); S. Chakravarty, Phys. Rev. Lett. 49, 681 (1982).
16. L. P. Kadanoff and G. Baym, Quantum Statistical Mechanics (W. A. Benjamin, New York 1962).
17. R. P. Feynman and F. L. Vernon, Jr., Ann. Phys. (NY) 24, 118 (1963).
18. A. Schmid, J. Low Temp. Phys. 49, 609 (1982).
19. A. B. Zorin, I. O. Kulik, K. K. Likharev and J. R. Schrieffer, Fiz. Nizk. Temp. 5, 1138 (1979). [Sov. J. Low Temp. Phys. 5, 537 (1979).]
20. J. Kurkijärvi, Phys. Lett. 88A, 241 (1982).
21. N. R. Werthamer, Phys. Rev. 147, 255 (1966).

NOTE:

A. J. Leggett has pointed out to me that to be consistent one should keep the next term coming from the slow $\phi$-variation expansion in the integral over $\beta$. When this is done, the $u^2$ moment of $\beta(u)$ enters in a manner analogous to Eq. (42), and Eq. (44) must be modified to read

$$S[\phi] = \int_0^\tau du\{\frac{\hbar}{8e^2} (\partial\phi/\partial u)^2[C+\delta C(1-\frac{1}{3}\cos\phi)]$$

$$-(I_J/2e)\cos\phi - (I/2e)\phi\} \qquad (44')$$

where $(\delta C/C)$ is given by Eq. (43). For small oscillations, this correction is thus reduced by a multiplicative factor of (2/3). For large oscillations the now phase dependent capacitance correction will change the harmonic content of the time dependence. In the quantum limit the analogue of the kinetic energy must then be an ordered product, i.e., $p^2/2M \to (1/8M(x))p^2+p(1/4M(x))p+p^2(1/8M(x))$. These points will be discussed in V. Ambegaokar, U. Eckern and G. Schön, in preparation.

# SMALL PARTICLES

B. Mühlschlegel

Institut für Theoretische Physik
Universität zu Köln
5000 Köln 41, W. Germany

## 1. INTRODUCTION

Small-particle physics deals with objects which neither belong to atomic physics nor to condensed matter. Small particles, aggregates or clusters with size as a variable quantity occupy a domain of great complexity between atoms and crystals. The two words MIE SCATTERING and CATALYSIS might indicate that this domain is in existence (and of importance) since the start of the century. However, only in recent years have the various research activities on small systems emerged into what may be called a new discipline. A look at meetings in the present decade is sufficient to demonstrate this. The 1980 Lausanne conference on Small Particles and Inorganic Clusters /1/ has produced a broad spectrum of results and has stimulated further research and interdisciplinary interaction; a similar conference is awaiting us in 1984 in Berlin. Entre l'atome et le cristal: Les agregats was the theme of a school in 1981 /2/; a Nato school on the Impact of Cluster Physics in Material Science and Technology was held in 1982 /3/, and several workshops took place during the last three years.

In addition we mention, without being complete, some reviews which have been written recently: Metal Colloids in Ionic Crystals, 1979 by Hughes and Jain /4/, Electronic Properties of Small Metallic particles, 1981 by Perenboom, Wyder and Meier /5/, Physics of Microparticles, 1982 by Baltes and Simanek /6/, Alkalihalide Cluster and Microcrystals, 1983 by Martin /7/. Already the titles of these reviews show that they cover only certain aspects of the properties of small particles and clusters. It is clear that also the following contribution will deal with some very selected topics and is far from being a systematic survey.

We confine our attention to electronic properties of small metal particles with emphasis on the approach from the bulk. This is in clear distinction to the true microcluster approach which begins with a few atoms and views the electrons in the small system as members of a giant molecule /8/. In the following we will describe a very simplified theoretical picture of free and interacting electrons in small metal grains which to a large extent has already been covered by the Perenboom-Wyder-Meier review paper (abbreviated here by PWM). This good article has also an exhaustive list of references to which we will refer for simplicity.

Isolated small metallic particles can, in a second step, be coupled in various ways to their surroundings. They are embedded in a medium, they can interact with outside atoms and molecules, they can be placed on a surface, and they can be coupled to each other forming a granular metal. Intrinsic properties of small particles are then of interest not only in themselves but are useful also for the understanding of the more complex behavior created by various couplings.

## 2. FREE ELECTONS IN EQUIDISTANT LEVELS

In a seminal paper in 1962 Kubo /9/ treated electrons in a fine metallic particle under three basic assumptions:

1. Due to the finite particle volume $\Omega$ the continuous spectrum of elementary excitations of the bulk metal becomes discrete with an average level separation near the Fermi energy $\varepsilon_F$

$$\delta = \frac{1}{\Omega N(0)} \sim \varepsilon_F/N \tag{2.1}$$

   where N(O) is the single-spin density of states per unit volume at the Fermi energy, and N is the number of conduction electrons in the small particle.
2. The electrostatic energy $U \sim e^2/a$ associated with the addition or removal of an electron to or from a small particle is extremely large compared to the thermal energy available. Therefore electric charge will not fluctuate (for well isolated particles) and low-temperature properties have to be calculated for N strictly kept fixed, that is canonically rather than grand canonically.
3. A collection of particles all of the same size $\Omega$ has the same average level separation $\delta$ near the Fermi energy $\varepsilon_F$ given by Eq. (2.1) but the level spectrum of two particles will be different due to uncontrollable surface irregularities. This requires a statistical description of level distributions.

In 1. it is tacitly assumed that surface irregularities will split apart degenerate levels so that the spectrum with average spacing $\delta$ is non-degenerate. Note that perfect spheres would have much broader spacings proportional to $N^{-2/3}$. A metallic particle of size of order 100 Å containing about $10^5$ conduction electrons would correspond to a level separation of roughly $\delta/k_B \sim 1K$.

To account for assumption 2, the canonical partition function Q(N) of N electrons is proficted out of the easier grand cononical form Q(z) by a contour integration in the complex z-plane

$$Q(N) = \frac{1}{2\pi i} \oint \frac{Q(z)}{z^{N+1}} dz \tag{2.2}$$

where $z = \exp(\beta\mu)$, $\beta = (k_BT)^{-1}$, $\mu$ = chemical potential. When we take out the ground state energy $E_0(N)$ of N electrons

$$Q(N) = e^{-\beta E_0(N)} Z \tag{2.3}$$

then Z accounts for all canonical thermal excitations of an equal number of electrons and holes. Of course, at temperatures $\beta^{-1}$ such that only the lowest excitations are important, it is easy to write out the partition function Z term by term. However, the precise counting of proper excitations becomes exceedingly complicated with rising temperature, and one has to rely on a saddle-point approximation. Only the case of equidistant levels $\varepsilon_n = n\delta$ is special because here the contour integration of Eq. (2.2) can be done exactly with the result for N = even or odd, resp. /10/

$$Z_{even} = \left[1 + 2\sum_{n=0}^{\infty} e^{-\beta\delta(n+1)^2} \cosh 2(n+1)h\right] Z_B^2$$

$$Z_{odd} = \left[2\sum_{n=0}^{\infty} e^{-\beta\delta n(n+1)} \cosh (2n+1)h\right] Z_B^2 \tag{2.4}$$

with

$$Z_B = \prod_{n=1}^{\infty} \left(1 - e^{-\beta\delta n}\right)^{-1}. \tag{2.5}$$

The influence of the magnetic field H enters through the ratio of Zeeman to thermal energy $h = 1/2\, \beta g \mu_B H$. $Z_B$ is the canonical partition function of spinless fermions. Note that $Z_B$ is identical with the partition function of bosons with energies $n\delta$, and this is a peculiar feature of equally spaced levels.

The exact counting of possible states leads to a clear distinction between the even and odd cases. The lowest terms of the partition function (2.4) are

$$\left.\begin{matrix} Z_{even} \\ \\ Z_{odd} \end{matrix}\right\} = \left\{\begin{matrix} 1 + 2e^{-\beta\delta}\cosh 2h \\ \\ 2(\cosh h + e^{-2\beta\delta}\cosh 3h) \end{matrix}\right\}(1 + 2e^{-\beta\delta} + 5e^{-2\beta\delta} + 10e^{-3\beta\delta}) + O(e^{-4\beta\delta}) \quad (2.6)$$

The partition functions (2.4) of canonical excitations are easily handled for all values of the two parameters $\beta\delta$ and h, and therefore allow the calculation of the specific heat and spin susceptibility in the whole range together with a comparison with the corresponding grand cononical results (Fig. 1).

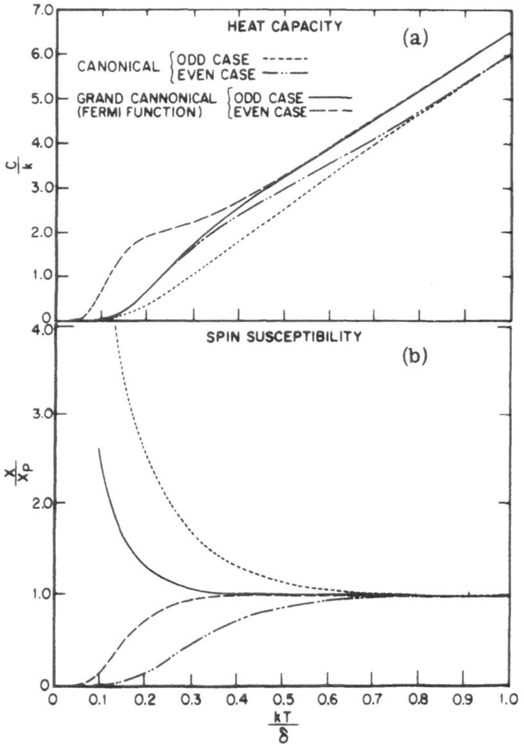


Fig. 1. (a) The electronic heat capacity C, and (b) the spin susceptibility $\chi$ normalized to the Pauli spin susceptibility $\chi_p$ as functions of temperature for a particle with an equal level spacing $\delta$, calculated with the canonical ensemble and with the grand canonical ensemble, respectively.

It is instructive to see the difference between the two thermodynamic ensembles in the spinless case for $T \to 0$, i.e. $\beta\delta \gg 1$. Then from (2.5)

$$Z_{canonical} \equiv Z_B = 1 + e^{-\beta\delta} + O\left(e^{-2\beta\delta}\right) \tag{2.7}$$

whereas for independent electron and hole excitations in the less restrictive grand canonical ensemble (Fermi energy in the middle between two levels) it follows

$$Z_{grand\ canonical} = \prod_{n=1}^{\infty} \left(1 + e^{-\beta\delta \frac{n}{2}}\right)^2 =$$

$$1 + 2e^{-\frac{\beta\delta}{2}} + 3e^{-\beta\delta} + 6e^{-\frac{3\beta\delta}{2}} + O\left(e^{-2\beta\delta}\right). \tag{2.8}$$

The susceptibility will exhibit for $T \to 0$ the Curie-law behavior in the odd case (due to one unpaired electron in the top-most occupied level) whereas it is exponentially attenuated for N even, as is always the case for the specific heat.

In the high-temperature limit, one has $\beta\delta \ll 1$, and there are many excited states. The exact counting becomes less important, and the even-odd distinction washes out:

$$\ln Z_{even} = \ln Z_{odd} = \frac{\pi^2}{\beta\delta} + \frac{1}{2}\ln\frac{\pi}{\beta\delta} - 2\int_{1/2}^{\infty} dx \ln\left(1 - e^{-\beta\delta x}\right) \tag{2.9}$$

The heat capacity and susceptibility are in this limit

$$C/k_B = \frac{2\pi^2}{3}\frac{k_B T}{\delta}\ , \quad \chi = 2\left(\frac{1}{2}g\mu_B\right)^2/\delta \tag{2.10}$$

and represent the bulk linear law and the usual Pauli spin susceptibility.

It is, of course, interesting to know when and how deviations from the bulk values (2.10) will occur for lower temperatures. Here Fig. 1 indicates that deviations for the susceptibility should be recognizable if $k_B T \lesssim 0.7\delta$. Experimentally, such deviations have been found first by Taupin /11/ in 1966 in small lithium particles by means of NMR, and subsequently by several other workers both with NMR and ESR techniques /12/.

## 3. ELECTRIC POLARIZABILITY

The approach to small particles coming from the bulk brings about a new energy $\delta$ . Quantum size effects are to be expected when relevant energy parameter as thermal energy $k_BT$, Zeeman energy, electromagnetic $\hbar\omega$ become comparable or smaller than $\delta$ . Of course, one has to realize that a collection of small particles is not uniform in size and consequently also not in $\delta$ , and this must be incorporated in a discussion of experimental results.

In Sec. 2 we have treated the $k_BT \lessgtr \delta$ - effect by means of the most simplified description of free electrons in equally spaced levels. Let us stick for a moment to this model and have a look at the $\hbar\omega \lessgtr \delta$ -effect (for the magnetic $h \gtrsim \beta\delta$ -effect see /10/). Here we begin with an even more hypothetical model namely a one-dimensional box of length a. It has sinus wave functions with energy eigenvalues ($\hbar$ = 1) $\varepsilon_n = \frac{1}{2m}(\frac{\pi n}{a})^2$. N electrons are filled into the N/2 lowest levels, and near the Fermi energy $\varepsilon_F = \varepsilon_{N/2}$ the difference between levels becomes

$$\varepsilon_{n'} - \varepsilon_n = \frac{1}{m}\left(\frac{\pi}{a}\right)^2 \frac{N}{2}(n'-n) = \delta \cdot (n'-n) \tag{3.1}$$

We have equidistant levels with spacings $\delta = 4\,\varepsilon_F/N$. At T = 0 one can probe the level discreteness by calculating the response to an electric field, given by the polarizability

$$\alpha_\omega = -e^2 \sum_{nn'} \frac{|\langle n'|x|n\rangle|^2 (f_{n'}-f_n)}{\varepsilon_{n'}-\varepsilon_n+\omega+i0^+} \tag{3.2}$$

where $f_n = (1,0)$ for $n \lessgtr N/2$. Using (3.1) and the expression for the dipole matrix element, one obtains

$$\alpha_\omega = \alpha_0 \frac{3}{y^2}\left[\frac{\tan(y+i0^+)}{y} - 1\right], \quad y = \frac{\pi}{2}\frac{\omega}{\delta} \tag{3.3}$$

with resonances at $\omega = n\delta$ as expected. The static polarizability is

$$\alpha_0 = \frac{e^2 a^2}{12\,\delta} = \frac{a^3}{12\pi\, a_0 k_F} \tag{3.4}$$

where the second expression on the r.h.s. is obtained by introducing the Bohr radius $a_0$ and the Fermi momentum $k_F = m\,\frac{a}{\pi}\,\delta$ .

It is amusing to compare (3.4) formally with the classical polarizability of a needle-shaped metallic body of length a and transversal radius $a_0 \ll a$

$$\alpha_{\text{classical}} = \frac{a^3}{24\,\ln(a/a_0)} \tag{3.5}$$

Taking $a_0k_F$ to be roughly of order one, we see that, due to the discrete level structure, the classical static polarizability of a quasi-one-dimensional small particle is enhanced by a factor $\ln(ak_F)$. A collection of such objects has been considered within another context /13/.

The above treatment is the trivial one-dimensional version of the case of a three-dimensional small particle which has been studied already in 1965 by Gorkov and Eliashberg /14/. Their paper was very stimulating, both experimentally and theoretically. For the static case it turns out that the above enhancement factor $\ln(ak_F)$ is replaced by $(ak_F)^2$ which would, at first glance, lead to a giant dielectric constant in artificial small-particle substances. However, here the difference between applied and inner electric field matters. The three-dimensional analogue of (3.2) is actually not the polarizability $\alpha$ but the susceptibility $\chi_e$ multiplied by the volume $\Omega$. Both quantities are related by

$$\alpha = \frac{\chi_e}{1 + L\chi_e}\Omega \tag{3.6}$$

with L being the depolarization factor. In the 1D case L is practically zero for parallel electric field; for 3D, however, screening leads to finite L and no enhancement in accordance with experiments.

Dynamically, three-dimensional small particles should exhibit marked effects when $\omega \sim \delta$. With $\delta/k_B \sim 1K$ such effects should show up in far-infrared absorption. In addition, Gorkov and Eliasberg /14/ have studied the influence of statistical level distribution (see the next section) which should lead to oscillations in the frequency dependence of the absorption coefficient. Unfortunately, this effect can be easily killed by deviations from equal size in a collection of particles. It appears that the whole field of infrared absorption does not give information about the quantum-size effect at present, and moreover, remains controversial regarding the magnitude of absorption /5/,/15/. A theorist's dream would be the scattering experiment on a single small particle!

## 4. LEVEL STATISTICS

We now turn to a brief discussion of Kubo's third assumption concerning the statistical description of level distributions in a collection of small particles all of the same size. A physical quantity C measured on the ensemble of particles with individual spacings is then given by

$$\overline{C} = \int C(\{\Delta_i\}) P(\{\Delta_i\}) \prod_i d\Delta_i \tag{4.1}$$

where $P\prod d\Delta$ is the probability to find spacings $\Delta_i$ between the various levels near the Fermi energy. The use of such level statistics assumes the absence of systematic surface perturbations on the particles. How many levels actually have to be averaged will depend on the special situation (e.g. temperature). The one-spacing function $P(\Delta)$ gives the probability that two adjacent levels are separated by an energy $\Delta$. Then

$$\int \Delta P(\Delta) d\Delta = \bar{\Delta} = \delta . \tag{4.2}$$

Note that the previous scheme of equally spaced levels is obtained with $P(\Delta) = \delta(\Delta - \delta)$, $\delta(x)$ = Dirac delta function.

The theory of eigen-value distributions of random matrices is fully developed /16/,/17/. As mentioned before, it was first applied to small particles by Gorkov and Eliashberg /14/. Fortunately, we don't have to burden ourselves with the whole apparatus of this theory when we confine our attention to the question of how the low-temperature properties of Sec. 2 are changed by level randomization.

Let us neglect all terms $O(e^{-2\beta\delta})$ in the partition function $Z(\beta\delta, h)$ given by Eq. (2.5). This is justified for $T \rightarrow 0$ where only the first level in a distance $\delta$ can be populated. Now, in a statistical description of a collection of small particles, one knows solely with the probability $P(\Delta)d\Delta$ that a level is a distance $\Delta$ away from the next one. Therefore

$$\overline{\ln Z} = \int_0^\infty \ln Z(\beta\Delta, h) P(\Delta) d\Delta \quad , \quad \bar{\Delta} = \delta . \tag{4.3}$$

For $T \rightarrow 0$ ($\beta \rightarrow \infty$) only small $\Delta$ will contribute, and here the spacing function P is assumed to behave like

$$P(\Delta) = a_n \Delta^n . \tag{4.4}$$

From

$$\overline{\ln Z} = \frac{a_n}{\beta^{n+1}} \int \ln Z(x, h) x^n dx \tag{4.5}$$

it then follows immediately that the specific heat $C/k_B = \beta^2 \partial^2 \overline{\ln Z} / \partial\beta^2$ is changed from exponential attenuation, seen in Fig. 1 to a power-law behavior

$$C = \gamma_n T^{n+1} \tag{4.6}$$

with a different coefficient $\gamma_n$ for the even and odd case. Similarly, the statistical average of the spin susceptibility is obtained. In the even case the relevant term in lnZ is $\ln(1 + 2e^{-x}\cosh 2h)$, $h = \frac{1}{2}\beta g \mu_B H$. This gives

$$\chi_{\text{even}} = \alpha_n T^n . \tag{4.7}$$

In the odd case, however, cosh(h) is a factor in the partition function which is not at all influenced by averaging. Therefore the Curie law

$$\chi_{\text{odd}} = \frac{(\frac{1}{2} g \mu_B)^2}{k_B T} \tag{4.8}$$

remains unchanged for $T \rightarrow 0$, a result which is plausible without any calculation.

According to Eq. (4.4) the probability P decreases with decreasing spacing $\Delta$ : energy levels like to repel each other. But with which power n? Random-matrix theory has an answer to this question: only the cases n = 1,2,4 are possible. These numbers originate from different symmetries imposed upon the random matrices. (orthogonal, unitary, symplectic ensemble).

With rising temperatures more levels are populated and have to be randomized. This leads to deviations from the limiting low-temperature behavior for $k_B T \gtrsim 0.1\,\delta$ . However at "high" temperatures $k_B T \gtrsim 0.5\,\delta$ the equal-level-spacing approximation is quite sufficient as comparison with the detailed statistical results shows /10/.

## 5. INTERACTIONS

We have so far considered a system with the Hamiltonian

$$H_0 = \sum_{ns} \varepsilon_{ns} c^+_{ns} c_{ns} \tag{5.1}$$

where $c^+_{ns} c_{ns}$ is the occupation number of a single-electron state n with spin s = $\uparrow$ or $\downarrow$ in the small particle. The $\varepsilon_{ns}$ form, in the simplest scheme, equidistant energy levels including Zeeman splitting. We may think that the main effects of interactions are incorporated in $H_0$ by viewing the free electrons as quasiparticle excitations near the Fermi energy. It is, however, of interest to treat additional interactions, and we shall indicate this here for a) short-range Coulomb interactions and b) spin-orbit coupling. The case of superconducting pairing will be discussed in more detail in Sec. 6.

a) A Coulomb-type interaction of the special form

$$H_1 = U \sum_{nn'} (c^+_{n\uparrow} c_{n\uparrow})(c^+_{n'\downarrow} c_{n'\downarrow}) = U N_\uparrow N_\downarrow \tag{5.2}$$

can be added to (5.1) in order to study the effect of exhange correlations within the framework of the canonical ensemble. The partition function of $H_0 + H_1$ is similar in form to the noninteracting case described before by Eq. (2.4). The only change necessary is the replacement of $\delta$ by

$$\hat{\delta} = \delta\left(1 - \frac{U}{\delta}\right) \tag{5.3}$$

in the exponential prefactors of the cosh-terms of Eq. (2.4). /10/

This result shows that the spin susceptibility is renormalized by the familar Stoner factor $(1 - U/\delta)^{-1}$ which enhances the bare density of states $\delta^{-1}$ to $\hat{\delta}^{-1}$. The heat capacity still depends on both $\delta$ and $\hat{\delta}$ .

b) The spin-orbit scattering of electrons on the surface of the small particle can be simulated by adding to (5.1) an interaction Hamiltonian of the type

$$H_1 = \sum_{\substack{nn' \\ ss'}} (\underline{V})_{nn'} \cdot \underline{\sigma}_{ss'} \, c^{+}_{ns} c_{n's'} \tag{5.4}$$

where $\underline{\sigma}$ is the Pauli spin vector. The quantity most sensitive to $H_1$ is the spin susceptibility $\chi$ . In the N even case $\chi$ vanishes at T = 0 for $H_1 = 0$ as seen in Fig. 1 because the two electrons in the top-most occupied level form a singlet. However, when $H_1$ is switched on, its spin flip allows a mixture with triplet-excited states which then produce a nonzero susceptibility $\chi_{even}$.

$H_0 + H_1$ was first studied by Shiba within the equal-level-spacings scheme, and grand cononically for temperature effects /18/. Since for $T \rightarrow 0$ level statistics should be included, one has to perform an average similar to Eq. (4.3), and this has been discussed by Buttet /19/. For further work on spin-orbit-coupling effects we again refer to the PWM-review article /5/. For completeness it should be mentioned that Shiba's main goal was to study the effect of spin-orbit interaction upon superconductivity in small particles; we will consider the interesting topic of BCS-pairing in the next section.

## 6. SUPERCONDUCTIVITY I

Not long after the microscopic theory of superconductivity was created in 1957 by Bardeen, Cooper and Schrieffer the pairing idea was applied to the finite system of a nucleus; strictly speaking a pairing force was considered in nuclear physics way back in 1943 by Racah /20/. In order to discuss superconductivity in small metallic particles it then appears advantageous to use the techniques developed for nuclear problems especially since these methods contain the projection on sharp particle number N, a condition we have stressed in the calculations of Sec. 2 by means of the canonical ensemble.

It seems, so far, that the similarities between nuclei and small-particle superconductors have seldom been exploited, with the exception, perhaps, of a paper by Kawabata /21/ who employs a projection method. We have to realize that nuclear structure is at zero temperature whereas experimental and theoretical studies of small superconductors were in the past more interested in the phase transition at finite temperature than in the $T \to 0$ behavior.

Also here we want to focus our attention on finite temperatures, and we begin with a simple reformulation of bulk BCS theory which eventually allows us to treat size effects. Consider the Hamiltonian

$$H = H_0 + H_1 = \sum_{ks} \varepsilon_k c^+_{ks} c_{ks} - \frac{1}{\Omega} \sum_q V_q B^+_q B_q \tag{6.1}$$

where

$$B^+_q = \sum_k c^+_{k+q\uparrow} c^+_{-k\downarrow} \tag{6.2}$$

creates pairs with momentum q. $H_0$ contains $-\mu N$, which means that $\varepsilon_k$ is measured relative to the Fermi energy; the k-summation in (6.2) is confined to a shell of thickness $2\omega_D$ around the Fermi surface. $V_q > 0$ simulates the pair attraction caused by phonons. In the following we will assume $V_q = V_0 \delta_q$ which represents the original BCS model ($V_q = V_0$ would describe the Gorkov interaction):

$$H = H_0 - \frac{V_0}{\Omega} B^+_0 B_0 \, , \quad B^+_0 = \sum_k c^+_{k\uparrow} c^+_{-k\downarrow} \tag{6.3}$$

The grand canonical partition function of the BCS superconductor

$$Z = \text{Trace}\, e^{-\beta H} = \text{Trace}\, e^{-\beta H_0 + \frac{\beta V_0}{\Omega} B^+_0 B_0} \tag{6.4}$$

is now calculated by using the identity

$$e^{-AB} = \int_{-\infty}^{\infty} dz \, e^{-\pi |z|^2 + \sqrt{\pi}(Az - Bz^*)} \, , \qquad \begin{array}{l} z = x + iy \\ dz = dx\,dy \end{array} \; . \tag{6.5}$$

In fact, the identification

$$A = \sqrt{\frac{\beta V_0}{\Omega}}\, B^+_0 \, , \quad B = -\sqrt{\frac{\beta V_0}{\Omega}}\, B_0 \tag{6.6}$$

leads immediately to

$$Z = \int dz\, e^{-\pi|z|^2} \text{Trace}\, e^{-\beta\left[H_0 - \sqrt{\frac{\pi V_0}{\beta\Omega}}\left(zB_0^+ + z^* B_0\right)\right]} \tag{6.7}$$

This is a two-dimensional Gaussian average of a much simpler trace. With the substitution

$$\psi = \sqrt{\frac{\pi V_0}{\beta\Omega}}\, z \tag{6.8}$$

and by inserting the expressions for $H_0$, $B_0^+$ and $B_0$ Eq. (6.7) becomes

$$Z = \frac{\beta\Omega}{\pi V_0} \int d\psi\, \text{Trace}\, e^{-\beta H(\psi)} \tag{6.9}$$

$$H(\psi) = \frac{\Omega|\psi|^2}{V_0} + \sum_k \left[\varepsilon_k \left(c_{k\uparrow}^+ c_{k\uparrow} + c_{k\downarrow}^+ c_{k\downarrow}\right) - \left(\psi\, c_{k\uparrow}^+ c_{-k\downarrow}^+ + \psi^* c_{-k\downarrow} c_{k\uparrow}\right)\right] \tag{6.10}$$

$H(\psi)$ contains the well-known influence of the pair field, and the trace under the integral in Eq. (6.9) is easily performed with the result ($Z_0$ partition function of free electrons)

$$Z/Z_0 = \frac{\beta\Omega}{V_0} \int_0^\infty d|\psi|^2\, e^{-\beta F(|\psi|^2, \beta)} \tag{6.11}$$

$$F(|\psi|^2, \beta) = \frac{\Omega|\psi|^2}{V_0} - \frac{2}{\beta} \sum_k \ln \frac{\cosh \frac{\beta E_k}{2}}{\cosh \frac{\beta \varepsilon_k}{2}}, \qquad E_k = \sqrt{\varepsilon_k^2 + |\psi|^2} \tag{6.12}$$

Let us here pause for a moment and reconsider the above steps. Eq. (6.7) was obtained by applying the c-number formula (6.5) to an operator exponential which contains non-commuting parts. Such a procedure is, of course, forbidden. The correct representation of the partition function is rather given by viewing Eq. (6.9) as a two-dimensional functional integral

$$Z = \int \delta\psi(\tau)\, \text{Trace}\, T_\tau\, e^{-\int_0^\beta d\tau\, H(\psi(\tau))} \tag{6.13}$$

over the trace of an ordered exponential. The $\beta$-periodic function $\psi(\tau)$, $0 < \tau < \beta$, describes the general "time"-dependent pair field. The physics behind Eq. (6.13) is that we have non-interacting electrons in this general pair field and find the partition function by a Gaussian functional average. When one neglects time dependence and replaces $\psi(\tau)$ by its zeroth Fourier component $\psi$ (we call this the static approximation), one is back to the normal integral of Eq. (6.9). For a more detailed discussion of functional-integral methods, also in connection with other solid-state models, we refer to an earlier Nato-ASI publication /22/.

The quantity F given in Eq. (6.12) is the BCS free-energy functional with its minimum determined by

$$0 = \frac{\partial F}{\partial|\psi|} = \frac{2\Omega|\psi|}{V_0}\left[1 - \frac{V_0}{2\Omega}\sum_k \frac{\tanh\frac{\beta E_k}{2}}{E_k}\right]. \tag{6.14}$$

In the bulk limit $\Omega \to \infty$ the extensive F will produce a sharp maximum of the integrand in Eq. (6.11) at the value $|\psi| = \Delta(T)$, determined by Eq. (6.14) - the gap equation. This allows one to replace the integral in (6.11) by the maximal integrand: we arrive at BCS thermodynamics $-\ln Z/Z_0 = \beta F(\Delta^2(T), \beta)$ for temperatures $k_B T \leq k_B T_c = 1.14\,\omega_D \exp[-(N(0)V_0)^{-1}]$.

## 7. SUPERCONDUCTIVITY II

The formulas (6.11) and (6.12) of the previous section suggest a very easy path to superconductivity in small particles. Also for small volume $\Omega$ we believe in a pairing model of the type (6.3); we replace the momentum k by the appropriate quantum number of the single-electron states and treat, for simplicity, the corresponding energies in the scheme of equidistant levels with spacings $\delta = [N(0)\Omega]^{-1}$. Then, the free energy given by Eq. (6.12) takes the form

$$F(|\psi|^2; \beta) = \frac{1}{\delta}\left[\frac{|\psi|^2}{g} - \frac{2}{\beta}\sum_n{}' \delta \ln \frac{\cosh\frac{\beta E_n}{2}}{\cosh\frac{\beta \varepsilon_n}{2}}\right]. \tag{7.1}$$

Here $g = N(0)V_0$ is the dimensionless coupling constant, the prime on the sum implies a cutoff for $|\varepsilon_n| = |n|\delta > \omega_D$, and $E_n$ is defined by $\sqrt{\delta^2 n^2 + |\psi|^2}$.

The formula (6.11) with F given by (6.12) and (7.1) has a very physical meaning. F is the mean-field approximation of the thermodynamic potential belonging to the pairing model. We might view F as a reduced energy in which all microscopic degrees of freedom have been taken into account except those associated with the order parameter $\psi$. The remaining integral over this classical variable then has the form of a classical partition function.

Superconductivity in bulk homogeneous material is characterized by a large intrinsic range of the pair coherence length $\xi_0$ (e.g. $\xi_0 \approx 2000$ Å in tin). The spatial variation of the order parameter, being possible only on a scale $\xi_0$, can be neglected for a small particle with diameter less than $\xi_0$. This justifies the simple model. It is, however, obvious that finite-size effects cannot be treated by retaining the saddle-point approximation of the previous section and using a gap equation (6.14) with discrete levels. With decreasing size of the system we have, rather, to integrate in (6.11) over a whole distribution of order parameters (fluctuations) and, consequently, the concept of a single energy gap becomes meaningless.

A theory of superconductivity in small particles based upon the static functional approximation (6.11) with F given by Eq. (7.1) was developed in 1972 /23/. The important size parameter is

$$\bar{\delta} = \frac{\delta}{k_B T_c} \tag{7.2}$$

with $k_B T_c = 1.14\,\omega_D \exp[-g^{-1}]$. For $T_c$ in the range 10K to 1K, $\bar{\delta}$ is of order 0.1 to 1 for a 100Å particle. A particle of diameter d $\approx$ 1000 Å and $T_c \approx$ 1K would have $\bar{\delta} \approx 0.001$.

The actual calculations were made both with the general F of Eq. (7.1) and, for comparison, with its Ginzburg-Landau truncation /24/

$$F_{GL} = \frac{1}{\delta}\left[\ln\frac{T}{T_c}|\psi|^2 + \frac{0.526}{(\pi k_B T_c)^2}|\psi|^4\right] = \Omega\left[a|\psi|^2 + \frac{b}{2}|\psi|^4\right] \tag{7.3}$$

which also can be used to estimate that $|\psi|$-fluctuations are important in a temperature region given by $|1-t| \leq 0.46\,\bar{\delta}^{1/2}$, $t = T/T_c$.

Detailed results can be found in ref. /23/ for the diamagnetic susceptibility, the specific heat and the spin susceptibility. As an illustration we show here only the behavior of the specific heat (Fig. 2). With decreasing particle size (increasing $\bar{\delta}$) a broadening of the bulk temperature dependence at the superconducting phase transition is seen in all cases. When $\bar{\delta}$ approaches unity, there is hardly a remnant of the peak of the specific heat. This is different for the spin susceptibility which shows the superconducting effect even for $\bar{\delta}$ as large as unity. Concerning the spin susceptibility $\chi$, spin-orbit coupling can produce important modifications. As was mentioned in Sec. 5, Shiba has added the interaction (5.4) to the pairing model (6.3), and studied its influence by perturbation theory /18/. The susceptibilitiy then increases for fixed $\bar{\delta}$ with increasing spin-orbit coupling.

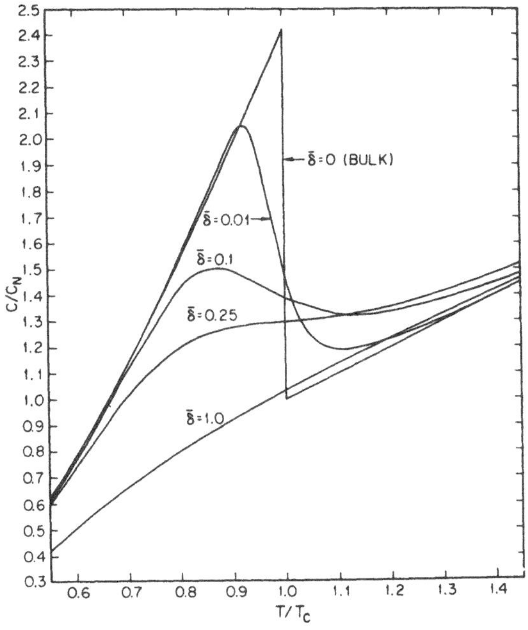


Fig. 2. Normalized specific heat calculated in the static functional approximation for several values of $\bar{\delta} = \delta/kT_c$. For reference the bulk BCS limit is also shown.

A number of experiments of all three quantities mentioned above were performed in recent years, and they confirm the predictions of the model in a qualitative manner /5/. We can say that increasing order-parameter fluctuations with decreasing size are by now established experimentally. They can be understood already with the simpler Ginzburg-Landau energy $F_{GL}$ of Eq. (7.3). Finer effects, however, such as contributions of terms beyond $|\psi|^4$ (which explicitly affect the specific heat) and the level discreteness contained in the full F of Eq. (7.1) are harder to detect. In addition to thermodynamic quantities also dynamical quantities such as nuclear-spin relaxation were studied both experimentally and theoretically /25/. Particle size influences here the order-parameter relaxation which enters the expression for the spin-lattice relaxation rate.

Let us finally make two comments: a) The simple model discussed above contains no theory of size dependence of phonons and of electron-phonon interactions. Therefore no prediction of the size dependence of the transition temperature can be made. $T_c$ is a phenomenological parameter which has to be fitted; b) We have stressed the restriction to fixed electron number N not only in Sec. 2 for normal metallic particles but also at the beginning of Sec. 6 for superconductivity in finite systems. The pairing interaction should make the N-fixed condition even more stringent for small superconducting particles. Clearly, the model described above is grand canonical and cannot account for this condition. We can, of course, project in a formal manner on the canonical partion function Z(N) by the contour integration prescribed in Eq. (2.2). However, it is very difficult to do an exact integration (as was possible for normal particles). At least for $\delta < 1$ and for $t = T/T_c$ not too small we can trust a saddle-point approximation which then produces unimportant modifications similar to the case of normal particles /23/.

## 8. CONCLUDING REMARKS

We have described some electronic properties of small metallic particles in a very simplified quantum statistical picture. A collection of particles is always necessary in order to do experiments, and we have indicated (in Sec. 4) that, besides size distributions, a special average over randomly distributed levels for particles of the same size modifies the low-temperature behavior of various quantities. It was further possible to determine the range in which these modifications are of importance. Level statistics is, of course, based upon the assumption that the particles are independent of each other.

For coupled particles it turns out that at least two things become relevant which already have appeared in our description of an isolated particle. This is first the large Coulomb energy of adding an electron to a small particle which led us to the canonical distribution of fixed N in Sec. 2. The same large Coulomb energy also plays a decisive role in normal and superconducting granular metals /26/. Secondly, the treatment of a superconducting particle in Sec. 6 and 7 in the static functional approximation has produced the complex order parameter $\Psi = |\Psi| \exp[i\varphi]$. The phase $\varphi$ is present in the small particle but does not participate in fluctuations since it was integrated away in passing from Eq. (6.9) to Eq. (6.11). This is changed when two particles come near to each other and their phases create Josephson coupling. A description of coupled small superconductors on a microscopic level has to use an appropriate generalization of the functional integral given by Eq. (6.13), and this was done by Efetov /27/.

Several properties of isolated small particles are rather well explained by applying the phenomenological laws of thermodynamics and electrodynamics to perfect geometries, e.g. spheres. The inclusion of surface energy in the thermodynamic treatment leads to results which sometimes agree surprisingly well with experiments /28/. An example is the decrease of the melting temperature of gold particles which can be fitted down to sizes smaller than 50 Å.

Plasmon excitations in small particles of perfect shape are in first approximation well treated by means of electrodynamics with a (bulk) dielectric function $\varepsilon(\omega)$. We regret that we cannot enter here the field of optical properties of small particles and refer instead to a recent review /29/. In a second approximation broadenings and shifts of plasmon resonances have to be considered, and quantum mechanics will come into the problem. With respect to these effects, the situation has been controversial in the past, both experimentally and theoretically; we mention in this context that shifts can also be produced by introducing surface roughness on the small sphere /30/.

At the end we find it useful to distinguish once more between three different categories of approaches to small particles:
A. Going up from the small end (atoms),
B. Coming down from infinity (bulk),
C. Using an ideal particle surface.
In A is certainly the most computational involvement. The structure of rare gas and ionic clusters can be studied /7/. The electronic structure of metal clusters is a domain of quantum chemistry but increasingly also methods originally developed for bulk solids are transferred to microstructures.

For approach C we have given some examples above. To this important category belongs, further, the transfer of experience with the infinite plane surface to the finite surface of a perfect particle (sphere, cube, etc.).

Approach B, eventually, deals with phenomena which show up when a bulk system is reduced in dimension. The description of a small particle given in the preceding sections belongs to this category. The approach is much less specific than A and C, and is confined to electronic properties of small metal particles. The surface of the system appears only in so far as it produces discreteness among the low lying excitations around the Fermi energy via boundary conditions. The advantage of such a procedure in spite of the loss of any structure influence: predictions of size effects for various intrinsic and extrinsic properties and the posssibility of handling a collection of small particles.

REFERENCES

/1/ Proceedings of the second international meeting on the small particles and inorganic clusters, Surface Sc. 106 (1981).
/2/ F. Cyrot-Lackmann (ed.),"Entre l'atome et le cristal: Les agregats", Les éditions de physique (1981).
/3/ Nato Advanced Study Institute on Impact of Cluster Physics in Material Science and Technology, Cap d'Agde, France 1982, to be published in 1983.
/4/ A.E. Hughes and S.C. Jain, Adv. Phys. 28:717 (1979).
/5/ J.A.A.J. Perenboom, P. Wyder and F. Meier, Phys. Reports 78:

173 (1981).
/6/ H.P. Baltes and E. Simanek, in: "Aerosol Microphysics II", W. Marlow, ed., Topics in Current Physics, Springer Berlin (1982).
/7/ T.P. Martin, Phys. Reports 95: 167 (1983).
/8/ See R.P. Messmer in /1/, D.R. Salahub in /2/.
/9/ R. Kubo, loc. cit. in PWM /5/.
/10/ R. Denton, B. Mühlschlegel and D. Scalapino, loc. cit. in PWM /5/.
/11/ C. Taupin et al., loc. cit. in PWM /5/
/12/ Compare the detailed discussion in Capter 3.5 of PWM /5/, and by J. Buttet in /2/.
/13/ R. Denton and B. Mühlschlegel, Solid State Commun. 11:1637 (1972); N.Grewe and B. Mühlschlegel, ibid. 14:231 (1974).
/14/ L.P. Gorkov and G.M. Eliashberg, loc. cit. in PWM /5/
/15/ D.M. Wood and N.W. Ashcroft, Phys. Rev. B25: 6255(1982).
/16/ See references /89/-/92/ in PWM /5/.
/17/ T.A. Brody, J. Flores, J. Bruce French, P.A. Mello, A. Pandey and Samuel S.M. Wong, Rev. Mod. Phys. 53:385 (1981).
/18/ H. Shiba, loc. cit. in PWM /5/
/19/ J. Buttet in /2/.
/20/ For pairing in nuclear physics see A.L. Fetter and J.D. Walecka, "Quantum Theory of Many-Particle Systems", McGraw-Hill, New York (1971).
/21/ A. Kawabata in /1/.
/22/ G.J. Papadopoulos and J.T. Devreese, eds. "Path Integrals and Their Applications in Quantum, Statistical, and Solid State Physics", NATO Adv. Study Inst. Ser. B 34, Plenum, New York (1978).
/23/ B. Mühlschlegel, D.J. Scalapino and R. Denton, loc. cit. in PWM /5/, see also B. Mühlschlegel in /1/.
/24/ The Ginzburg-Landau integral was studied also in earlier contributions, see References in /23/ and PWM /5/.
/25/ See the References in the article of Baltes and Simanek /6/.
/26/ B. Abeles, Ping Sheng, M.D. Coutts, and Y. Arie, Adv. Phys. 24: 407 (1975); B. Abeles, Phys. Rev. B 15:2828 (1977).
/27/ K.B. Efetov, Zh. Eksp. Teor. Fiz. 78:2017 (1980), (Sov. Phys. JETP 51:1015 (1980)).
/28/ For a review see the article by Châtelain and Borel in /2/.
/29/ U. Kreibig in /3/.
/30/ A.A. Maradudin et al., to be published.

# PARTIALLY CONNECTED SYSTEMS

P.G. de Gennes

Ecole de Physique et Chemie
Paris, France

## INTRODUCTION

Many physical structures are made of random "islands", and in certain conditions, among these islands, one macroscopic continent emerges. This idea was displayed first in a historical article of Broadbent and Hammersley.[1] It was rediscovered soon after (in connection with substitutional alloys) by Lafore, Millot, and the present writer.[2] From this point on, most numerical studies on this "percolation phenomenon" were performed on periodic lattices where a fraction p of the sites (or of the bonds) was active, and a fraction 1-p was inactive - different sites (or bonds) being uncorrelated. But the results are more general: the statistics of islands near threshold is very universal.

Our aim, in the present talks, is first to define the main features of the islands and of the continent near threshold (section II). Then we proceed to transport properties, with special emphasis on the problems of one "ant" moving on a percolation lattice (section III). These geometrical and dynamical questions have many applications in physics. Here, in section IV, we restrict our attention to one of these: the magnetic behavior of granular superconductors.

None of this is original: the main ideas on "ants", and their transposition to the superconductor problem, have been invented by S. Alexander and his coworkers. I very much regret that he could not present them himself at this workshop. I also do not give a complete list of references, but restrict myself to a few recent reviews.

II. PERCOLATION CLUSTERS

We see on Fig. 1 a typical example of site percolation, with a fraction p of active sites. Following Ref. 2, we focus our attention on one particular lattice site 0, and construct the probabilities $P_n$ for this site to belong to an n-cluster. For example with the square lattice of Fig. 1, we have

$$\begin{aligned} P_0 &= q \\ P_1 &= pq^4 \\ P_2 &= 4p^2q^6 \\ P_n &= \sum_m C_{nm} p^n q^m \end{aligned} \qquad (11.1)$$

where $C_{nm}$ is the number of clusters containing 0 with n sites and m adjacent sites (m is what in Ref. 2 we called the "adherence" of the clusters). It is of particular interest to consider the sum

$$S(p) = P_0 + P_1 + \ldots + P_n + \ldots \qquad (II.2)$$

At low p this sum is identically equal to unity. But when p is beyond a certain "percolation threshold" $p_c$ this is not true any more: there is a finite probability $P_\infty(p)$ for 0 to belong to the infinite cluster, and

$$S(p) = 1 - P_\infty(p) \qquad (II.3)$$

Using a function S(p) approximated by a finite number of terms one can locate $p_c$ and construct $P_\infty(p)$. This has been carried out in great detail. A recent summary of numerical results on percolation transitions is found in Ref. 3. Just above threshold ($p = p_c + \Delta p$) the weight of the infinite cluster $P_\infty(p)$ increases with a certain power $\beta$

$$P_\infty(p) = \Delta p^\beta \qquad (II.4)$$

The exponent $\beta$ is the same for all 3 dimensional systems $\beta_3 = 0.45$ and is also the same for all 2 dimensional systems $\beta_2 = 0.14$. This is our first encounter with universality: all percolation systems (defined with short range links, in d dimensions) have the same critical behavior near threshold.

Just below threshold ($p = p_c - \Delta p$) some clusters become very large, but some others remain quite small. The "weight average" of the distribution is defined by

$$n_w = \sum_n nP_n \qquad (II.5)$$

and diverges like $\Delta p^{-\gamma}$ near the threshold. Here $\gamma$ is another critical exponent ($\gamma_3 = 1.74$; $\gamma_2 = 2.39$).

The complete distribution $P_n$ has been studied by D. Stauffer.[4] It is very broad distribution, with a cut off $n = N^*$, which can be idealized as follows

$$P_n \cong \begin{cases} n^{-a} & (n < N^*) \\ 0 & (n > N^*) \end{cases} \qquad \text{(II.6)}$$

where $a = 1 + \frac{\beta}{\beta+\gamma}$. The cut off $N^*$ itself varies with the distance to threshold

$$N^* \cong \Delta p^{-(\beta+\gamma)} \qquad \text{(II.7)}$$

An important feature is the spatial size of the clusters r (defined, for instance, through their radius of gyration). The size of the largest clusters ($n = N^*$) is called the correlation length $\xi_p(p)$ and diverges with a certain exponent

$$\xi_p(p) \cong a(\Delta p)^{-\nu} \qquad \text{(II.8)}$$

where a is the mesh size of the network. $\nu$ is not a new exponent, but is related to $\beta$ and $\gamma$. In d dimensions $\nu d = 2\beta + \gamma$.

It is important to realize that the average size of a cluster is smaller than $\xi$. For instance, if we define the average according to: $\langle r^2 \rangle = \sum_n P_n \langle r_n^2 \rangle$ we find, following Stauffer, that

$$\langle r^2 \rangle = \Delta p^{\beta} \xi_p^2 = a^2 \Delta p^{-2\nu+\beta} \ll \xi_p^2 \qquad \text{(II.9)}$$

(This is very similiar to the law for the weight average $n_w \cong \Delta p^{\beta} N^*$ which can be derived from Eq. (11.6).)

One remark concerning the finite clusters: we find them both below and above threshold. All formulas such as Eqs. (II.6, II.7) can be applied in both regimes (although the numerical prefactors are different above and below $p_c$).

Let us end up this section by a few specific properties of the <u>infinite cluster</u> above threshold. The infinite cluster is a random network, with a distribution of mesh sizes. The largest mesh size present is equal to the correlation length $\xi_p(p)$. But there are smaller mesh pieces. There are also a number of dangling ends belonging to the network. From a practical point of view the essential property of the infinite cluster is the electrical conductivity $\Sigma$ (corresponding to active bonds which are conducting, and inactive bonds which are insulating). Near threshold:

$$\Sigma = \Sigma_0 \, \Delta p^{\bar{t}} \tag{II.10}$$

The exact value of $\bar{t}$ in 3 dimensions is still a matter of discussion - early work led to $\bar{t} \sim 1.7$ but recent data suggest much higher values ( ~ 2). In any case the conductance $\Sigma$ increases much more slowly than the weight of the infinite cluster. Part of this is due to the dangling ends: the weight of the "backbone" (obtained by removing the dangling ends) is

$$P_b = \Delta p^{\beta_B} \ll P_\infty \tag{II.11}$$

since $\beta_B$ ( ~ 0.9 in 3 dimensions) is larger than $\beta$. But apart from this there is also a reduction in $\Sigma$ due to the contorted paths which the currents must follow.

## III. THE ANT PROBLEM

To define local dynamics on the percolation clusters, let us consider an "ant" jumping from site (i) to site (j), with a distribution $f_i(t)$ ruled by

$$\frac{\partial f_i}{\partial t} = \sum_j W_{ij}(f_j - f_i) \tag{III.1}$$

where $W_{ij} = 0$ if the neighboring sites (i,j) are not connected, and $W_{ij} = W$ otherwise. Consider first an ant which has started <u>on the infinite cluster</u>, at one point O, and which moves during time t by a distance x(t). By symmetry we have $\langle x(t)\rangle = 0$ but the square is more interesting. In a macroscopic limit, when $x \gg \xi$, we expect

$$\langle x^2(t)\rangle = 2Dt \tag{III.2}$$

where D is a certain diffusion constant for the ant. D is related to the conductance $\Sigma$, as follows. The ant current $J_{ij}$ between neighboring sites is $J_{ij} = W(f_i - f_j)$ and this is similar to the electrical current in the conductance problem, if $f_i$ and $f_j$ become local voltages. Thus we expect the macroscopic ant current to be

$$\bar{J} = W\,\Delta p^{\bar{t}}(-\nabla f) \qquad \text{(III.3)}$$

Let us now write the conservation law for the number of ants, all moving on the infinite cluster. In a volume $\Omega$ the number of sites is $\Omega P_\infty$, and the number of ants is $\Omega P_\infty f$. Thus we may write

$$\frac{\partial}{\partial t}(\Omega P_\infty f) = \int \bar{J}.dS = W\Delta p^{\bar{t}}\Omega\nabla^2 f \qquad \text{(III.4)}$$

and we see that $\frac{\partial f}{\partial t} = d\nabla^2 f$ where

$$D = W\Delta p^{\bar{t}-\beta} \qquad \text{(III.5)}$$

Knowing Eq. (III.5) we have a description of the long time motions via Eq. (II.2). This is valid provided $\langle x^2(t)\rangle > \xi^2$ or $t > \tau$ where $\tau$ is a characteristic time

$$\tau = \xi_p^2/D \sim \Delta p^{-2\nu+\beta-t} \qquad \text{(III.6)}$$

Let us now focus on earlier times $t < \tau$ (always assuming that our ant is on the finite lattice).[4] Then the space traveled during time t is $\langle x^2(t)\rangle = \ell^2(t)$. This $\ell(t)$ must be independent of $\Delta p$: the spacial structure of the clusters at scales $< \xi_p$ has a well-defined critical limit just as the Stauffer distribution (Eq. (11.6)) is independent of $\Delta p$ for $n < N^*$. Imposing this condition, and noting the boundries

$$\ell(t = W^{-1}) = a$$
$$\ell(t = \tau) = \xi_p \qquad \text{(III.7)}$$

(where a is the unit cell of the underlying lattice), we are unequivocally led to

$$\ell(t) = a(Wt)^{1/(2+\theta)} \qquad \text{(III.8)}$$

with $\theta = (\bar{t} - \beta)/\nu$ ($\theta_3 \sim 1.4$). The law (Eq (II.8)) is not of the diffusion type, but it is often convenient to visualize it as the effect of a scale dependent diffusion coefficient $\ell^2 = D(\ell)t$

$$D(\ell) = Wa^2 \left(\frac{a}{\ell}\right)^\theta \qquad (\ell < \xi_p) \qquad \text{(III.9)}$$

Eq. (III.9) will be the starting point of our later discussion on granular superconductors.

One final remark: all the above discussion was restricted to ants moving on the infinite cluster. If we put an ant on a finite cluster, things are usually more complex, because the ant may be trapped in small clusters, and never travel over distances such as $\xi_p$: this is discussed in detail in Ref. 4.

## IV. GRANULAR SUPERCONDUCTORS UNDER FIELDS

### Loops and pendent chains

We start with a system of grains, a fraction p of these being superconducting. Near the percolation threshold, we can idealize the superconducting clusters as a network of superconducting wires: the wires are very thin (diameter smaller than the penetration depth) and they are branched. On this structure we impose a homogeneous magnetic field H. The screening effect of the superconducting currents is neglected: we consider only an extreme type II limit.

The first observation is that the field has effects only through the closed loops in the network:[5] it is only in these regions that the vector potential A cannot be removed by a trivial gauge transformation. A second observation is that, although the field couples through the closed loops, it's effects are propagated elsewhere. The simplest example to show this is a loop with one pendent chain (a "lasso") for which one can discuss the upper critical field in detail:[6] one finds that, under fields, the order parameter is also depressed in the pendent chain.

It has been known for a long time that the solutions of the Landau-Ginsburg equations for the superconductor are deeply related to the diffusion properties of the carriers in the normal state.[7] Thus it is natural to think that the ant plays a role here. This has been worked out in detail by S. Alexander, and we shall describe a simplified version of his ideas in what follows.

### Characteristic lengths in zero field

A) If all grains are superconducting (p = 1) we have a superconductng correlation length $\xi_s$ which (in the dirty limit) is simply given by

$$\xi_s^2 = D_1 \tau_s \tag{IV.1}$$

Here $D_1$ is the diffusivity of one electron at the Fermi surface and may depend sensitively on the tunneling between adjacent grains. $\tau_s$ is a pair lifetime $\tau_s = \hbar/k_B(T_{co} - T) = \hbar/k_b \Delta T)$ where

$T_{co}$ is the superconducting transition point. We restrict our attention to the regime where $\xi_s$ is much larger than the size (a) of the grains ($\Delta T$ small).

B. <u>In the percolation regime</u> (p close to $p_c$) we have to define two lengths.

a) One is purely geometrical: this is the largest size of the clusters (or the mesh size of the infinite network): i.e. the correlation length for percolation $\xi_p$ (Eq. (11.8)).

b) The second one $\xi$ describes the superconductivity. If we are above $p_c$, and in the macroscopic limit where $\xi > \xi_p$, this is simply defined by the ant diffusivity D:

$$\xi^2 = D\,\tau_s \qquad (\tau < \tau_s) \tag{IV.2}$$

where $D = D_1\,\Delta p^{\bar{t}-\beta}$. On the other hand, if $\xi$ is smaller than $\xi_p$ (or $\tau_s$ smaller than $\tau$) we must replace D by the scale dependent diffusivity $D(\ell)$ defined in Eq. (III.9). The resulting correlation length has been called $\lambda$ by Alexander. (Note: this is <u>not</u> a penetration depth!) The equation for $\lambda$ is

$$\lambda^2 = D(\lambda)_s = \left(\frac{a}{\lambda}\right)^{\theta} \xi_s^2 \tag{IV.3}$$

and holds provided that $\lambda < \xi_p$. Note the temperature dependences $\xi \sim \Delta T^{-1/2}$ but $\lambda \sim \Delta T^{-1/(2+\theta)}$.

To summerize: we have two lengths, one geometric ($\xi_p$) one associated with the superfluidity ($\lambda$ or $\xi$). All scaling laws will involve the ratio x of these two lengths. In early work by Toulouse, Rammal, and Lubensky,[8] x was defined as

$$x_{TRL} = \frac{\xi s}{\xi_p} \tag{IV.4}$$

This, however, is not physical, because it compares an $\xi_s$, defined for p = 1, with an $\xi_p$ defined for $p = p_c$. In other words, we expect a change in behavior when the diffusion length of the pairs <u>in the medium</u> ($\xi$ or $\lambda$) becomes equal to $\xi_p$. Thus the scaling parameter is

$$x_A = \frac{\xi}{\xi_p} \tag{IV.5}$$

and not Eq. (IV.4). This will be essential in what follows.

The macroscopic limit under fields

We now select the case where $\xi >> \xi_p$, and discuss the magnetic behavior of the infinite cluster. This regime is conceptually simple; since the granularity of the percolation medium occurs only at small scales, we can apply standard discussions for homogeneous superconductors.

A. Let us start with the upper critical field of the infinite cluster (the field which destroys macroscopic superconductivity):

$$H_{C2} = \frac{\emptyset_0}{\xi 2} = \frac{\emptyset_0}{D\tau_s} \quad (\xi > \xi_p) \tag{IV.6}$$

where $\emptyset_0$ is the flux quantum. Returning to Eq. (IV.8) we see that $H_{C2}$ is linear in $\Delta T$ and proportional to $\Delta p^{-t+\beta}$ ($\sim \Delta p^{-1.4}$ in 3 dimensions).

B. It is also of interest to compute the contribution of the infinite cluster to the low field dimagnetism - always assuming strong type II behavior: the field H is still above $H_{C1}$, but small. Then we expect vortices, with a core of radius $\xi$, and a 2 dimensional density $\frac{H}{\emptyset_0}$. The modification in core energy per $cm^3$ is of order $N(0)\ \Delta_0^2 P_\infty$ where N(0) is the density of states in the metal, $\Delta_0$ the gap parameter in zero field and $P_\infty$ selects the infinite cluster part. Then the free energy per unit volume is

$$F - F_0 \cong \frac{H\xi^2}{\emptyset_0} N(0)\ \Delta_0^2 P_\infty \tag{IV.7}$$

and the magnetic moment $M = -\frac{\partial F}{\partial H}$ is independent of H

$$-M = P_\infty \frac{2\ N(0)\Delta_0^2}{\emptyset_0} \sim \frac{D_1\hbar\ N(0)\Delta_0}{\emptyset_0} \Delta p^{\bar{t}} \tag{IV.8}$$

where we have used Eq. (IV.2) (and also written $\Delta_0 \sim k_B \Delta T$). Note that in this regime we cannot define a susceptibility, since $\frac{\partial M}{\partial H} = 0$.

Finite Clusters Near Threshold

Let us discuss first the qualitative structure of the Landau-Ginsburg free energy. The gradient terms, when evaluated on a scale $\ell$ (smaller that $\xi_p$) take the form

$$F\Big|_{grain} \tilde{=} C\ D(\ell)\ [(i\nabla - A/\emptyset_0)\ \Delta]^2 \qquad (IV.9)$$

where $C \sim N(0)\ a^3/k_B\ T_{c0}$ is a normalization factor. We may estimate the vector potential A by setting A = H .

A. In low fields, the diamagnetic currents are spread over the whole cluster. If our cluster has n grains, and a size $r_n$, we must take $\ell = r_n$, and we see that the H dependent terms in Eq. (IV.9) are of order

$$\Delta F = C\ D(R_n)\ (\frac{H}{\emptyset_0})^2\ r_n^2\ \Delta_0^2 \qquad (IV.10)$$

This corresponds to a diamagnetic moment linear in field, or to a susceptibility per cluster.

$$x_n = -nC\ \Delta_0^2\ D(r_n)\ r_n^2 = -nC\ \Delta_0^2\ D_1\ a^{\theta}\ r_n^{2-\theta} \qquad (IV.11)$$

(The factor n comes from our counting per cluster.) Finally the susceptbility averaged over all finite clusters is[8]

$$x = \sum_n P_n\ x_n = -\ C\ \Delta_0^2 D_1\ a^{\theta}\ \xi_p^{\ 2-\theta}\ \Delta p^{\beta} \qquad (IV.12)$$

The last factor $\Delta p^{\beta}$ comes from the integration over the Stauffer distribution of clusters (Eq. (II.6)). Making use of Eqs. (III.8) for $\theta$ and (II.8) for $\xi_p$ we find that $x \sim \Delta p^{t-2\nu}$. This coincides with an unpublished conjucture which I had made in connection with Ref. 5, but disagrees with a proposal of Ref. 9. Monte Carlo data of Rammel and coworkers tend to support Ref. 9: it is hard, however, to see where the Alexander argument[8] could go astray.

B. Upper critical field. Here we face two different possibilities, depending on the size of the cluster ($r_n$) when compared to the superconducting correlation length ($\lambda$):

a) for large clusters ($r_n > \lambda$) we optimize the energy by building a wave packet $\Delta(r)$ such that

$$\frac{1}{\ell} = \Delta^{-1}\nabla\Delta = \frac{H\ell}{\emptyset_0} \qquad (IV.13)$$

and the Landau-Ginsburg equation imposes $\ell^2 = D(\ell)\tau_s$ on $\ell = \lambda$. Thus the upper critical field is

$$H_{C2} \cong \frac{\emptyset_0}{\lambda^2} \tag{IV.14}$$

Note that $H_{C2}$ is not a linear function of temperature: returning to Eq. (IV.3) we see that $H_{C2} \sim \Delta T^u$ with $u = 1/(1 + \theta/2)$;

b) for small clusters ($r_n < \lambda$), the order parameter $\Delta$ is nearly constant inside the cluster, even at the upper critical field (just as it is in the more familiar case of thin films in tangential fields). The characteristic length is $\ell = r_n$ and the L-G equation gives

$$\frac{1}{D(r_n)\tau_s} = \left(\frac{H_{C2} r_n}{\emptyset_0}\right)^2 \tag{IV.15}$$

Here $H_{C2}$ is proportional to $\Delta T^{1/2}$ and to $r_n^{-1+\theta/2}$. Note that (since $r_n < \lambda$) the value of $H_{C2}$ which is derived from Eq. (IV.15) is higher than Eq. (IV.14). Thus, at fields H below $\emptyset_0/\lambda^2$, all clusters are superconducting. At fields $H > \emptyset_0/\lambda^2$ certain large clusters are normal, but the small clusters are still superconducting. We need to approach $H \rightarrow \infty$ (in this model with thin wires) to make all clusters normal.

Summary of Various Regimes

All of these results are summarized in the Alexander plot[8] of Fig. 1. This diagram is constructed at fixed T (fixed $\lambda$), variable H and variable percolation levels, with $\xi_p^{-1}$ as the abscissa and H as the ordinate. A first curve of importance is the parabola

$$H = \frac{\emptyset_0}{\xi_p^2}$$

Below this parabola, even the largest finite cluster (of size $\xi_p$) has less than one flux quantum: the vortices do not enter, and all finite clusters are superconducting. Above the parabola, some clusters are large enough to contain many vortices, but they remain superconducting up to a field $\emptyset_0/\lambda^2$. Some smaller clusters remain superconducting even in very high fields. The diagram also shows the fate of the infinite cluster: this has an upper critical field $H_{C2}$ which is small at high p, larger for $p \rightarrow p_c$. (Eq. (IV.6). $H_{C2}$ has a finite value (of order $\emptyset_0/\lambda^2$ at $p = p_c$.

All this description has been restricted to scaling laws: a lot of work remains to be done to transform it into a quantitative theory, with precise coefficients and crossover shapes. But the general picture is relatively simple, and probably correct.

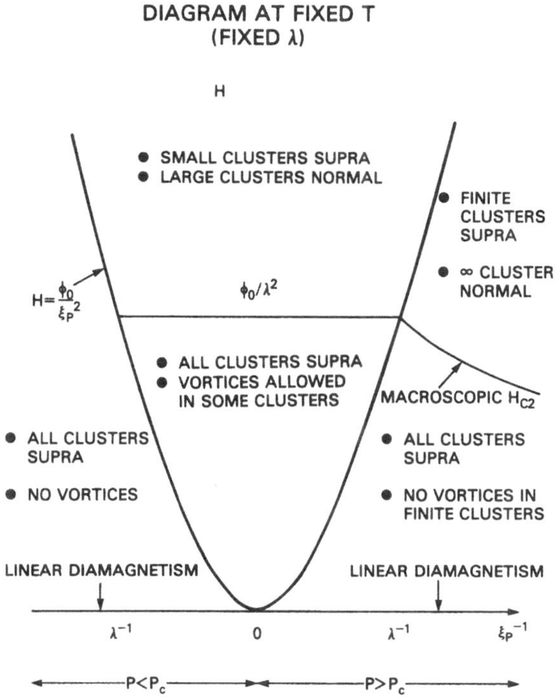


Figure 1

ACKNOWLEDGMENTS - This paper owes a lot to G. Deutscher, who introduced me to the experimental facts which he describes elsewhere in these proceedings; to G. Toulouse, T. Lubensky, and R. Rammal for stimulating discussions; and above all to S. Alexander, for his patient explanations on the magnetic behavior of a Stauffer distribution of clusters.

REFERENCES

1. S.R. Broadbent and J.M. Hammersley, Proc. Cambridge Phil. Soc. 53, 629 (1957).

2. P.G. de Gennes, P. Lafore, and J.P. Millot, J. Physique (Paris) 20, 624 (1959).

3. J.P. Clerc, G. Giraud, J. Rossenq, R. Blanc, J.P. Carton, E. Guyon, H. Ottavi, and D. Stauffer, Annales de Physique 8, 1 (1983). See also S. Kirkpatrick, Rev. Mod. Phys. 45, 574 (1973).

4. Y. Gefen, A. Aharony, and S. Alexander, Phys. Rev. Lett. 50, 77 (1983); R. Rammal and G. Toulouse, J. Physique Lett. 44L, 13 (1983).

5. P.G. deGennes, Comptes Rend. Acad. Sci. Paris 292, 9 (1981).

6. P.G. deGennes, Comptes Rend. Acad. Sci. Paris 292, 279 (1981).

7. See for instances, P.G. de Gennes, "Superconductivity of metals and alloys", Benjamin, NY (1964).

8. S. Alexander, to be published.

9. R. Rammal, T. Lubensky, and G. Toulouse, J. de Physique Lett. 44L, 65 (1983).

# PERCOLATION AND SUPERCONDUCTIVITY

Guy Deutscher*

Department of Physics and Astronomy
Tel Aviv University
Ramat Aviv, Tel Aviv 69978, Israel

## I. INTRODUCTION

It has been known for many years[1] that it is possible to prepare by vacuum deposition metal-insulator mixtures which display interesting structural properties. When the metal and the insulator are insoluble in each other - which is the only case we will be interested in in these lectures - codeposition of the constituents onto a room temperature substrate (by reactive evaporation, sputtering or thermal evaporation) usually results in the formation of a granular structure. Diffraction experiments indicate that the metal is crystalline and the insulator amorphous. The metal forms well characterized and separate grains. Dark field electron microscopy[2] shows that they have in general a random orientation; in particular, there is no indication that neighbouring grains have correlated crystallographic orientations. On the other hand, the grain size can be fairly uniform, with a distribution width of the order of the mean value or even smaller in some cases such as Aℓ-Ge.[3] Typical average grain sizes are of the order of 100 A. The insulator forms a thin amorphous coating around the metallic grains, of the order of a few atomic layers. Hence, we can consider such granular mixtures as composed of well characterized metallic regions separated by tunnel junctions. If the inelastic mean free path is of the order of the grain size or smaller, we can ascribe to each of these junctions a well defined resistance. Since this resistance varies exponentially with the thickness of the insulator,

*The Oren Family Chair of Experimental Solid State Physics.

we have essentially a random resistor network (if the thicknesses are randomly distributed), whose macroscopic resistance can usually be described by percolation exponents.[4] (However, the critical metal volume fraction p at the metal insulator transition is usually close to 50%[1], much higher than the value of 15% predicted for a random continuum). When the metallic grains are superconducting, some of the thin junctions may allow Josephson tunneling, while in the thicker ones single particle tunneling will dominate. At low temperatures, the junctions' resistances will tend either to zero or to infinity: here again, we have a well-defined percolation problem, as first suggested by Straley.[5]

More recently, it has been found that metal-insulator mixtures do not always form in a granular structure. If the constituents have relatively low melting points (such as In-Ge)[6], or if the film is deposited on a hot substrate,[7] or if the granular structure is annealed so that the amorphous insulator recrystallizes[8] - i.e. in all known cases where both the metal and the insulator are crystalline, the mixture shows a random topology. The insulator does not coat systematically the metal grains. Both the metal and the insulator form clusters of grains. In that case, the concepts of percolation in a random continuum can be applied in a straightforward fashion to the physical properties of the mixture, i.e. there exists a close relationship between geometrical percolation, electrical conduction and superconducting properties. For instance the value of $p_c$ is indeed equal to 15% within experimental accuracy.[6]

The third type of structure that has been investigated is that of thin, discontinuous metal films. The nucleation and growth processes of thin films have been the subject of numerous studies,[9] which have mainly concentrated on the early stages of these processes. Relatively little attention has been paid to the labyrinthine structure observed when the film has almost reached continuity. While the early growth is dominated by interactions that are not universal in character (such as interaction with the substrate, surface tension, etc. ...) it turns out that the very large structures (clusters) observed near continuity do obey certain universal scaling relationships identical to those of percolation clusters.[10,11] This opens up the possibility of a quantitative characterization of these structures and use of theoretical models to predict their normal and superconducting properties.[12]

We first review some of the relevant structural aspects of metal-insulator mixtures before discussing their superconducting properties.

## II. PERCOLATIVE STRUCTURES

We review in this section the experimental structural evidence for the relevance of percolation concepts to metal-insulator mixtures. Direct evidence exists only for random mixture films and discontinuous films, because the main element of randomness in granular films (the insulating barrier thickness) is not amenable to direct observation.

## II.A. Random Mixtures

Random mixture films can be prepared by codeposition at room temperature for some low melting point metals (In, Pb) mixed with Ge.[6] Some mixtures that form in the granular structure when deposited at room temperature can also be formed in the random phase by deposition on a hot substrate[7] or by annealing of the granular phase. Such is the case for Aℓ-Ge for which a detailed study of the "structural phase diagram" has been performed.[8] On the other hand, annealing of other granular structures such as $W$-$SiO_2$[14] only produces grain growth but no qualitative change in the structure (probably because $SiO_2$ is not recrystallized in these experiments). For granular Aℓ-Ge prepared at room temperature, the annealing temperature required to obtain the random phase depends on the concentration of the mixture. It decreases as the Aℓ content is increased, reaching a plateau of about 160°C at high Aℓ concentration (Fig. 1). There the annealing transformation is between the metallic granular phase and the random phase, while at low Aℓ concentration the transformation is between an amorphous insulating Aℓ-Ge phase directly into the random phase. It is clear from these experiments that both the amorphous and the granular phase are metastable, the random phase being the stable one.

Little is understood about the thermodynamics of these transformations and their kinetics. There seems to be some correlation between the occurrence of the random phase and percolation of the metal in the mixture. When Aℓ-Ge is deposited on a 180°C substrate it forms in the random phase down to the percolation threshold, and in the amorphous phase below that concentration. A similar result was reported for room temperature evaporated In-Ge.[6] A related observation is the occurrence of coherent clusters in the random mixtures. Figure 2 shows a bright field and a dark field picture of the same area of an In-Ge sample, the dark field showing only In grains close to a given orientation. It is clear that the large scale In cluster is composed of grains with strongly correlated orientations. By slightly tilting the sample stage, nearby parts of the In (infinite) clusters become visible in the dark field picture. This shows that the infinite cluster may grow in a quasi-coherent fashion over large distances. In contrast, dark field pictures of Ge show uncorrelated grains. These observations suggest a relation between percolation as a growth process and

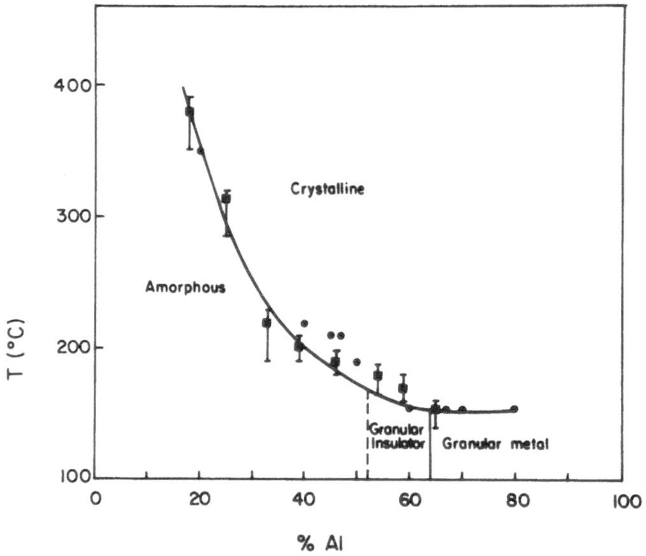


Fig. 1. Recrystallization temperature of granular Aℓ-Ge for formation of the random phase.

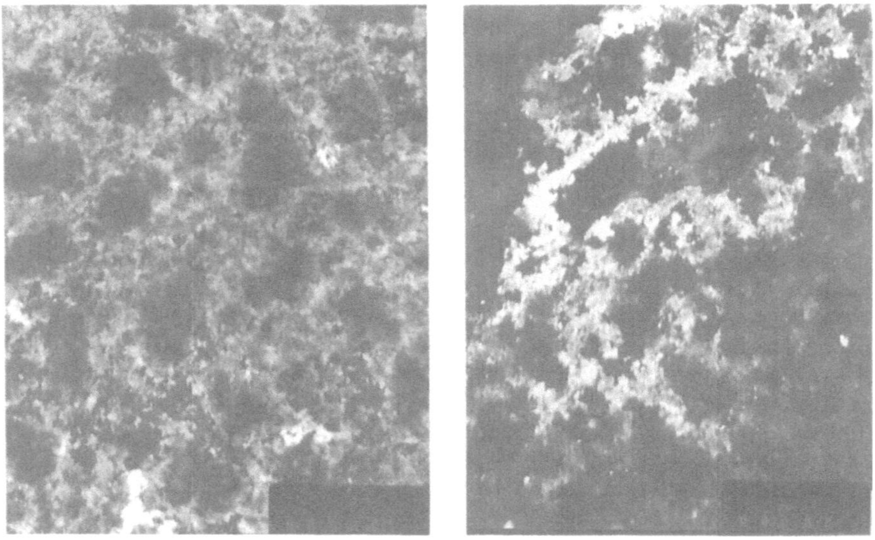

Fig.2a. Fig.2b.

Fig. 2. Bright field (Fig.2a) and dark field (Fig.2b) of random In Ge showing coherent orientation of In grains over long distances.

and percolation as a conduction process. More experiments are needed to study the growth kinetics of the random phase.

Near threshold, the random character of the mixtures leads to self-similarity over a significant range of length scales, as predicted by the scaling theory of percolation. A series of micrographs of In-Ge[13] taken at varying magnifications shows qualitatively similar pictures, i.e. the observer is unable to tell in what order of magnifications they were taken. However, below a certain magnification gross features become apparent and the order of magnifications is easily established. According to our current understanding of percolation[4], this occurs when the frame size is of the order of a characteristic length $\xi_p$, which diverges at $p_c$

$$\xi_p = a \; |p - p_c|^{-\nu} \qquad (1)$$

where a is a length of the order of the typical grain size and is a critical index $\nu_3 = 0.85$ in three dimensions (film thickness $d>\xi_p$) and $\nu_2 = 1.33$ in two dimension ($d<\xi$ ). Near $p_c$, $\xi_p$ can become quite large, values of the order of 1 μm are easily achieved experimentally. The property of self-similarity is then observable over a sizeable range of scales L, i.e. $a < L < \xi_p$.

II.B. Discontinuous Films

The nucleation and growth processes in thin films before they reach continuity has been the subject of numerous studies.[9] Most theoretical efforts concentrated on understanding the initial stages of the process, which are not universal in character as they depend on the interaction between the metal and the substrate, the substrate temperature and other parameters. Although the existence of a labyrinthine structure near continuity was well documented, it was only recently realized that this structure has the universal features predicted by percolation theory:[10,11 15]

i) Below the continuity (percolation) threshold, the finite clusters present a broad size distribution. For cluster sizes L such that $a<L<\xi_p$ the number $n_s$ of clusters of area s varies as $s^{-\tau}$, where $\tau$ is an exponent close to 2. This behaviour, first reported for Aℓ films[15] was confirmed for Pb and Au films.[10] It is in excellent agreement with Monte-Carlo simulations.[16 17]

ii) Above threshold, the study of the infinite cluster shows that it has an anomalous scale dependent density.[10] Measuring the density $\rho$ as a function of length scale L around randomly selected sites of the infinite cluster, and averaging over all sites, one obtains a power law density decrease, $\rho(L) \propto L^{-(0.1\pm0.02)}$.
In other words, the mass is not proportional to the area $L^d$, where

d is the usual dimensionality (here d=2), but rather varies as $L^D$ where D is an anomalous, or fractal,[18] dimensionality, here $D=1.9\pm0.02$. This behaviour is a straightforward result of the scaling theory of percolation which predicts that $D=d-\beta/\nu$, where $\beta$ is the mass exponent of the infinite cluster and $\nu$ the exponent of its correlation length. In 2D, $\beta/\nu=0.14/1.33$, in excellent agreement with the experimentally measured value of D. The prediction $D=d-\beta/\nu$ is a direct result of the assumption that there is only one relevant length scale $\xi_p$ in the problem, i.e. on scales smaller than $\xi_p$ there is no observable characteristic length scale,[19] and the object must therefore be self similar.

This self similarity property can be visually identified by the fact that the infinite cluster has in it holes (and therefore loops) of all sizes up to $\xi_p$ (Fig. 3). At $L>\xi_p$ the self similarity property breaks down. $\xi_p$ appears as the characteristic length scale of the cluster, the density is size independent[10] and changes in magnification can be detected. It will turn out that these properties are quite important for the discussion of superconducting properties near the percolation threshold.

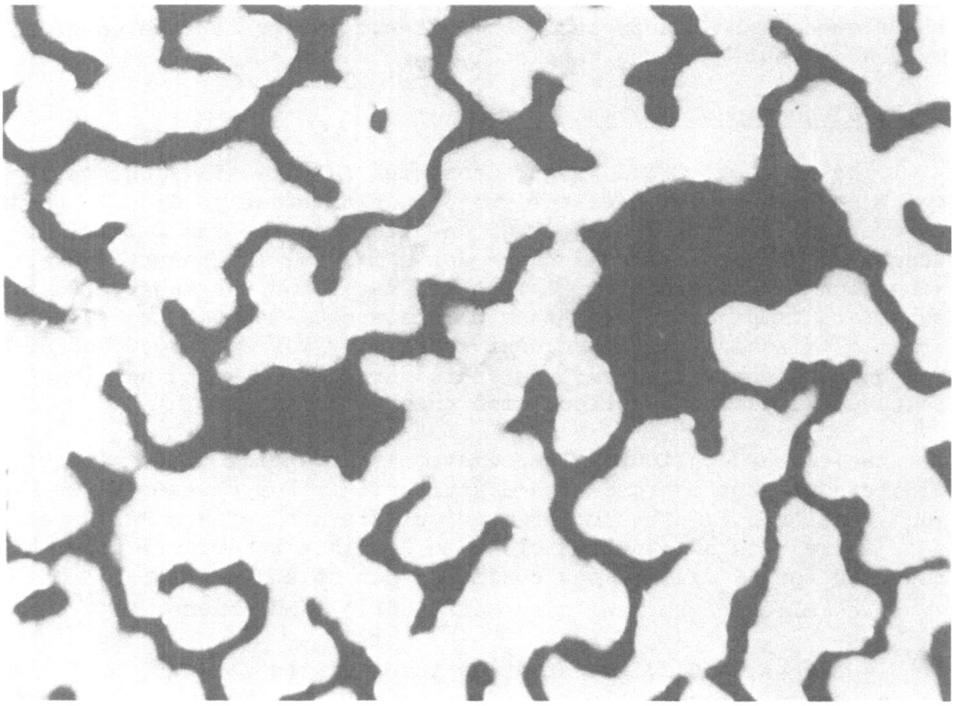

Fig. 3. Structure of infinite cluster of a Pb film close to threshold.

## III. TRANSPORT IN PERCOLATIVE FILMS

A problem closely related to percolation is that of the random resistor network.[7] The macroscopic conductance $\Sigma$ of a network composed of conductances $\sigma_o$ randomly placed on a lattice with probability p has a power law variation

$$\Sigma = a\sigma_o (p - p_c)^{\mu} \qquad (2)$$

where a is a coefficient of order unity. The best Monte-Carlo results are $\mu$=1.3 in 2D and $\mu$=2 in 3D. The Hall coefficient has a power law divergence in 3D, $R_H \propto (p-p_c)^{-0.3}$, while in 2D an exact result is that $R_H$ must remain constant near $p_c$.[20]

A power law behaviour of the conductivity was first reported for granular W films by Abeles et al.,[14] with an exponent equal to 1.9. Later on $\mu$=1.75 was reported for granular Aℓ-Ge[6], $\mu$=1 for 2D Pb-Ge films[21] and $\mu$=1.2 in 2D Aℓ-Ge.[22]

### III.A. Discontinuous Films

Very recently, the value of $\mu$ has been measured for discontinuous films of Au and In by Rappaport.[23] In Au films, for which the existence of a percolative structure is well documented,[11] it turns out that the average thickness $\bar{t}$ of the discontinuous film is proportional to its surface coverage p over a broad thickness range near continuity, hence $(\bar{t}-\bar{t}_c) \propto (p-p_c)$. This allows for a very accurate measurement of $\mu$ in an experimentally pure 2D system. The result of Rappaport $\mu=1.25 \pm 0.05$, in excellent agreement with the best Monte Carlo results,[4] was obtained by a direct in situ measurement of the film's conductivity during deposition, as a function of its average thickness as read by a thickness monitor.

While the critical surface coverage of Au films is close to 50%, as predicted by percolation theory for a random 2D continuum, the situation in In films is quite different. Because of their low melting point, In films tend to agglomerate and the critical surface coverage is above 90%. Nevertheless, the critical behaviour of the conductivity near $p_c$ is the same as that of the Au films.[23] This experimental result is very important, because it lends strong support to the assumption of universality of the critical indices: while the value of $p_c$ depends on the short range correlations in the system and may vary a lot, the value of the critical indices does not.

The Hall coefficient was measured in 2D thin Au films and found to be essentially constant near $p_c$, for a range of samples with resistivities varying over more than two orders of magnitude.[24] This verifies the theory of Bergman.[4] (Hall effect measurements on thick granular $A\ell-A\ell_2O_3$ are compatible with the theoretical 3D result.[25]).

III.B. Dimensionality Cross-over in Mixture Films

Mixture films have the interesting property that they can be either 3D or 2D depending on the ratio of their thickness d to the percolation correlation length $\xi_p$. Percolation theory makes a number of interesting predictions with respect to dimensionality crossover. Evidently, when $d/\xi_p < 1$ the critical indices should be 2D, and 3D in the opposite case. Since $\xi_p$ diverges at $p_c$, all films become eventually 2D near the transition. This produces a change in the critical behaviour, and also of the threshold value which becomes thickness dependent:

$$p_c(d) - p_{c3} \propto d^{-1/\nu_3} \tag{3}$$

where $p_{c3}$ designates the bulk, 3D value of the threshold.

Experimentally, cross-over effects can basically be studied in two different ways:

i) by measuring the conductivity of the film as a function of thickness at a fixed concentration. Such measurements have been performed by measuring the conductivity of the film in-situ during deposition.[25] For a given concentration, this measurement lends the critical value of the thickness at which the film becomes conducting, $d_c(p)$. Equation 3 was checked by comparing the values of $d_c$ obtained at different concentrations. From the curves $\sigma(d)$ it is also possible to infer the value of the thickness at which the 2D to 3D crossover occurs. Roughly speaking, it is the thickness beyond which the conductivity (not the conductance) becomes constant. Finally, the variation of this limiting conductivity as a function of concentration gives the 3D value of $\mu$. As can be seen, a wealth of information on percolation can be obtained by such in-situ measurements. The values obtained for $\nu_3$ (from $d_c(p)$ through eq. 5 and independently from the cross-over thickness $d=\xi_p$), $\mu_2$ and $\mu_3$ are in good agreement with theoretical values for Pb-Ge and In-Ge films.

ii) another method is to study a series of samples of fixed thickness covering a range of concentrations near $p_c(d)$. This method was used for the study of $A\ell$-Ge films.[22] Here one makes use of the concentration dependence of $\xi_p$ to reach the condition $\xi_p=d$. The results clearly show the thickness dependence of $p_c$, which varies from 18% at $d \simeq 5000$ A up to 35% at $d \simeq 700$ A. (Below that thickness structural changes in the layer prevent the observation of the predicted increase of $p_c$ up to 50%). They also show a power law behaviour $\sigma \propto (p-p_c(d))^{1.2}$ characteristic of a 2D system near $p_c(d)$, and a crossover towards $\sigma \propto (p-p_{c3})^2$ at a concentration where $\xi_p \simeq d$.

Both sets of experiments illustrate the importance of the correlation length $\xi_p$ for the description of random mixture films. The measurement of $p_c(d)$, and of the concentration where the 2D to 3D crossover occurs, allow an experimental determination of $\xi_p$ within a numerical constant of order unity. The value of $\nu_3$ obtained from these experiments (0.9) is in very good agreement with the best Monte Carlo results (0.88 [4] ).

In conclusion of sections II and III, it may be said that the existence of a characteristic length $\xi_p$ in discontinuous films and in random mixtures is now well established. The self similarity property at scales $L < \xi_p$ has been confirmed by detailed density measurements in the discontinuous films, and by direct observation in mixture films. The homogeneity at scales $L > \xi_p$ has been established directly by measurements of the conductivity as a function of thickness at a fixed concentration,[26] and indirectly by the observation of a transition towards a bulk 3D behaviour at concentrations such that $\xi_p < d$.[22] This body of experimental results provides the justification for applying percolation concepts to the discussion of the superconducting properties, with special emphasis placed on the importance of the length $\xi_p$.

## IV. SUPERCONDUCTING PROPERTIES

There are essentially two classes of superconducting composites: the superconductor - normal metal, and the superconductor - insulator mixtures. Although the first class - such as the Tsuei wire[27] - has some very interesting properties which may be related to percolation through the proximity effect, we shall concentrate here on the superconductor - insulator mixtures for which the relevance of percolation has been independently established, as shown in the preceding sections.

A number of superconducting properties of these mixtures have been studied experimentally: the zero field resistive transition,[28] the heat capacity transition,[29,30,31] the London penetration depth,[32] the critical currents[21,33] and critical fields.[34,35,36,37]

### IV.A. Granular Composites

Here again we must distinguish between granular and random composites. A percolation description of granular superconductors is based upon two related assumptions: that the resistance $R_N$ of the intergrain Josephson junction provides the element of randomness in the composite, resulting in a distribution of Josephson coupling energies; and that the critical temperature of the grains is uniform, and independent of the average value of $R_N$.

With such a model it is possible to explain[30] in some detail the heat capacity transition of granular Aℓ-$A\ell_2O_3$.[29] The basic feature of the experimental data is a lowering of the transition accompanied by a progressive diminuation and rounding of the heat capacity jump, as the normal state resistivity is increased. The percolation model states that the reduced transition temperature is due to the lowering of the Josephson coupling energy ($\propto 1/R_N$), and <u>not</u> to a lowering of the critical temperature of the grains. The validity of this model was recently confirmed by critical field measurements in the Pauli limit, where the critical field $H_p$ is proportional to $T_c$.[38] It was shown that $H_p$ - and hence the $T_c$ of the grains - indeed remain constant over a broad range of resistivity values (Fig. 5).

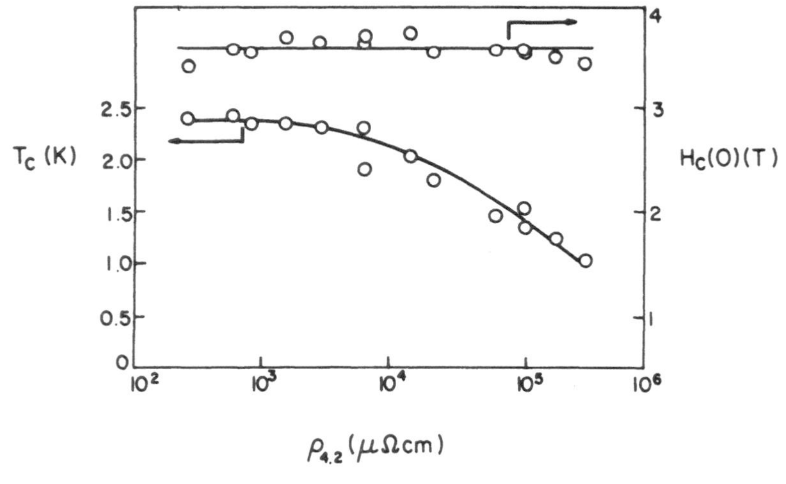


Fig. 5. Critical temperature and low temperature critical field $H_p$ of granular Aℓ (after Ref. 38). The field $H_p$ is independent of the resistivity, showing that the critical temperature of the grains is constant.

The percolation model also explains the penetration depth measurements in $A\ell\text{-}A\ell_2O_3$.[32] In agreement with the argument of de Gennes,[39] the normal state conductivity is proportional to the superfluid density and hence to the square of the London penetration depth. This is precisely what is observed experimentally over a range of more than 2 decades in the conductivity.

Finally, the model explains some of the resistive transitions in zero field of granular composites with fairly large (100 A) grain sizes.[28] Here a significant drop in resistance is observed at a temperature which is independent of the resistivity (Fig. 6). This drop is followed by a tail which gets more important at high resistivities. The proposed interpretation of this experiment is as follows: the initial drop occurs at the $T_{co}$ of the grains, and signals the transition of <u>clusters</u> of well coupled grains. The

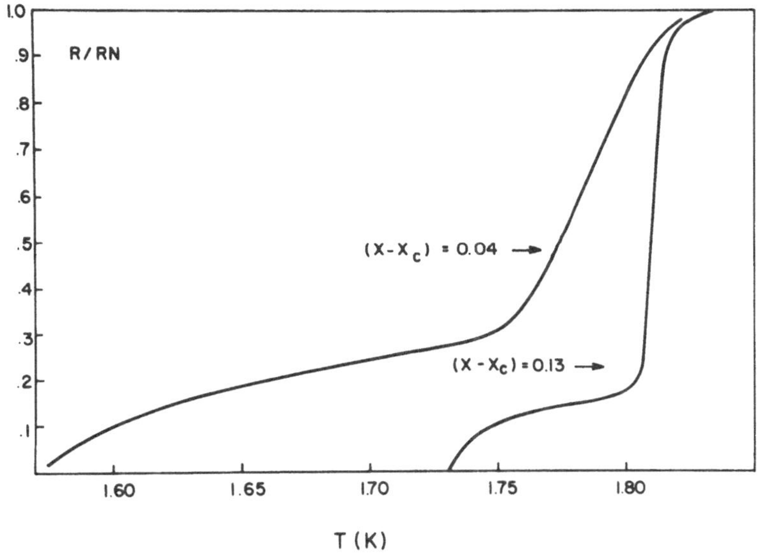


Fig. 6. Resistive transition of $A\ell$-Ge films showing percolation tail developing as $p_c$ is approached.

tail reflects the progressive increase in the Josephson coupling energy between weakly coupled grains, until the fraction p of effectively coupled grains (i.e. for which $E_c > kT$) becomes large enough to allow superconducting percolation. Assuming that $(p(T) - p_c) \propto (T - T_c)$, the shape of the curves is analysed to lend the exponent s of Straley. The value found (s=0.7) is in good agreement with 3D Monte Carlo simulations. A similar interpretation was given to the transition of Pb-Ge 2D films.[28]

It is interesting to note that some properties of granular superconductors can be interpreted by the existence of the grains properties ($T_c$, size), with disorder effects (intergrain coupling) playing only a minor role. This is the case for the upper critical field in the Pauli limit, as mentioned above. It is also the case when the effective superconducting coherence length $\bar{\xi}_s$ is smaller than the characteristic grain size. There is then a characteristic upturn in the temperature dependence of the critical field, which becomes that of the isolated grains.[36] If d is a characteristic average grain size, the upturn should occur at the temperature where $H_{c2} \simeq \phi_0/2\pi\, d^2$. The data of Ref. 36 on Aℓ-Ge then give d ≃ 400 A, significantly larger than the grain size observed in the electron microscope (∿ 100 A). A possible explanation for this discrepancy is the existence of clusters of grains. Such clusters have indeed been observed in $A\ell$-$A\ell_2O_3$ by Petit and Silcox,[40] by the study of low angle electron diffraction, with a cluster to grain size ratio of about 5.

## IV.B. Random Composites

Superconducting percolation effects are more directly established in random composites than in granular ones, as the random element in the latter (the intergrain barrier thickness distribution) cannot be directly observed experimentally.

We model superconducting random composites as a network of thin superconducting wires connected at nodes. The properties of such networks are reviewed in this volume by de Gennes.

Our understanding of these materials is based upon the distinction between the homogeneous and the inhomogeneous limit, as first introduced by Deutscher et al.[37] The boundary between the two regimes is defined by the condition $\bar{\xi}_s = \xi_p$, where $\bar{\xi}_s$ is the effective superconducting coherence length of the medium, and not the superconducting coherence length of the pure superconductor. As explained by de Gennes in this volume, the erroneous condition $\xi_s = \xi_p$ leads to incorrect theoretical results.

It is clear that the condition for inhomogeneity, $\bar{\xi}_s < \xi_p$, is always realized close enough to $p_c$ because of the divergence of $\xi_p$

and because $\bar{\xi}_s$ can only be reduced by the increased "dirtiness" of the system (in reality, what happens is that $\bar{\xi}_s$ saturates near $p_c$). In practice, the condition $\bar{\xi}_s = \xi_p$ is met fairly far away from $p_c$. Taking $\bar{\xi}_s \simeq 500$ A as a typical order of magnitude, and $\xi_p \simeq 100$ Å x $(p-p\ )^{-0.9}$ (3D behaviour) we see that the crossover occurs roughly at $p-p_c \simeq 0.2$. So that, in the well defined percolative critical region where scaling applies, we expect essentially all percolating superconductors to be in the inhomogeneous limit. Hence, predictions relevant to the behaviour in this limit should be easily observed.

IV.B.1. Critical currents. According to the scaling approach to the structure of the percolative network, which has been clearly experimentally substantiated as reviewed at the beginning of this paper, the infinite cluster contains loops of all sizes up to $\xi_p$. A corollary statement is that it has bottlenecks a distance $\xi_p$ apart. Hence, the number of independent conducting paths per unit cross section is $\xi_p^{-(d-1)}$, where d is the percolative dimensionality of the sample. Based on that argument, and assuming all bottlenecks to have the same critical current, Deutscher and Rappaport[21] predicted the law:

$$j_c = \propto (p - p_c)^{(d-1)\nu} \tag{4}$$

which they observed experimentally in 3D A -Ge and 2D Pb-Ge films. More recent measurements by S. Alterovitz and T. Duby[40] have confirmed this result for thick (3D) In Ge films. The results for $\nu$ are: $\nu_2=1.35$ and $\nu_3=0.9$. The interesting point here is that the superconducting critical current is a much simpler property than the normal state conductivity, because it is not sensitive to the details of the local structure of the infinite cluster, but only to the length scale $\xi_p$ and the dimensionality.

The assumption made in deriving eq.(4) is that the bottlenecks are independent from each other, i.e. that the superconducting order parameters at different bottlenecks are uncorrelated. This assumption is verified if the typical distance between neighbouring bottlenecks - $\xi_p$ - is larger than $\bar{\xi}_s$, i.e. in the inhomogeneous regime. As discussed above, this is indeed the case near $p_c$ due to the divergence of $\xi_p$, if T is not too close to $T_c$. These are the conditions under which eq.4 was verified ($(p-p_c)<0.1$; $(T/T_c)\simeq 0.5$). Very near $T_c$, typically at $(T_c-T)/T_c<$ a few $1 \times 10^{-2}$, the homogeneous regime $\xi_p<\xi_s(T)$ must be reached due to the divergence of $\xi_s(T)$. The concentration dependence of $j_c$ is then quite different

$$j_c \propto (p - p_c)^{(\mu+\beta)/2} \tag{5}$$

as shown recently.[42] This result is close to the bulk behaviour $j_c \propto \sigma^{1/2} \propto (p-p_c)^{\mu/2}$, with a correction due to the density variation of the infinite cluster. Notice that the critical indices predicted by eq.4 and eq.5 are quite different. In 2D, $(d-1)\nu_2 = 1.35$, $(\mu+\beta)/2 = 0.65$; in 3D, $(d-1)\nu_3 = 1.8$, $(\mu+\beta)/2 = 1.25$.

In practice, eq.4 is easily and well verified: near $p_c$ the critical current densities are small so that one does not have to worry about heating effects. Experiments in the homogeneous limit are much more difficult, because one has to perform them very close to $T_c$, or else one has to use fairly concentrated samples and in that case the critical current densities are high and heating effects are important. However, the experiments of ref.(40) do indicate that near $T_c$ the concentration dependence of $j_c$ is weaker than predicted by eq.4, as expected.

The temperature dependence is also different in the homogeneous ($\propto (T_c-T)^{3/2}$) and inhomogeneous ($\propto (T_c-T)^{4/3}$) limits, as shown in ref.42. However, the difference between the predicted exponents is not large enough to allow clear experimental verification. Moreover, close to $T_c$ where these predictions hold, the inhomogeneous limit can only be obtained for p very close to $p_c$. Experimentally one observes, for a sample very close to $p_c$ ($(p-p_c) \simeq 0.01$), a temperature dependence definitely weaker than $(T_c-T)^{3/2}$. A linear behaviour could be compatible with the data, reminding of weak link behaviour. More concentrated samples fit a $(T_c-T)^{3/2}$ variation up to $T_c$.

IV.B.2. Critical fields. Interest in the critical fields of random composites[37] stems from the fact that the superconducting state probes the properties of the infinite cluster (for instance) on a time scale of its own, i.e. the characteristic time for the relaxation of the order parameter $\tau \simeq h/k_B(T_c-T)$. In dirty homogeneous superconductors, the superconducting coherence length is the diffusion length over that time scale, $\xi_s \simeq (D\tau)^{1/2}$ where D is the coefficient of diffusion related to the conductivity and the density of states through Einstein's relation. In percolating superconductors we may still use the same definition provided we take into account the time dependence of D in the inhomogeneous regime $\bar{\xi}_s < \xi_p$ (we use the notation $\bar{\xi}_s$ for percolating superconductors).

As noted above, the inhomogeneous regime is easily reached near $p_c$. Through the relation $H_{c2} = \phi_0/2\pi\ \xi_s(T)^2$, a measurement of the upper critical field then allows a determination of the time dependence (or scale dependence) of D. Both the short time, anomalous diffusion regime on scales smaller than $\xi_p$, as well as the long time homogeneous regime can be probed - simply by measuring the concentration dependence of the composite at some fixed reduced temperature.

Critical field experiments have been carried out on In-Ge[37,43] and Pb-Ge.[41] Their mean features can be characterized as follows.

1) At low resistivities ($\rho$ < a few $10^{-5}\Omega$cm) the behaviour is very similar to that of a dirty type II superconductor: $H_{c2}(T)$ has a linear behaviour near $T_c$ and the slope $(dH_{c2}/dT)_{T=T_c}$ has a value close to that predicted for a homogeneous superconductor, in particular it is proportional to the normal state resistivity $\rho_n$ within experimental accaracy.[37,42] The critical field anisotropy $H_{/\!/}/H_{\perp}$ is close to the value 1.69 = $H_{c3}/H_{c2}$. These samples are in the homogeneous limit $\xi_s > \xi_p$, for which theory predicts[12] that $H_{c2}$ should vary linearly near T and be inversely proportional to the coefficient of diffusion D $\propto (p-p\ )^{\mu-\beta}$. The studied In Ge samples being in the 2-dimensional percolation limit,[37] the small value of $\beta$ (0.14) does not allow to distinguish experimentally between $\mu$ and $(\mu-\beta)$. However, measurements on the presumably three dimensional Pb-Ge[41] give $H_{c2} \propto \rho_n^{0.7}$, in fair agreement with the theoretical prediction $H_{c2} \propto \rho_n^{(\mu-\beta)/\mu}$.

At low temperatures T << $T_c$ and for samples in the intermediate range of resistivities $\rho_n \sim$ a few $10^{-5}\Omega$cm, it has been noted that $H_{c2}(T)$ tends to show an upward curvature,[43] unlike homogeneous type II superconductors. It is not clear at the moment whether this is related to the existence of localization effects,[44] to the approach to the inhomogeneous limit, or to some macroscopic inhomogeneities in the samples.

2) At resistivities $\rho_n \simeq 10^{-4}\Omega$cm, In Ge samples show a change in behaviour. $H_{c2}(T)$ still has a linear variation near $T_c$, but the slope $dH_{c2}/dT$ varies more slowly with $\rho_n$.[37] The critical field anisotropy decreases with $\rho_n$.

3) At high resistivities $\rho_n > 3.10^{-4}\Omega$cm there is a drastic change in behaviour: $H_{c2}(T)$ approaches $T_c$ with an infinite slope (Fig.7), and the low temperature value of $H_{c2}(\rho_n)$ reaches saturation (Fig.8). This new behaviour is in qualitative agreement with the predictions of theory. These samples are now fully (i.e. essentially at all temperatures, except very close to $T_c$) in the inhomogeneous limit $\xi_s < \xi_p$. The critical field should then be independent of $(p-p_c)$ - which explains the saturation effect - and its temperature dependence is $H_{c2}(T) \propto \xi_s^{-2} \propto (\Delta T)^{1/(1+\theta)/2}$, where $\theta = (\mu-\beta)/\nu$.

If fitted to a power law $H_{c2}(T) \propto (T_c-T)^u$, the data on In Ge[43] gives u=1 up to $\rho_n \simeq 3 \times 10^{-4}\Omega$cm. Beyond that resistivity, u decreases continuously as $\rho_n$ increases, approaching a value close to 0.5 near the threshold. The continuous variation of u with $\rho_n$ is probably due to the fact that strictly speaking at $T_c$ all samples are in the homogeneous limit, hence the fit to a single power law

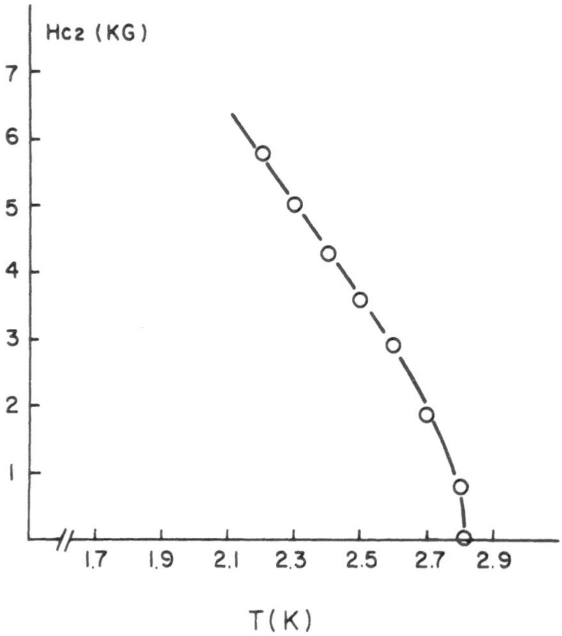


Fig. 7. Temperature dependence of $H_{c2}$ near $T_c$ in the inhomogeneous limit (after Ref.42). Sample is In Ge with $\rho_n \simeq 5.10^{-4}\Omega$cm, thickness 2000 A.

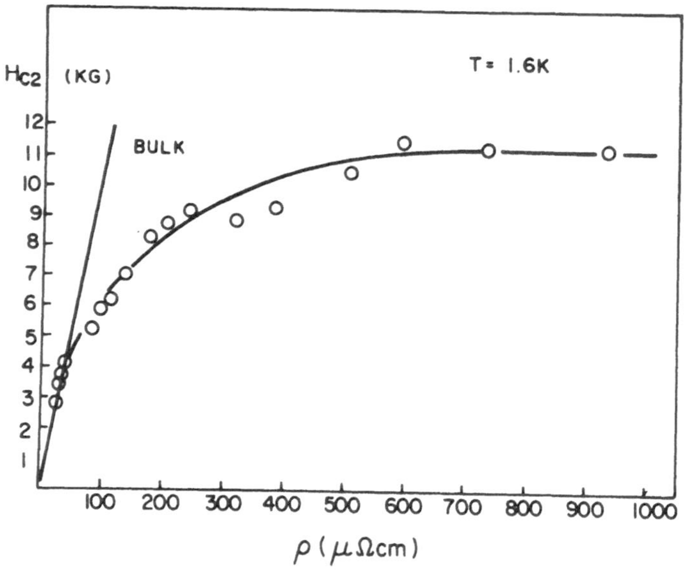


Fig. 8. Saturation of low temperature $H_{c2}$ at high resistivities. Samples are In Ge, thickness 2000 A.

up to $T_c$ is not justified (although the fit looks rather good).

To summarize the results on the upper critical field of the infinite cluster, it can be said that both the homogeneous $\bar{\xi}_s > \xi_p$ and inhomogeneous $\bar{\xi}_s < \xi_p$ limits have been observed. However, a number of questions remain open for further study. The transition between the two regimes appears to be quite broad, unlike the sharp crossover suggested by Fig.1 in de Gennes article in this volume. Of course, scaling theory only predicts the two limiting behaviours and where - but not how - it occurs. Numerical work is probably necessary in the crossover region. Another point which requires further theoretical investigation is the critical field anisotropy, which goes down from about 1.7 in the homogeneous limit (as expected) to a value of the order of 1.2 in the inhomogeneous limit. Qualitatively, the reduction of the anisotropy can be understood by the appearance of "internal" surfaces near $p_c$. It is tempting to conjecture that the value of the anisotropy near $p_c$ must somehow reflect the fractal nature of the infinite cluster on the scale $\bar{\xi}_s$, and provides some measure of this anomalous dimensionality. Finally, we do not have yet experimental results on the diamagnetism and critical field of finite clusters, which deserve a detailed study.

Acknowledgements: It is a pleasure to thank S. Alexander and P. G. de Gennes for numerous stimulating discussions on the properties of percolating superconductors, in particular for pointing out at an early stage the importance of anomalous diffusion and suggesting further experiments. Critical field experiments were performed by I. Grave and A. Gerber. The understanding of the structure and transport properties, as summarized in this paper, owes a lot to the contributions of M. Rappaport, A. Kapitulnik and A. Palevski. Discussions on the Hall effect with D. Bergman and fractals with Y. Gefen and A. Aharony are gratefully acknowledged. Finally, I wish to thank T. Duby and S. Alterovitz for communication of their critical current data prior to publication. This work was partially supported by the Israel National Council for Research and the Karlsruhe Nuclear Research Center.

REFERENCES

1. B. Abeles, Adv. Phys. 24:407 (1975).
2. G. Deutscher, H. Fenichel, G. Gershenson, E. Grunbaum and Z. Ovadyahu, J. of Low Temp. Phys. 10:231 (1973).
3. Y. Shapira and G. Deutscher, Phys. Rev. B27:4463 (1983).
4. For reviews on percolation see "Percolation, Structure and Processes", Annals of the Israel Physical Society, Vol. 5, Eds. G. Deutscher, R. Zallen and J. Adler (1983).
5. J. Straley, Phys. Rev. B15:5733 (1977).

6. G. Deutscher, M. Rappaport and Z. Ovadyahu, Solid State Commun. 28:593 (1978).
7. A. Kapitulnik, M. Rappaport and G. Deutscher, J. Phys. Lett. 42:L-541 (1981).
8. M. Rappaport, A. Kapitulnik, G. Deutscher and Y. Lareah, to be published.
9. See for instance K. Chopra, Thin Film Phenomena, McGraw Hill, New York, 1969.
10. A. Kapitulnik and G. Deutscher, Phys. Rev. Lett. 49:1444 (1982).
11. R. Voss, R. Laibowitz and E. Alessandrini, Phys. Rev. Lett. 49:1441 (1982).
12. S. Alexander, Phys. Rev. B27:1541 (1983) and in "Percolation, Structures and Processes" op.cit. p.149.
13. G. Deutscher, A. Kapitulnik and M. Rappaport in "Percolation, Structures and Processes", op.cit. p.207.
14. B. Abeles, H. L. Pinch and J. I. Gittleman, Phys. Rev. Lett. 35:247 (1976).
15. R. B. Laibowitz, E. I. Alessandrini and G. Deutscher, Phys. Rev. B25:2965 (1982).
16. J. Kertesz, D. Stauffer and A. Coniglio in "Percolation, Structures and Processes", op.cit. p.121.
17. D. Stauffer, Physics Reports 53:3759 (1980).
18. B. B. Mandelbrot, Fractals: Form, Chance and Dimension (W. H. Freeman, San Francisco, 1977).
19. It is now recognized that the relation $D = d-\beta/\nu$ is the correct one. For a review of the fractal structure in percolation, see H. E. Stanley and A. Coniglio in "Percolation, Structures and Processes" op.cit. p.101.
20. See for instance D. Bergman in "Percolation, Structures and Processes" op.cit. p.297.
21. G. Deutscher and M. Rappaport, J. Physique Lett. 37:L-9 (1976).
22. A. Kapitulnik and G. Deutscher, J. Phys. A16:L-255 (1983).
23. M. Rappaport, private communication and to be published.
24. A. Palevskii, M. Rappaport, A. Kapitulnik, A. Fried and G. Deutscher, submitted to J. Physique Lett.
25. B. Bandyopadhyay, P. Lindenfeld, W. McLean, and H.K. Sin, Phys. Rev. B26:3476 (1982).
26. M. L. Rappaport and O. Entin-Wohlman, Phys. Rev. B27:6152 (1983).
27. C. C. Tsuei, Science 180:57 (1973).
28. G. Deutscher and M. L. Rappaport, J. de Physique Colloque C-6:581 (1978).
29. R. L. Filler, P. Lindenfeld, T. Worthington and G. Deutscher, Phys. Rev. B21:5031 (1980).
30. G. Deutscher, O. Entin-Wohlman, S. Fishman and Y. Shapira, Phys. Rev. B21:5041 (1980).
31. Y. Shapira, Ph.D. Thesis, Tel Aviv (1981).
32. D. Abraham, G. Deutscher, R. Rosenbaum and S. Wolf, J. de Physique Colloque C-6:586 (1978).

33. S. Alterovitz and T. Duby, private communication and to be published.
34. B. Abeles, R. W. Cohen and W. R. Stowell, Phys. Rev. Lett. 18:902 (1967).
35. G. Deutscher and S. Dodds, Phys. Rev. B16:3936 (1977).
36. G. Deutscher, O. Entin-Wohlman and Y. Shapira, Phys. Rev. B22:4264 (1980).
37. G. Deutscher, I. Grave and S. Alexander, Phys. Rev. Lett. 48:1497 (1982).
38. T. Chui, P. Lindenfeld, W. L. McLean and K. Mui, Phys. Rev. B24:6728 (1981).
39. P. G. de Gennes, J. Physique Lett. 37:L-9 (1976).
40. R. B. Pettit and J. Silcox, J. Appl. Phys. 45:2858 (1974).
41. S. Alterovitz, to be published.
42. O. Entin-Wohlman, A. Kapitulnik, S. Alexander and G. Deutscher, to be published.
43. A. Gerber, M.Sc. Thesis, Tel Aviv 1983.
44. S. Maekawa, H. Ebisawa and H. Fukuyama, Technical Report ISSN 0082 - 4798, Ser. A. 1273 (1982), Institute for Solid State Physics, the University of Tokyo.

# ARTIFICIALLY-STRUCTURED SUPERCONDUCTORS

M. R. Beasley
Department of Applied Physics
Stanford University
Stanford, California 94305

## I. INTRODUCTION

Superconductivity is perhaps the best understood cooperative phenomenon. Very successful theories (both microscopic and phenomenological) are available with which to help interpret experiments. At the same time superconductivity occurs in a wide variety of material systems, has many unique features and is influenced by a wide range of factors. As a result, superconductivity is increasingly being used as a tool with which to study other phenomena and/or as a probe of the properties of materials in general, as well as for its own sake. The topics of interest at this NATO ASI are a reflection of this fact. They are listed in Table I.

As seen in the table the phenomena of interest relate to fundamental questions of quantum mechanics (macroscopic quantum tunneling), statistical physics (phase transitions, commensurate-incommensurate transitions and percolation) and basic solid-state physics (small particles, disorder and localization). However, as also seen from the table, because dimensionality and/or randomness play a large role in these issues, the pertinent superconducting systems are in general complex. Indeed, it is just this complexity that gives these systems their interesting properties. Most of the systems can be classified in terms of the scheme illustrated in Fig. 1. They consist of basic superconducting elements (e.g., Josephson junctions, small particles, filaments or films) either in isolation or coupled together to form higher dimensional structures. The couplings may be regular or random. Note that even when the individual elements

TABLE I

Some Current Topics in Superconductivity Research

| Phenomena | Systems |
| --- | --- |
| Macroscopic Quantum Tunneling and Coherence Effects | Tunnel Junctions and Arrays |
| Percolation | SNS Junctions and Arrays |
| Disorder, Localization and Coulomb Interactions | Amorphous Superconductors |
| Phase Transitions (x-y model) | Granular Superconductors and Randum Metal/Insulator Mixtures |
| Commensurate-Incommensurate Transitions (vortex pinning) | Quasi 3-, 2-, and 1-Dimensional Materials |
| Microstructure Science | Small Particles |

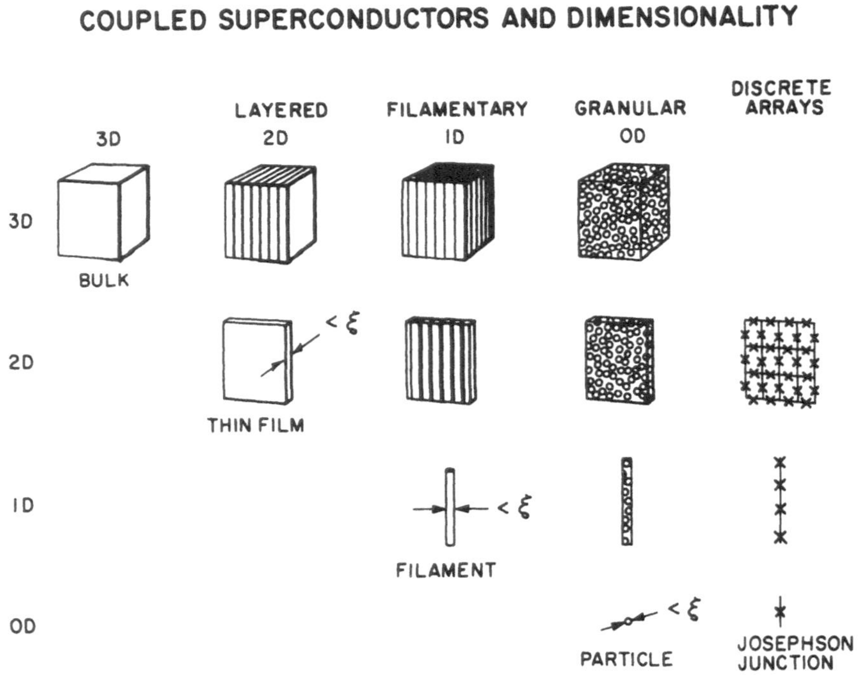


FIG.1--Schematic matrix showing various inhomogeneous and artificially-structured superconducting systems of interest.

themselves are of a reduced dimensionality (0-, 1-, or 2-D), they can nonetheless be of macroscopic dimensions because of the relatively large coherence lengths typical of superconductivity. The characteristic dimension of the element need only be less than the temperature-dependent Ginzburg-Landau coherence length. One is also interested in materials disordered on an atomic scale (e.g., amorphous superconductors) or serpentine percolative structures in 2- and 3-dimensions.

Because of the inherent complexity of these systems, in trying to establish the connection between theory and experiment, it is desirable to have model systems. By model systems we mean systems in which the particular desired structure is present in a well-characterized way and in which ideally it can be varied systematically. Once such model systems are understood, they provide the basis upon which the properties of more complicated (and probably more typical) material systems can then be understood.

Model systems can be obtained in two ways. The first, more traditional method is to seek such systems from materials occurring naturally and to vary the properties of the material, say, by chemical substitution. For example, in the study of quasi-two-dimensional superconductivity, the intercalated layered transition-metal dichalcogenides have played such a role. On the other hand, thanks to recent advances in thin-film vapor deposition and micropatterning, a second approach, increasingly referred to as artificially-structured materials, is now possible. In this approach the desired structure on the characteristic length scale of interest (e.g., $\xi(T)$ or $\lambda(T)$) is either built into the material in a controlled way during its formation or carved out of a larger piece of material using microlithography. For example, artificially-layered quasi-two-dimensional superconductors with easily varied interlayer coupling have recently been made with great success using the "build-in" approach, and planar dot-matrix SNS Josephson arrays for studies of the Kosterlitz-Thouless transition have been made with comparable success using the "carve out" approach. Which approach one uses depends on the geometry desired and the characteristic length scales of interest. At the atomic length scale one must use the build in approach, but as the length scale increases, microlithography becomes possible. These relationships are illustrated in Fig. 2. Note that using advanced techniques microlithography can be carried out down to the ~ 500 Å level.

Of course the natural and artificial approaches are not mutually exclusive. For example, in studying macroscopic quantum tunneling one would like to have an ideal tunnel junction as assumed in the detailed microscopic theories. In general the native oxides of most superconductors with convenient transition

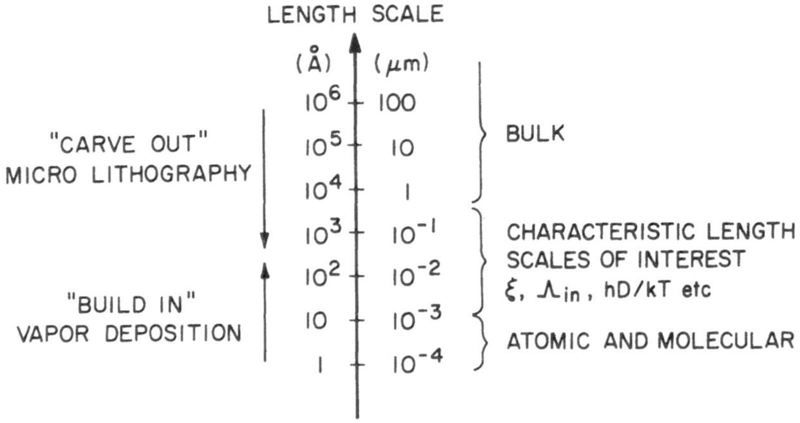


FIG. 2--Comparison of characteristic physical length scales of interest and the range of lengths over which materials can be artificially structured.

temperatures don't qualify, e.g., Nb. Consequently, it may be preferable to use one of the artificial barriers such as oxidized Al overlayers. Here the barrier is obtained artificially but utilizes the excellent native oxide of Al.

This artificial route to model systems, and to new materials generally, is an increasingly effective approach. It can be expected to become even more important as the most powerful vapor deposition tools (e.g. metallic MBE) and micropatterning techniques are brought to bear. In this chapter, however, we deal only with the vapor deposition half of the approach to artificially-structured materials, that is we concentrate on the build-in approach. The micropatterning approach is discussed in detail in the chapters by Prober and Laibowitz. Specifically, in Section II we survey those vapor deposition techniques that have proved the most versatile and successful to date. In Section III we show some examples of the systems that have been produced. The objective of these two sections is not to be definitive or encyclopedic - that would be inappropriate at the present stage of development of this field. Rather it is to introduce the reader to the various approaches that have been successfully applied with the aim of illustrating the general principles involved and stimulating the reader to invent new and better approaches and applications.

## II. Advanced Vapor Deposition Tools for Artificially-Structured Superconductors

Vapor deposition of metal films is hardly new. What is new are the improved sources, rate control and vacuum quality that now allow much more complex materials and structures to be produced via this approach. Also, some very clever ideas about how to use vapor deposition in new and imaginative ways have emerged recently. For example, using codeposition, alloys and compounds can now be made almost as readily as elemental films.[1] Careful sequencing of the depositions can be used to make a variety of multilayer thin-film structures and even (in some cases) metallic superlattices.[2,3,4] Phase separation of the constituents can be used to form granular or percolative structures,[5] as can reactive depositions.[6,7] The principal techniques are thermal evaporation (particularly electron beams), sputtering and the use of ion beams - the latter being useful as both a primary and a secondary (e.g. ion beam sputtering) source. General reviews and references are available in the literature for each of these techniques.[8,9,10] Our interest here lies in their use for superconducting materials.

### A. Materials of Interest

As indicated above these new, more powerful techniques open up the range of materials that can be prepared in complex thin-film structures. Specifically, the transition metals are now on an almost even par with the simple s-p metals in their utility for the fabrication of model systems. Indeed, they may be even more interesting because of their generally superior superconducting properties, and, more importantly, because of their ruggedness and relative lack of interdiffusion when deposited in nonequilibrium configurations. Indeed, their relative advantage here is for precisely the same reasons that the transition metals, their alloys and their compounds are the ultimately preferred materials for superconducting electronic applications.[11] This advantage increases as the scale of the non-equilibrium structure being formed decreases.

Figure 3 shows the materials that have received the most attention to date. In addition to the familiar transition-metal and s-p superconductors, also shown are various semiconductors used in alloying, compound formation and/or for artificial tunnel barriers. Carbon, nitrogen and oxygen are frequently used in reactive depositions and/or in forming native tunnel barriers.

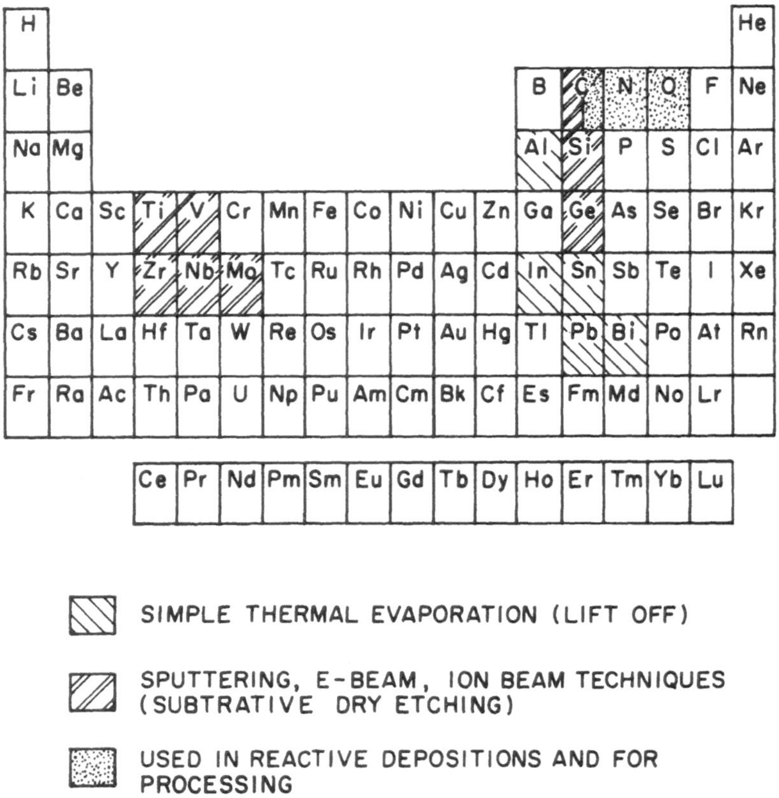


FIG. 3--Materials that have been used to make artificially-structured superconductors. Typical deposition and patterning techniques are indicated.

Note also that as interest shifts from the s-p metals to the transition metals, the preferable method of patterning goes from standard lift off to so-called substractive etching in which the desired structure is formed out of a previously deposited thin film or multilayer using dry (e.g., plasma or sputtering) etching. Wet etching can be used in some cases but is not a generally useful approach as the size of the desired structures decreases [see chapters by Prober and Laibowitz].

B. Thermal Evaportion (Electron Beams)

Because of the importance of the transition metals with their high melting temperatures, electron beam heating is the preferred method of thermal evaporation for superconducting artificially-structured materials. The ease with which sources can be changed in-situ using turret sources adds to the flexibility of this approach. Two typical configurations are shown in Fig. 4. Figure 4(a) shows a dual e-beam source. Such a system can be used for coevaporation or layering (using sequenced shutters). As indicated in the figure this technique also has the feature that there is a compositional spread across the substrate due to the physical separation of the sources. This so-called phase-spread is often extremely useful. It allows a series of samples over a particular desired region of the phase diagram to be prepared in a single evaporation.[1] Similarly if phase separation occurs one can, for suitable system such as Aℓ-Ge, deposit a series of samples that go progressively through the percolation threshold.[5] This approach is not only convenient but has the added advantage that all other variables (known and unknown) are held constant for all samples. Finally, in all these cases it is possible to deposit various overlayers in-situ in order to protect the samples.

Thermal evaporation is line of sight, which means that shaddowing can be used to create structures on a very fine scale without using fine lithography [see chapter by Prober]. Such shadowing can also occur at bumps on the surface due either to dirt (e.g. dust) or natural fluctuations in the film growth process. This is usually undesirable and can be minimized by rotating the substrates. Thermal evaporation has the drawback that the substrates are directly exposed to the high-temperature blackbody radiation of the source. This is particularly severe in the case of high melting temperature materials such as the transition metals. The resultant heating of the substrate can be a problem, for example when using photoresist lift off. In this case substractive patterning or refractory metal lift off masks[12] must be used.

Figure 4(b) shows a reactive evaporation system incorporating an ion beam source in conjunction with thermal evaporation. This approach is particularly useful when one of the species being incorporated is a gas, such as N in the formation of NbN.[13] Of course the gas could be bled directly into the vacuum system. With an ion beam, however, the gaseous species enter as an ion of controlled kinetic energy and direction. Also, with the ion beam, phase-spread depositions are possible. The ion source also provides the possibility of presputtering (ion milling) the substrate and/or processing the surface of the film, for example to form a tunnel barrier.

One final advantage of thermal evaporation is that since no gaseous plasmas are required to operate the source, it can be carried out under high-vacuum conditions. This permits the use of in-situ structural and chemical analytical instrumentation, to some degree even during deposition itself. Indeed the first

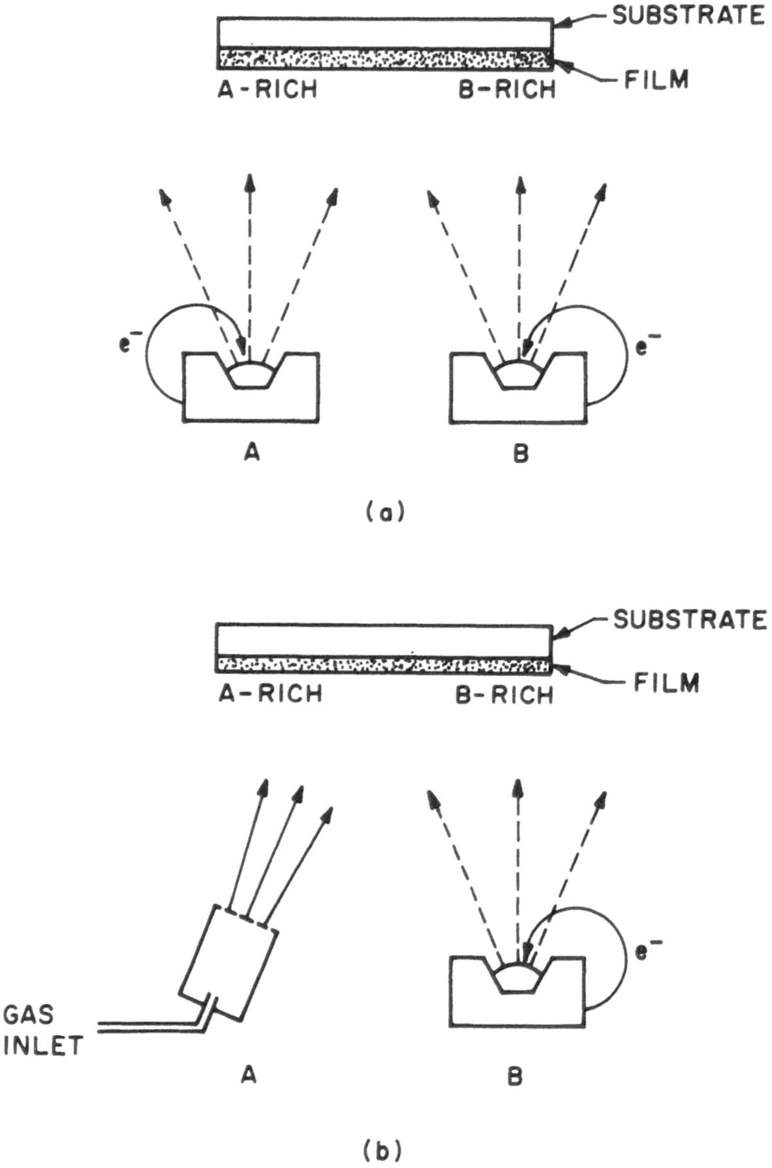


FIG. 4--Schematic of advanced electron beam deposition techniques.

epitaxial metallic superlattice (Nb/Ta) was produced using thermal (e-beam) evaporation in a UHV vacuum system.[14] This is essentially metallic " molecular beam epitaxy" (MBE), and it will be extremely interesting to see what interesting structures can be produced as this approach is increasingly applied to metals.

## C. Sputter Deposition

Sputter deposition is also not new. However, the introduction of magnetron sputtering has led to qualitative improvements over more traditional diode sputtering. Specifically, by confining the plasma of sputtering ions close to the source, magnetron sputtering is cleaner and allows higher rates. Higher rates in turn ease base vacuum and gas purity requirements. Moreover, the substrate is no longer part of the sputtering electrical circuit. Deposition systems incorporating multiple magnetron sputtering heads and a rotating substrate table (see Fig. 5) have provided a simple but highly effective tool for making new materials in general as well as artificially structured materials in particular (see Ref. 2).

The configuration illustrated in Fig. 5 can be used in various ways. With the baffle removed and the substrate table stationary, it can be used to produce samples in a double (or triple) phase spread configuration. If the substrate table is rotated and the rates set so as to deposit less than one monolayer per pass, very homogeneous alloy films result. This approach has

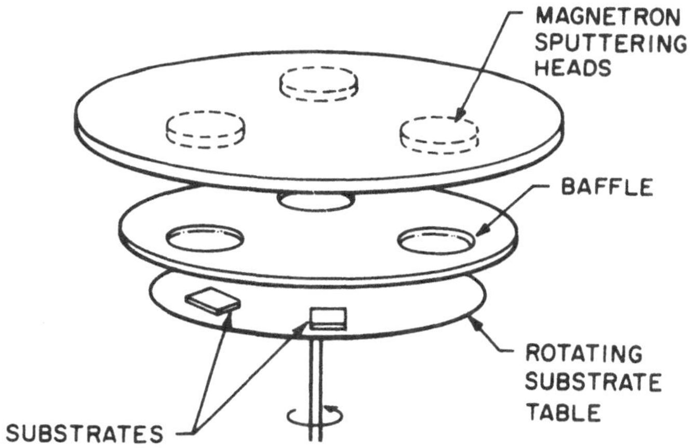


FIG. 5--Schematic of advanced sputtering technique including rotating substrate table.

been extremely effective in making amorphous alloys, for example.[15] Finally, if the baffles are inserted and the table rotated or sequentially positioned by a stepping motor, multi-layered films can be deposited.[2] This approach has been used to produce a variety of superconducting multilayers with great success.[3] This type of system can also be used to deposit protective under- or over-layers of a third material with a very short delay. The oxidized Al overlayers serving as tunnel barriers developed by the Bell group are an example of this possibility.[16] Finally reactive sputtering, in which the reactive gas is bled into the vacuum system or used to form the sputtering plasma itself, is a very common technique. It has been used in the case of superconductors most notably in the formation of granular NbN films.[6]

In contrast to the case of thermal evaporation, in sputtering the source is cool and produces negligible blackbody radiation. On the other hand, sputtered atoms have very high energies unless thermalized by a background gas in the system. If unthermalized the impinging beam of high energy atoms can heat the substrate substantially. Also, in this case the atoms arrive line of sight, whereas if thermalized they arrive isotropically. The trade offs involved here as a function of system background pressure have been discussed by Schuller and Falco.[17] Finally because of the need for a plasma in-situ structural and chemical analysis cannot be carried out during sputter deposition. Note, however, that metallic superlattices of Nb-Ta of quality nearly comparable to those produced by thermal evaporation have been produced using sputtering and rotation on a heated substrate.[18]

### D. Ion Beam Deposition and Processing

The newest approach to vapor deposition and film processing involves the use of ion beam sources.[10] Some examples are illustrated in Fig. 6. The simplest of these is ion beam sputter deposition (Fig. 6(a)). This is similar to conventional sputtering but where the plasma is even more fully confined (allowing lower system pressures during deposition) and physically separated from the target. Also, the energy of the sputtering ions can be readily controlled. Target preparation is similar to conventional sputtering but more complicated than for thermal sources. Binary alloys and compounds can be made from a single target of the correct composition, but deposition from dual sources is more flexible.

An interesting variant on dual ion beam sputtering is shown in Fig. 6(b). Here the second ion beam can be used to introduce a desired reactive gas as in Fig. 4(b), or it can be used with a nonreactive species (e.g., Ar) to affect the resultant growth morphology of the film being deposited. Reactive ion beam sputter deposition can also be carried out as shown in Fig. 6(c). In this case, however, the reaction takes place both at the target and on the substrate. Under these conditions a rich variety of film structures can be produced that involve deposition of both the basic target element itself and some reacted complex (e.g., an oxide) of this element. For example, as found by Nakahara and Hebard,[7] the physical structure of In films can range from amorphous to granular to columnar to agglomerated as the $O_2$ background pressure is varied. Other materials studied (Al, Cu, Pb, and Sn) behaved similarly but did not necessarily exhibit this full progression.

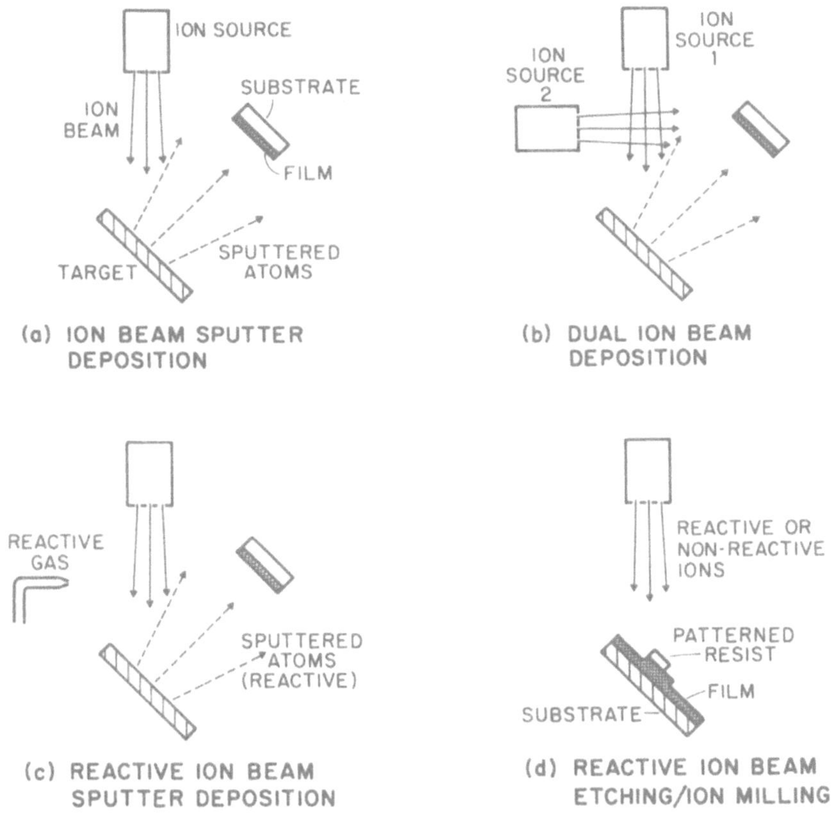


FIG. 6--Schematic of various uses of ion beams useful in making artificially-structured superconductors.

Finally when the substrate is placed in the target position, the ion beam can be used to pattern samples as shown in Fig. 6(d). If unreactive gases are used (e.g., Ar), this is called ion milling. If reactive gases are used (e.g., $CF_4$), it is called reactive ion beam etching. In the latter case the removal of material is believed to be accomplished in large part by chemistry (reaction to form a volatile compound) and in part by gentle sputtering (to remove residues, etc.). In any event, it is much gentler than ion milling and is especially useful when the materials are damage sensitive (e.g., many high-Tc superconductors). Because the ions arrive line of sight, very sharp edges and high aspect ratio structures are possible with this dry etching approach (see chapter by Prober).

## III. EXAMPLES OF ARTIFICIALLY-STRUCTURED SUPERCONDUCTORS

### A. Layered Superconductors

Clearly the simplest type of artificially-structured material one can make using vapor deposition is a multilayered material formed by means of sequential deposition. One special attraction for such multilayers in the case of superconductivity (compared with other cooperative phenomena) is the relatively long characteristic lengths typical of this phenomenon. The point is that one does not have to layer on an atomic scale in order to

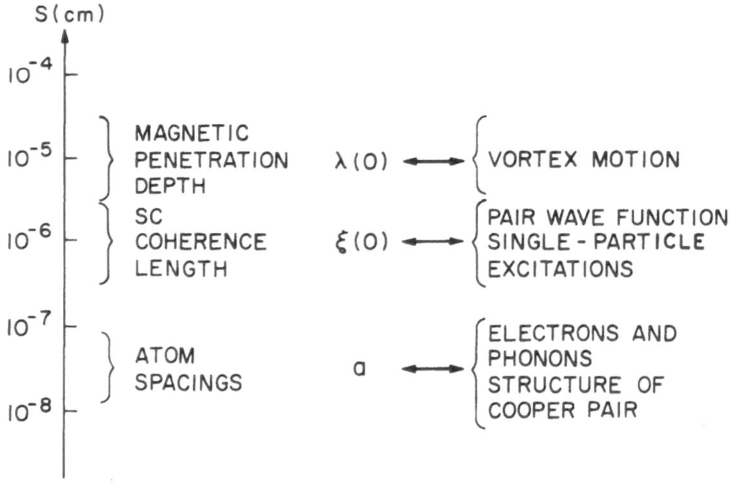


FIG. 7--Comparison of the various length scales of interest in superconductivity and the physical properties affected by layering on that length scale.

achieve interesting effects, although such fine scale layering is certainly of considerable interest. The effects of layering on the properties of superconductors has been reviewed recently in detail by Ruggiero and Beasley.[3] The general picture is summarized in Fig. 7, which shows the layering periodicity required typically in order to modulate the material on the various length scales of interest in superconductivity. Also shown are the physical properties that would be strongly affected at these various length scales. Layered superconductors are also usefully characterized by whether the layering involves two superconductors (S/S' layering), a superconductor and a normal metal (S/N layering), or a superconductor and an insulating barrier of some kind (S/I layering).

Layered superconductors of each type and on all the various length scales indicated have been produced using the techniques described in Section II. The most carefully studied examples are listed in Table II. Note that with only two exceptions the superconducting materials involved are transition metals for the reasons that we have already discussed in Section II-A.. Also, essentially all of these materials showed high-quality layering in the direction of the layering itself, i.e., they have a high degree of one-dimensional structural order. On the other hand, the degree of chemical order, i.e., lack of interdiffusion and the degree of lateral structural coherence is very system dependent. By lateral coherence we mean the degree to which the interfaces between the layers are coherent and this coherence is maintained throughout the entire layer.

The issues involved in defining and characterizing order in multilayered materials, and the x-ray techniques needed to establish this order quantitatively have been reviewed recently by McWhan.[25] As noted by McWhan, the degree of lateral coherence in any given multilayer system is governed by the balance between the strain energy present due to lattice mismatch at the interface and the energy required to introduce misfit dislocations. For sufficiently thin layers the strain energy, which decreases with layer thickness, is small enough that the interfaces will be coherent. But, as the layer thickness increases, a commensurate-incommensurate phase transition occurs in which misfit dislocations enter the system and coherence is lost. This transition begins at a critical wavelength $\lambda_c$ , which depends on the degree of lattice mismatch.

The degree of lattice mismatch for potential layering combinations can be gauged usefully by comparing the in-plane density of atoms in the close-packed plane of the two materials of interest. A convenient plot of these densities due to J. Rowell is shown in Fig. 8. Note that the crystal structures indicated are those of the elements in their bulk equilibrium form.

TABLE II

Multilayered Superconductors Studied to Date

| Material System | Fabrication Approach | Nominal Type | Periodicities [Å] | Structure & Properties Studied | Ref. No. |
|---|---|---|---|---|---|
| Al/Ge | e-beam | S/I | 60-500 | Al layers granular; $T_c$, $H_{c2}$. | 19 |
| Nb/Al | sputter | S/N | 40-300 | Limited ⊥ coherence; $T_c$, Tunneling density of states $N_s(E)$ | 16 |
| Nb/Cu | sputter | S/N | 10-10,000 | No ⊥ coherence; $T_c$, $H_{c2}$. | 20 |
| Nb/Ge | sputter | S/I | 10-200 | Ge amorphous, Nb polycrystalline; $T_c$, $H_{c2}$. | 21 |
| Nb/Ta | MBE*/sputter | S/S' | 20-200 | Single-crystal epitaxial growth; $T_c$, Tunneling density of states $N_s(E)$. | 14 |
| Nb/Ti | sputter | S/N | 10-200 | Ti layers possibly metastable bcc phase; $T_c$, $H_{c2}$. | 22 |
| Nb/Zr | sputter | S/S' | 10-200 | Zr layers metastable bcc phase; $T_c$. | 23 |
| Pb/Bi | e-beam | S/S' | 700-800 | Modulated alloy; $J_c$. | 24 |

*By MBE we mean Thermal Evaporation in an UHV Environment.

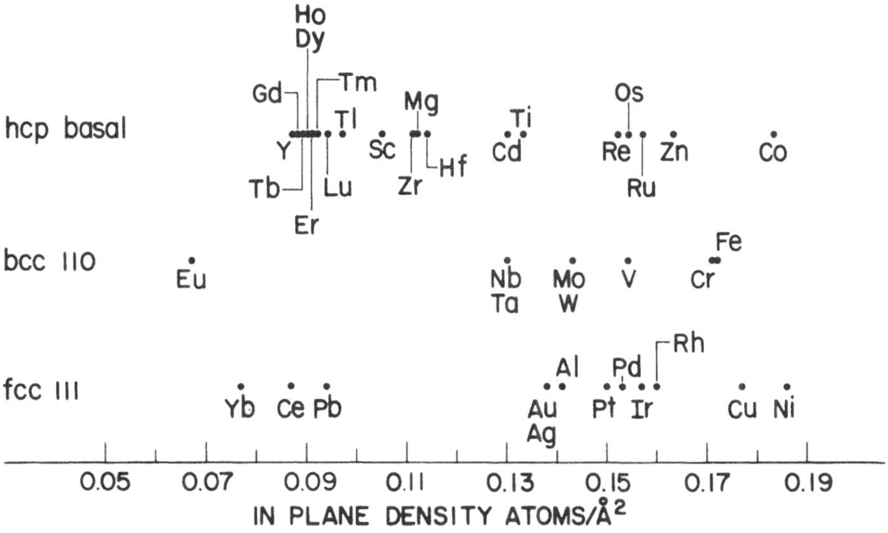


FIG. 8--Comparison of in-plane densities of various elements of interest in metallic superlattice formation. (Courtesy of J. M. Rowell)

On the basis of this plot the structural behavior observed for the various multilayers listed in Table II can be readily understood. Among the crystalline materials Nb/Ta is clearly the best. Nb/Aℓ and Nb/Cu are progressively less ideal. The excellent quality of the layering in the Nb/Ta system results from its near perfect lattice match (both in spacing and symmetry). For Nb/Aℓ and Nb/Cu there is a symmetry change and progressively larger lattice mismatches. On the other hand, the relative immiscibility of Cu into Nb compared with that of Ta results in a much more perfect chemical modulation of the layers. Whether a high degree of structural order or chemical order is better depends on which particular physical properties of the system are of greatest interest.

Another interesting feature of note is that whereas Nb and Zr are not well matched when in their equilibrium structures, coherent Nb/Zr multilayers have in fact been made for Zr layers thinner than 30 Å.[23] This arises because the Zr layers are in metastable bcc phase, which lattice matches the symmetry of the Nb layers. Here we have an obvious case of epitaxy producing a metastable material. The same situation may arise in the Nb/Ti system.[22] The potential of this technique for producing whole new classes of metastable materials has only just begun to be exploited.

A dramatic demonstration of the perfection of layering and coherence that can be achieved is illustrated in Fig. 9, which shows TEM micrographs of thin sections of a W/a-C multilayer and a Nb/bcc-Zr superlattice. (The Nb/a-Ge multilayers listed in Table II are believed to be very similar to the W/a-C case illustrated.) Note that the lateral coherence along (and between) the layers of the Nb/Zr system can be seen directly in the lower right-hand section of Fig. 4(b).

(a)

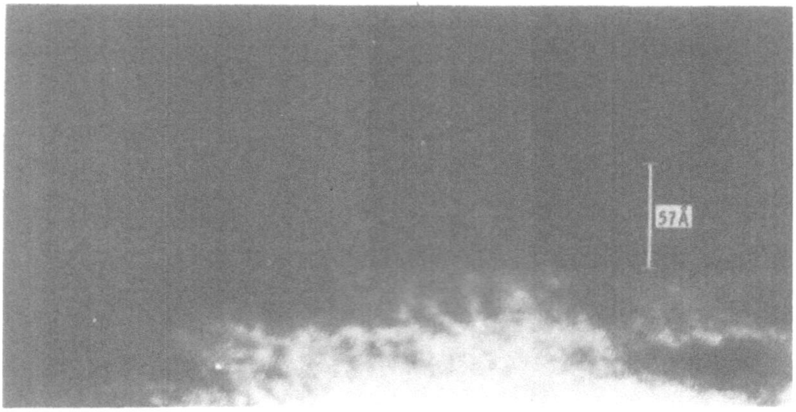


(b)

FIG. 9--(a) TEM of a section through a W/a-C multilayer. Light areas are amorphous carbon. Dark areas are the tungsten. (Courtesy of T. W. Barbee, TEM micrography by Y. Lepetre and A. Charai). (b) TEM of a Nb/bcc-Zr superlattice. Note coherence shown in lower right corner. (Courtesy of T. W. Barbee, TEM micrograph by R. Byers.)

The physical properties of superconducting multilayers that have been studied to date are indicated in Table II and were reviewed in Ref. 3. From the point of view of the theme of this chapter--artificially-structured superconductors as model systems--the results obtained to date on the Pb/Bi, the Nb/a-Ge and the Nb/Cu systems are of the greatest interest. In the work of Raffy and co-workers[24] on Pb/Bi, periodicities on the scale of the penetration depth $\lambda$ were introduced into the material with the expected impact on vortex pinning. Coherent pinning of the vortex lattice was demonstrated and studied. Interdiffusion of the Pb and Bi (at room temperature) made it necessary to maintain these samples at low temperatures, however. Also only a modest compositional modulation could be achieved, well short of 100%. All of this illustrates the problems of working with the soft superconducting materials. Nonetheless this was probably the first really successful example of an artifically-structured superconductor.

In the Nb/a-Ge[21] and Nb/Cu[20] systems very much finer layering has been achieved, as has essentialy 100% chemical modulation. Neither of these materials is even close to being perfectly ordered from the structural point of view, however. In Nb/a-Ge the Ge layers are amorphous and the Nb is polycrystalline. Also there is evidence for structure across the Nb layers themselves. In Nb/Cu both layers are polycrystalline and there is essentially no lateral coherence as already noted. Nevertheless from the point of view of quasi-two-dimensional superconductivity, both of these systems have proved highly successful. In each case the non-superconducting layers act as barriers in the material through which the superconducting (Nb) layers are Josephson coupled, of course the mechanism of the Josephson coupling is different in the two cases (i.e., tunneling versus the proximity effect). Most importantly, in both cases this Josephson coupling can be readily varied and controlled.

The quasi-two-dimensionality of these materials is illustrated for the case of Nb/Ge in Fig. 10.[21] Here we show the parallel $H_{c2\parallel}$ and perpendicular $H_{c2\perp}$ upper critical fields of a series of Nb/Ge multilayers in which the thickness of the Ge layers $D_{Ge}$ has been systematically varied. Note first that the samples are clearly anisotropic ($H_{c2\parallel} > H_{c2\perp}$) . The dimensionality of the superconductivity can be established from the $H_{c2\parallel}$ behavior. For 3D behavior one expects the bulk result $H_{c2}(T) \propto (T_c-T)$ near $T_c$, whereas for 2D one expects $H_{c2} \propto (T_c-T)^{1/2}$ as found in very thin films. Clearly as $D_{Ge}$ increases, there is a progression from 3D to 2D behavior in $H_{c2\parallel}$.

More interestingly for $D_{Ge}$ = 35 Å there is a 3D to 2D crossover in $H_{c2\parallel}$ as a function of temperature. Near $T_c$ where the coherence length perpendicular to the layers is large, 3D behavior is obtained. At lower temperatures the coherence length decreases and becomes comparable to the layer separation producing a dimensional crossover to 2D. This crossover is clearly seen

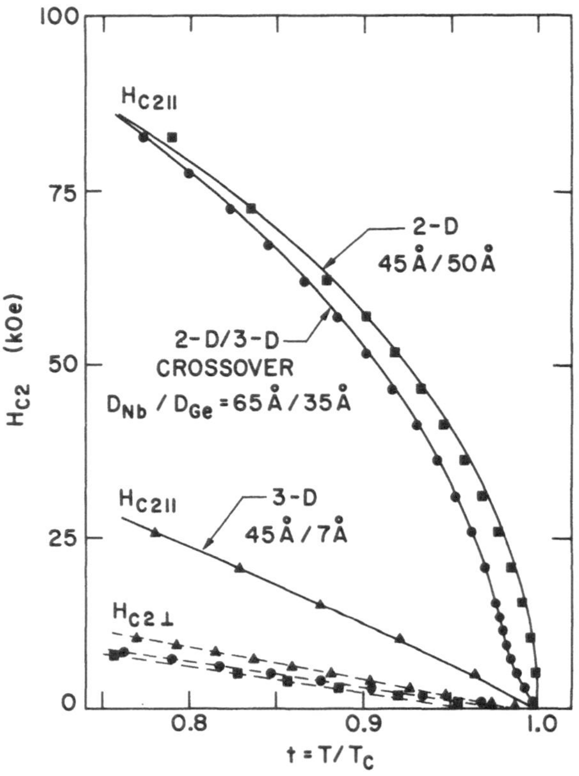


FIG. 10--Parallel and perpendicular upper critical fields for a series of Nb/Ge multilayers illustrating the 3D to 2D dimensional crossover.

where the $H_{c2\parallel}$ curve shows a marked upturn from a linear temperature dependence and joins the 2D behavior at lower temperatures. One can physically interpret this result in terms of the vortex cores "fitting" between the layers. This interpretation also emphasizess the fact that anomalies in $H_{c2}$ are expected when periodicities of any kind are introduced into a superconductor on the scale of the coherence length.

From these $H_{c2}$ measurements both the ratio of the Ginzburg-Landau effective masses M/m and the Josephson coupling energy $E_J$ can be determined. The dependence of these quantities on $D_{Ge}$ is shown in Fig. 11, which also illustrates the remarkable degree of control that has been achieved. As seen in this figure as $D_{Ge}$ increases there is a sharp rise in the effective mass ratio. Moreover, over the entire range shown the Josephson coupling energy ($E_J \propto r$ in the figure) goes exponentially with $D_{Ge}$ demonstrating that the coupling is via tunneling. Similar anomalies in $H_{c2\parallel}$ have been seen in the Nb/Cu system.

B. <u>Artificial Tunneling Barriers</u>

Another case where layered thin film structures have been highly successful is in the formation of artificial tunneling barriers. The deposition techniques involved are closely related to those used to make the multilayers described above. The idea here is to form a very thin layer (or sequence of layers) of a material that, either in itself or when subsequently processed (e.g., oxidized), forms a good tunnel barrier. This approach is most useful when the material of interest does not produce a good native oxide barrier, or when undesirable proximity effects are present. Interestingly, attempts to make artificial tunnel barriers are not at all new.[26] Even some of the recent approaches were tried in the past. The current rash of successes probably lies in the improved deposition techniques now available, along with (in several cases) an innocent ignorance of what had been tried and found to fail in the past.

Basically there have been three approaches. These are illustrated in Fig. 12. Two are based on the use of amorphous silicon (a-Si)[27,28] and the other involves a deposited metal.[29] Nevertheless, as can be seen in the figure in each case there is a strong similarity in the final structure of the barrier. Specifically, there is a buffer layer between the actual barrier itself ($SiO_x$, a - Si:H or $A\ell_2O_3$) and the base electrode. This buffer layer serves to chemically protect the base electrode. When a chemically sensitive counterelectrode is used (e.g., Nb as opposed to say, Pb) a similar buffer layer is used above the actual barrier. The bilayer (trilayer) nature of these barriers appears to be an essential part of their success.

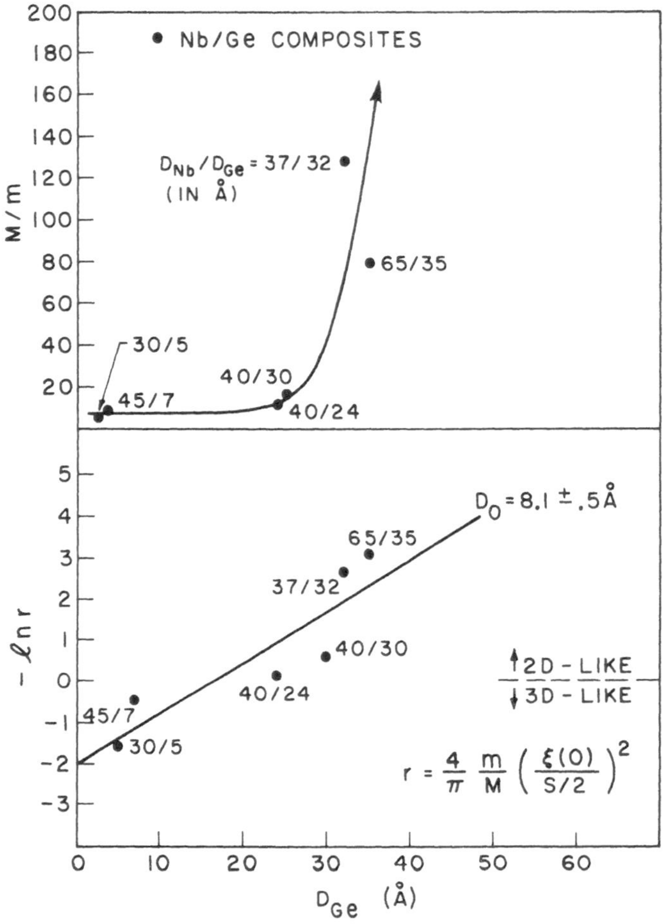


FIG. 11--(a) Dependence of GL effective mass ratio on $D_{Ge}$. (b) Dependence of the Josephson coupling energy $E_J \propto r$ on $D_{Ge}$.

The oxidized a-Si barriers are formed as follows. First, a very thin layer (~ 20 to 30 Å) of a-Si is deposited followed by subsequent oxidation to both form the barrier ($SiO_x$) and possibly block pin holes. Additional silicon is then deposited if an upper buffer layer is desired. The a-Si has been deposited both by

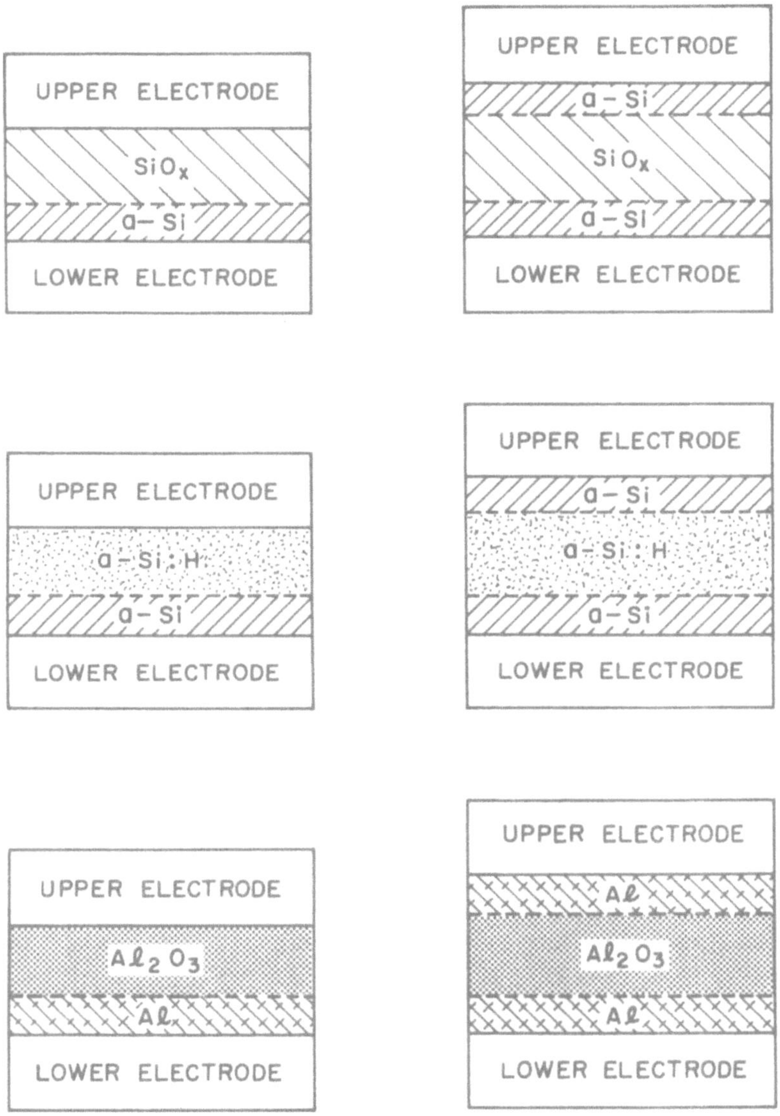


FIG. 12--Schematic representation of various successful artificial tunnel barriers.

e-beam evaporation and sputtering. The detailed electrical behavior that results from these two approaches are not equivalent, however, probably reflecting differences in the structure of the resultant a-Si. As shown explicitly by

Meservey et al.,[30] a-Si by itself does not form a particularly good barrier. The role of the oxygen appears to be to reduce the density of localized electron states in the a-Si. The detailed physics of the tunneling process in these barriers is not well understood and is an interesting problem of disordered materials in its own right.

To date the hydrogenerated a-Si barrier (a-Si:H) has only been formed using sputtering. A layer thickness of ~ 30-50 Å is typical. As stressed by Kroger et al.,[31] the role of the hydrogen clearly is to reduce the density of localized states. The buffer layers serve to prevent the hydrogen from entering the Nb electrodes where it degrades the superconductivity of the Nb. The electrical properties of these a-Si:H barriers are also discussed in Ref. 31.

From the point of view of artificially structured materials, the most interesting of these artificial tunnel barriers are those involving oxidized metal layers. This approach was originally demonstrated with Aℓ but has subsequently been extended to various other materials, Mg,[32] various rare-earth metals,[32,33] Ta[34] and Zr.[35] Only Nb electrodes have been used to date. In the original work[29] a very thin ($\gtrsim$ 20 Å) metallic Aℓ layer was deposited on a Nb base electrode and subsequently oxidized to form the barrier ($A\ell_2O_3$). Hence, like the oxidized a-Si barrier, this is a hybrid (artificial/natural) approach in which the artificial barrier actually incorporates a native oxide. Sound metallurgical and chemical principles appear to be underlying the success of the deposited metal barriers in that the Aℓ evidently wets the Nb surface without being incorporated into the bulk and also reduces any oxygen near the surface, thereby also serving a cleaning function.[32] Indeed it was just these principles that were used in the selection of many of the other successful barriers. These remarkable results have been complimented recently by the observation that comparably good barriers can be obtained on Nb films that are essentially single crystals.[36] The implication is that grain boundaries are somehow responsible (at least in part) for the traditional problems encountered with Nb tunnel barriers.

In short, cleverly structured artificial tunnel barriers are now available for use in tunnel junctions on previously difficult materials, and for those applications where highly uniform barriers are desired (e.g. large Josephson junction arrays). In particular, it appears that Nb can now be used as a reliable material in model systems incorporating tunnel junctions. The possibilities for model studies of the tunneling process itself using such structured barriers are yet to be exploited. Single-crystal, epitaxial growth of oxides or other barriers on these materials are also a possibility as MBE-type systems become available.

## C. Deposited Planar SNS Josephson Junction Arrays

Appropriate depositions on pre-configured substrates can be used to make arrays of SNS Josephson junctions. Here the substrate is prepared using lithography, but the actual array is formed insitu via the deposition process itself. Hence this is an example of artificial structure that is in some sense intermediate between those formed wholly by deposition and those formed via lithography. The approach is based on the concept of step-edge patterning first introduced by Feuer and Prober.[37] It has been exploited to make SNS Josephson junctions by de Lozanne et al.[38] The approach is illustrated in Fig. 13. The idea is to use the step to determine the length of the bridge, which must be submicron and is the critical length determining the properties of the bridge. Given the step, first the superconductor is deposited at an angle such that the step edge is not covered. Then, rapidly, so as to preserve a clean interface, a normal metal is deposited from a complementary angle so as to cover both the superconducting banks and the step edge, which then forms the bridge connecting the two banks. The Josephson properties of these bridges have been studied in detail and are found to be quite good.[38]

To adapt this technique to the formation of linear arrays is straight forward. Arrays containing up to ~ 1000 Nb/Cu/Nb SNS bridges have been made using substrates patterned produced using holographic techniques to define the photoresist pattern. An actual array is shown in Fig. 14. Several independent linear arrays can be seen going from the upper right to the lower left of the figure. The orthogonal lines running from the upper left to the lower right are the step edges used to form the bridges. Two-dimensional arrays should also be possible. This approach to SNS array fabrication should be contrasted with the dot matrix approach of Resnick et al.[39] and Abrahams et al.[40] The principal advantage of the step-edge approach is that higher resistance junctions can be formed, which is desirable in certain experiments, for example in studying quantum fluctuations, although obtaining high enough resistances to observe these particular effects will be difficult. Nonetheless this example does serve to illustrate another approach to artificially structuring materials.

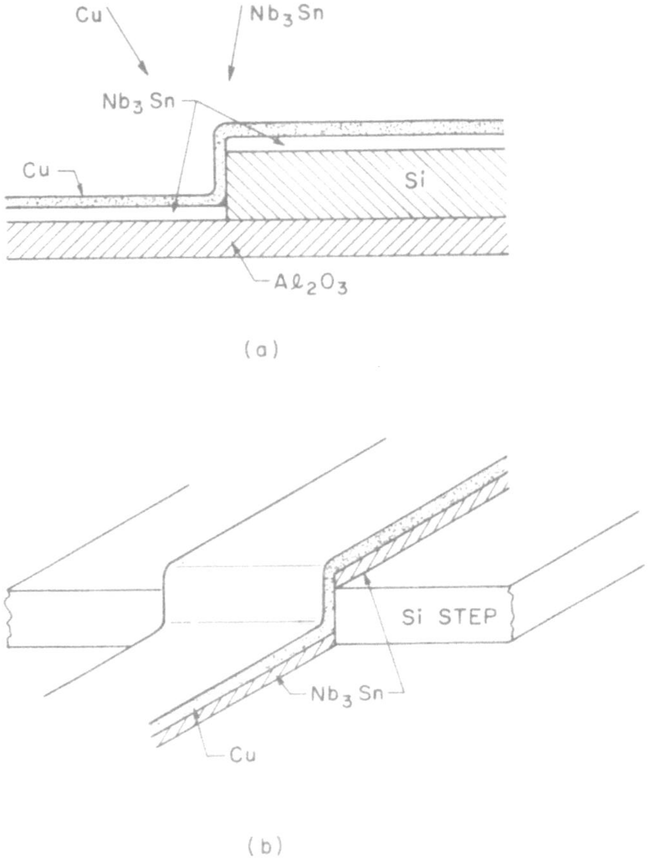


FIG. 13--Schematic of SNS planar bridge fabrication procedure.

## D. Some Related Thin-Film Model Systems

Finally let us turn to some examples that while not artificially-structured materials in the strictest sense of term, were made possible by advanced vapor deposition techniques of the sort we have been considering. The first of these are the discontinuous films and the granular systems discussed in the

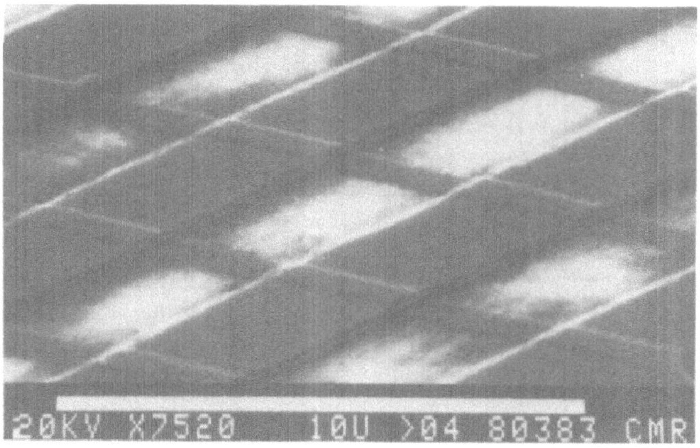


FIG. 14--SEM picture of several arrays of SNS bridges. (Courtesy W. Anklam and A. deLozanne.)

chapter by Deutcher in this volume. Here percolative structures have been made by deposition of Pb on Ge substrates and by Aℓ-Ge codepositions. In the case of discontinuous Pb films, the self-similarity required of a true random percolative network has been demonstrated explicitly from direct study of the structure as well as from the electrical properties.[5] The Aℓ-Ge mixtures can be made amorphous or granular depending on a combination of the relative compositions and the substrate temperature.

A similar example are the In-In oxide films produced by Nakahara and Hebard[7] using reactive depositions (as in Fig. 6c). As we have already discussed a wide variety of structures, from amorphous to aglomerated films are possible. The structure of these films has been carefully studied in Ref. 7, but not from the point of view of self-similarity as in the case of the Pb films above. Very homogenous, high sheet resistance films have been produced for studies of the Kosterlitz-Thouless transitions in superconductors, however. (See chapter by Mooij.)

Two well-known granular systems are Aℓ-Aℓ oxide[41] and anodized NbN.[42] These systems have been discussed at length in the past and will not be reviewed further here.

By way of a final example that illustrates some entirely new oportunities, we mention the recent work of Graybeal[43] in which ultra-thin amorphous superconducting films were produced by means of the same sputtering techniques used previously to make fine multilayers. By cosputtering on a rapidly rotating substrate in conjunction with carefully selected over- and under-layers, Graybeal has produced a-(Mo-Ge) films whose specific resistivity appear to be constant (except for the expected localization effects) down to ~ 20 Å. Hence there is no evidence that the films are breaking up. The sheet resistances of these homogeneous films can be as high as ~ 1500 Ω. To achieve this striking result, amorphous substrates (a-$Si_3N_4$) coated with a 30 Å a-Ge underlayer were employed. Also for chemical protection from the ambient atmosphere (and photolithography chemicals) an a-Si overlayer was likewise employed. This a-Si layer was also used to form a tunnel barrier for tunneling studies of these films. Thus the desired function of passivation and a tunnel barrier are integrated into a single layer. Studies of the superconductivity in these thin films demonstrates a substantial reduction of $T_c$ due to localization effects. In any event, the samples illustrate how advanced vapor deposition and layering techniques can be combined to produce a model system that is not otherwise available in nature.

Acknowledgements

The author would like to acknowledge formative discussions with T.W. Barbee, S.T. Ruggiero, J. Graybeal, D.E. Prober, R. Hammond, and T.H. Geballe from which he learned most of what he knows about artificially-structured superconductors. This chapter was prepared with the support of the U.S. National Science Foundation.

References

1. See, for example: R.H. Hammond, IEEE Trans. MAG-11, 201 (1975).

2. For a general reference on multilayered materials see: Synthetically Modulated Structure Materials, L. Chang and B.C. Giessen, Eds., Academic Press, 1984 (in press).

3. For a general review of multilayered superconductors, see: S.T. Ruggiero and M.R. Beasley, "Synthetically Layered Superconductors," in Ref. 2.

4. I.K. Schuller and C.M. Falco, "Metallic Superlattices," in Microstructure Science and Engineering/VLSI Vol. 4, N.G. Einsbruch, Ed., Academic Press (1982).

5. See chapter by G. Deutscher. Also see A. Kapitulnik, Ph.D. Thesis, Tel Aviv University, 1983.

6. J. R. Gavaler, M. A. Janocko and C. K. Jones, Appl. Phys. Lett. 19, 305 (1971).

7. See, for example: S. Nakahara and A.F. Hebard, to appear in Thin Solid Films.

8. Unfortunately no recent comprehensive reviews of electron-beam thermal evaporation are available. Useful information can be found, however, in the Handbood of Thin Film Technology, L. I. Maissel and R. Glang, Eds., McGraw Hill, 1970; and The Physics of Microfabrication, I. Brodie and J. J. Murray, Plenum, 1982.

9. For a review of the type of sputtering systems found most useful to date for artificially-structured superconductors, see: H.A. Huggins and M. Gurvitch, J. Vac. Sci. Technol. A1, 77 (1983).

10. For ion beams, see J.M.E. Harper, J.J. Cuomo, and H.R. Kaufman, J. Vac. Sci. Technol. 21, 737 (1982).

11. For a general review see: M.R. Beasley and C.J. Kircher, "Josephson Junction Electronics: Materials Issues and Fabrication Techniques," in Superconductor Materials Science, NATO ASI, B68, S. Foner and B.B. Schwartz, Eds., Plenum, 1981.

12. R.E. Howard, Appl. Phys. Lett. 33, 1034 (1978).

13. J. Cuomo, J.M.E. Harper, C.R. Guarnieri, D.S. Yee, L.J. Attanasio, J. Angilello, C.T. Wu, and R.H. Hammond, J. Vac. Sci. Technol. 20, 349 (1982); T. Goto and R.H. Hammond, to be published.

14. S.M. Durbin, J.E. Cunningham, M.E. Mochel, and C.P. Flynn, J. Phys. F: Metal Phys. 11, L223 (1981); S.M. Durbin and C.P. Flynn, J. Phys. F: Metals Phys. 12, L75 (1982).

15. See article by T.W. Barbee, Jr., in Ref. 2.

16. J. Geerk, M. Gurvitch, D.B. McWhan, and J.M. Rowell, Physica B & C 109 and 110, 1175 (1982).

17. See Ref. 4.

18. G. Hertel, D.B. McWhan, and J.M. Rowell, in Superconductivity in d- and f-Band Metals, Kernforschungszentrum, Karlsruhe, 299 (1982). Also see Ref. 16.

19. T.W. Haywood and D.G. Ast, Phys. Rev. B 18, 2225 (1978).

20. I. Banerjee, Q.S. Yang, C.M. Falco, and I.K. Schuller, Solid State Commun. 41, 805 (1982).

21. S.T. Ruggiero, T.W. Barbee, Jr., and M.R. Beasley, Phys. Rev. B 26, 4894 (1982).

22. Y.J. Qian, J.Q. Zheng, B.K. Sarma, Q.H. Yang, J.B. Ketterson, and J.E. Hilliard, J. Low Temp. Phys. 49, 279 (1982).

23. W.P. Lowe and T.H. Geballe, to appear in Phys. Rev. B.

24. H. Raffy, J.C. Renard, and E. Guyon, Solid State Commun. 11, 1679 (1972); H. Raffy, E. Guyon, and J.C. Renard, Solid State Commun. 14, 427 (1974), ibid p.431; H. Raffy and E. Guyon, Physica 108B, 947 (1981).

25. See chapter by McWhan in Ref. 2.

26. For a review of this early work, see: P. Cardinne, J. Nordman, and M. Renard, Rev. Phys. Appliquee 9, 167 (1974).

27. D.A. Rudman and M.R. Beasley, Appl. Phys. Lett. 36, 1010 (1980).

28. H. Kroger, L.N. Smith, and D.W. Jillie, IEEE Trans. MAG-15, 488 (1979).

29. J.M. Rowell, M. Gurvitch, and J. Geerk, Phys. Rev. B 24, 2278 (1981); M. Gurvitch, M.A. Washington, and H.A. Huggins, Appl. Phys. Lett. 42, 472 (1983).

30. R. Meservey, P.M. Tedrow, and J.S. Brooks, J. Appl. Phys. 53, 1563 (1982).

31. H. Kroger, L.N. Smith, D.W. Jillie, and J.B. Thaxter, IEEE Trans. MAG-19, 783 (1983).

32. J. Kwo, G.K. Wertheim, M. Gurvitch, and D.N.E. Buchanan, IEEE Trans. MAG-19, 795 (1983).

33. C.P. Umbach, A.M. Goldman, and L.E. Toth, Appl. Phys. Lett. 40, 81 (1982).

34. S.T. Ruggiero, D.W. Face, and D.E. Prober, IEEE Trans. MAG-19, 960 (1983).

35. S. Celaschi, R.H. Hammond, T.H. Geballe, and W.P. Lowe, Bull. Am. Phys. Soc. 28, 423 (1983).

36. S. Celaschi, T. H. Geballe and W. P. Lowe, Appl. Phys. Lett. 43, 794 (1983).

37. M.D. Feuer and D.E. Prober, Appl. Phys. Lett. 36, 226 (1980).

38. A. deLozanne, M.S. diIorio, and M.R. Beasley, Appl. Phys. Lett. 42, 541 (1983).

39. D.J. Resnick, J.C. Garland, J.T. Boyd, S. Shoemaker, and R.S. Newrock, Phys. Rev. Lett. 47, 1542 (1981).

40. D.W. Abraham, C.J. Lobb, M. Tinkham, and T.M. Klapwijk, Phys. Rev. B 26, 5268 (1982).

41. See, for example: W.L. McLean, P. Lindenfeld, and T. Worthington, AIP Conf. Proc. 40, 403 (1977); W.L. McLean, T. Chui, B. Bandyopadhyay, and P. Lindenfeld, AIP Conf. Proc. 58, 42 (1980).

42. See, for example, D.U. Gubser, S.A. Wolf, T.L. Francavilla, and J.L. Feldman, AIP Conf. Proc. 58, 159 (1980).

43. J. Graybeal and M.R. Beasley, submitted for publication.

# CLUSTERING IN THIN Au FILMS NEAR THE PERCOLATION THRESHOLD

R. B. Laibowitz, R. F. Voss and E. I. Alessandrini

IBM Research Center
PO Box 218
Yorktown Heights, NY 10598

## INTRODUCTION

A random mixture of conducting (fractional concentration p) and insulating (fractional concentration 1-p) material will abruptly exhibit electrical conduction at a critical concentration, $p_c$. While this idealized type of percolation problem is ideally suited for computer simulation studies and as a model for conductivity measurements in real systems, efforts have recently been made to examine these mixtures in thin metal films via transmission electron microscopy and to analyze the resulting micrographs using digital processing. In this way, detailed comparisons between the cluster formation in the metal film and the predictions of percolation theory can be directly made. We have made such comparisons in our studies of the thickness dependence of the resistance and microstructure of very thin metal films. While changes in the conductivity properties as a function of metal to insulator concentration have been interpreted in terms of percolation[1], these measurements are complicated by the presence of other conductivity mechanisms unrelated to percolation. Leakage between clusters either on the surface or through the bulk including quantum mechanical tunneling[2] can smear the percolation transition and add a temperature dependence to the conductivity. Detailed analysis of the TEM micrographs has been shown to be a reliable and informative way to compare the geometric properties of the clustered film with the theoretical predictions of percolation and scaling near the insulator-metal transition.

In the vapor deposited Au films reported on in this paper, we have observed the formation of the 'infinite' cluster at a critical area coverage of about 0.73 (= $p_c$) The substrate in this case was amorphous $Si_3N_4$. The percolation threshold was identified and statistical and geometrical studies performed directly using computer analysis of the digitzed micrographs. In addition, computer enhancement of the micrographs can also be used to study metallurgical features in the clustered grains.

EXPERIMENT

The thin Au films were vapor deposited[3] at room temperature using electron beam heating in a system with a background pressure of about $10^{-7}$ torr. The deposition rate was kept at about 2 nm/sec and the system shutters were programmed so that adjacent samples had nominal thicknesses that differed by 10 nm as measured by quartz crystal monitors. The samples or substrates consisted of sections of a Si (single crystal) wafer containing a thin window or membrane supported by an etch-defined portion of the Si[4,5]. An example of such a window substrate is shown in Fig. 1. Typical window sizes range from 50 microns on a side to several mm. The membrane thickness is an important parameter as it must be transparent to the electron beam in a transmission electron microscope (TEM). The membrane thickness was always less than 200 nm and for high resolution microscopy, 20 nm thick membranes could be used. The $Si_3N_4$ films were amorphous and very smooth and hence did not contribute any images to the subsequent micrographs.

The substrates were placed in the vacuum system in a rectangular array and the deposition thickness was varied systematically over

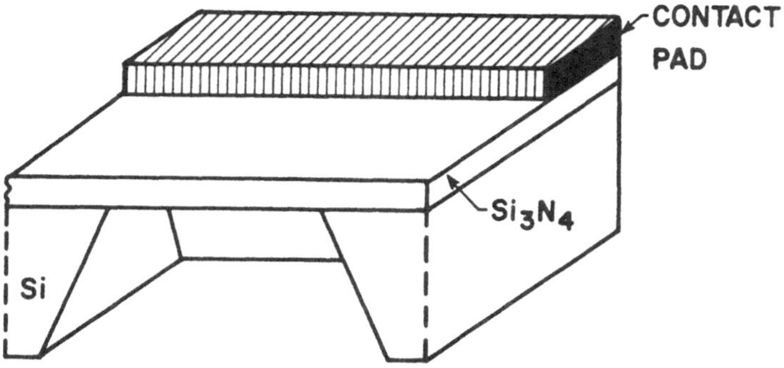


Figure 1. A schematic representation of a window substrate which consists of a etched hole in a section of a silicon wafer covered by a thin membrane of $Si_3N_4$. The windows were typically 0.1 mm on a side while the membrane thicknesses were generally between 20 and 200 nm.

the entire array in order to maximize the the number of samples around the interesting region near $p_c$ In general, a very narrow thickness range was desired and thus the average thickness variation over the array was from 60 to 90 nm. The number of samples in the array varied from run to run with the maximum number of samples in a single run of about 150. The sample sheet resistances at room temperature varied from a few Ω/□ for the thicker films to greater than $10^9$ Ω/□ for the thinnest samples. In this way the entire insulator-metal transition was covered in a single deposition. In order to measure the resistance of the films, Au connecting pads were predeposited on the substates in several configurations. In-situ Ar plasma cleaning was used for about 1 min prior to the film deposition to insure a clean reproducible substrate surface and provide for good electrical contact between the pads and the thin gold film.

With the above procedures, substrates and thicknesses, the subsequent Au films grew in the form of irregularly shaped clusters as shown in Fig. 2. In this micrograph the light areas are the insulating substrate material and the dark areas are the metal. Structure can also be seen within each cluster which is due both to the different crystal orientations of the Au grains that make up a given cluster and to crystalline defects such as dislocations and twin boundaries. Such micrographs are the principal data for the results described below although room temperature resistance was also used. In order to analyse such complex data, the micrographs were first digitized[6] on a scanning densitometer on a 512 by 512 grid. Each image recorded required scanning for about 8 hours. An example of a digitized micrograph is shown in Fig. 3. The digitizing process stores the intensity of the light transmitted through the negative micrograph and by choosing the proper threshold it was possible with high contrast images to automatically distinguish between between metal and insulator in the digitized image. After the data has been stored in the computer further analysis can be accomplished by operating on the stored information. Knowledge of the cluster area distribution as a function of Au thickness and cluster perimeter give direct evidence for percolation processes and scaling theory.

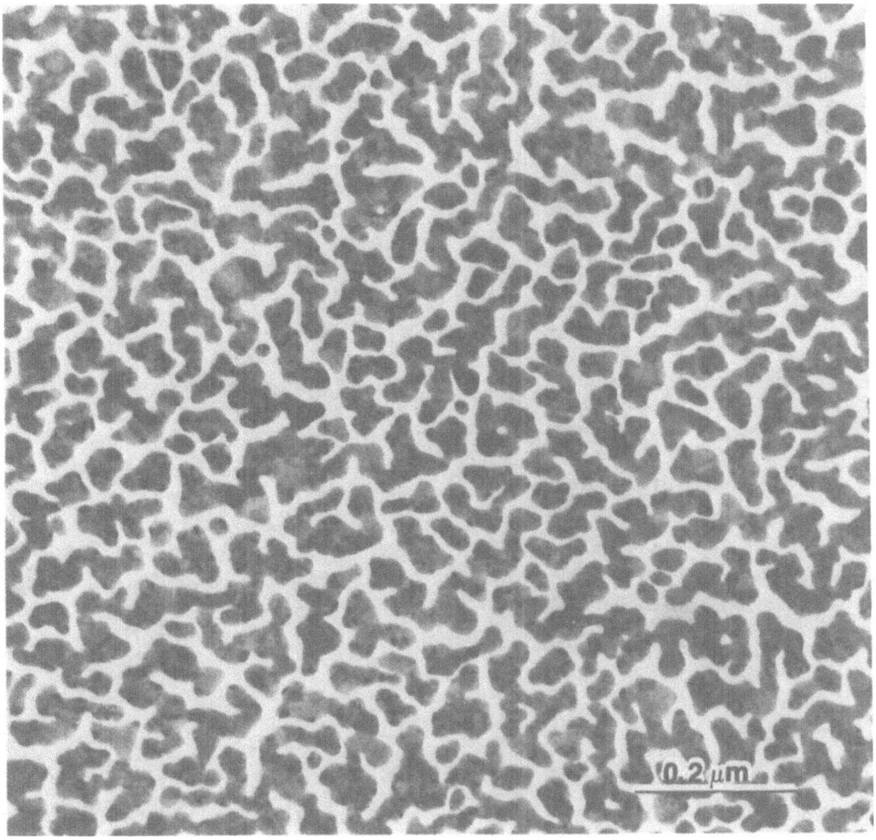


Figure 2. TEM micrograph of a thin Au film on an amorphous $Si_3N_4$ substate. The Au thickness as determined during the deposition by a crystal monitor was about 7 nm.

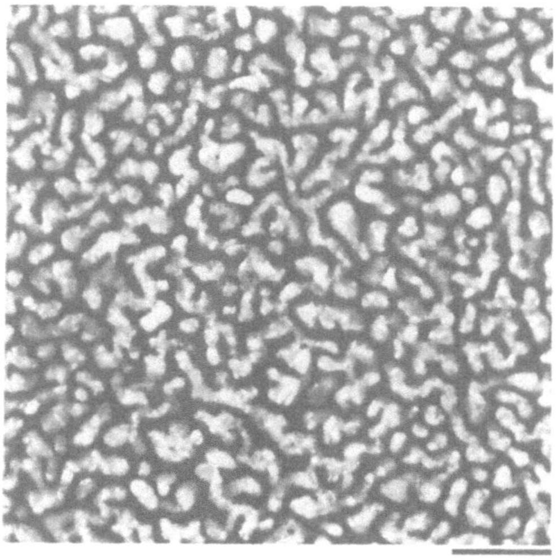


Figure 3. Digitized TEM micrograph in which the non-metal regions are shown in darkest shade. Microscopic structure within the clusters such as grain boundaries, twin boundaries and dislocations which are observed in the original micrographs are also clearly visible in the digitzed image.

The digitized images also provide a unique method of identifying the percolation threshold ($p_c$) and the appearence of the infinite cluster.

RESULTS AND DISCUSSION

A great deal of the previous work[1] on thin metal films exhibiting cluster formation concentrated mainly on conductivity measurements due in part to the complexity of the information contained in a TEM micrograph. In the micrographs as shown in Figs 2 and 3 it is difficult to identify the insulator-metal transition or to find the longest or infinite cluster. However, operating on the stored data which contains the areas of all the clusters in a given view, we can enhance the micrographs such that the largest clusters are shown in the darkest shade, the next largest cluster in a lighter shade and so on. Color enhancement has also been used to great advantage in this regard. An example of such an enhancement is shown in Fig. 4 where the images span the range from insulating (a) to conducting (b) and (c). The various shapes that the clusters assume and their variation as p crosses $p_c$ is now easy to observe. The actual percolation threshold for the samples as shown is about 0.73. One can also observe in these enhanced micrographs the actual appearence of the infinite cluster. In fact, it is a more direct way to identify the percolation threshold in the Au samples than the conductivity measurements at room temperature. Not all metals show the formation of such all-metallic, infinite clusters[2] particularly those metals such as Al where the clusters may be separated by thin, possibly conducting, oxide barriers. However, the appearence of such an infinite cluster is a basic prediction of percolation theory and is observed in the computer enhanced micrographs of Fig. 4. This is also an advantage when one considers that the actual transition can complicated by several 2 and 3-D processes such as surface tension and cleanliness.

In Fig. 5 (a) we show the results of measuring the room temperature conductance for a series of Au samples as a function of the frac-

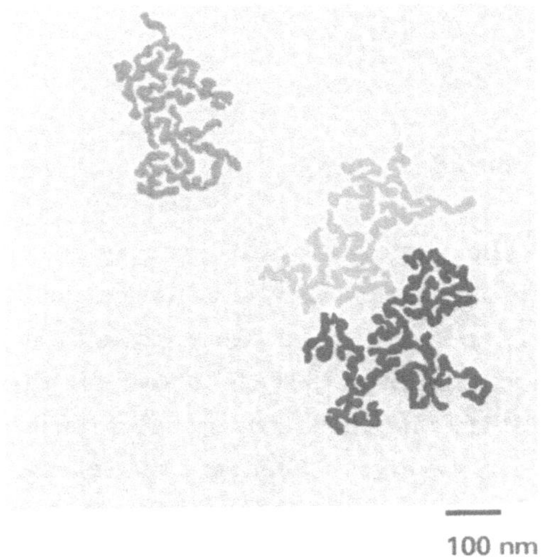


metal P = 0.659

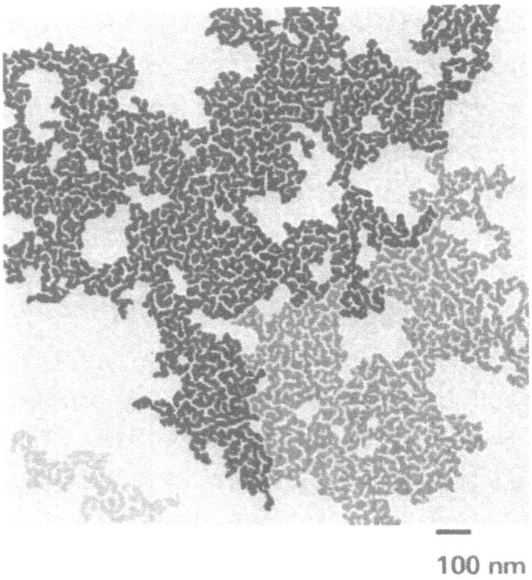


metal P = 0.707

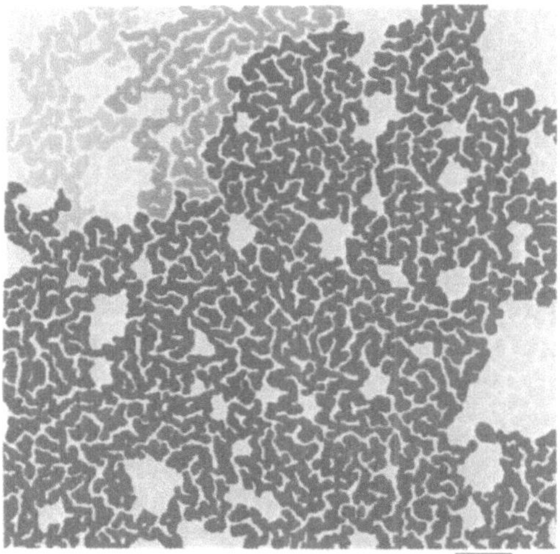


Figure 4. Cluster analysis from a series of Au films that span the percolative transition. In each case, the three largest clusters are shown in dark shades.

metal p = 0.659

metal p = 0.707

metal p = 0.752

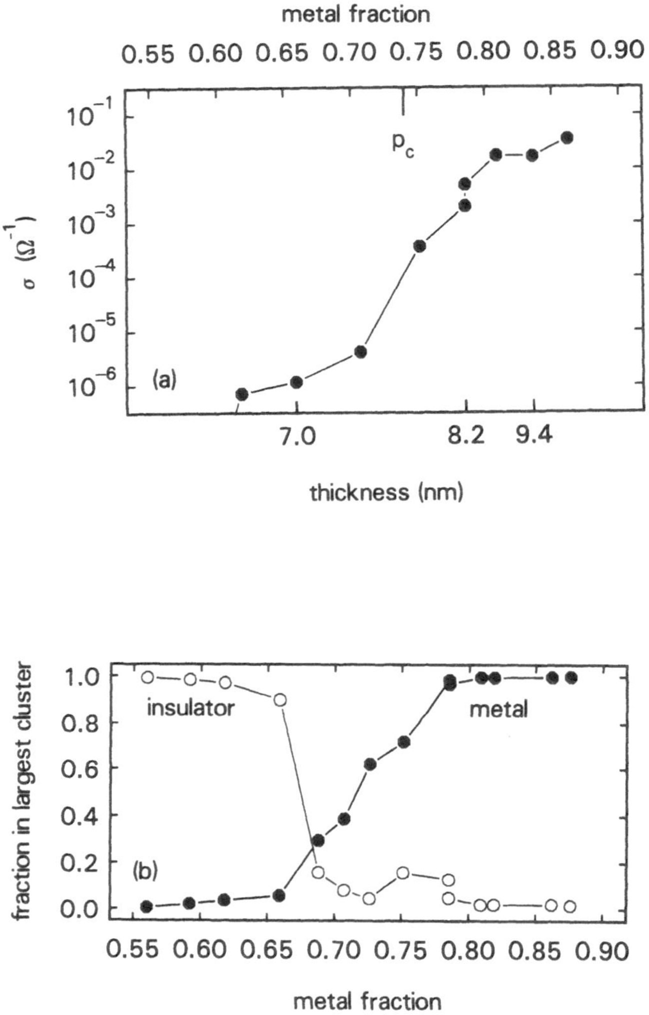


Figure 5. (a) Measured room temperature conductivity. (a) Fraction of image in the largest cluster as p is varied. The plots for both metal and non-metal in the same image are shown.

tional coverage p. The different values of p are derived from the micrographs which are taken as function of thickness in this case. Clearly an insulator to metal transition is taking place as the thickness is increased. Comparison of the conductance data with the actual digitized micrographs shows that the transition corresponds to the appearance of the infinite cluster with a $p_c$ of about 0.73. The conductivity measurements sample a much larger area than the TEM data. Due to this large area sampling and the possiblity of other conduction mechanisms (as mentioned above), there is no sharp theshold in the conductivity data at $p_c$. The two curves plotted in Fig. 5 (b) show for the first time that in a classical percolation system which the Au clusters closely resemble, both the metal and the absence of metal clusters exhibit percolation. We have plotted in Fig. 5 (b) the fraction of material in the largest cluster as p is varied from the insulator side to the high conductivity metal side. As the infinite cluster grows, it is expected that the fraction of material in each image will corresponingly increase. This is observed in the figure in both the insulator and in the metal coverages.

The percolation problem is one of the simplest realizations of a 2nd order phase transition[7]. The scaling theory of percolation is based upon computer simulations and renormalization group arguments in which typical studies are centered around the quantity n(A), the average number of clusters (per lattice site) of area A as a function of p. At $p_c$ there is no characteristic size and n(A) is proportional to $1/A^{\tau}$. As shown by Voss et al[6] this is equivalent to the Korcak-Mandelbrot formalism (developed for islands on the Earth's surface) which states that the number of clusters with area greater than some size A, N(area>A) is proportional to $1/A^{\tau-1}$. $\tau$ is known from theoretical estimates at the percolation threshold to be ~2.05. From the data of Fig. 6 we get that near $p_c$ N(area>A) decays with the expected $1/A^{\tau-1} = 1/A^{1.05}$ in good agreement with theoretical predictions for a large range of A.

As has been shown by numerical simulation[8] and later proven rigorously[9] for percolation clusters, P (the cluster perimeter) is

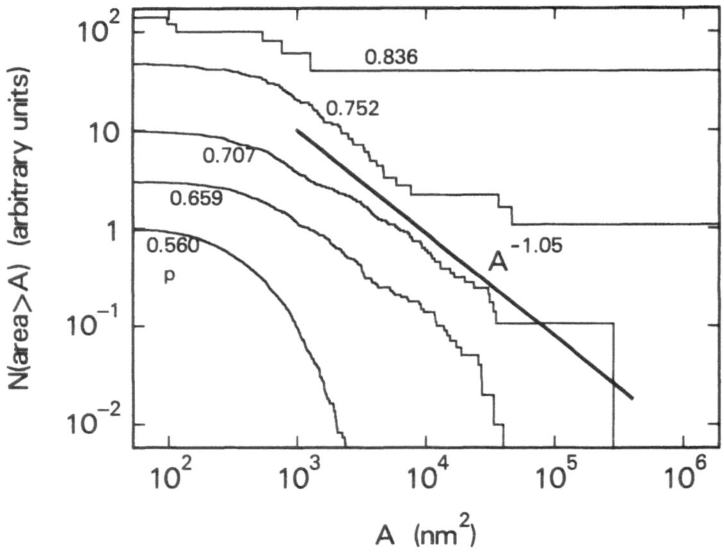


Figure 6. The cumulative distribution N(area>A) vs A at different fractional Au coverages p. The solid line shows the expected $1/A^{1.05}$ dependence at $p_c$.

proportional to A for large A at all p. The smallest clusters are almost circular in shape and have a P proportional to $A^{1/2}$ These predictions are verified by the data as shown in Fig. 7 which is a scatter plot of P (defined as the number of unoccupied sites on the digitzed grid adjacent to a given cluster) as a function of A for the Au films. Each cluster is represented by a single point. It should be observed that for A greater than about 600 $nm^2$ P is proportional to A both above and and below $p_c$. The crossover between P proportional to A and to $A^{1/2}$ is the same for all p and appears to be related to the Au-substrate interaction. A reasonable growth mechanism for the large clusters is that clusters combine by developing small connecting necks to their neighbors. In this way the larger and longer clusters are weak connections of N smaller clusters. Each of these clusters has a perimeter of $P_o$ and area of $A_o$ giving a total P $\sim NP_o$, $A \sim NA_o$ and ,therefore, P/A = constant.

Mandelbrot's fractal geometry[10] provides an alternate formalism for understanding cluster shapes and their scaling behavior. This formal formalism is covered in the references and only a few conclusions will be included in this report.The cluster boundaries which are topologically 1D are characterized by two fractal dimensions D and $D_c$ It has been shown[6] that $D_c$ has a value of about 1.9 for the Au infinite cluster and that this number is directly related to the universal exponents of percolation theory. Taken together the collection of all boundaries at large sizes are characterized by D=d=2.

## CONCLUSIONS

Computer analysis of TEM micrographs of thin Au films has been used to show that at the percolation threshold an infinite cluster appears and begins to fill the entire plane. Near $p_c$ the Au clusters have the scaling properties expected from percolation theory. Local Au-substrate interactions appear to raise $p_c$ to about 0.73. The equivalence between the analytical scaling theories of percolation with its universal exponents and a geometric interpre-

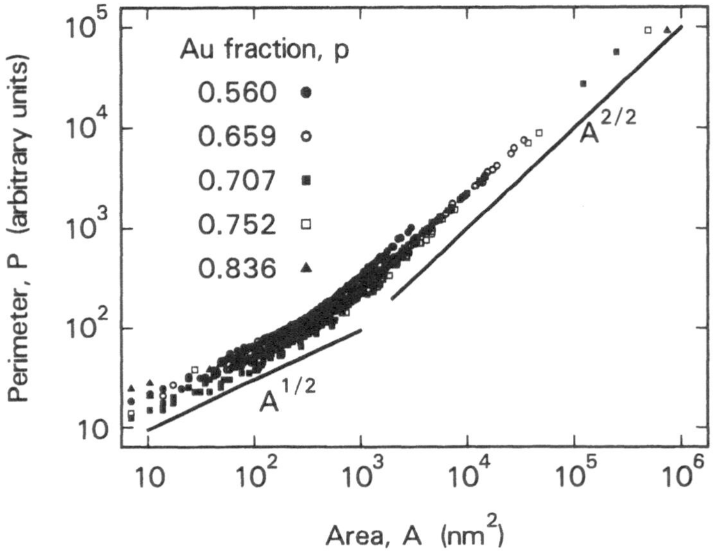


Figure 7. Scatter plot of the perimeter P vs area A for the gold clusters from samples as shown for example in Fig. 2. Each cluster produces one point. The solid line shows the expected P ∝ A behavior at large A.

tation based on fractals has been demonstrated. We have also shown that computer analysis of the complex micrographs can yield high quality statistics that were previously only available in simulations.

ACKNOWLEDGEMENTS

The authors would like to thank C. R. Guarnieri for expert sample development and are grateful to J. M. Viggiano, M. D'Agostino, J Speidell, K. Nichols and N. Us for their continuing assistance. We also acknowledge helpful discussions with B. B. Mandelbrot, Y. Gefen, S. Kirkpatrick, C. Umbach and Y. Imry.

REFERENCES

1. See for example B. A. Abeles in Applied Solid State Science, "Granular Metal Films", edited by R. Wolfe (Academic, New York, 1976) **6**, 1; B. A. Abeles, H. L. Pinsh, and J. I. Gittleman, "Percolation Conductivity in W-$Al_2O_3$ Granular Metal Films", Phys. Rev. Lett. **35**, 247 (1976); or C. J. Lobb, M. Tinkham, and W. J. Skocpol, "Percolation in Inhomogeous Superconducting Composite Wires", Solid State Comm. **27**, 1253 (1978).

2. R. B. Laibowitz, E. I. Alessandrini and G. Deutscher, "Cluster-size distribution in $Al_2O_3$ films near the metal-insulator transition", Phys. Rev. **B25**, 2965 (1982).

3. E. I. Alessandrini, R. B. Laibowitz, C. R. Guarnieri, R. F. Voss and D. S. MacLachlan, "Cluster Formation in Thin Au Films near the Metal-Insulator Transition", Proc. Electron. Microscopy Soc. Amer., **40**, 732 (1982) or R. B. Laibowitz, E. I. Alessandrini, C. R. Guarnieri and R. F. Voss, "Cluster Formation and the Percolation Threshold in Thin Au Films", **A1**, 438 (1983); see also Ref. 6.

4. E. Bassous, R. Feder, E. Spiller and J. Topalian, "High Transmission X-ray Masks for Lithographic Applications", Sol. St. Tech. **19**, 55 (1976).

5. R. B. Laibowitz and A. N. Broers, "Fabrication and Physical Properties of Ultra-Small Structures", in Treatise on Materials Science and Technology, (Academic Press, New York, 1982), **24**, 237 (1982).

6. R. F. Voss, R. B. Laibowitz, and E. I. Alessandrini, "Fractal (Scaling) Clusters in Thin Gold Films near the Percolation Threshold", Phys. Rev. Lett. **49**, 1441 (1982) and R. F. Voss, R. B. Laibowitz, "Percolation and Fractal Properties of Thin Au Films", Proc. Workshop on the Math. and Physics of Disordered Media", U. of Minn., Feb 1983, Springer- Verlag.

7. See the excellent review by D. Stauffer, "Scaling Theory of Percolation Clusters", Phys. Reports **54**, 1, (1979) and references therein.

8. P. L. Leath, "Cluster Size and Boundary Distribution near Percolation Threshold", Phys. Rev. **B14**, 5046 (1976).

9. H. Kunz and B. Souillard, "Essential Singularity in Percolation Problems and Asymptotic Behavior of Cluster Size Distribution", J. Stat. Phys. **19**, 77 (1978) and A. Coniglio and L. Russo, "Cluster Size and Shape in Random and Correlated Percolation", J. Phys. **A 12**, 545 (1979).

10. For a general discussion of fractals see B. B. Mandelbrot, The Fractal Geometry of Nature (Freeman, San Francisco 1982) and references therein. Chapter 13 deals specifically with percolation.

# THEORY OF THE WEAKLY LOCALIZED REGIME

Hidetoshi Fukuyama

Institute for Solid State Physics
The University of Tokyo
Roppongi, Minato-ku, Tokyo 106

## INTRODUCTION

Superconductivity and localization are two characteristic and extreme quantum transport properties of electrons. The former, as is well known, was observed experimentally first by Kamerlingh Onnes in 1911, and its microscopic explanation has been given by Bardeen, Cooper and Schrieffer in 1957 after the accumulation of the experimental data. Since then microscopic investigations on the superconductivity has been explored extensively both experimentally and theoretically and the present status of the understanding is in general very satisfactory though there exist still nowadays interesting and fundamental problems. Such development is mainly due to the validity of the mean field theory for the interacting electrons in essentially clean systems and the comparison between theory and experiment can be pursued to a rather sophisticated degree.

On the other hand the localization caused by the randomness, the Anderson localization, has clearly been realized by Anderson [1] almost at the same time as BCS theory appeared, i.e. in 1958, when there were only few experiments in doped semiconductors to suggest the phenomenon. The physical implications of this monumental paper have been explored by Mott [2] in 1960's. Mott proposed such concepts as the minimum metallic conductivity, the mobility edge and the variable range hopping particular to the Anderson localization, and stimulated the various experimental works in disordered systems. Microscopic theoretical investigations, however, have been very difficult due to the randomness intrinsic to the problem. In such a circumstance computer simulations are very powerful as has been worked out by Thouless and

Table 1.

| SUPERCONDUCTIVITY | | | ANDERSON LOCALIZATION |
|---|---|---|---|
| Kamerlingh Onnes | 1911 | | |
| Meissner | 1933 | | |
| F. & H. London | 1935 | | |
| | | 1955 | Feher et al. |
| BCS | 1957 | | |
| | | 1958 | ANDERSON |
| Josephson | 1962 | | |
| | | 1967 | Mott<br>Thouless |
| | | 1979 | AALR |

coworkers [3,4], which supplemented the physical arguments by Mott. Simulations, however, have intrinsic limitations and the detailed understanding of the localization has been impossible.

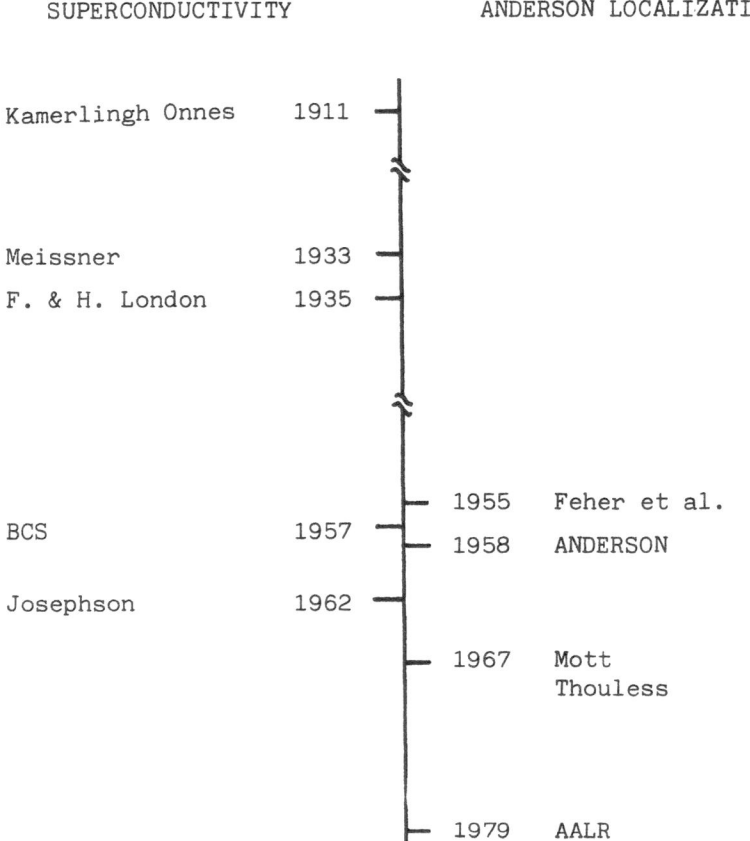

In 1979 Abrahams, Anderson, Licciardello and Ramakrishnan (AALR) [5] made a breakthrough. By elaborating scaling theories which had been discussed before by Thouless [3] and Wegner [6] they not only predicted new results but also gave a birth to a systematic analytical methods to treat the localization problem. Namely in their theory for the non-interacting systems they assumed a one-parameter (i.e. dimensionless conductance) scaling, which has later been confirmed numerically by MacKinnon and Kramer [7], and they showed that the scaling function can be calculated perturbatively from the metallic limit in two- and three- dimensions. Once this is known, the ordinary diagrammatical investigations familiar in the many-body problem and the critical phenomenon can be employed as was first shown by Abrahams and Ramakrishnan [8], and by Gorkov et al. [9]. Hence various scattering mechanism are shown to have their characteristic effects on the localization. Above all the nature of mutual interactions in the disordered systems, which play important roles but can not be treated by the computer simulations, are now being clarified since then [10,11].

In these progress the existence of the weakly localized regime (WLR) is realized which is playing a vital role. The WLR, which can be termed as "dirty metals," is essentially a metallic regime where the weak precursor effects of the complete localization are seen. Theoretically the first order perturbation with respect to randomness is valid in this regime and then the theoretical predictions can be unambiguous once the model is fixed and the detailed comparisons between theory and experiment are possible and actually have been pursued with great success [12]. Such knowledge of WLR will be inevitable for the better and complete understanding of the true localization.

In this note brief survey of the present status of WLR will be given.

We take units of $\hbar=k_B=1$.

## 2. THE WEAKLY LOCALIZED REGIME (WLR)

In the scaling theory by Abrahams et al. [5] the dimensionless conductance g obeys the following equation as a function of the system size L,

$$\frac{\partial \ln g}{\partial \ln L} = \beta(g) \tag{2.1}$$

The initial condition, $g(L_0)=g_0$, together with eq.(2.1), determines uniquely $g(L)$ where $L_0$ is the minimum length for the validity of the scaling relation. The conductivity, $\sigma(L)$, is related to $g(L)$ as $\sigma(L)\propto e^2 g(L)L^{2-d}$. Abrahams et al. [5] indicated that $\beta(g)$ depends on g as shown in Fig.1, which yielded the surprizing results for d=2 and d=3. The metallic regime in the sense of Boltzmann corresponds to the region of $g\to\infty$ in Fig.1, where $\beta(g)$ can be considered as a constant, $\beta=1$ in d=3 and $\beta=0$ in d=2. In this case the conductivity of a macroscopic system, $\sigma_0$, is given as follows.

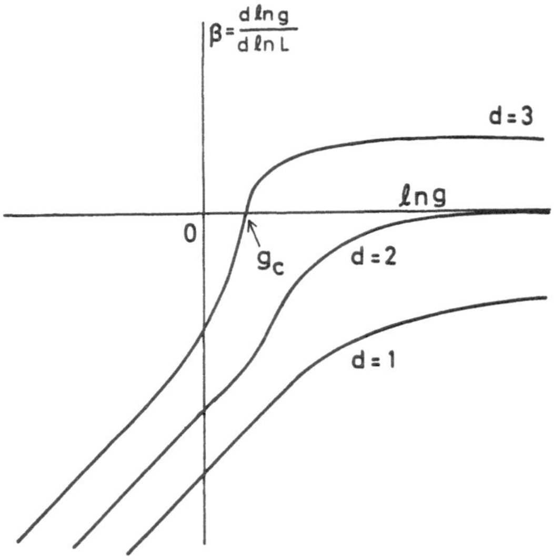


Fig. 1. Scaling function in each d.

$$\sigma_0 = \frac{ne^2}{m} \tau_0 \quad , \tag{2.2a}$$

$$= e^2 \frac{2}{3\pi^2} k_F \varepsilon_F \tau_0 \quad , \quad (d=3) \quad , \tag{2.2b}$$

$$= e^2 \frac{\varepsilon_F \tau_0}{\pi} \quad , \quad (d=2) \quad , \tag{2.2c}$$

where e, m, n, $\tau_0$, $\varepsilon_F$ and $k_F$ are electronic charge, the effective mass, electronic density, the life time, the Fermi energy and the Fermi momentum, respectively.

As is easily understood the first correction to this classical result will thus be represented as that to $\beta(g)$ in the region of large g, i.e.

$$\beta(g) \sim d - 2 - \frac{g_a}{g} \quad , \tag{2.3}$$

where $g_a$ is a constant of order unity. Equation (2.1) together with (2.3) result in

$$g = \frac{L}{L_0}(g_0 - g_a) + g_a \quad , \quad (d=3) \quad , \tag{2.4a}$$

$$g = g_0 - g_a \ln L/L_0 \quad , \quad (d=2) \quad , \tag{2.4b}$$

or, correction to the conductivity, $\sigma'$,

$$\sigma'/\sigma_0 = -3\pi^4 \lambda^2 g_a \quad , \quad (d=3) \quad , \tag{2.5a}$$

$$\sigma'/\sigma_0 = -2\pi^2 \lambda g_a \ln L/L_0 \quad , \quad (d=2). \tag{2.5b}$$

where $\sigma_0$ is defined by eq.(2.2a). In eqs.(2.5a) and (2.5b), $\lambda$ is the small expansion parameter from the metallic limit and is defined by (recovering $\hbar$ to stress the quantum nature of this parameter)

$$\lambda = \hbar / 2\pi\epsilon_F\tau_0 \quad . \tag{2.6}$$

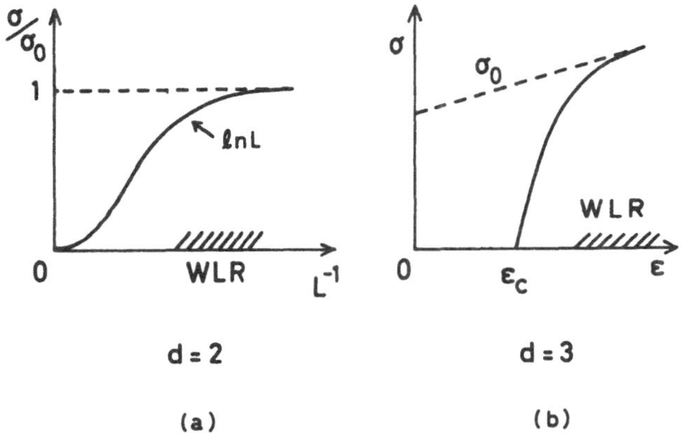


Fig. 2. WLR in d=2(a) and d=3(b).

In Figs.2(a) and (b) we have schematically shown the region where quantum correction, $\sigma'$, is small compared to $\sigma_0$ and then perturbation with respect to $\lambda$ is reliable. The validity of the perturbation theory from the metallic limit implies that the electronic state can be viewed as essentially metallic and then this regime corresponds to that of "dirty metals." In Fig.2(b), $\varepsilon_c$ is the mobility edge, which is given as a solution of $g_0(\varepsilon_c)=g_c$, $g_c$ being defined in Fig.1. These shaded regions are WLR in each dimension.

The WLR is the regime where the interference effect between different plane waves, being treated as independent in the Boltzmann transport equation, is expected to start to play a role. This is actually demonstrated in the following. Since this interference is sensitive to the types of perturbation, $\sigma'$, i.e. $g_a$ in eq.(2.3), also depends distinctly on scattering processes, as we will discuss in detail.

As is expected from the $\lambda$ dependences in eqs.(2.5a) and (2.5b), the situation is drastically different in d=1 where such perturbational approach by $\lambda$ is not valid. We will then discuss d=2 and d=3 exclusively in this lecture.

## 3. NON-INTERACTING SYSTEMS

### 3.1 Interference Effect

In order to see explicitly the interference effect which starts to play a role in WLR we adopt the following model;

$$H = \sum_i \frac{p_i^2}{2m} + \sum_i U(r_i) \quad , \tag{3.1a}$$

$$= \sum_{k,s} \varepsilon(k)\, a^+_{ks} a_{ks} + \sum_{k,q,s} u_q\, a^+_{k+qs} a_{ks} \quad , \tag{3.1b}$$

where $\varepsilon(k)=k^2/2m$ and $U(r)$ is the random potential which is assumed to satisfy

$$\langle U(r)U(r')\rangle \propto \delta(r-r') \quad , \tag{3.2}$$

or in terms of the Fourier transform

$$\langle u_q\, u_{q'}\rangle = 2\pi N(0)\tau_0^{-1}\, \delta_{q+q',0} \quad . \tag{3.3}$$

In eq.(3.3) N(0) is the density of states at the Fermi energy per spin and $\tau_0$ is the life time. Actually the self-energy correction to the one-particle Green function is given by the process shown in Fig.3, where the dotted line is the average over the random potential, eq.(3.3). Fig.3 results in ($\delta>0$)

$$G(k,\varepsilon\pm i\delta) = [\varepsilon - \varepsilon(k) + \varepsilon_F \pm \frac{i}{2\tau_0}]^{-1} \quad , \tag{3.4a}$$

$$\equiv G_{R(A)}(k,\varepsilon) \quad . \tag{3.4b}$$

The sign in front of $i/2\tau_0$ indicates either retarded (incoming) or advanced (outgoing) wave. By use of $G_{R(A)}(k,\varepsilon)$ the classical conductivity, $\sigma_0$, is given diagrammatically by the process shown in Fig.4 as the current-current correlation function for the Kubo formula where solid lines are $G_{R(A)}$.

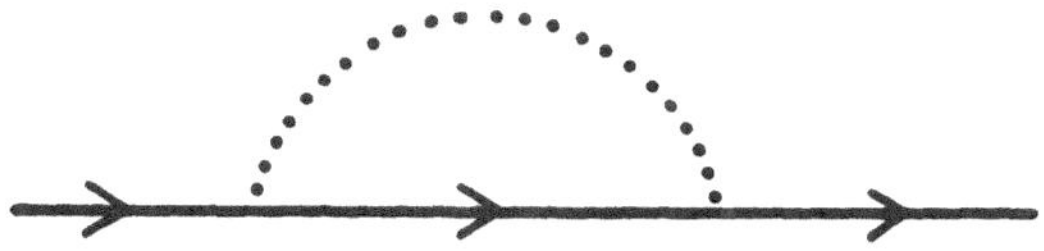

Fig. 3. Self-energy correction due to randomness.

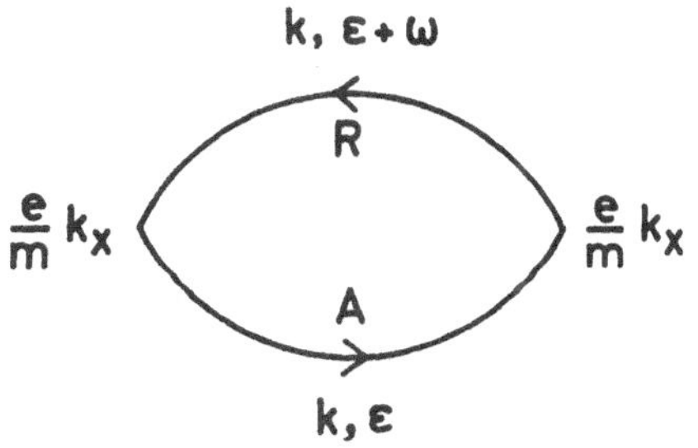


Fig. 4. Process leading to classical conductivity, $\sigma_0$.

Next we are concerned with the interference between the outgoing wave with the momentum, k, and the energy, $\varepsilon$ (i.e. $\varepsilon-i\delta$), and the incoming wave with k' and $\varepsilon'$ ($\varepsilon'+i\delta$). This is conveniently given by the process shown in Fig.5. The function defined by Fig.5 depends on momentum and energy only in the forms of $k+k'\equiv q$ and $\varepsilon'-\varepsilon=\omega$ and is given explicitly as follows;

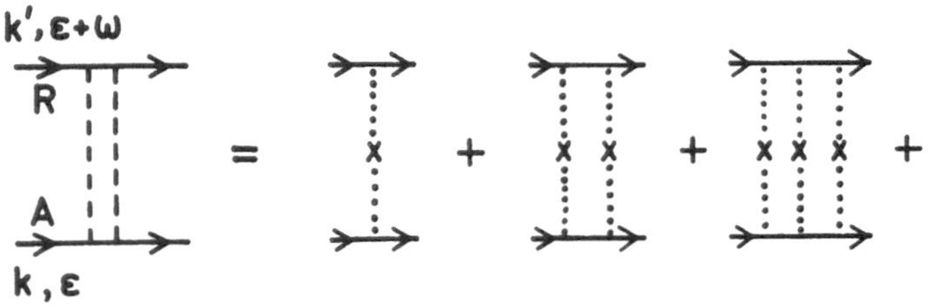


Fig. 5. Cooperon describing the interference effect.

$$D_\varepsilon(q,\omega) = \frac{1}{2\pi N(0)\tau_0}\,[1 + X + X^2 + \cdots] \quad ,$$

$$= \frac{1}{2\pi N(0)\tau_0}\,\frac{1}{1-X} \quad , \tag{3.5}$$

where X is defined as

$$X(q,\omega) = \frac{1}{2\pi N(0)\tau_0}\sum_k G_R(-k+q,\ \varepsilon+\omega)\ G_A(k,\varepsilon) \quad . \tag{3.6}$$

For small q and ω, i.e. if $Dq^2$, $\omega<\tau_0^{-1}$, $D=2\varepsilon_F\tau_0/md$ being the diffusion constant, X is given as

$$X = 2\ N(0)\ n_i\ u^2\ \tau_0\ [1 - Dq^2\tau_0 + i\omega\tau_0] \quad , \tag{3.7a}$$

$$= 1 - Dq^2\tau_0 + i\omega\ \tau_0 \quad . \tag{3.7b}$$

By use of eq.(3.7b), $D_\varepsilon(q,\omega)$ is given as

$$D_\varepsilon(q,\omega) = \frac{1}{2\pi N(0)\tau_0^{\ 2}}\,\frac{1}{Dq^2-i\omega} \quad . \tag{3.8}$$

In some cases the momentum q of interest is not so small as $Dq^2<\tau_0^{-1}$, or $q\ell<1$, $\ell=v_F\tau_0$ being the mean free path, but can be $q\ell>1$ (still $q<<2k_F$, however), in which case eq.(3.8) should be modified as follows;

$$D_\varepsilon(q,\omega) = \frac{1}{2\pi N(0)\tau_0} \frac{\sqrt{(1-i\omega\tau_0)^2+q^2\ell^2}}{\sqrt{(1-i\omega\tau_0)^2+q^2\ell^2}-1} \quad , \quad (d=2) \ , \qquad (3.9a)$$

$$D_\varepsilon(q,\omega) = \frac{1}{2\pi N(0)\tau_0} \frac{1}{q\ell + \frac{i}{2}\ell n\left[\frac{1-i\omega\tau_0+iq\ell}{1-i\omega\tau_0-iq\ell}\right]} \quad , \quad (d=3) \ . \qquad (3.9b)$$

This function $D_\varepsilon(q,\omega)$ is called as particle-particle diffusion propagator, or the Cooperon. The divergence as $\omega\to 0$ and $q=k+k'\to 0$, implies the existence of strong correlation between k and -k states both on the Fermi sphere. This implies that the backward scattering is overwhelming so that localized states start to be formed. It is to be noted that the interference effect is absent in the limit of large energy $D\to\infty$ as it should be but that once D is finite there exist important effects for sufficiently small q, $Dq^2\to 0$, or in the long distance.

The effect of such interference on the conductivity can be expressed as a correction term σ', shown in Fig.6.

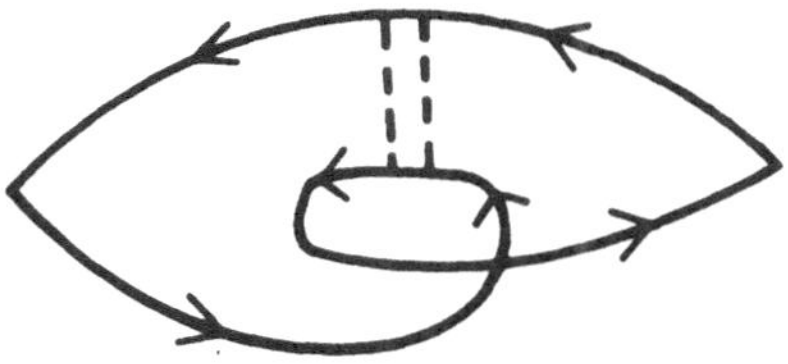

Fig. 6. Process for σ'.

By use of eq.(3.8) and by noting that dominant contributions are coming from the region of $q\to 0$, the correction to conductivity, $\sigma_L'$ is evaluated as follows (Here the suffix L indicates that this is due to so-called localization effects to be contrasted with the interaction effects to be discussed later.);

$$\sigma_L' = -\frac{2e^2}{\pi m}\varepsilon_F\tau \sum_q \frac{1}{Dq^2-i\omega} \quad , \qquad (3.10a)$$

$$= -\frac{e^2}{2\pi^2}\ln\frac{1}{-i\omega\tau_0} \quad , \quad (d=2) \quad , \qquad (3.10b)$$

$$= -\frac{3\sqrt{3}e^2}{2\sqrt{2}\pi^2}\sqrt{\frac{m}{\varepsilon_F\tau_0^2}}\left[1-\frac{\pi}{2}\sqrt{-i\omega\tau_0}\right] , \quad (d=3) \quad . \qquad (3.10c)$$

These results for an infinite system, $L=\infty$, but for a finite frequency, $\omega\neq 0$ can easily be transcribed to those of $\omega=0$ but $L<\infty$ by replacing $-i\omega$ by $DL^{-2}$.

## 3.2 Temperature Dependence

So far we considered the case at T=0. Once $T\neq 0$ the inelastic scattering, which makes the electron hop between states with different energies, results in the reduction of the interference effect. This effect is represented by the finite life time, $\tau_\varepsilon$, in $D_\varepsilon(q,\omega)$ as was indicated by Anderson et al. [13];

$$D_\varepsilon(q,\omega) = \frac{1}{2\pi N(0)\tau_0^2}\frac{1}{Dq^2-i\omega+\frac{1}{\tau_\varepsilon}} \quad , \qquad (3.11)$$

This $\tau_\varepsilon$ is distinctly different from $\tau_0$ and is due solely to the inelastic scattering. Hence it is expected that $\tau_\varepsilon \propto T^{-p}$ with p>0. Substituting eq.(3.11) into eq.(3.10a) we obtain for static conductivity, $\omega=0$,

$$\sigma_L'(T) = -\frac{e^2}{2\pi^2}\ln\tau_\varepsilon/\tau_0 = \frac{e^2}{2\pi^2}p\ln T \quad , \quad (d=2) \quad , \qquad (3.12a)$$

$$\sigma_L'(T) = C\, e^2\sqrt{\tau_0/\tau_\varepsilon} \propto T^{p/2} \quad , \quad (d=3) \quad , \qquad (3.12b)$$

where $C=(3\sqrt{3}/4\sqrt{2})\ (m/\varepsilon_F\tau_0^2)^{1/2}$.

These results imply that electrons can only diffuse a characteristic length, $L_\varepsilon=(D\tau_\varepsilon)^{-1/2}$ (called as a Thouless length [14]), before they are affected by the inelastic scattering and hop to different energy state and then $L_\varepsilon$ plays a role of a system size for an electron in an eigenstate.

The unique $\ln T$ dependence in d=2 has been confirmed in various experiments nowadays [11].

In the case where the effective mass of electrons is anisotropic and if each principal value is given by $m_\alpha$, eqs.(3.12a) and (3.12b) in the α-th direction are to be multiplied by $\sqrt{m_1 m_2/m_\alpha^2}$ and $(m_1 m_2 m_3/m_\alpha^3)^{1/3}$, respectively, where m in C of eq. (3.12b) is to be understood as $m=(m_1 m_2 m_3)^{1/3}$.

As an intermediate case between d=2 and d=3 we can think of metallic films with thickness d such that the electronic state is three dimensional in the sense that the quantized level spacing in the direction perpendicular to the film is much smaller than the Fermi energy, i.e. $\frac{1}{2m}(\frac{2\pi}{d})^2 << \varepsilon_F$, and then $D=2\varepsilon_F\tau/3m$ but $L_\varepsilon >> d$, in which case the classical sheet conductance is given

$$\sigma_\square = \frac{ne^2\tau_0}{m} d = \frac{\varepsilon_F^*\tau}{\pi} e^2 \quad , \tag{3.13}$$

where $\varepsilon_F^*=\varepsilon_F(2k_F d/3\pi)$ but $\sigma_L'$ is still given by eq.(3.12a).

## 3.3 Magnetoresistance (MR)

As has been discussed so far $\sigma_L'$ is essentially determined by the interference effect between two waves k and -k. Then the magnetic field affects such interference effect of orbital motions and results in magnetoresistance (MR), which is anomalous in the sense that it is purely quantum mechanical in contrast to the ordinary MR treated by the Boltzmann transport equation. If the field is weak enough, $\omega_c\tau_0<<1$, $\omega_c=eH/mc$ being the cyclotron frequency, the modification of the electronic states is negligible but the momentum q in the Cooperon is to be replaced by $\hat{q}=q-2eA/C$, A being the vector potential [15], since each momentum, k, an electron carries is shifted by k-eA/c and q is the sum of momenta of two electrons. Quantum mechanical treatment of $D\hat{q}^2$ results in the Landau quantization $D\hat{q}^2=a(N+\frac{1}{2})$ where $a=4D\ell_H^{-2}=4\varepsilon_F\tau_0\omega_c$ and N is integer. Here $\ell_H=\sqrt{c/eH}$ is the Larmor radius. Consequently $\sigma_L'$ in the presence of the field is given by

$$\sigma'(H) = -\frac{2e^2}{\pi m}\varepsilon_F\tau_0 \sum \frac{1}{D(q-2eA/C)^2+Dk_z^2+\frac{1}{\tau_\varepsilon}} \quad , \tag{3.14a}$$

$$= -\frac{4e^3H}{\pi mc}\varepsilon_F\tau_0 \sum_{k_z}\sum_{N=0} \frac{1}{a(N+\frac{1}{2})+Dk_z^2+\frac{1}{\tau_\varepsilon}} \quad , \tag{3.14b}$$

or the change of $\sigma_L'$ due to finite H, $\Delta\sigma_L = \sigma_L'(H) - \sigma_L'(0)$, for each d is given as follows

$$\Delta\sigma_L = \frac{e^2}{2\pi^2 \ell_H^{d-2}} F_d(h) \quad , \tag{3.15}$$

where

$$F_2(h) = \psi\left(\frac{1}{2} + \frac{1}{h}\right) - \psi\left(\frac{1}{2} + \frac{1}{a\tau_0}\right) + \ln \tau_\varepsilon/\tau_0 \quad , \tag{3.16}$$

$$F_3(h) = \sum_{N=0} \left\{2\left(\sqrt{N+1+\frac{1}{h}} - \sqrt{N+\frac{1}{h}}\right) - \frac{1}{\sqrt{N+\frac{1}{2}+\frac{1}{h}}}\right\} \quad . \tag{3.17}$$

Here $h = a\tau_\varepsilon$ and $\psi(z)$ is the di-gamma function. These results for $\Delta\sigma_L$ have been derived by Hikami et al. [16] and by Altshuler et al [17] for d=2 and by Kawabata [18] for d=3.

Since the field works to destroy the interference causing the localization, the present MR is negative.

The explicit field dependences of $F_2(h)$ and $F_3(h)$ are shown in Fig.7(a) and (b). The parameter to characterize these field dependences is $h = 4\varepsilon_F \tau_0 \tau_\varepsilon \omega_c$ which is much larger than that in the classical transport, $\omega_c \tau_0$. Then the present anomalous MR is expected in a relatively weak fields, i.e. $\tau_\varepsilon^{-1} < a < \tau_0^{-1}$, where $F_2(h) \sim \ln h$ and $F_3(h) \to 0.605 \equiv C_0$, or

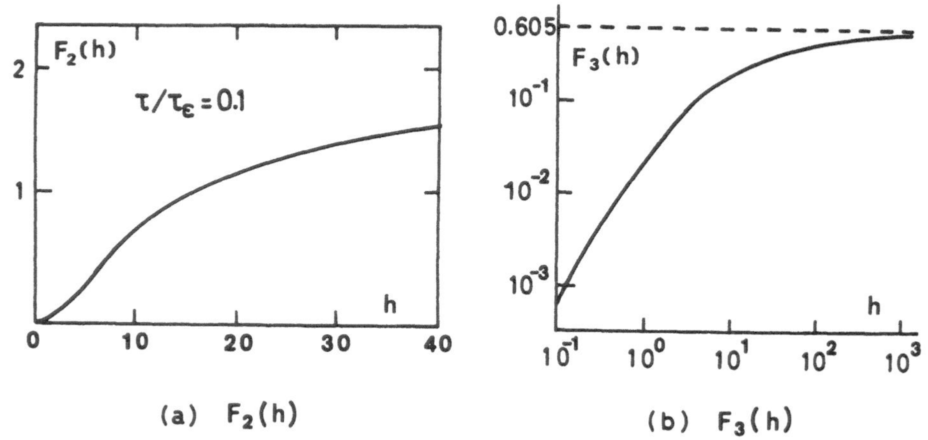


Fig. 7.

$$\Delta\sigma = \frac{e^2}{2\pi^2} \ln H \quad , \tag{3.18}$$

$$\Delta\sigma = \frac{C_0}{2\pi^2 \ell_H} e^2 \propto \sqrt{H} \quad . \tag{3.19}$$

These characteristic field dependences have been observed in various systems, such as Si-MOS, heterostructure, doped semiconductors and metallic films.

The MR so far discussed is due to the orbital effect and the spin Zeeman effect does not play a role for $\sigma_L'$.

### 3.4 Aharonov-Bohm Effect

The fact that $\sigma_L'$ is governed by the interference effect leads to a unique possibility of observing the Aharonov-Bohm [19] effect in dirty metals as has been predicted by Altshuler et al [20]. If a metal film of a cylindrical form is placed in a magnetic field parallel to its axis as in Fig.8, the vector potential in eq.(3.14a) is constant on the film and, as the short perimeter of the cylinder causes a size effect, $\sigma_L'$ oscillates according to the change of magnetic flux in the cylinder. This prediction has already been confirmed experimentally [21].

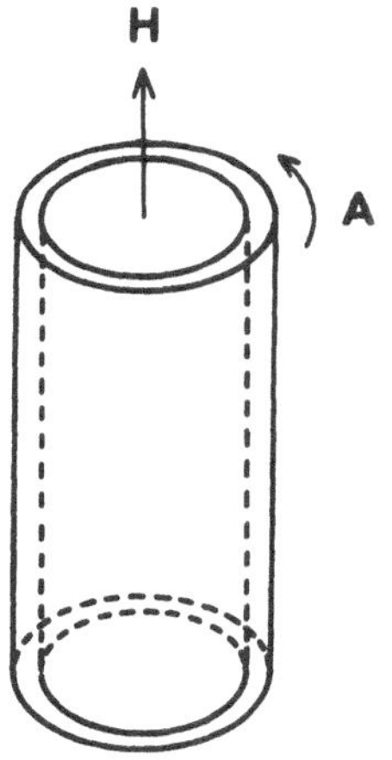


Fig. 8. Geometry to observe Aharonov-Bohm effect in $\sigma'$.

## 4. INTERACTION EFFECTS

The mutual interactions between electrons play important roles [10]. In this section effects of Coulomb interaction in WLR will be discussed first in its lowest order and then including the infinite order.

### 4.1 Lowest Order

In the clean system the lowest order correction to the self-energy is given by the Hartree and the Fock processes shown in Fig.9, where wavy lines are Coulomb interactions. In the first process the momentum transfer, q, via interaction is q=0 and the contribution is cancelled by the uniform background, while the second process with the proper screening effect included results in essentially constant shift in the chemical potential without any physical importance. In the presence of randomness, or once $\lambda$ is finite, these self-energy corrections are modified as in Fig.10 where the double broken lines are Cooperons defined in Fig.5 and the broken lines are the process shown in Fig.11, i.e. the particle-hole propagators is called a diffuson. We call the processes in Fig.10 as the $g_i$-process ($i=1\sim4$), respectively. (The way the $g_2$-process is represented is different in Fig.10(b) from that in e.g. [12].)

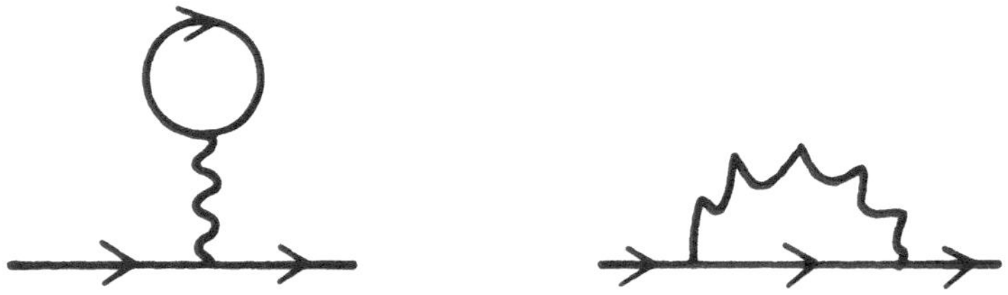

Fig. 9. Hartree-Fock self-energy correction in clean systems.

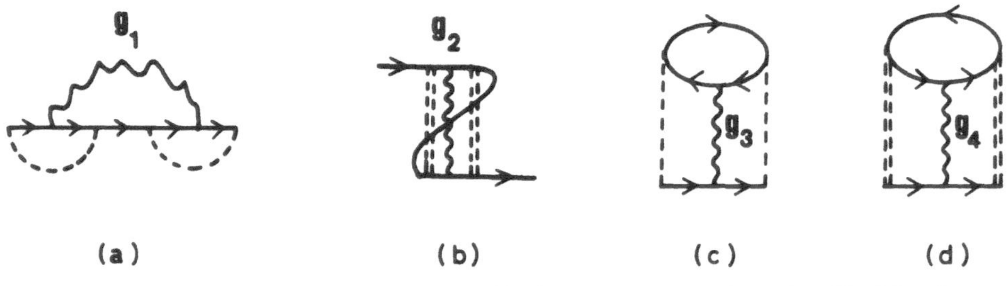


Fig. 10. Corrections to Hartree-Fock processes due to finite $\lambda$.

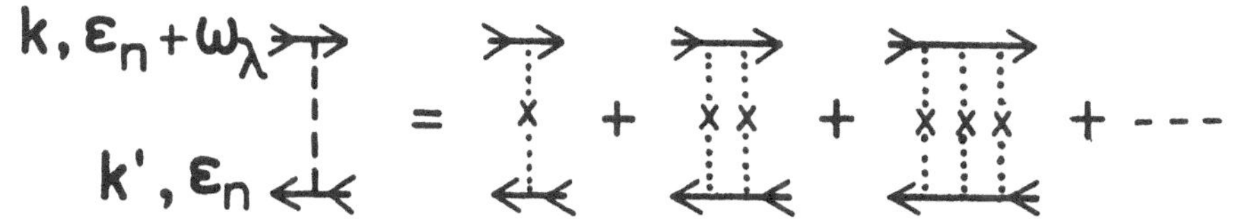


Fig. 11. The diffuson.

The diffuson, $\Lambda(q, \omega)$, as defined in Fig.11 is a function of k-k' (not k+k' as in Cooperon) and is given as follows for small $q=k-k'$ and $\omega$

$$\Lambda(q,\omega_\ell) = \frac{1}{2\pi N(0)\tau_0^2[Dq^2 + |\omega_\ell|]} , \tag{4.1}$$

where D is the same diffusion constant as in the Cooperon. The diffuson $\Lambda(q,\omega_\ell)$ is not affected by the inelastic time, $\tau_\varepsilon$, even at $T\neq 0$, since it is the density-density correlation function which always has singularity as $q\to 0$ and $\omega_\ell\to 0$ as far as the scatterings conserve the total number of electrons.

In order to see the typical effects of interactions we examine the $g_1$-process. The self-energy correction $\Sigma_1(k,i\varepsilon_n)$, from this process is given as follows;

$$\Sigma_1(k,i\varepsilon_n) = -T \sum_{\omega_\ell}{}' \sum_q (2\pi N(0)\tau_0^2\Lambda(q,\omega_\ell))^2 v(q,\omega_\ell)\, \mathcal{G}(k+q,i\varepsilon_n+i\omega_\ell), \tag{4.2}$$

where only singular contributions are retained and then the summation over $\omega_\ell$ is restricted to the region of $\varepsilon_n(\varepsilon_n+\omega_\ell)<0$. For $|k|\sim k_F$ and $\varepsilon_n>0$ ($\varepsilon_n\to 0$), $\mathcal{G}(k+q, i\varepsilon_n+i\omega_\ell)$ can be approximated by $2i\tau_0$ since the dominant contributions in the integrations are due to small q and $\omega_\ell$ and then

$$\Sigma_1(k,i\varepsilon_n) = -2i\tau_0^{-1}T \sum_{\omega_\ell<-\varepsilon_n} \sum_q v(q,\omega_\ell) \frac{1}{(Dq^2+|\omega_\ell|)^2} . \tag{4.3}$$

From this equation it will be obvious that $\Sigma_1$ has singular temperature dependence ; if q- and $\omega_\ell$- dependences of $v(q,\omega_\ell)$ are ignored, $\Sigma_1 \propto \ell n T$ in d=2 and $\sqrt{T}$ in d=3, as $\varepsilon_n\to 0$.

If q- and $\omega_\ell$-dependences in each region of $g_i$-process are ignored, the life time of the plane wave and the density of states are modified as follows;

$$\frac{1}{\tau} = \frac{1}{\tau_0}\,[1 + K_d] \quad , \tag{4.4}$$

$$\frac{\delta N}{N(0)} = -\,2K_d \quad , \tag{4.5}$$

where $K_d$ for each d is given by

$$K_2 = \lambda g\ 2\pi T \sum_{\omega_\ell < -\varepsilon_n} \frac{1}{|\omega_\ell|} \equiv \lambda g \phi_2(T) \quad , \tag{4.6}$$

$$K_3 = C_3\lambda^2 g\ 2\pi T\,\sqrt{\tau_0} \sum_{\omega_\ell < -\varepsilon_n} \frac{1}{\sqrt{\omega_\ell}} \equiv \lambda^2 g \phi_3(T) \quad . \tag{4.7}$$

Here $g=g_1+g_2-2(g_3+g_4)$, $g_i=N(0)v(q,\omega_\ell)$ being characteristic coupling constant in each process and $C_3=3\sqrt{3}\pi^2/4$. In the pionering work by Altshuler and Aronov [22], T-dependence of $\delta N$ in d=3 has been derived by considering the $g_1$-process. Similarly there exists a correction to the conductivity, $\sigma_I'$; [23,24]

$$\sigma_I' \,/\, \sigma_0 = -\,K_d \quad . \tag{4.8}$$

Consequently various physical quantities have singular dependences of $\ell nT$(d=2) and $\sqrt{T}$ (d=3), given by eqs.(4.6) and (4.7).

In eqs.(4.4), (4.6) and (4.8) the q- and $\omega_\ell$-dependences of the interaction are ignored and interactions are parametrized by $g_i$. The actual values of $g_i$ depend on the types of interactions. Among $g_i$'s the $g_1$-process has a particular feature that it is concerned with the small momentum, q, and energy, $\omega_\ell$, transfer via interactions, i.e. $q \lesssim q_0 \equiv (D\tau_0)^{-1/2} = \sqrt{d/2}\ (k_F/\varepsilon_F\tau_0)$ and $|\omega_\ell| < 1/\tau_0$. By this reason $g_1$ is sensitive to the choice of bare interactions. On the other hand the momentum transfer, $q_L$, is large, $q_L \lesssim 2k_F$, in other processes though the energy transfer is small as well.

In the case of dynamically screened Coulomb interaction, $v(q,\omega_\ell)$ in RPA is given by

$$v(q,\omega_\ell) = \frac{v_B(q)}{1 + v_B(q)\,\chi^0(q,\omega_\ell)} \quad , \tag{4.9}$$

where $v_B(q)$ is the bare Coulomb interaction, $v_B(q)=2\pi e^2/\varepsilon_0 q$ (d=2) or $4\pi e^2/\varepsilon_0 q^2$ (d=3) and $\chi^0(q,\omega_\ell)$ is the polarization function in the absence of interaction but in the presence of impurity scattering. In the limit of small q and $\omega_\ell$ this $\chi^0(q,\omega_\ell)$ is given by

$$\chi^0(q,\omega_\ell) = 2N(0)\,\frac{Dq^2}{Dq^2 + |\omega_\ell|} \quad , \tag{4.10}$$

By use of eq.(4.10) the screened Coulomb interaction in the region of small q is given as follows for both d=2 and 3

$$2N(0)v(q,\omega_\ell) = \frac{Dq^2+|\omega_\ell|}{Dq^2} \quad . \tag{4.11}$$

If eq.(4.11) is employed, $g_1$ in the conductivity turns out to be $g_1=1$ in d=2 and $g_1=2/3$ in d=3.

As regards $g_2$, $g_3$ and $g_4$ on the other hand they will conveniently be characterized by the constant, F, defined by

$$g_2 = g_3 = g_4 = N(0)\,\langle v(k-k',\,0)\rangle \equiv F/2 \quad . \tag{4.12}$$

where the dynamics of $v(q,\omega_\ell)$ is ignored, i.e. $\omega_\ell=0$, and $\langle \quad \rangle$ is meant to take average over the Fermi surface, $|k|=|k'|=k_F$. Here the effect of impurity scattering on the screening can be ignored since for $q \gg q_0$ the existence of finite $\tau_0^{-1}$ does not result in any important consequence.

## 4.2 Higher Order Effect

So far discussions are for the lowest order both in $\lambda$ and the interaction. These parameters are independent; $\lambda$ can be varied externally whereas the mutual interaction is intrinsic to each system. There now exist various investigations on higher order effects in $\lambda$ and/or interaction [25-31]. In the following we discuss the higher order interaction effect in the WLR, i.e. in the linear order of $\lambda$ [31,32].

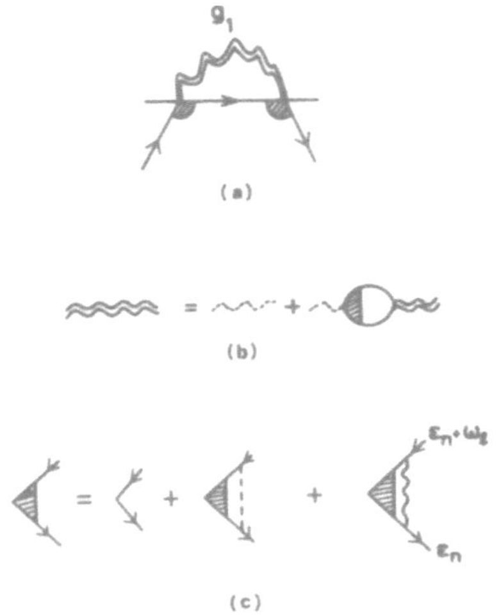

Fig. 12. Higher order corrections in the $g_1$-process.

We first note that the contributions linear in $\lambda$ result from processes with as many as either diffusons or Cooperons carrying the same momentum. Consequently the $g_1$-process is determined by the processes shown in Fig.12, where the vertices to the Coulomb interaction have exchange corrections. The net result of the higher order corrections corresponds to modify eq.(4.11) as

$$2N(0)v(q,\omega_\ell) = \frac{D'q^2 + |\omega_\ell|}{Dq^2} \quad , \tag{4.13}$$

where $D'=D(1-F/2)$. The correction to the conductivity is now given as

$$\sigma_1' / \sigma_0 = - g_1(F)\ \phi_d(T) \quad , \tag{4.14}$$

where $\phi_d(T)$ is defined either by eq.(4.6) or by (4.7) and $g_1(F)$ is

$$g_1(F) = \begin{cases} \dfrac{2}{F}\,\ell n\,\dfrac{2}{2-F} \quad , & (d=2) \quad , \qquad (4.15a) \\ \dfrac{8}{3F}\left(\dfrac{1}{\sqrt{1-\frac{F}{2}}} - 1\right) \quad , & (d=3) \quad . \qquad (4.15b) \end{cases}$$

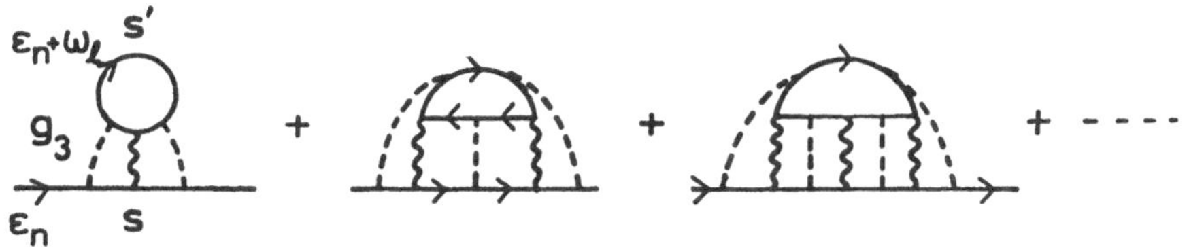


Fig. 13. Higher order corrections in the $g_3$-process.

The contributions from the $g_3$-process are given by those in Fig.13, where s is the spin index. These are known as the paramagnon type of processes [33-35]. In this case the effective interaction, $g_3(q,\omega_\ell)$, is modified from F/2 as [35]

$$g_3(q,\omega_\ell) = \frac{F}{2} \frac{Dq^2 + |\omega_\ell|}{D'q^2 + |\omega_\ell|} \quad . \tag{4.16}$$

The correction to the conductivity is

$$\sigma_3' / \sigma_0 = 2g_3(F)\ \phi_d(T) \quad , \tag{4.17}$$

where $\phi_d(T)$ is the same as in eq.(4.14) and $g_3(F)$ is given by

$$g_3(F) = \begin{cases} 2(\frac{2}{F}\ \ln \frac{2}{2-F} - 1) = 2g_1(F) - 2 \quad , \quad (d=2) \quad , & (4.18a) \\ \frac{8}{3F}\ (\frac{2}{\sqrt{1-\frac{F}{2}}} - 2 - \frac{F}{2}) = 2g_1(F) - \frac{4}{3} \ , \ (d=3). & (4.18b) \end{cases}$$

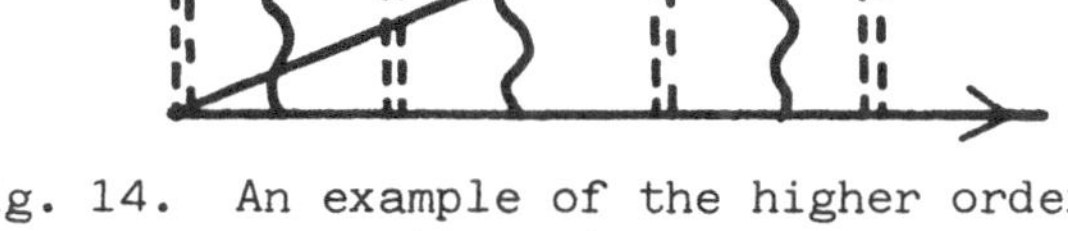

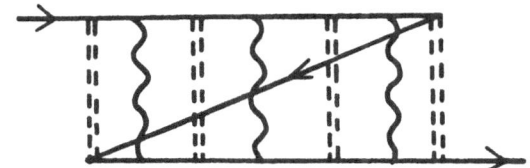

Fig. 14. An example of the higher order corrections in the $g_2$-process.

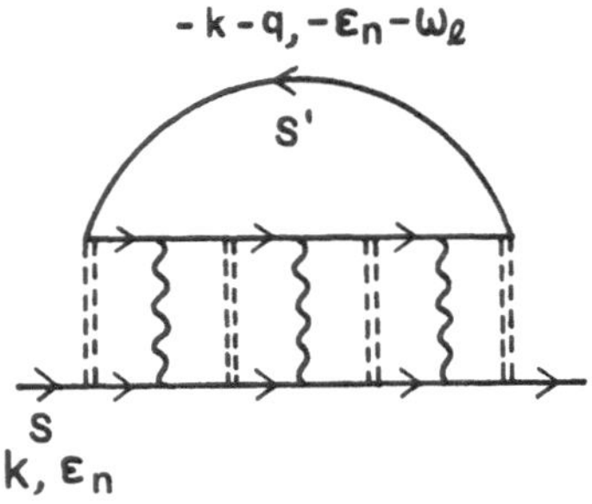


Fig. 15. An example of the higher order corrections in the $g_4$-process.

As regards the $g_2$- and $g_4$- processes coupled with Cooperon the situation is different. The higher order contributions of the $g_2$-process consist of processes of the type shown in Fig.14, which are totally cancelled by those of the $g_4$-process shown in Fig.15 with parallel spin s'=s. The net results are those of Fig.15 between antiparallel spins, s'=-s, which are given as follows for $Dq^2+|\omega_\ell|>>4\pi T$

$$g_2-2g_4 = -\frac{F}{2}\frac{1}{1+\frac{F}{2}\ln\frac{2\varepsilon_F}{Dq^2+|\omega_\ell|+\frac{1}{\tau_\varepsilon}}}, \tag{4.19a}$$

$$\equiv -g_4(q,\omega_\ell) \quad . \tag{4.19b}$$

Here it is to be noted that q and $\omega_\ell$ in eq.(4.19a) are defined as in Fig.15, which are not the same as the momentum and the energy the Cooperon carries, i.e. the latters are 2k+q and $2\varepsilon_n+\omega_\ell$, respectively. These differences result in the complicated entanglement of energy variables which are carried by $g_4(q,\omega_\ell)$, eq. (4.19b), and by the Cooperon corrections. If the dependences of $g_4(q,\omega_\ell)$ on q and $\omega_\ell$ are totally ignored as the most naive approximation and if $g_4(q,\omega_\ell)$ is characteized by its smallest value $g_4(0,0)$ given by

$$g_4(0,0) = \frac{F}{2}\frac{1}{1+\frac{F}{2}\ln\frac{1.13\varepsilon_F}{T}}, \tag{4.20}$$

the correction to the conductivity is

$$(\sigma_2' + \sigma_4') / \sigma_0 = g_4(0,0)\ \phi_d(T) \quad . \tag{4.21}$$

In this approximation the logarithmic divergences of $\phi_d(T)$ and $g_4(0,0)$ are cancelled in d=2 yielding a finite value of $(\sigma_2'+\sigma_4')$ as $T\to 0$. However if the above-mentioned entanglement of $g_4(q,\omega_\ell)$ and the Cooperon are properly taken into account, the result is [32]

$$(\sigma_2' + \sigma_4') / \sigma_0 = \lambda\ \ln\left[\frac{1 + \frac{F}{2}\ \ln \frac{2\varepsilon_F}{T_M}}{1 + \frac{F}{2}\ \ln 2\varepsilon_F\tau_0}\right] , \tag{4.22}$$

where $T_M$=Max $(2\pi T,\ 1/\tau_\varepsilon)$. There remains a divergence though weak.

In the present case of the Coulomb interaction, which is of course repulsive, the effective values of $g_1$ and $g_3$ are seen to be enhanced whereas those of $g_2$ and $g_4$ are reduced [25,26]. If the mutual interaction is dominantly attractive, the situation is reversed. The interaction constant, F, so far used is to be understood as the effective interaction between two electrons in clean systems. Once this interaction gets strong, it is no longer equal to the bare interaction but has to be determined by the t-matrix approximation given in Fig.16 as is known in the theories of itinerant magnetism [36] and correlations in the electron gas [37]. This t-matrix approximation results in the effective interaction, F*, given by

$$F^* = \frac{F}{1 + F/2}$$

and this F* is to be used in eqs.(4.15) and (4.18) in place of F, i.e.

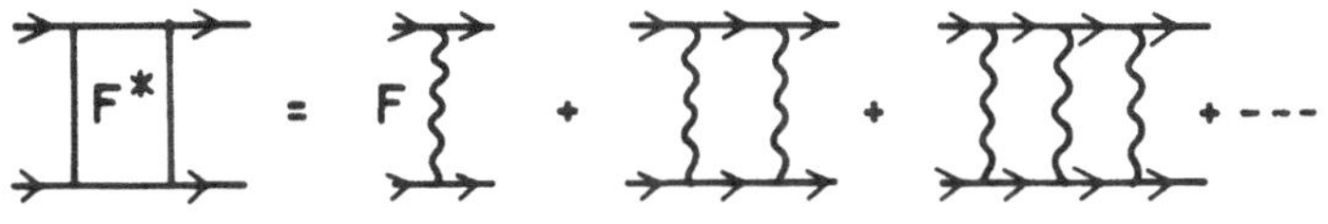


Fig. 16. The t-matrix approximation to the effective interaction.

$$g_1(F) = \begin{cases} \frac{2+F}{F} \ln \frac{2+F}{2} \quad , \quad (d=2) \quad , & (4.23a) \\ 4 \frac{2+F}{3F} \left( \sqrt{\frac{2+F}{2}} - 1 \right) \quad , \quad (d=3) \quad , & (4.23b) \end{cases}$$

and the relationship between $g_1(F)$ and $g_3(F)$ is the same as in eqs.(4.18a) and (4.18b).

The present F-dependence of $g_1(F)-2g_3(F)$ for d=2 is the same as the corresponding quantity calculated by Finkelstein [38] by a different method with use of a replica.

## 4.3 Magnetoresistance

The $\sigma_I'$ is affected by both orbital and spin Zeeman effect [26,39]. The existence of the former [12] is clear since the $g_2$- , $g_4$- processe are determined by Cooperons as in $\sigma_L'$. The latter effect is due to the fact that both Cooperons and diffusons between antiparallel electrons are affected by the spin Zeeman splitting [40]; i.e. $Dq^2$ in eqs.(3.8) and (4.1) is modified as $Dq^2 \pm ig\mu_B H$, where g and $\mu_B$ are the effective g-factor and the Bohr magneton, respectively. Such spin effects are consequently associated with the $g_3$- and $g_4$- process. These classifications of the MR are listed in Table 2 [41].

The field dependence of MR due to spin effect is characterized by $g\mu_B H/2\pi T$ [26,39], which is in contrast to $a\tau_\varepsilon = 4\varepsilon_F \tau_0 \omega_c \tau_\varepsilon$ in the case of the orbital effect.

Table 2. Classification of interaction-induced magnetoresistance

| | | |
|---|---|---|
| $g_1$ | | none |
| $g_2$ | | orbital |
| $g_3$ | $g_3^{\uparrow\uparrow}$ | none |
| | $g_3^{\uparrow\downarrow}$ | spin |
| $g_4$ | $g_4^{\uparrow\uparrow}$ | orbital |
| | $g_4^{\uparrow\downarrow}$ | orbital+spin |

If the mutual interaction is dominantly repulsive, the absolute magnitude of MR from the orbital effect is small since effective values of $g_2$ and $g_4$ are reduced. In this case the temperature dependence in the absence of the field is determined by $g_1(F)-2g_3(F)\equiv G_d(F)$, and those in the region of the field $g\mu_B H/2\pi T>1$ are given by $g_1(F)-g_3(F)\equiv H_d(F)$. The F-dependences of $G_d(F)$ and $H_d(F)$ for d=2 and 3 are shown in Fig.17. It is seen from this figure that the signs of $G_d(F)$ and $H_d(F)$ can be different for wide regions of parameter, F. Such change of the sign of the temperature dependences has actually been observed by Ootuka et al. [42] in the Sb-doped Ge in the uniaxial stress, though the experimental value of $g_3$ appears to be appreciably larger than the theoretical value.

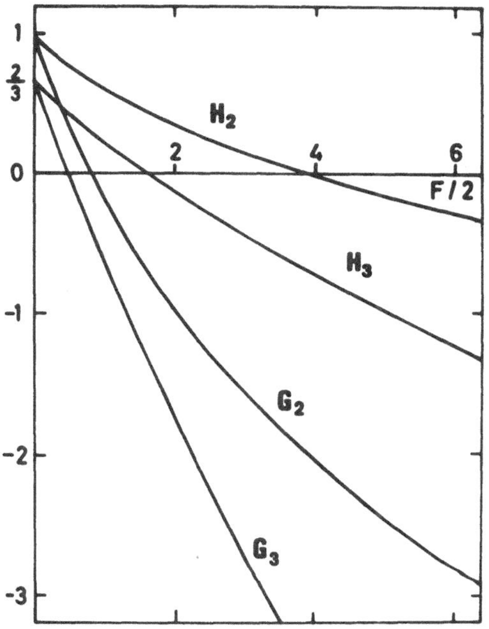


Fig. 17. The F-dependence of $g_1-2g_3=G_d$ and $g_1-g_3=H_d$.

## 4.4 Inelastic Scattering Time

So far we treated the inelastic scattering time, $\tau_\varepsilon$, phenomenologically. Usually this $\tau_\varepsilon$ is identified with the energy relaxation time, $\tau_e$, and $\tau_e^{-1}\propto T^2/\varepsilon_F$ for the Coulomb interaction in the clean system. Effects of the smearing of the Fermi surface by the randomness on $\tau_e$ have been investigated some time ago by

Schmid [43] who found $\tau_e^{-1} \propto T^{d/2}$. The case of d=2 has recently been examined in detail by Abrahams et al. [44] and the result is

$$\tau_e^{-1} = \lambda \pi T \ln T_1 / T \quad , \tag{4.24a}$$

$$T_1 = D\kappa_2^2 / \pi^2 \lambda^2 \quad , \tag{4.24b}$$

where $\kappa_2 = 2me^2/\varepsilon_0$.

It is to be noted, however, that $\tau_\varepsilon$ determining the conductivity is defined as the life time of the Cooperon and that it will not be necessarily equal to $\tau_e$. From this viewpoint it is worthy to evaluate $\tau_\varepsilon$ by its definition, which has been diagrammaticaly pursued quite recently [45]. The result of detailed self-consistent calculation has confirmed eqs.(4.24a and b), $\tau_\varepsilon = \tau_e$.

The values of $\tau_\varepsilon$ deduced from the fitting of the field dependence of MR in various experiments are almost two orders of magnitude longer than the theoretical value. This discrepancy is an unresolved problem.

## 5. COMPETITION BETWEEN LOCALIZATION AND SUPERCONDUCTIVITY

In the light of new development in the understanding of the effects of randomness, the theories of the dirty superconductors have to be elaborated.

It has been well known by the investigation by Anderson [46] and by Gorkov [15] that the critical temperature, $T_c$, is not affected by static and non-magnetic disorder. In their theory of dirty superconductors the combined effects by randomness and dynamical interactions (i.e. effects in the order of $\lambda g_i$) has not been taken into account. The quantum corrections to Gorkov theory, however, indicate the existence of effect of randomness on $T_c$. To examine this we consider the following Hamiltonian.

$$H = \tilde{H} + g_{BCS} \int dr \, [\Delta^+ \psi_\uparrow(r)\psi_\downarrow(r) + \Delta\psi_\downarrow^+(r)\psi_\uparrow^+(r)] \quad , \tag{5.1}$$

where $\psi(r)$ is the electron field operator, and $g_{BCS}$ is the BCS interaction. Here $\Delta$ is the order parameter given by

$$\Delta = \int dr \, \mathrm{Tr}\, e^{-\beta H} \psi_\uparrow(r)\psi_\downarrow(r) / \mathrm{Tr}\, e^{-\beta H} \tag{5.2}$$

The Hamiltonian, $\tilde{H}$, in eq.(5.1) is

$$\tilde{H} = H_0 + H_{int} \quad , \tag{5.3}$$

where $H_0$ is given by eq.(3.1) and $H_{int}$ represents interactions besides the BCS interaction. By use of the eigenstates, $\phi_\alpha(r)$, of $H_0$, i.e. $H_0\phi_\alpha(r)=\epsilon_\alpha\phi_\alpha(r)$ and $\psi_s(r)=\sum_\alpha C_{\alpha s}\phi_\alpha(r)$, $T_c$ is given as a solution of the following equation,

$$|g_{BCS}|^{-1} = \sum_{\alpha,\beta} \int_0^\beta du \langle T\, C^+_{\alpha\downarrow}(u) C^+_{\alpha^*\uparrow}(u)\; C_{\beta^*\uparrow}(0) C_{\beta\downarrow}(0)\rangle \ , \tag{5.4}$$

where $C^+_{\alpha s}(u)=e^{uH}C^+_{\alpha s}\; e^{-uH}$ and $\alpha^*$ is the time-reversed state of $\alpha$ and then $\epsilon_{\alpha^*}=\epsilon_\alpha$.

In the absence of $H_{int}$, eq.(5.4) results in

$$\begin{aligned} g_{BCS}^{-1} &= T \sum_n \sum_\alpha \mathcal{G}_\alpha(i\epsilon_n)\, \mathcal{G}_{\alpha^*}(-i\epsilon_n) \\ &= \pi\ N(0)\ T \sum_n \frac{1}{|\epsilon_n|} \quad . \end{aligned} \tag{5.5}$$

As is seen, equation for $T_c$ is identical to that of clean system if N(0) is not changed by randomness. This is the Anderson's theorem [46].

As is easily expected from the discussions in the preceeding section, there exist corrections to eq.(5.5) in the presence of $H_{int}$. Consequently $T_c$ is affected in the order of $\lambda g_i$,

The results for $T_c$ obtained by Maekawa et al. [47] and by Takagi et al. [48] for d=2 are as follows,

$$\begin{aligned} \ln \frac{T_c}{T_{c0}} &= -\frac{\lambda}{2} (g_1 - 3g') \left(\ln \frac{1}{T_c\tau_0}\right)^2 \\ &\quad - \frac{\lambda}{3} (g_1 + g') \left(\ln \frac{1}{T_c\tau_0}\right)^3 \quad , \end{aligned} \tag{5.6}$$

where $g_2=g_3=g_4=g'$ is assumed and $T_{c0}$ is the critical temperature in the absence of interaction effects.

The numerical results of $T_c$ given by eq.(5.6) are shown in Fig.18 for several choices of $T_{c0}\tau_0$ and g'=0. As seen the reduction of $T_c$ can be appreciable. Experimentally the reduction of $T_c$ in proportion to the sheet resistance ($\propto\lambda$) has been observed in various metallic films.

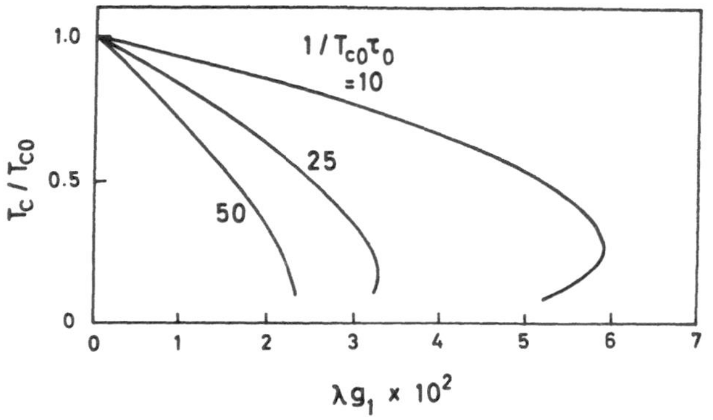


Fig. 18. The reduction of $T_c$ due to $\lambda g_1$.

As regards the pair breaking parameter, $\gamma$, which plays a crucial role in the Maki-Thompson process of fluctuations and for which there already exist some possible explanations [50-52] one can expect its close relationship to the inelastic scattering time, $\tau_\varepsilon$, discussed in Sec.5. As is analyzed in detail by Ebisawa et al. [53], $\gamma$ in the Maki-Thompson processes is seen to be identical to $\tau_\varepsilon^{-1}$. Then the result of Sec.4.4 indicates the existence of $\gamma$ essentially in proportion to $\lambda$, or the sheet resistance, which is in accordance with the observation in Al films by Crow et al. [54] and by Kajimura et al. [55].

Although intresting effects of localization on the transport properties of superconductors above $T_c$ are expected, and have actually been observed as will be discussed in this Summer Institute there does not exist to the knowledge of the author any systematic investigations, which are comparable with those of the repulsive force discussed in Sec.4.

REFERENCES


1. P.W. Anderson : Phys. Rev. **102** (1958) 1008; Rev. Mod. Phys. **50** (1978) 191.
2. N.F. Mott : Rev. Mod. Phys. **50** (1978) 203.
3. D.J. Thouless : Phys. Rept. **13C** (1974) 93.
4. D.C. Licciardello and D.J. Thouless : J. Phys. C **8** (1975) 4157.
5. E. Abrahams, P.W. Anderson, D.C. Licciardello and T.V. Ramakrishnan : Phys. Rev. Lett. **42** (1979) 673.
6. F.J. Wegner : Z. Phys. **25** (1976) 327.
7. A. MacKinnon and B. Kramer : Phys. Rev. Lett. **47** (1981) 1546 ; A. MacKinnon : Anderson Localization ed. by Y. Nagaoka and H. Fukuyama (Springer Verlag, 1982) p.54.
8. E. Abrahams and T.V. Ramakrishnan : J. Non-Crystalline Solids : **35** (1980) 15.
9. L.P. Gorkov, A.I. Larkin and D.E. Khmelnitzkii : JETP Lett. **30** (1979) 248.
10. For review P.A. Lee : Anderson Localization ed. by Y. Nagaoka and H. Fukuyama (Springer Verlag, 1982) p.62 ; H. Fukuyama, ibid. p.89.
11. P.W. Anderson, Physica **117 & 118B** (1983) 30.
12. For review, H. Fukuyama : Surface Science **113** (1982) 489 Physica **117 & 118B** (1983) 673.
13. P.W. Anderson, E. Abrahams and T.V. Ramakrishnan : Phys. Rev. Lett. **43** (1979) 718.
14. D.J. Thouless : Phys. Rev. Lett. **39** (1977) 1167.
15. L.P. Gorkov : JETP **36** (1959) 1364.
16. S. Hikami, A.I. Larkin and Y. Nagaoka : Prog. Theor. Phys. **63** (1980) 707.
17. B.L. Altshuler, D.E. Khmelnitzkii, A.I. Larkin and P.A. Lee : Phys. Rev. **B22** (1980) 5142.
18. A. Kawabata : Solid State Commun. **34** (1980) 432 ; J. Phys. Soc. Jpn. **49** (1980) 628.
19. Y. Aharonov and D. Bohm : Phys. Rev. **115** (1959) 485.
20. B.L. Altshuler, A.G. Aronov and B.Z. Spivak : JETP Lett. **33** (1981) 101.
21. D. Yu. Sharvin and Yu. V. Sharvin : JETP Lett. **34** (1981) 285.
22. B.L. Altshuler and A.G. Aronov : Solid State Commun. **30** (1979) 115.
23. B.L. Altshuler, A.G. Aronov and P.A. Lee : Phys. Rev. Lett. **44** (1980) 1288.
24. H. Fukuyama : J. Phys. Soc. Jpn. **48** (1980) 2169.
25. B.L. Altshuler, A.G. Aronov, D.E. Khmelnitzkii and A.I. Larkin : Soviet Physics-JETP **54** (1981) 4111.
26. P.A. Lee and T.V. Ramakrishnan : Phys. Rev. **B26** (1982) 4009.
27. W.L. McMillan : Phys. Rev. **B24** (1981) 2739.
28. R. Oppermann : Solod State Commun. **44** (1982) 1297.

29. C. Castellani, D. Di Castro, G. Forgacs and E. Tabet : J. Phys. C **16** (1983) 139 ; and preprint.
30. G.S. Grest and P.A. Lee : Phys. Rev. Lett. **50** (1983) 693.
31. H. Fukuyama, Y. Isawa and H. Yasuhara : J. Phys. Soc. Jpn. **52** (1983) 16.
32. Y. Isawa and H. Fukuyama : to be submitted to J. Phys. Soc. Jpn.
33. N.F. Berk and J.R. Schrieffer : Phys. Rev. Lett. **17** (1966) 433.
34. S. Doniach and S. Engelsberg : Phys. Rev. Lett. **17** (1966) 750.
35. P. Fulde and A. Luther : Phys. Rev. **170** (1968) 570.
36. J. Kanamori : Prog. Theor. Phys. **30** (1963) 275.
37. H. Yasuhara : Solid State Commun. **11** (1972) 1481 ; J. Phys. Soc. Jpn. **36** (1974) 361.
38. A.M. Finkelstein : preprint.
39. A. Kawabata : J. Phys. Soc. Jpn. **50** (1981) 2461.
40. S. Maekawa and H. Fukuyama : J. Phys. Soc. Jpn **50** (1981) 2516.
41. Y. Isawa, K. Hoshino and H. Fukuyama : J. Phys. Soc. Jpn. **51** (1982) 3262.
42. Y. Ootuka, S. Katsumoto, S. Kobayashi and W. Sasaki : in preparation.
43. A. Schmid : Z. Physik **271** (1974) 251.
44. E. Abrahams, P.W. Anderson, P.A. Lee and T.V. Ramakrishnan : Phys. Rev. **B24** (1981) 6783.
45. H. Fukuyama and E. Abrahams : Phys. Rev. **B27** (1983).
46. P.W. Anderson : J. Phys. Chem. Solids **11** (1959) 26.
47. S. Maekawa and H. Fukuyama : J. Phys. Soc. Jpn. **51** (1982) 1380.
48. H. Takagi and Y. Kuroda : Solid State Commun. **41** (1982) 643.
49. M. Strongin, R.S. Thompson, O.F. Kammerer and J. E. Crow : Phys. Rev. **B1** (1970) 1078.
50. B.R. Patton : Phys. Rev. Lett. **27** (1971) 1237.
51. J. Keller and V. Korenman : Phys. Rev. **B5** (1972) 4367.
52. B. Keck and A. Schmid : Solid State Commun. **17** (1975) 799.
53. H. Ebisawa, S. Maekawa and H. Fukuyama : Solid State Commun. **45** (1983) 75.
54. J.F. Crow, R.S. Thompson, M.A. Klein and A.K. Bhatnager : Phys. Rev. Lett. **24** (1970) 371.
55. K. Kajimura and N. Mikoshiba : J. Low Temp. Phys. **4** (1971) 331.

# LOCALIZATION AND SUPERCONDUCTIVITY

Yoseph Imry

Department of Physics and Astronomy
Tel Aviv University
Tel Aviv 69978, Israel

## I INTRODUCTION, BASIC CONCEPTS, ORDERS-OF-MAGNITUDE

The Bloch (Boltzmann) theory of electronic transport in lightly disordered conductors has been quite successful in describing the impurity and temperature dependence of the conductivity in ordinary relatively pure conductors. Further transport properties such as magnetoconductivity, Hall effect, thermal conductivity and thermopower can also be handled with general success. However, when the amount of disorder (or impurity concentration) becomes very large, novel phenomena occur which are unexplainable within the weak-scattering theory. In particular, the temperature dependence of $\rho$ becomes much weaker and eventually changes sign for high enough disorder (Wiesmann et al., 1979). The Mathiessen rule, according to which the disorder and temperature-contributions to $\rho$ are additive, thus breaks down. An extremely interesting correlation was found by Mooij (1973), namely that in a large number of metallic conductors, $d\rho/dT$ becomes negative when $\rho$ becomes larger than a value ranging around 80-180 $\mu\Omega$ cm in something like a hundred disordered systems. This almost "universal" trend must have an explanation which is only weakly dependent on material properties, and we remember that $d\rho/dT > 0$ always in weak-scattering theory. More recently (and, in fact, following ideas from localization theory - for general references on localization see, e.g. articles in the books edited by Friedman and Tunstall, 1978; Balian et al., 1979, Stern, 1980; Castellani et al., 1981; Nagaoka and Fukuyama, 1982; as well as the book by Mott and Davis, 1979 and lecture notes by Lee, 1981) it has been found that "restricted geometry" systems i.e. thin films and wires appear to always have $d\rho/dT < 0$ at low enough temperatures. This behavior of effectively one dimensional (1D) and 2D systems is also quite universal. A growing amount of evidence is indicating (Dolan and Osheroff, 1979; Giordano et al., 1979; Bishop et al., 1980; Pepper,1980) that these latter systems are always (Thouless, 1977; Abrahams et al., 1979) insulators, i.e. $\rho \to \infty$, as $T \to 0$.

Ordinary bulk "3D" systems will also become insulators in the above sense when the disorder is strong enough. (Anderson, 1958) Thus, disorder is another important mechanism for the transition to the insulating state. Other mechanisms being (Mott, 1974) electronic (band) structure effects, electron-electron interactions (Mott-Hubbard,1967; Hubbard, 1964) or excitonic (Kohn,1967) and, possibly, self-trapping of the electron by the phonons (Toyozawa 1961). These mechanisms may strongly couple and influence each other, as we shall see, but we would like to start with the problem of non-interacting electrons in a given, static, aperiodic potential.

The solution for the electronic transport for this problem is called "localization theory". It appears to be a good candidate for the new discipline needed for discussing the above-mentioned problems (Jonson and Girvin, 1979; Imry, 1979; 1981). It explains much of the temperature and magnetic field (B) (Hikami, et al., 1980; Kawaguchi and Kawaji,1980) dependence of $\rho$, predicts that "1D and 2D systems are not truly metals" and gives useful indications on the metal-insulator transition in 3D. Interesting insights are obtained on disordered magnetic metals, superconductivity in disordered metals, etc. The electron-electron (and electron-phonon) interaction is also important and should be considered (Schmid, 1974; Abrahams et al., 1981) once the pure "localization" part is understood.

The models usually considered for the aperiodic potential are: A random potential $V(\vec{r})$ such that

$$< V(\vec{r}) > = 0, \quad < V(\vec{r})V(\vec{r}\,') > = C(|\vec{r}-\vec{r}\,'|) \tag{I-1}$$

where the range of C, called a, is the microscopic length in the problem and the size of C (C(o) > 0) is related to the strength of the potential fluctuations. The case $C(x) = \delta(x)$ is referred to as the white-noise potential (since the Fourier transform, $\hat{C}_k$ is constant). Another useful model potential is a periodic one with a random modulation, which is called the Anderson model (Anderson, 1958) when taken as a nearest neighbor tight-binding model, i.e.

$$\mathcal{H} = \sum_i \varepsilon_i C_i^+ C_i + \sum_{<ij>} t_{ij} C_i^+ C_j + \text{h.c.} \qquad \text{(II-2)}$$

where $C_i^+$ creates an electron on the "atomic" state of the $i^{th}$ site of a simple lattice $\varepsilon_i$ (diagonal disorder) or $t_{ij}$ (non-diagonal disorder), or both, can be taken to be random. In the former case, if $t_{ij} = V$ and the width of the $\varepsilon$ distribution is W, W/V is a convenient measure for the disorder. Solving for the eigenvalues of (II-2) amounts to diagonalizing a random matrix. Obviously, this will solve a number of other physical problems such as phonons in a disordered crystal, etc.

In his pioneering paper of 1958 (preceded only by Landauer's work of 1957 to be discussed in Section III) Anderson considered the Hamiltonian (III-2). A strong enough disorder can localize a state, as obviously happens when $\varepsilon_i$ is very large or very small, with respect to typical values of $\varepsilon_j$ analogous to the formation of a bound state or a local vibration. This means that the envelope of the corresponding wavefunction $\psi$, decays strongly (say, exponentially) at large distances from the localization center (e.g. site 1). More general types of localization, $\psi$ extending over a characteristic length $\xi$ and sometimes having bulges and oscillations after being small for a while, also exist. The formation of localized states, taking orthogonalization with different states into account, has presumably some, as yet not understood, similarities to bound state formation (for example, both happen very easily in 1D and 2D). Following a lot of analytical arguments and numerical work, (Licciardello and Thouless, 1978; Stein and Krey, 1980; Domany and Sarker, 1979) it is generally agreed that for a large enough disorder all (or almost) all states are localized. For an intermediate disorder the situation (at dimensionality d > 2) is thought to be as follows (Mott, 1966). Due to the disorder, states are created in the gap. The states in the middle of the band (shaded area) are not localized (i.e. they are extended), while the states near the band extremities can be localized. The extended and localized states are separated by the mobility edges Em1 Em2 (Mott, 1966). The existence of these follows from the physically reasonable increase of the tendency for localization as E gets further from the band center and from an intuitive argument by Mott according to which, extended and localized states cannot coexist at the same energy since they will be mixed by any interaction, however small. This argument is not rigorous, since for a system with linear size L the characteristic difference in energies of the states that are close in energy is $0(L^{-d})$ while if the extended and localized states reside in different parts of the system, the interaction may be $0(e^{-0(L)})$. This can probably happen in very inhomogenous, e.g. percolating systems, but is assumed not to happen if the system is homogenous enough. When W/V is increased, the mobility edges approach each other and coalesce at some limiting value $(W/V)_c$, where all states become localized, and the Anderson transition has occurred. We emphasize that the density of states (DOS) at the Fermi energy is only slightly depressed, and is not expected to develop any singular structure due to the localization process alone (Kirkpatrick, 1981).

Some of the physical relevance of localization follows from the observation that if all the physically relevant states - such as the states near the Fermi energy $E_F$ - are localized the system will be insulating at T = 0. This physically obvious assertion can be proven by showing that the diffusion constant D vanishes. From the Einstein relation, (Kubo, 1957 ) at $k_BT << E_F$

$$\sigma = e^2(dn/dE)D \tag{I-3}$$

it will then follow that $\sigma$ vanishes. To demonstrate the vanishing of D, form a narrow minimal wavepacket at $t = 0$, $\vec{r} = 0$; $\psi(t = 0) =$

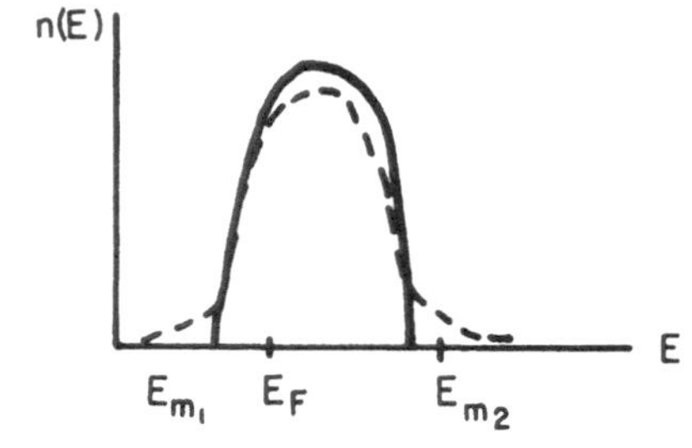


Fig.I-1: The density of states with and without disorder and the mobility edges in the former case.

$\Sigma a_i \psi_i$, from the available states. $a_i$ will be exponentially small for states $|i\rangle$ localized many localization lengths $\xi$ from r=0. Thus,

$$\psi(t) = \Sigma a_i e^{-i(E_i/h)t} \psi_i \qquad (I-4)$$

will also decay exponentially in r at any time, so that $\langle r^2 \rangle \equiv 2Dt$ is never (even when $t \to \infty$) larger than 0 ($\xi^2$), so that D=0.

This observation gives us a very simple mechanism for the metal-insulator (henceforth abbreviated as M-I) transition. $E_F, E_{mi}$ can be changed by changing the electron density or disorder, respectively. Whenever $E_F$ passes from the extended to the localized range the system will go from the metallic to the insulating phase. Since in the insulating phase $\rho(T=0)=\infty$ and $\rho$ decreases with temperature, it is not surprising that $d\rho/dT$ can be negative near the transition, certainly in the "poor" insulator - and, by continuity, in the "poor" metal. One thus already sees that localization theory may be useful in explaining the anomalous properties of disordered conductors as well as the disorder induced metal-insulator transition.

When the disorder, e.g. W/V, becomes very small, the usual weak scattering theory should apply. A convenient dimensionless parameter to express this is $k_F \ell$ or $E_F \tau$, which are of the same order of magnitude. $\tau$ is the mean (elastic) free time and $\ell$ the mean free path, $\ell = v_F \tau$. The small parameter in the weak-scattering theory is $1/(k_F \ell)$. Note that the order-of-magnitude of the diffusion constant is $D \sim v_F^2 \tau = v_F \ell$ and the usual weak-scattering conductivity is given by

$$\sigma = \frac{ne^2 \tau}{m} = \frac{1}{3\pi^2} \frac{e^2 k_F}{\hbar} (k_F \ell), \text{ in 3D} \qquad (I-5)$$

To understand this, note that $(e^2/\hbar)$ is a conductance, which happens to be equal to $\sim(4K\Omega)^{-1}$ in MKS. (This should not be surprising, since $\alpha = e^2/\hbar c \simeq 1/137$, and $c^{-1} \sim 30\Omega$, related to the "free space impedance".) Thus, $(e^2/\hbar) k_F$ is a conductivity unit appropriate to the microscopic, interelectron distance scale conductivity for a conductance of $e^2/\hbar$. $\sigma$ is multiplied by $k_F \ell$ as well. The situation in 2D is more interesting, since $\sigma$ (or the conductance of a thin film) $\propto e^2/\hbar\ (k_F \ell)$ - which is just a universal constant times $k_F \ell$. Of course, the weak-scattering theory will breakdown when the parameter $k_F \ell$ is no longer >> 1. This defines the concept of the minimum metallic conductivity (Yoffe-Regel, 1960; Mott, 1966), obtained when $k_F \ell \sim 1$. In 3D:

$$\sigma_{min} = C \frac{e^2}{\hbar} k_F \qquad (I-6)$$

where C is a constant on the order of .01-.05. For metallic systems where $k_F$ is a few inverse A, this yields resistivities on the order of $10^{-3}$ $\Omega$ cm, somewhat larger but of the same order of magnitude as the Mooij value of $1\text{-}2\times10^{-4}$ $\Omega$ cm. (Thus $d\rho/dT$ becomes negative when $k_F\ell\sim5\text{-}10$.) In 2D, this yields a "maximum universal metallic resistance" of $\sim30$ K$\Omega$. $\sigma_{min}$ is clearly the range where localization is very relevant. Mott has argued that no metals can exist with $\sigma<\sigma_{min}$. This will be discussed in some detail later in these lectures.

## II THERMALLY ACTIVATED CONDUCTION IN THE LOCALIZED REGIME

If the states at the Fermi energy, $E_F$ are localized, $E_F<E_m$ (we assume, for definiteness, that $E_F$ is in the lower half of the band, $E_m \equiv E_{m1}$), then the quantum conduction of T=0 is zero. At low but finite temperatures the electron can gain thermal energy (typically from other excitations, e.g. phonons) to perform a number of possible processes (Mott and Davis, 1979).

1. Activation to (and above) the mobility edge, which will yield

$$\sigma_1 \propto e^{-(E_m-E_F)/k_BT} \qquad \text{(II-1)}$$

2. Activation to neighboring localized states. If the localization length is $\xi$ and the density of states (DOS) at the Fermi level $n(0)$ then the number of states in a volume of linear dimension in D dimensions is $n(0)\xi^D$, hence the typical energy separation between such states is

$$W_\xi \sim [n(0)\xi^D]^{-1} \qquad \text{(II-2)}$$

which will yield a "nearest neighbor" activated conductivity of the form

$$\sigma_2 \propto e^{-W_\xi/k_BT} \qquad \text{(II-3)}$$

However, as suggested by Mott, (1969), it pays sometimes for the electron to hop a larger distance, thereby reducing the necessary inelastic energy transfer. This introduces the next type of activated conductivity:

3. Variable range hopping, we assume that the hopping conductivity to a state localized a length $L\gg\xi$ away is proportional to the overlap matrix element squared, which goes like $I^2e^{-2L/\xi}$, where I is a characteristic energy. On the other hand, the energy needed now, $W_L$, is obtained by generalizing the argument leading to (II-2) and noting that $W_L$ decreases with L

$$W_L \sim (n(o)L^d)^{-1} \sim W_\xi \left(\frac{\xi}{L}\right)^D \quad (L\gg\xi) \qquad \text{(II-4)}$$

The hopping over a length L is controlled by $e^{-2L/\xi - W_L/k_BT}$. Sometimes it pays to jump with $L>\xi$. The optimal $L$, $L_0$, for such jumps is given by minimizing the exponent

$$L_o \sim \left(\frac{\xi}{n(o)k_oT}\right)^{1/(D+1)} \tag{II-5}$$

and this mechanism is relevant as long as $L_o \gtrsim \xi$ i.e. when the temperature is low enough so that $T \ll T_o$, where

$$k_BT_o \sim W_\xi \tag{II-6}$$

At such low temperatures,

$$\sigma_3 \propto e^{-C\left(\frac{T_o}{T}\right)^{1/(D+1)}} \tag{II-7}$$

where C is a dimensionless constant. For $T > T_o$ nearest neighbor hopping, $\sigma_2$, is obtained, which has a simple activation form. One has then to consider the competition between $\sigma_2$ and $\sigma_1$. It appears that $\sigma_2$ usually wins, especially near the transition, when $W_\xi$ vanishes probably faster than $(E_m - E_F)$. For $(E_m - E_F) < W_\xi$, $\sigma_1$ will be dominant in the simple activated region and the crossover to $\exp.(-C'/T^{1/(D+1)})$ will occur at a temperature below $T_o$.

We also remark that in the variable range hopping range, $L_o$ is an important length scale which determines, for example, the effective dimensionality of a thin film or wire (Fowler, Hartstein and Webb, 1982).

Mott (1970) has also made an argument showing that the T=0 frequency-dependent conductivity, $\sigma(\omega)$ should behave like

$$\sigma(\omega) \propto \omega^2 \, \ell n^{d+1}(I/\omega) \tag{II-8}$$

## III THE LANDAUER FORMULA

In 1957 Landauer made a fundamental contribution to the understanding of conduction in disordered systems, to what turned out to be localization (Mott and Twose, 1961) in 1D and to clarification of concepts associated with quantum phase. One should hope that it is not typical to the ability of physicists to grasp really new ideas that it took this work 13 years to be properly published (Landauer, 1970) and 10 more years for its signficance to be appreciated. The digestion processes for Anderson's fundamental paper (Anderson, 1958) were, in fact, comparably slow.

Landauer considered the simplest transport problem - transfer of a 1D wave through a finite reflection (R) barrier. Instead of appealing to transport theory formalism he derived his result using seemingly straightforward but really a rather delicate argument. A wave of amplitude unity impinges from the left on the barrier, a fraction r is reflected and a fraction t transmitted. t and r are

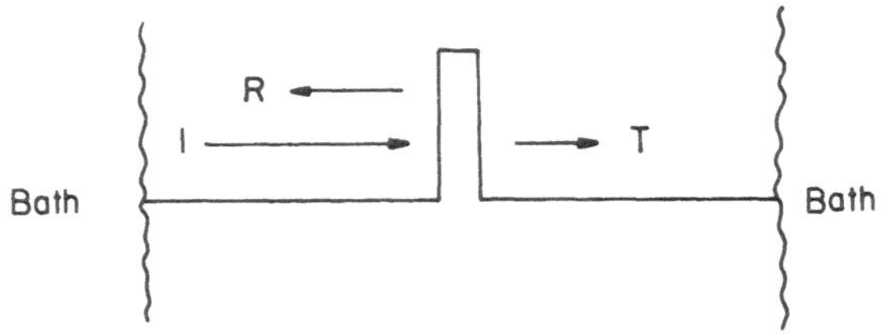


Fig.III-1: The Landauer geometry for transport through a barrier.

complex numbers. The reflection coefficient is $R=|r|^2$ and the transmission one is $T=|t|^2=1-R$, thus

$$|t|^2+|r|^2 = 1 \qquad \text{(III-1)}$$

The average electron density, n, on the left-hand side of the barrier is 1+R and on the right-hand side it is T, the current transmitted is Tv where v is the wave velocity. Thus, the diffusion constant D is given by (noting $\Delta n=1+R-T=2R$)

$$D = \frac{\text{current}}{\text{density gradient}} = \frac{aT}{2R}\, v, \qquad \text{(III-2)}$$

where a is the extent of the barrier.

The conductivity is given in terms of D by an Einstein relation (I-3), for degenerate electrons

$$\sigma=e^2\, D\, n(0) \qquad \text{(III-3)}$$

and $v=v_F$. Thus, the conductance $G=\sigma/a$ is given by

$$G = \frac{e^2\, T}{R}\, v_F\, n(0) = \frac{e^2}{\hbar\pi}\, \frac{T}{R} \qquad \text{(III-4)}$$

i.e., the conductance is units of $e^2/\hbar$ is just T/R, it is given in terms of the scattering properties of the barrier (see also Erdos and Herndon, 1982).

However, as discussed by Landauer, (1970,1975) the picture is still incomplete. Consider the situation time-reversed to that of Fig.III-1. The wave of amplitude t impinges now from the right, one with amplitude 1 ($\geq|t|$) is transmitted and, in addition, the left-hand side of the barrier is hit with an incoming wave of amplitude r (and a specified phase). This situation is pure nonsense, the current goes from lower to higher density and D is negative. Physically, of course, we know that if a wave impinges on the barrier from the right, a fraction r' of it will be reflected and t' transmitted, where t=t' and (Landau and Lifshitz, 1976)

$$-t/t'^* = r/r'^* \qquad \text{(III-5)}$$

yielding the same values of G and D. Thus, it is important to avoid the possibility of electrons with correlated phases to arrive from the two sides. This will eliminate the unphysical time-reversed situation. It is not surprising, in fact, that one needs to break time reversal symmetry in order to obtain real transport. How to do it was in fact suggested by Landauer: We assume electron baths on the two sides, which can send only electrons that are incoherent - with no phase correlations. The baths can also be used to make the electron electrical potential uniform, so that the current in Eq.III-2 is due solely to the density change. [Remember that to prove the Einstein relation, III-2, one uses a different arrangement - an open circuit where the total I vanishes because of the cancellation of the diffusion current with the ohmic one due to the electrical potential difference generated by the open circuit. In the linear case these two effects add.]

The above considerations suggest that very interesting effects may be obtained if the reservoirs are eliminated and the two sides of Fig.1 are connected to form a ring (Büttiker et al., 1983).

The Landauer formula can also be understood (Imry, 1981) by employing known aspects of the conduction through a tunnel junction. We start with a small transmittance, $T \ll 1$. Take the lid going into the junction as an ideal conductor of length L, where the electron moves with a velocity v. Each time the electron is reflected from the barrier it goes back to the reservoir and then shot back at the barrier. It takes, on the average $T^{-1}$ such attempts until the barrier is crossed, so that the average time the electron stays on that side of the barrier before making it to the other side is:

$$\tau_L = 2L\, v\, T^{-1} \qquad \text{(III-6)}$$

To get a current I one thus needs $I\tau_L$ excess electrons on the left-hand side of the junction, the density change is $\Delta n = I\tau_L/L = 2I/vT$, and the diffusion constant is thus

$$D = \frac{I}{\Delta n/a} = \frac{v\,Ta}{2} \qquad G = \frac{e^2}{\pi\hbar}\, T \qquad \text{(III-7)}$$

which is indeed Eq.(III-2,4) with $R \simeq 1$. To generalize this for large transmission, one may combine n barriers of $T \ll 1$, in series, (see next Section) so that the total transmission is not too small, and recover the order-of-magnitude of Eq.(III-2). (See Imry, 1980.)

## IV SERIES ADDITION OF OBSTACLES, 1D LOCALIZATION

We now consider, following Landauer, the effect of two obstacles in series. We denote the phase change on the wave passing through the constant potential (ideal conductor) between the obstacles, by $\phi$. The waves in the region between the obstacles are the sum of all the multiply scattered waves yielding a total left-

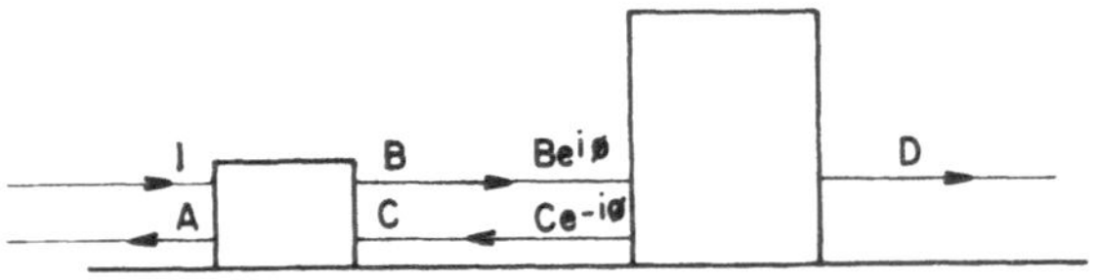


Fig.IV-1: Two obstacles in series

going wave and a right-going one, an amplitude A is reflected and D transmitted through the whole device. A,B,C,D are complex numbers. The wave just emerging from obstacle 1 is $Be^{i(kx-\omega t)}$ and acquires a phase $\phi$ upon impinging on obstacle 2. The wave C suffers a similar phase change from 2 to 1. The barrier equations are

$$A = r_1 + Ct_1 \qquad B = t_1 + Cr_1'$$
$$Ce^{-i\phi} = Be^{i\phi}r_2 \qquad D = Be^{i\phi}t_2 \tag{IV-1}$$

Solving these equations, we find $D = \dfrac{e^{i\phi}t_1t_2}{1-e^{2i\phi}r_2r_1'}$, which yields the total transmittance $T_{12}$ of the device:

$$T_{12} = \frac{T_1T_2}{1 + R_1R_2 - 2\sqrt{R_1R_2}\cos\theta} \tag{IV-2}$$

where $\theta=2\phi+\arg(r_2r_1')$ and

$$\frac{R}{T} = \frac{R_1 + R_2 - 2\sqrt{R_1R_2}\cos\theta}{T_1T_2} \tag{IV-3}$$

Assume now that we have an ensemble of systems prepared with similar $R_1$ and $R_2$ but such that the "optical" phase difference, $\phi$, in different members of the ensemble spans many times $2\pi$ with a uniform probability. The average of the inverse of the dimensionless conductance, g,

$$g \equiv G/(e^2/\pi\hbar) \tag{IV-4}$$

is thus given by

$$(g^{-1})_{av} = \frac{R_1 + R_2}{(1 - R_1)(1 - R_2)} \tag{IV-4}$$

This result is already quite surprising. Ohm's law of series addition of resistances, $g^{-1}=g_1^{-1}+g_2^{-1}=R_1/(1-R_1)+R_2/(1-R_2)$ is not valid in general! It is only valid in the limit of good transmission, or small resistance, $R_i \ll 1$. This has very serious and important consequences that were not easy to accept in 1957. In fact, if one combines good transmittances ($R \ll 1, T \cong 1$) in series, the resulting, resistance $g^{-1}$, first increases linearly with n as it should, but once n is so large that the total transmittance is smaller than unity, then

$$(g^{-1})_{av,n+1} = \frac{R_n + R}{T_n} = (g^{-1})_{av,n} + \frac{R}{T_n} \qquad \text{(IV-5)}$$

One may form what is called these days a Renormalization Group (RG) equation for the length-scale (n) dependence of $(g^{-1})_{av,n}$

$$\frac{d}{dn}(g^{-1})_{av,n} = R[(g^{-1})_{av,n} + 1] \qquad \text{(IV=6)}$$

so that the (dimensionless) resistance, after having increased linearly with n to O(1), will then increase exponentially with the length n. This is our first encounter with 1D localization.

The above is not entirely satisfactory, however, as remarked already by Landauer. The distribution of the resistances in the ensemble is not narrow and thus, as emphasized by Abrahams et al (1981), the results depend on what quantity is being averaged. Abrahams et al. also pointed out what is the proper way to average for large n. One needs something that will behave like an ordinary extensive quantity with both average and mean-square average increasing linearly with n. Such a quantity is $\ell n(1+g^{-1})$. This is because $1+g^{-1}=1+R/T=1/T$ so that $\ell n(1+g^{-1})= -\ell nT$. $-\ell nT$ plays the role of the extinction exponent and one would expect it to be additive for two scatterers if the relative phase is averaged Indeed, from (IV-2) one finds, using (Abrahams, et al., 1981)

$$\int_0^{2\pi} d\,\theta\, \ell n(a + b\cos\theta) = \pi\, \ell n\, \frac{1}{2}\,[a + (a^2 + b^2)^{1/2}] \qquad \text{(IV-7)}$$

that $\langle \ell n\, T_{12}\rangle = \ell n\, T_1 + \ell n\, T_2$. Thus, the exact scaling of the 1D resistance with n is given by

$$\ell n\,(1 + g_n^{-1}) = \rho_1 n \qquad \text{(IV-8)}$$

with $\rho_1$ the resistance, in units of $\pi\hbar/e^2$ of a single obstacle. This indeed increases first linearly and then exponentially with n, differing only quantitatively from the results of (IV-6). This establishes the 1D localization, manifest in a measurable quantity, i.e. the resistance. The localization of all eigenstates in 1D (Mott and Twose, 1961; Borland, 1963 ) is now well known and has been rigorously proven. (All this is, of course, at "zero" or appropriately low temperatures.)

Interesting effects may exist on top of the average behavior, as discussed by Azbel (1982). For a given, finite, system if the energy (and hence the optical path difference between scatterers is varied, T can vary and, in fact, will show sharp "transmission resonances". These may show up as sharp oscillations, at low temperatures, of the resistance as function of the electron density (which is variable in a MOSFET device (Ando et al., 1982)) and perhaps magnetic field or temperature.

Now, we are also in a position to generalize the derivation of (III-7) to larger transmission coefficients (Imry, 1981). Given a T=1, R << 1 we combine n obstacles so that $T_n$ is a number C smaller but on the order of unity. Thus, $|\ell nC| = n|\ell nT| \cong nR$, but if C is still small, Eq.III-6 still roughly holds for $g_n \sim T_n = C$. Since Ohm's law roughly holds as long as $g \gtrsim 1$ we can now obtain $g_1$- the conductance of a single obstacle as $ng_n$, i.e.

$$g_1 = nC = O(1)\,\frac{T}{R} \qquad \text{(IV-9)}$$

which agrees within an order of magnitude with Eqs.(III-2,4).

We conclude this section with further remarks on the tunnel junction picture of conduction which also pave the way to the material of the next section. The lifetime $\tau_L$ (Eq.III-6) for an electron on one side against a transition to the other side is given by the (Landau and Lifshitz, 1976) Fermi golden rule (when tunneling is a weak perturbation, i.e. T << 1):

$$\tau_L^{-1} = \frac{2\pi}{\hbar}\,\overline{v^2}\,N_L\,(E_F) \qquad \text{(IV-10)}$$

where $\overline{v^2}$ is the average of the tunneling matrix element squared and $N_L(E_F)$ the density of states on the final side. Assuming the two sides to be equivalent, when a voltage V is applied $eVN_L(E_F)$ states are available, each decaying with a time constant $\tau_L$, so that the current is $I = e\,N\,(E_F)\tau_L^{-1}V$ and the conductance is

$$G = e^2\,N_L(E_F)/\tau_L = \frac{2\pi e^2}{\hbar}\cdot\overline{v^2}\,[N_L(E_F)]^2 \qquad \text{(IV-11)}$$

which is a useful result, the second equality is a well-known formula in the tunnel junction theory. Note that Eqs.IV-10,11 are valid in any number of dimensions.

## V THE THOULESS PICTURE, LOCALIZATION IN THIN WIRES AND FINITE TEMPERATURE EFFECTS

The first equality in Eq.(IV-11) is very general. Let us use it for the following scaling picture (Thouless, 1977): Divide a large sample to (hyper) cubes or "blocks" of side L. The typical level separation at the relevant energy (say, the Fermi level), w , is given by the inverse of the density of states (per unit energy) $N_L(\varepsilon_F)$. Defining an energy associated with the transfer of electrons between two adjacent such systems by $v_L \equiv \pi\hbar/\tau_L$ ($\tau_L$ is the lifetime of an electron on one side against transition to the other side) the dimensionless conductance $g_L \equiv G_L/(e^2/\pi\hbar)$ is:

$$g_L = v_L/w_L \tag{V-1}$$

I.e. $g_L$ is the (dimensionless) ratio of the only two coupling energies in the problem. The way Thouless derived this relation is by noting that the electron's diffusion on the scale L is a random walk with a step L and characteristic time $\tau_L$, thus

$$D_L \sim L^2/\tau_L \tag{V-2}$$

For metals the conductivity, $\sigma_L$, on the scale of the block size L is given by the Einstein relation (I-3), and the conductance in d dimensions is given by $G_L \sim \sigma_L L^{2-d}$. Putting these relations together and remembering that $N_L(E_F) \sim L^d dn/d\mu$, yields Eq.V-1. To get some physical feeling for the energy $h/\tau_L$ we note that, at least for the weak coupling case, the Fermi golden rule yields Eq.(IV-10) or:

$$v_L = 2\pi^2 \overline{v^2}/w_L \tag{V-3}$$

Thus, $v_L$ is defined in terms of the interblock matrix elements. Clearly, (V-3) is also related to the order of magnitude of the perturbation theory shift of the levels in one block by the interaction with the other. For a given block this is similar to a surface effect - the shift in the block levels due to changes in the boundary conditions on the surface of the block. Indeed, Thouless has given appealing physical arguments for the equivalence of $v_L$ with the sensitivity of the block levels to boundary conditions. This should be valid except perhaps for $g_L \ll 1$, where the latter should at least be an upper bound for the former. This can be used also for numerical calculations of g(L), which gives us most of the physics of the problem, for non-interacting electrons, as we shall see. Alternatively, Eq.(V-3) as well as generalizations of the Landauer formula (Fisher and Lee, 1981) can and have been used for numerical computations. It is important to emphasize that $g_L \gg 1$ means that states in neighboring blocks are tightly coupled while $g_L \ll 1$ means that the states are essentially single-block ones. Thus, if $g_L \to 0$ for $L \to \infty$, then the range of scales L where $g_L \sim 1$ gives the order of magnitude of the localization length, $\xi$.

The analysis by Thouless (1977) of the consequences of Eq.V-I for a long thin wire has led to extremely important results. First, it showed that 1D localization should be manifest not only in "mathematically 1D" systems but also in the conduction of realistic thin wires, demonstrating also the usefulness of the block-scaling point of view. Second, the understanding of the effects of finite temperatures (as well as other experimental parameters) on the relevant scale of the conduction clarifies the relationships between g(L) and experiment in any dimension. Thouless simply analyses what happens in a wire of a given cross-section A as function of L. For a real wire $\sqrt{A}$ is many atomic distances a, but still much less than "macroscopic" sizes (the precise requirements will become clear later). The usual Ohm's law $G_L \propto L^{-1}$ can at best, hold only for a limited range of L's. Indeed, suppose $G_L \propto L^{-1}$ for some range of L's, once $L > L_c$ where $L_c$ is defined by $G_{L_c} = e^2/2\hbar$ one would obtain $g \ll 1$ for $L \gg L_c$. This means that localization occurs of the scale $L_c$, which is therefore the localization length $\xi$ in this case. Stated simply, Ohm's law (which does appear to hold for ordinary thin wires) can only hold as long (for $T \to 0$) as the T=0 resistance of the wire is less than about 10 kΩ. For lengths larger than this length, $\xi$, the resistance should increase exponentially with L (see discussion following Eq.I-4), like in the mathematically 1D case (see Eqs.IV-6,IV-8). $\xi$ is easily estimated assuming $G_L \sim L^{-1}$ for $a < L < \xi$, for a resistivity $\rho$, $2\hbar/e^2 \simeq \rho\xi/A$ or

$$\xi \simeq \frac{2\hbar}{e^2} A\sigma \simeq \frac{2}{3\pi^2}(Ak_F^2)\,\ell \tag{V-4}$$

where Eq.(I-5) was used to obtain the second approximate equality. Thus, the order-of-magnitude of the length $\xi$ is given by the elastic mean free path $\ell$ times the number of electrons in the wire's cross-section. For a cross-section of atomic dimension this yields just $\ell$, in agreement with the "purely 1D" case. The assumption of $G \sim L^{-1}$ for $L \ll \xi$ at least agrees with our intuition on wires. Theoretically it means that $v^2 \sim 1/L$, since $w \sim 1/L$, which is not unreasonable for a surface effect, as long as $L \ll \xi$. Of course, $g_L \sim L^{-1}$ for $L \ll \xi$ in the purely 1D case as well.

Obviously, we know that the resistance of thin wires does not ordinarily increase exponentially with their length. However, one never measures zero temperature resistance. For the localization result to hold, T must be low enough so that the electron should not feel the effects of temperature during its motion on scales that are even larger than $\xi$. Thouless' work indicates the following very plausible physical consideration. When T goes to zero the characteristic time, $\tau_T$, between inelastic (or, more generally, any phase breaking) events, becomes very large. For example, in many cases one can write

$$\tau_T \propto T^{-p} \tag{V-5}$$

where much will have to be said later on both the exponent p and the prefactor. Consider a diffusing electron, after a time t, it covers a length $\sqrt{Dt}$. To feel strongly the localization effects one needs that the nominal length, defined by the diffusion coefficient, $D_1$, in the non-localized regime,

$$\ell_T \equiv \sqrt{D_1 \tau_T} \tag{V-6}$$

be much larger than $\xi$. In the opposite case, of course, the electron will perform many inelastic collisions before it feels the localization and one would expect that the effects of the latter will not be very strong. Thus, we are led to viewing $\ell_T$ as an important length-scale for the electron's motion. For the wire to be effectively 1D, one is likewise led to the analogous necessary condition,

$$\ell_T >> \sqrt{A} \quad , \qquad \text{(V-6)}$$

$\sim\sqrt{A}$ being the thickness and width of the wire.

Thouless has also made a rather complete analysis of the low temperature transport in the thin wire. Once T is so small that $\tau_T > \xi^2/D$, (we shall denote the cross-over temperature, where $\tau_T = \xi^2/D$, by $T_\ell$), the motion of the electron starts to be diffusive with a step $\xi$ and characteristic time $\tau_T$, thus:

$$D \sim \xi^2/\tau_T \qquad \sigma \propto \tau_T^{-1} \propto T^p \qquad \text{(V-7)}$$

I.e. a vanishing of the conductivity for small T like a power of T! This very important result has still not been noticed or appreciated by all localization practitioners. Note that, (V-7) assumes that an inelastic event gives the electron enough energy to move from one block of size $\xi$ to the next. The condition for this is that $\xi < T$. This defines a further cross-over temperature, $T_0$, to exponential temperature dependence (which can be, e.g. simple or variable-range one, see Section II) given by

$$k_B T_0 = [n(0)A\xi]^{-1}.$$

Thus, the condition for observing the $T^p$ behavior is $T_\ell >> T_0$. This analysis is valid for localized states in any dimension.

Let us concentrate now on the range $T > T_\ell$, where the effects of localization are weak. Here the length $\ell_T$ is physically meaningful - it is the scale to which the electron diffuses quantum mechanically. From then on the motion is classical and controlled by the inelastic scattering. Thus, even if we know the quantum mechanical g(L) (at T=0), it is only relevant for $L \lesssim L_T$. Since one believes in the classical intuition in the appropriate range, (which simply asserts that for a given current the voltages along consecutive segments add in 1D) it is suggested that the macroscopic conductivity of the sample will be determined by Ohm's law using the T=0 one on scale $\ell_T$. This important observation also follows from the Landauer picture - where the electron becomes incoherent in the baths. The length $\ell_T$ over which this happens in the long system defines the maximum length over which the T=0 theory is valid. At larger scales classical conduction sets in. This forms the basis for obtaining the so called "weak localization" effects which occur when $\ell_T < \xi$.

# VI THE SCALING THEORY OF LOCALIZATION AND ITS CONSEQUENCES

## 1. General

The conductance of a (hyper) cube of size L, at T=0 can be calculated, in principle, numerically using the Thouless relation (V-1) in any number of dimensions, employing a variety of methods to determine $V_L$. Using generalizations of the Landauer formula provides another method which appears to be more effective (Fisher and Lee, 1981). Knowing how g scales with L, for the appropriate range of L values, and understanding that $L_T$ (or $\xi$ and $\tau_T$ in the localized regime, for $L_T \gg \xi$) determines the relevant scale for the temperature dependence of g, enables us to obtain the temperature dependence of the conductivity, $\sigma$:

$$\sigma(T) \cong \frac{e^2}{\pi\hbar} g(L)\, L^{2-d} \Big]_{L=L_T} \tag{VI-1}$$

i.e. $\sigma$ evaluated from G using the geometry and the scale $L=L_T$. This is valid for $L_T < \xi$ and also in the whole metallic range. In the localized regime, for $L_T > \xi$, Eq.(V-7) and its appropriate counterparts at low temperatures have to be used, as discussed in Sections II, V.

We shall now present and discuss the scaling theory of localization by Abrahams, Anderson, Licciardello and Ramakrishnan (1979). This theory is really a clever guess based on an interpolation between the limits of a good conductor, $g_L \gg 1$, and a localized insulator, $g_L \ll 1$. It is consistent with the first correction to the good (weak scattering) conductor, with all the presently available numerical work (those numerical results that contradicted it have been superceded by more reliable ones), with the 1D and thin wire cases and with some analytical approximations. Being an interpolation picture it should be qualitatively correct (except possibly for the details inside the interpolation range, some of which can be quite important). While the behavior of real systems may be sensitive to other effects too (notably, electron-electron interactions), the scaling theory does explain better than qualitatively a large amount of data on many systems, and had had made surprising (at the time) predictions that have been confirmed by experiment. Moreover, concerning reports about disagreement with the scaling theory predictions one should make sure that real predictions of the scaling (e.g. Eq.(VI-1) is not valid in the insulator) are tested and that electron-electron interactions do not play a role.

In the limit of a good conductor one expects the usual Ohm's law to hold, i.e. $\sigma(L)$ = const. = $g(L) \propto L^{d-2}$. In the opposite limit, one expects (See Section L) both $\sigma$ and g to decrease exponentially with L. Thus, in these two limits, the logarithmic derivative of g;

$$\beta \equiv \frac{d\ \ell n g}{d\ \ell n_L} = \begin{cases} d-2 & g>>1 \\ \text{const} + \ell n g & g<<1 \end{cases} \qquad \text{(VI-2)}$$

(where small corrections due to a possible power law prefactor for g << 1 have been neglected) is independent of L and of the details of the system. Since β can be obtained from computations on finite systems, it must be analytic and it can be expected not to decrease with g. Assuming that β stays a function of g i.e., using RG language, that g is the only "relevant" variable and that L is large enough so that the "irrelevant" ones vanish only in te whole range (not only for $g \lessgtr 1$), one arrives at the picture of β(g) changing monotonically and smoothly between the two limits of VI-2, as shown in the figure where β is schematically given as function of log g.

2. The Case d < 2

For d < 2, β is thus always negative. This means that if we know $g(L_0)=g_0$ for some small $L_0$, then, obtaining g(L) for every L by solving $d\ell ng/d\ell nL=\beta(g)$ with $g(L_0)=g_0$, we find that always $g(L)\sim\exp(-\alpha L)$ as $L\to\infty$. This can be visualized by noting that $g_0$ is represented by some point on the β(g) graph and that g(L) will simply flow down that curve until it reaches the linear range at small g (large negative ℓng). The above procedure for this very simple case is called "solving the RG equations" and the motion of the point along the β(g) a "RG-flow" in the theoretical jargon. The localization length, ξ, is the L above which the flow has reached the linear range in ℓng. Since β(g) becomes very flat when g>>1, ξ becomes larger when $g_0$ is increased. For g>>1 we expect, from analyticity in $g^{-1}$,

$$\beta(g) \sim d-2 - \frac{c}{g} . \qquad \text{(VI-3)}$$

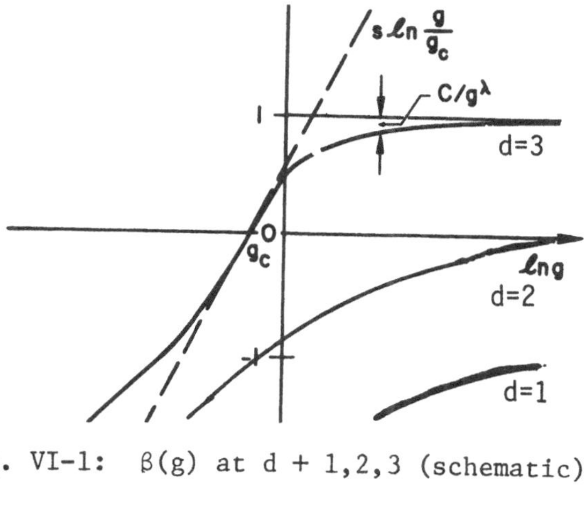


Fig. VI-1: β(g) at d + 1,2,3 (schematic).

This has been confirmed and the numerical constant C computed as a function of d by perturbation theory. $\xi$ is roughly defined as the scale at which $\beta$ decreases substantially from d-2 and approaches the linear range. This can be estimated using the rough approximation VI-3 or even $\beta \sim d-2$, extrapolated to $g \gtrsim 1$, and is consistent with taking

$$g(\xi) = \text{a numerical constant.} \tag{VI-4}$$

We note that this agrees with (V-4) for the effectively 1D case, in 2D, $\xi$ will be exponentially large for large $g_o$ (i.e. certainly larger than the distance to the sun for $R\ (L_o) \lesssim 10^{-3}\Omega$).

Note that VI-3 yields a correction to the ohmic behavior, $\sigma(L)$=const, for $L \ll \xi$, i.e.

$$\begin{aligned} \sigma(L) &= g_o L_o - CL \qquad d=1 \\ g(L) &= g_o - C\ \ell n(L/L_o) \qquad d=2 \end{aligned} \tag{VI-5}$$

Thus, as function of L, when L is increased, one should first obtain the weak localization correction VI-5, until $L \sim \xi$ and then one gets into the (strongly) localized range with the behavior discussed before. Most interesting is the 2D behavior: The system is, in principle never a metal. Non-Ohm's law logarithmic corrections will be seen at relatively high temperatures $\ell_T \ll \xi$ (when the temperature is too high so that $\ell_T$ is smaller than the smallest microscopic $L_o$ allowed - $L_o \sim \ell_{el}$ - this theory will break down). Upon <u>decreasing</u> T, g will decrease and R <u>increase</u>; until $R \sim h/e^2$ (the numerical constants yield R, critical $\sim$ 30 k$\Omega$), one crosses over into strong localization. Thus "2D metals are not really metals." A necessary condition for 2D behavior is: thickness of film $\ll \ell_T$, as before. These surprising predictions are now confirmed in the weakly localized range by many experiments. (e.g. Dolan and Osheroff, 1979; Biship et al., 1980; Pepper et al., 1980; Ovadyahu and Imry, 1981.) It is also known that $\sigma\ (T \to 0) \to \infty$ for 2D samples with R > 30 k$\Omega$. There is a recent experiment by Ovadyahu and Imry (1982), where the <u>same</u> sample crosses-over from weak lnT to a stronger increase with decreasing temperature around $R \cong 30$ k$\Omega$, which is a remarkable confirmation of the above surprising prediction.

For convenience we shall briefly discuss the theoretical predictions in the weak localization regime. The constant C in Eq.(VI-3,5) as $1/\pi^2$. This means that the weak localization correction is

$$\Delta G = -\frac{e^2}{\pi^2\hbar}\ \ln L,\ \Delta G(T) = +\frac{e^2 p}{2\pi^2\hbar}\ \ln T \tag{VI-6}$$

(using Eqs. 3-5). In terms of the resistance per square, $R_\square$,

$$\Delta R_\square / R_\square = - \frac{e^2 p}{2\pi^2 \hbar} R_\square \ln T \tag{VI-7}$$

i.e. $R_\square$ decreases when T increases, the relative effect is increasing like $R_\square/(\hbar/e^2)$. This explains qualitatively many experiments with complications related to uncertainties in p, effects like many - valley conduction and mainly electron-electron interaction, which was found to also lead to a correction similar to Eq.(VI-6,7), as will be discussed later. This makes discrimination between the localization and interaction effects rather difficult.

Measurements of the magnetoconductance (Kawaguchi and Kawaji, 1979) may clarify this distinction. While the interactions do lead to non-trivial magnetoresistance its two contributions (spin and orbital, see e.g. Fukuyama, 1981,1982) appear to be smaller or of the same order of magnitude than the localization one. The latter is large, positive in the usual cases, and rather well defined. The easiest way to understand it is as follows: Assume a magnetic field H perpendicular to the 2D layer. Over a range $\ell_H$ in space with

$$H\ell_H^2 = \phi_0, \quad \phi_0 = hc/e = \text{flux quantum} \tag{VI-8}$$

one flux quantum pierces the system. The usual gauge transformation on the $\hat{P}+eA/c$ term in the Hamiltonian yields that the electron acquires a phase $\sim 2\pi$ by motion on a scale $\ell_H = \sqrt{\hbar c/eH}$ in this field. This $\ell_H$ is a candidate for the physical length determining the "phase coherence" scale or that on which the motion is T=0, H=0 one. Now we have two cases: a) $\ell_H \ll \ell_T$ ("strong fields") and the characteristic physical length is $\ell_H$, and

$$\Delta\sigma(H) \sim \frac{e^2}{e\pi^2\hbar} \ln H \tag{VI-9}$$

in the weak localization regime in 2D.

b) $\ell \gg \ell_T$ ("weak fields"), here $\ell_T$ is the relevant length. The field yields a small $O(H^2)$ magnetoconductance.

There are now several cases in which the measured magnetoresistance agrees quantitatively with the detailed predictions in the weak localization regime (Imry and Ovadyahu 1981; Wheeler, 1981; Bergman, 1981). (See also Uren et al., 1981, Kaveh, et al., 1982.) In some cases, there may be complications due to spin-orbit interactions, magnetic impurities, etc.

A further measurement that can distinguish between the localization and interaction contributions is the Hall constant $R_H$. In the pure localization theory it should not have any $\ell n T$ temperature term (this may be very roughly interpreted as confirming that the density-of-states (DOS) does not change for non-interacting electrons). On the other hand, the interaction terms yield a $\Delta R_H(T)/R_H$ which is twice $\Delta R(T)/R$. Experimentally, inversion layers appear to behave like the former or in an intermediate manner at small fields and like the latter at very large fields (where, unfortunately, both theories do not strictly hold). In one well defined system, $In_2O_{3}$-x, no $\ell n T$ term is $\Delta R_H$ was found at small fields (Ovadyahu and Imry, 1981). We are not aware of conclusive $R_H$ data or on dirty 2D metallic systems. A scaling theory for $R_H$ was given by Shapiro and Abrahams (1981).

We conclude the 2-D discussion by mentioning a development due to Kaveh & Mott (1981) who claimed that in a weakly disordered system one need not always have expenential localization but one can have the weaker power-law decay, which will become exponential at strong enough disorder. Their argument becomes simple in terms of a formula which we shall develop later. For a diffusing electron, one finds for the following matrix elements of its exact eigenstates:

$$|\langle m \,|\, e^{i\vec{q}\cdot\vec{r}} \,|\, n \rangle|^2 = \frac{1}{\pi n(0)} \frac{Dq^2}{(Dq^2)^2 + \omega^2}, \quad \omega \equiv \frac{E_n - E_m}{\hbar} \qquad \text{(VI-10)}$$

For weak scattering, diffusion may be assumed for small scales. The wave function $\psi$ are given perturbatively by the zero disorder plane wave $\psi_n^o$ plus a small correction $\Delta\psi_n$. Using Eq.(VI-10) for m=n, we find to lowest order, for $q \neq 0$,

$$|\langle \psi_n{}^o | e^{i\vec{q}\cdot\vec{r}} | \Delta\psi_n \rangle|^2 = \frac{1}{\pi n(0) Dq^2} \qquad \text{(VI-11)}$$

This means that the $|$Fourier coefficients$|^2$ of $\Delta\psi_n$ with a wave-vector differing from that of $\psi_n{}^o$ by q is

$$|a(q)|^2 = \frac{1}{\pi n(0) Dq^2} \qquad \text{(VI-12)}$$

So that the Fourier transform of $\Delta\psi_n$ goes like 1/q and $\Delta\psi_n$ decays at large r like 1/r (in 2D). Thus

$$|\Delta\psi_n(r)|^2 = \frac{1}{\pi n(0) Dr^2} \qquad \text{(VI-13)}$$

The diffusing electron has a wavefunction which decays like a power-law. This extremely interesting observation does not prove, however that the electron can diffuse on every scale and that Eq. (VI-13) is always valid. While Kaveh and Mott were able to derive the weak-localization correction (VI-6) when (VI-13) holds, it is clear that the picture does not hold at arbitrarily large r. One difficulty is that the normalization integral for $\psi_n$ diverges logarithmically with L. The cut-off length needed is easily seen to be equal to the localization length, $\xi$, found before from the scaling theory! The remedy to this is that D should somehow decrease on scales larger than $\xi$, the contribution of $\psi_n^o$ should become unimportant and Eq.(VI-11-13) become invalid. The simplest self-consistent scenario for this is the exponential localization one. This does not rule out more interesting power-law behaviors - which have yet to be demonstrated conclusively (M.Kaveh , unpublished results). Different arguments for power-law localization were also given by Azbel (1982). However, the initial numerical evidence for the power-law localization by Pichard and Sarma (1981) did not persist in larger samples (Pichard and Sarma, 1982). There is an experimental evidence for power-law localization, (Davies et al., 1982) consisting of power-law fits to $\sigma(\tau)$. It has to be checked whether these will not be explainable by the Thouless $e^2n(0)\xi^2/\tau_{in}$ mechanism.

The simplest way to derive Eq.(VI-10), for which we will have many uses later, is to note (Azbel, 1981) that it follows from the classical approximation for a diffusing particle. In this approximation the transition probabilities - the LHS of Eq.(VI-10) in the quantum case - are equal to the Fourier transforms of the appropriate time-dependent classical quantity with frequencies $(E_n-E_m)/\hbar$. For $e^{iqr(t)}$ (taking $r(0)=0$) the classical average is given by:

$$\langle e^{i\,q\,r(t)}\rangle = e^{-1/2q^2\langle r(t)^2\rangle} = e^{-q^2Dt}, \qquad \text{(VI-14)}$$

the Fourier transform of which yields the well-known Lorentzian in the RHS of Eq.(V-10). Another derivation of Eq.(VI-10) (Kaveh and Mott, 1981; Abrahams et al., 1981; McMillan, 1981; Imry et al., 1982) uses the appropriate dynamic structure factor, $s(q,\omega)$, for diffusion, which is proportional to the RHS of Eq.(VI-10) and given in terms of the desired matrix elements squared.

### 3. The Case d>2, The Metal-Insulator (M-I) Transition

The new feature which appears for d>2 is the occurrence of a metal-nonmetal transition, associated with the fact that $\beta$ vanishes at some $g=g_c$. This zero of $\beta(g)$ follows because $\beta$ is positive for $g\to\infty$ and negative for $g\to 0$. The value of $g_c$, as well as the slope, s,of $\beta$ at $g_c$ - which will have an important role to play - are numerical constants that can be obtained from approximations (Wollhardt) and Wolfle, 1981) or from simulations (Stein and Krey, 1981; McKinnon, 1982). It is agreed that s is not far from unity and $g_c$ is perhaps 2-3 for d=3 - which is the case of most interest. We note that if g on any scale is $>g_c$ - the conductance will "flow" to the Ohmic, conducting, limit as $L\to\infty$. Likewise, if g is sometimes $<g_c$ - it will "flow" to the insulating range, $g\sim e^{-\alpha L}$, when $L\to\infty$. $g=g_c$ is a "fixed point" of the RG transformation" - if $g=g_c$ on some scale then $g=g_c$ on any scale, including $L\to\infty$ and thus $\sigma\sim g_cL^{2-d}\to 0$ in the macroscopic limit. According to this simple theory all materials (where an important necessary condition for the applicability of the

theory is their being homogenous on the scales of interest) "sit" on the same universal $\beta(g)$ curve and can be distinguished from each other, say, by their conductance, $g_0$, on some microscopic scale, $L_0$. Clearly, all materials with $g_0 > g_c$ are conductors and all those with $g_0 < g_c$ insulators. It is interesting to find out what happens when the transition is approached when $\varepsilon = |\ell n g_o - \ell n g_c| \cong |g_0 - g_c|/g_c \ll 1$. For a range of scales L from $L_0$ on, the behavior can be approximated by $\beta(g) \sim s \ \ell n \ g/g_c$, until g changes enough to get into the macroscopic range (where the limiting forms of Eq.(V-2) are valid). The scale beyond which the macroscopic laws are valid is denoted by $\xi$, in analogy with the usual correlation length in other phase transactions. By integrating the linear approximation to $\beta(g)$ from $L_0$ to $\xi$ we find

$$\frac{\ell n \ g/g_c}{\ell n \ g_0/g_c} = \left(\frac{\xi}{L_0}\right)^s \qquad \text{(VI-15)}$$

so that

$$\xi \sim L_0 \frac{\text{const}}{\varepsilon^{1/s}}. \qquad \text{(VI-15)}$$

Thus, the critical exponent of the diverging $\xi$ is $\nu = 1/s$ ($\cong$ 1 at d=3), employing the usual notation. In the "macroscopic", $L \gg \xi$, regime, $g \propto e^{-L/\xi}$ in the insulating and $g \propto L^{d-2}$ in the conducting phase where the first correction to the latter is given by Eq.(VI-22) below In both cases, in the whole range $L_0 \lesssim L \lesssim \xi$ g does not change by much more than an order of magnitude within the cross-over regime from the linear ("critical" or "microscopic") to the macroscopic range. Thus, the macroscopic conductivity for the metal is given by

$$\sigma_\infty(L \to \infty) \sim \frac{\text{const}}{\xi^{d-2}} \qquad \text{(VI-17)}$$

Note that the M-I transition in this theory is second order and there is no "minimum metallic conductivity" ($\sigma \to 0$ continuously as the transition is approached). The minimum metallic conductivity $\sigma_m$ still gives one an estimate of when non-trivial things start to happen.

We note that for $L \ll \xi$, the two phases are qualitatively similar (in both of them g does not change by more than an order of magnitude in the range from $L_0$ to $\xi$ as mentioned above) thus (apart from the variation in g, which is relatively unimportant for large $\xi$):

$$\sigma(L) \sim \sigma_\infty \left(\frac{\xi}{L}\right)^{d-2} \qquad (L \lesssim \xi). \qquad \text{(VI-18)}$$

I.e. the conductivity (and therefore the diffusion constant) is scale-dependent for $L \ll \xi$. The physical meaning of $\xi$ in the insulating phase is obvious - it is the localization length. The above considerations suggest that $\xi$ in the conducting phase is the length below which the behavior is roughly the same as in the insulating phase. For $L \ll \xi$ the wavefunctions and various correlation-and Green's-functions behave similarly in the two phases. It is only for $L > \xi$ where the differences between exponential decay and a non-zero average value is apparent. The order-parameter describing this transition can probably be determined using these considerations. The "anomalous" diffusion in the microscopic regime $L < \xi$, given by Eq.(VI-18), has interesting consequences for the relationship between time- and length-scales in this regime. Instead of $L^2 \sim Dt$ for usual diffusion, here D is renormalized as the scale is changing, and $dL^2/dt = D_L$. This implies

$$L^d \sim \xi^{d-2} D_\infty t \qquad \text{(VI-19)}$$

where $D_\infty$ is the macroscopic diffusion constant in the metallic phase for the given $\xi$. Thus, in terms of an inelastic scattering time $\tau_{in} \propto T^{-p}$, as before the appropriate length which is given by $L_T \sim \sqrt{D\tau_{in}} \sim T^{-p/2}$ (Eq.(V-6) in the macroscopic regime, is given here by:

$$L_T \sim (\xi^{d-2} D_\infty \tau_{in})^{1/d} \propto T^{-p/3}, \qquad \text{(VI-20)}$$

in the microscopic regime (d=3).

In both the macroscopic metal and in the microscopic regime of both metal and insulator the large-sample conductivity function of temperature is given by $g(L_T)\, L_T^{\;2-d}$ as discussed earlier. In the microscopic regime g stays a constant within about an order of magnitude and

$$\sigma(T) \sim L_T^{\;2-d} \sim T^{(d-2)p/3} = T^{p/3} \;\text{(at d=3)} \qquad \text{(VI-21)}$$

In the macroscopic conducting regime $\beta(g)$ is given by Eq.(VI-3), integrating which yields $g(L_T)$ and hence

$$\sigma(T) = \text{const} + C\, L_T^{\;2-d} = \text{const} + O\,(T^{p(d-2)/2}) \qquad \text{(VI-22)}$$

where the correction ("weak localization") increases with T like an appropriate power. There exists now experimental evidence for both Eq.(VI-21) and Eq.(VI-22) where Eq.(VI-21) is the relevant correction to the normal behaviour in the microscopic regime. In the range where these considerations are valid ($L_T \gg \ell_{el}$ is an important condition necessitating not too clean samples and low temperatures) we are getting a negative TCR which is a universal attribute of dirty conductors! Moreover, to get a negative TCR at temperatures around room temperature, where $\ell_T$ is very small, one needs a $\ell_{el}$ comparable to the interelectron distance and $\sigma$ just somewhat larger than $\sigma_{min}$. We believe that this constitutes a valid qualitative explanation for the Mooij correlations discussed in Section 1.

The smallest permissible value of the length $L_0$ is usually taken as $\sim \ell_{el}$. However, there exist many cases where inhomogeneity of some sort exists in the system and it may be viewed as homogenous only on scales larger than some homogenization length $\ell_{ho}$. For granular metals, $\ell_{ho}$ should be on the order of the grain size, d; near the percolation threshold $\ell_{ho}$ will be on the order of the percolation correlation length. $\ell_{ho}$ values of $10^3$ A are not uncommon. It is well known experimentally that in these systems the effective elastic mean free path near the M-I transition is very small ($10^{-2}$-$10^{-1}$A is possible) and hence the associated conductivity is much smaller than any appropriate $\sigma_m$. In granular metals, (for a review, see Abeles et al., 1975), these conductivity values characterizing the M-I transition are found to go like 1/d. These facts are easily understood noting that $\ell_{ho}$ (or d in granular metals) is the appropriate scale $L_0$. Thus, from the scaling theory the conductivity around which localization is important should be (Imry, 1980)

$$\sigma_0 \sim \frac{e^2}{\hbar \ell_{ho}} \tag{VI-23}$$

in good agreement with both the order of magnitude and grain size dependence (Adkins,1976 ) mentioned above. These facts are hard to understand using naive $\sigma_m$ Yoffe-Regel considerations.

In the presence of a magnetic field, the relevant length scale becomes $\ell_H$ once it is $<<\ell_T$. This yields a large negative MR, in analogy to the 2D case (except that now the MR behaves like $\sqrt{H}$ for large H in the weak localization limit - Eq.(VI-22). This may be the explanation for many cases of "anomalously large" negative MR in dirty systems. Further complications, such as spin-orbit scattering may change the sign of the MR, and there is now a large body of work on these aspects, including also the possible effects of electron-electron intractions.

A concise presentation of the results of the scaling theory for $\sigma(L)$ in the 3D metallic range not too far from the transition is provided by the following figure.(VI-2).

The lower curve depicts $\sigma(L)$ at the transition, it simply goes as $\sigma_c L_0/L$ where $\sigma_c = g_c e^2/\hbar L_0$. The upper curve is for $\sigma(L_0)$ somewhat above the transition. Here $\sigma(L)$ goes like 1/L in the microscopic range $\xi > L > L_0$, and like $\sigma_m + e^2 C/\hbar \ell_T$ the macroscopic regime $L >> \xi$. The macroscopic $\sigma_m$ is $\sim Ae^2/\hbar\xi$ where A is the order of magnitude of g where $\beta(g)$ becomes close to unity, i.e. $A \sim 20$. The correction to $\sigma$ at $L=\xi$ is on the order of $e^2/\hbar\xi$. Thus $\sigma_m$ is still important for $L \lesssim \xi$ , this is relevant for the interpretation of experiments in the microscopic regime that have supported Eqs. (IV-20-21) (Ovadyahu and Imry, 1982).

As mentioned before, many experiments on 2D and 3D systems

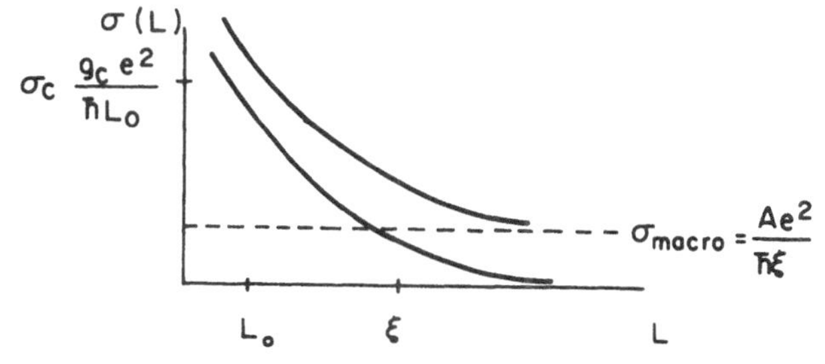


Fig.VI-2: σ(L) in 3D (schematic)

are in very good qualitative agreement with this picture, provided that some new ideas on the mechanisms for inelastic scattering are accepted. One can approach now a quantitative understanding of the experiments, especially if the effects of electron-electron correlations are understood. The two general new features that the experiments have revealed are:

1. $\tau_T$ is typically shorter by a few orders of magnitude in dirty systems than expected for otherwise similar pure ones.

2. The temperature dependence of $\tau_T$ is weak, usually p=1-1.5 in 1D, 2D and 3D.

We are now in the process of gaining an understanding of these effects. There is still some difficulty in the disorder-dependence of $\tau_T$ as well as in the 1D case, although there are interesting ideas about these as well. Before embarking on all this we have to review two further subjects - screening and the effects of Coulomb interactions.

## VII DIELECTRIC PROPERTIES

The ability to screen the long-range part of the Coulomb interaction is as important an attribute of the conducting state of matter as the finite conductivity itself. In addition to the static screening, the frequency dependencies of σ and the dielectric constant ε determine the non-trivial optical properties (including the microwave and infrared ranges)of the conductor. Important anomalies in all these properties also exist around the M-I transition and in the insulating phase near the transition.

We start by reviewing briefly the Thomas-Fermi picture (see e.g. Kittel, 1963) for static screening in a degenerate electron system. We discuss the response of the system to a small external potential $V^{ext}(x)=V_k^{ext}e^{ikx}$ (more Fourier components can be handled by linear superposition). The system will reach equilibrium generating a screened potential $V^{tot}{}_k e^{ikx}$ so that the total electrochemical potential become a constant. Thus, the electron density n(x) should change, causing a change in the Fermi level (denoted by μ(x), so that the electrochemical potential, $-eV^{tot}(x)+\mu(x)$), should stay constant. Since $\Delta\mu=(\partial\mu/\partial n)\,\Delta n$, we find

$$\Delta n(x) = e \frac{\partial n}{\partial \mu} V_k^{tot} \quad e^{ikx} \tag{VII-1}$$

which yields an induced potential, via Laplace's equation

$$k^2 V_k^{ind} = -4\pi e^2 \frac{\partial n}{\partial \mu}(V_k^{ext} + V_k^{ind}) = -4\pi e^2 \frac{\partial n}{\partial \mu} V_k^{tot} \tag{VII-2}$$

so that the total potential energy becomes

$$V_k^{tot} = V_k^{ex} + V_k^{ind} = V_k^{ex}[1 - \frac{1}{1+\Lambda^2k^2}] = V_k^{ex}\frac{\Lambda^2k^2}{1+\Lambda^2k^2} \tag{VII-3}$$

where the Thomas-Fermi screening length is Λ, defined by:

$$1/\Lambda^2 = 4\pi\, e^2 \frac{\partial n}{\partial \mu} \tag{VII-4}$$

$V^{tot}$ is given by $V_{ex}$ divided by the dielectric constant, so that

$$\varepsilon_k = 1 + \frac{1}{\Lambda^2k^2}, \tag{VII-5}$$

which yields e.g. the well-known screened field of a point charge. We note that no specific properties of the conductor have been used here, except that the electrons can move to keep the equilibrium (constant electrochemical potential) condition. Thus, we expect the dirty conductor to also display static Thomas-Fermi type screening.

One can establish this (Imry, Gefen and Bergman, 1982) for non-interacting electrons in a random potential by using the linear response expression for the complex wavevector-and-frequency-dependent dielectric function:

$$\tilde{\varepsilon}(q,\omega) = \varepsilon(q,\omega) + \frac{4\pi i}{\omega}\sigma(q,\omega), \tag{VII-6}$$

$$\tilde{\varepsilon}(q,\omega) = 1 - \lim_{\eta\to 0}\frac{4\pi e^2}{q^2}\int d\,\varepsilon_i\, N(\varepsilon_i)\int d\,\varepsilon_f N(\varepsilon_f) \; |\langle f|e^{iq\cdot r}|i\rangle|^2 \; \frac{f(\varepsilon_i)-f(\varepsilon_f)}{\varepsilon_i-\varepsilon_f+\hbar\omega+i\eta} \tag{VII-7}$$

We shall factor out the densities of states at $\varepsilon_F$, use Eq.(VI-10) for the matrix elements, the well-known expression:

$$\lim_{\eta\to 0} \frac{1}{x+i\eta} = P(\frac{1}{x}) - i\pi\delta(x), \tag{VII-8}$$

and the approximation $f(\varepsilon_i)-f(\varepsilon_f)/(\varepsilon_i-\varepsilon_f) = f'(\varepsilon_f)$, which is valid for $Dq^2 << kT << E_F$, to get, for $\omega=0$

$$\varepsilon(q,\omega=0) = 1 + \frac{1}{\Lambda^2 q^2} P\int_{-\infty}^{\infty} \frac{Dq^2}{(Dq^2)^2+\omega^2} d\omega = 1 + \frac{1}{\Lambda^2 q^2}, \tag{VII-9}$$

as expected! Actually, here $1/\Lambda^2 = 4\pi n(0)e^2$, the equality to Eq. (VII-4) follows because for free electrons $n(0)=\partial n/\partial\mu$.

In fact, this establishes the Thomas-Fermi result quite generally, even for more complex types of diffusion, such as the "anomalous" one in the microscopic range. The reason being that the integrand in Eq.(VII-9) which is the "dynamic structure factor" of a diffusing electron is the Fourier transform of $\exp(-q^2\langle r^2(t)\rangle)$ Since $\langle r^2(t)\rangle$ vanishes at t=0, by definition, the integral is equal to unity (a well-known sum rule) so that the Thomas-Fermi result is thus established quite generally.

In fact, this result seems to be too general - what about the insulating phase, that does not screen the $1/q^2$ part of the Coulomb interaction, and thus has a finite $\lim_{q\to 0} \varepsilon(q,0)=\varepsilon(0,0)$? It turns out that if the electron is localized on a scale $\xi$, and $\langle r^2\rangle$ is thus bounded - then the integrand in Eq.(VII-9) acquires a $\delta$ function component, which does not contribute to the principal part. A simple analysis yields

$$\varepsilon_{ins}(0,0) \sim \xi^2 \tag{VII-10}$$

in agreement with what might be expected in a scaling picture (Rosenbaum et al., 1980; Imry et al., 1982). Thus, the static dielectric constant diverges in the insulator as the M-I transition is approached. Such a divergence, with the appropriate critical behavior has been found in the Si:P system (Rosenbaum et al., 1980; Capizzi et al., 1980).

It is almost trivial now to obtain the imaginary part of Eq.(VII-7) as well, which yields for simple diffusion (valid for $q << 1/\xi$):

$$\sigma(q,\omega) = \frac{e^2 n(0) D\omega^2}{(Dq^2)^2+\omega^2} = \frac{\sigma(0,0)\omega^2}{(Dq^2)^2+\omega^2} \tag{VII-11}$$

This is valid as long as Eq.(VI-10) can be used, i.e. at small q and small $\omega$. At q=0, this yields $\sigma(0,\omega) = \sigma(0,0)$, which is appropriate for the small $\omega$ limit.

It is possible to derive from Eq.(VII-7) and the appropriate expressions for the dynamic structure factor, governing the matrix elements in Eq.(VI-10), the behaviors of $\sigma(q,\omega)$ and $\varepsilon(q,\omega)$ (Wegner, 1976) in all the regimes in both insulating and conducting phases (Imry, Gefen, Bergman, 1982; Shapiro and Abrahams, 1982). These determine, for example, the optical properties in the relevant frequency range ($\omega\,\tau_{el} < 1$). In particular, at low frequences $\sigma(0,\omega)$ in the insulating phase goes like $\omega^2$ (with a coefficient that diverges like $\xi^4$ when the transition is approached). A very interesting behavior is obtained in the microscopic regime ($q>1/\xi$ and/or $\omega> D/\xi^2$ in both metallic and insulating phases). For q=0 one finds

$$\varepsilon(0,\omega) \propto \left(\frac{D\xi/\Lambda^3}{\omega}\right)^{2/3}, \quad \sigma(0,\omega) \propto (D\xi/\Lambda^3)^{2/3}\,\omega^{1/3}, \qquad \text{(VII-12)}$$

which can be checked by optical experiments in the (far) infrared range.

## VIII EFFECTS OF ELECTRON-ELECTRON INTERACTIONS

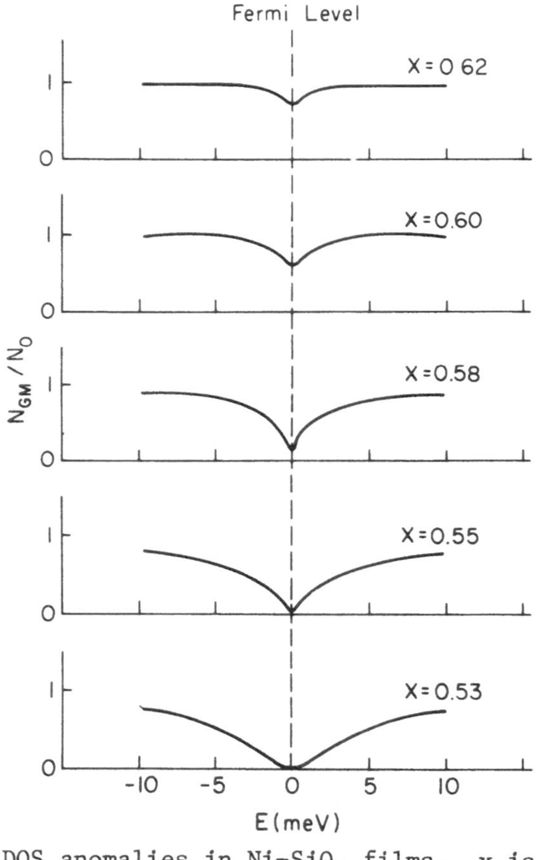


Fig.VIII-I: DOS anomalies in Ni-$SiO_2$ films. x is the Ni fraction, E is measured from $E_F$ (after Abeles and Sheng, 1974).

1. Changes in the Density of States (DOS)

It is well known, (see e.g. Kittel, 1963), that for the usual electron gas with Coulomb interactions in the Hartree-Fock approximation, the exchange terms yield anomalies in the self-energy and in the resulting DOS (a logarithmic vanishing of the latter at $E_F$). These logarithmic singularities are spurious; for example, they are eliminated if the screened Coulomb interaction is used. It is found, however, that in the electron gas with even a weak disorder, a weak singularity of n(E) at $E_F$ should really exist (Altshuler and Aronov, 1979, see also McMillan, 1981), and it is , in fact, found experimentally (Abeles et al., 1974 (Fig. VIII-1); Mochel, 1981, Dynes and Garno, 1981; Imry and Ovadyahu, 1982).

We remind ourselves of the Hartree-Fock theory, (see e.g. Kittel, 1963), the self-consistent approximate Schrodinger equation reads:

$$\frac{\hat{p}^2}{2m} \phi_m + \int d^d y\, V(x-y) \sum_l f_l\, \phi_l^*(y)\phi_l(y)\phi_m(x) - \int d^d y V(x-y)$$

$$\sum_l{}' f_l \phi_l^*(y)\phi_l(x)\phi_m(y) = E_m \phi_m(y), \qquad \text{(VIII-1)}$$

where $E_m$ and $\phi_m$ are the one electron energy levels and orbitals respectively. $f_l$ are the occupations, $\sum_l'$ signifies summation over spins parallel to that of m, V is the electron-electron interaction, the second term is the (direct) Hartree contribution and the third is the (exchange) Fock one. The energies are given by

$$E_m = \varepsilon_m + \Sigma_m \qquad \text{(VIII-2)}$$

$\varepsilon_m$ being the unperturbed non-interating electrons energies and $\Sigma_m$ is the self energy given in the Hartree-Fock approximation by

$$\Sigma_m = \sum_l f_l \langle ml|V|ml\rangle - \sum_l{}' f_l \langle ml|V|lm\rangle \qquad \text{(VIII-3)}$$

where $\langle ml|V|ml\rangle = \iint d^d x d^d y\, \phi_m^*(x)\phi_m(x)V(x-y)\phi_l^*(y)\phi_l(y)$ and $\langle ml|V|lm\rangle$ is the same with x and y interchanged in the arguments of $\phi_l$ and $\phi_m$.

In the usual electron gas without disorder, the Hartree term just cancels exactly the contribution of the uniform positive background. This cancellation will not occur in our disordered system. We shall nevertheless concentrate on the exchange terms $\Sigma_m^{ex}$. More complete calculations including the direct term and using dynamic screening multiply the result we shall get by important factors, but do not change the nature of the singularity.

We use the Fourier representation of V.

$$V(r) = \frac{1}{(2\pi)^3} \int d\vec{q}\, V_q\, e^{iq\cdot r}, \quad \text{thus:} \qquad \text{(VIII-4)}$$

$$\langle m\ell|V|\ell m\rangle = \frac{1}{(2\pi)^3}\int d\vec{q}\; |\langle m|e^{iq\cdot r}|\ell\rangle|^2 V \tag{VIII-5}$$

In the simplest and least sophisticated approximation, we use the statically screened Coulomb interaction for $V_q$ and Eq.(VI-10) for the matrix elements squared. The interesting behavior follows from the small q limit where $V_q=4\pi e^2/q^2 \cdot \Lambda^2q^2=1/n(o)$ (note that the electron charge e has been cancelled). Thus as function of the unperturbed energy, $\varepsilon$

$$\sum^{ex}(\varepsilon) = -\frac{1}{(2\pi)^3}\int_{\varepsilon'<\varepsilon_F} d\varepsilon'\int d^dq\,\frac{1}{\pi n(0)}\,\frac{D\,q^2}{(Dq^2)^2+(\varepsilon-\varepsilon')^2/\hbar^2} \tag{VIII-6}$$

We note that the change in the DOS is given by:

$$\frac{dn}{dE} = \frac{dn}{d\varepsilon}\,\frac{d\varepsilon}{dE} = \frac{dn}{d\varepsilon}\Big/\left(1+\frac{d\Sigma}{d\varepsilon}\right), \tag{VIII-7}$$

where $\frac{dn}{d\varepsilon} = n(\varepsilon)$ is the unperturbed DOS. Taking the derivative of VII-6, $\sum'^{ex}$ is seen to have a mild singularity at $\varepsilon=\varepsilon_F$ in 3D. It is conveniet to accentuate the singularity by taking another derivative. We first note that changing the $\varepsilon$ integration in Eq.(VII-6) to the variable $(\varepsilon-\varepsilon_F)$, then for $\varepsilon > \varepsilon_F$, the upper limit on $\varepsilon-\varepsilon'$ is$(\varepsilon-\varepsilon_F)$. Thus, the first derivative is just the integrand:

$$\sum^{ex'}(\varepsilon) = \frac{1}{8\pi^4 n(0)}\int d\vec{q}\,\frac{Dq^2}{(Dq^2)^2+\omega^2} \tag{VIII-8}$$

where here $h\omega\equiv\varepsilon-\varepsilon_F$. We thus find

$$\sum^{ex''}(\varepsilon) = \frac{1}{4\pi^4 n(0)}\int d^dq\,\frac{Dq^2\omega}{[(Dq^2)^2+\omega^2]^2} = \frac{\text{const}}{D^{3/2}\omega^{1/2}} \tag{VIII-9}$$

Thus, $\sum'$ will have a singular contribution which is a numerical constant, C times $\omega^{\frac{1}{2}}/hn(0)D^{3/2}$.

$$n(E)=n_o(E)[1 + C_1\omega^{1/2}/\hbar n(0)D^{3/2}] \tag{VIII-10}$$

Thus, the DOS has a square-root singularity at $\varepsilon_F$, whose total relative amplitude (for $\hbar\omega\sim\varepsilon_F$, say) is of the order of $1/(k_F\ell)^{3/2}$. The calculation we presented here (Lee, 1981, McMillan, 1981) is a simplification of the low-order systematic one by Altshuler and Aronov (1979). That calculation is only valid when the DOS correction is small (or $k_F\ell >>1$). In this case, there is a good agreement with the experiments. One may try to extrapolate this type of argument to the critical region near the M-I transition using self-consistent scaling methods (Gefen and Imry, 1982) or carrying the perturbation calculation to higher order and using a RG extrapolation (Lee and Grest 1983; Castellani et al., 1982; Finkelstein, 1983). The results of all these theories are that the dip in the DOS increases as the transition is approached and that, in fact, n(0) vanishes at the transition. There is still no theoretical agreement on the details of this critical behavior, but the above-mentioned experiments clearly show the expected qualitative features. A recent experiment by Hertel et al. (1983) found, for example, that n(0) vanishes near the M-I transition in Au:Si systems linearly with the distance from the transition.

An interesting aspect of the DOS anomalies is their dimensionality dependence. For a system which is thin enough to be effectively 2D, the $\sqrt{\omega}$ singularity is replaced by a stronger, logarithmic, one. To be effectively 2D, it turns out that the thickness, d, of the film should be smaller than the characteristic length $\ell_\omega \sim \sqrt{D/\omega}$ (we take $\omega \gg kT, \tau_T^{-1}$). Both this crossover and the logarithmic behavior have recently been observed experimentally (Imry and Ovadyahu, 1982).

## 2. On The M-I Transition Due to Interactions

In the last section we discussed briefly the relevance of the electron-electron interaction near the localization transition. Electron-electron interactions may induce a M-I transition on their own, even if no disorder is present. A simplified model for this transition is the Hubbard one defined on a lattice, in which the long range Coulomb interaction is replaced by a simpler, very short range, on-site repulsive interaction U (Hubbard, 1964). U operates only between electrons with antiparallel spins, since electrons with parallel spins cannot occupy the same site. For N sites with nearest neighbor transfer matrix element t; a band with a width proportional to t is obtained for U=0. This band has 2N states, including spin. On the other hand, for large U/t, there is a gap due to the U interaction between the N lowest states (lower Hubbard band) and the N highest states (upper Hubbard band). In the half-filled case one thus has an insulator for large U/t, where the lower Hubbard band is full and the upper one empty. We thus expect a transition as a function of U/t from a single band to two split bands of some critical value $(U/t)_c$. This will yield a M-I transition in the half-filled case. This is a special case of the Mott transition - a M-I transition due to electron correlations, even without disorder.

Clearly, then, both interactions and localization can cause M-I transitions on their own. When both are present, one would expect them to be relevant and change the critical behavior from the two separate cases. The problem of a Hubbard model with a disordered site energies (such as in Eq.(I-2) was recently studied, employing an approximate real space RG method, by Ma (1982). In 1D, an Anderson insulator is found for any finite disorder, making a transition to a Hubbard insulator (with a gap) for a large enough U/t. In 3D, a metallic phase is also obtained in addition to the two insulating phases. Therefore, there also exists a triple point where the three phases meet. The critical behavior at the transition from the Anderson insulator to the metal is modified due to the interactions. An intriguing and unexpected result is that a small U turns out to provide delocalization, in the sense that more disorder is needed to get the localized phase. This is probably due to some type of an effective "screening" in this short range interaction system. Namely, the existence of an electron near a deep potential site makes, through the U, this site less attractive for opposite spin electrons. The subject of the M-I transition for disorder with both long- and short- range interactions deserves further study.

## 3. Inelastic Scattering Due to Electron-Electron Interactions

As mentioned at the end of Section 6, the large number of experiments that followed the initial suggestion by Thouless (1977) on localization effects in thin wires (Giordano et al., 1979 Giordano, 1981, Fowler et. al., 1982) and the scaling theory predictions for thin films (Dolan and Osheroff, 1979; Bishop et al., 1980; Markiewicz, 1981, Pepper, 1980, Uren et al., 1981; Ovadyahu and Imry, 1981) and 3D systems (Imry and Ovadyahu, 1981; Hertel et al., 1983) yielded results that were in only very rough qualitative agreement with the theory. To explain the experiments more quantitatively, one has had to assume that the inelastic scattering mechanisms, yielding $\tau_T$ and hence the characteristic temperature-dependent length scale $L_T$, were very different from what one has been used to in clean metals. The general features are that $\tau_T$ is typically much shorter than in clean systems, and its temperature dependence is much weaker (i.e. like a small power of T, not like $T^{3-4}$ as in usual electron-phonon scattering). In fact, Schmid has already found in 1974 that in dirty systems the electron-phonon scattering is much weaker but the electron-electron scattering is much stronger than in clean metals and may well be dominant. The latter fact is due to the relaxation of the momentum conservation which greatly inhibits the scattering in a degenerate Fermi system. The calculated $\tau_T$ in the various regimes (Schmid, 1974; Abrahams et al., 1981, Gefen and Imry, 1981) makes the localization contribution to be in a much better, but not perfect agreement with experiment. On the other hand, one also has a competing theory - direct interaction contributions to the transport - (Altshuler and Aronov, 1979; Altshuler et al., 1980, Fukuyama, 1983) which may also be relevant to some of the experiments - but again does not explain all the trends fully. We shall briefly review these contributions in the next subsection. It appears that a quantitative understanding of experiments will also necessitate the localization theory, with a better understanding of the inelastic scattering.

We now review briefly the calculation of the lifetime, $\tau(E)$, of a quasiparticle having an energy E above the Fermi level, due to electron-electron scattering. At a temperature T, this lifetime, with $E \sim kT$, will yield the appropriate $\tau_T$. Clearly, the lifetime is determined by the imaginary part of the self-energy:

$$\tau(E) = \hbar/\mathrm{Im}\Sigma(E) \qquad \text{(VIII-11)}$$

To obtain $\mathrm{Im}\Sigma$, we use the dynamically screened interaction $4\pi e^2/q^2\,\tilde{\varepsilon}(q,\omega)$ (for $\omega=(\varepsilon-\varepsilon')/\hbar$) in Eq.(VIII-6), instead of the static one, $1/n(0)$. From the imaginary part of $1/\tilde{\varepsilon}$ one can get, in 3D,

$$\mathrm{Im}\Sigma(\varepsilon) = \frac{1}{\pi(2\pi)^3}\int_{\varepsilon'<\varepsilon_F} d\varepsilon' \int d^3q\, \frac{4\pi e^2}{q^2}\,\mathrm{Im}\left(\frac{1}{\varepsilon(q,\omega)}\right)\frac{Dq^2}{(Dq^2)^2+\omega^2} \qquad \text{(VIII-12)}$$

From VII-6 and evaluating $\mathrm{Im}\,1/\tilde{\varepsilon}$, using Eq.(VII-11) and the generalization (Imry, et al., 1982) of Eq.(VII-9), for small q-$\mathrm{Re}\varepsilon(q,\omega)=\Lambda^{-2}D^2q^2/[(Dq^2)^2+\omega^2]$ we find $\mathrm{Im}(1/\tilde{\varepsilon}) = -\omega/4\pi\sigma(o)$. Thus:

$$\mathrm{Im}\Sigma(\varepsilon) = \frac{1}{\pi(2\pi)^3}\int^{\varepsilon_F} d\varepsilon' \int d^3q\, \frac{e^2\omega}{\sigma(o)q^2}\,\frac{Dq^2}{(Dq^2)^2+\omega^2} =$$

$$= \frac{\sqrt{3}}{4}\,\frac{1}{(k_F\ell)^{3/2}}\,\frac{\varepsilon^{3/2}}{\sqrt{\varepsilon_F}} \qquad \text{(VIII-13)}$$

Where the singular structure for small $\omega$ was analyzed in the same way as done before for the real part and the resulting DOS. Thus (Schmid, 1974, Abrahams et al., 1981),

$$\tau_T \sim \frac{1}{T^{3/2}} \quad \text{in 3D} \qquad \text{(VIII-14)}$$

The corresponding results being $1/T$ in 2D and $1/T^{1/2}$ in 1D, where the effective dimensionality is again determined by the ratio of the appropriate dimension of the film to $\sqrt{D/T}$. In strictly 2D systems (i.e. atomic thickness), more sophisticated calculations have yielded $T\ln T$ instead of $T$ (Abrahams et al., 1981).

Another interesting modification of the above results in 3D occurs in the "microscopic" regime (where $L_T < \xi$, see discussion around Eq.(VI-21). Here $\varepsilon(q,\omega)$ and $\sigma(q,\omega)$ acquire somewhat different powers of $q$ and $\omega$, due to the anomalous (scale-dependent) diffusion (cf. Eq.(VII-12) and Imry et al., 1982). Repeating the above calculations with the modified expressions for $\varepsilon$ and $\sigma$ yields (Gefen and Imry, 1981).

$$\tau_T \sim \hbar/kT, \qquad \text{(VIII-15)}$$

in the microscopic regime.

This result means that at higher temperatures, once $L_T < \xi$, one should get (since p=1)

$$\Delta\sigma \sim T^{1/3} \qquad \text{(VIII-16)}$$

Such a cross-over to $T^{1/3}$ behavior has indeed been observed in $In_2O_{3-x}$ samples (Ovadyahu and Imry, 1981) and also in some Si:P experiments. However, there are many cases where p=1 seems to occur in the macroscopic 3D regime, as well as in "1D" thin wires. One might be tempted to conclude that this is due to the interaction effects of the type mentioned in the next subsection. However, the agreement with these contributions is also not full. For example, it has been found empirically in several cases that not only is p equal to unity but that the effective $\tau_T$ is proportional to $\rho$ (or to $(k_F \ell)^{-1}$), (Giordano, 1981; Markiewicz, 1982; Ovadyahu, 1983), which is not in accord with the pure interaction theory. This cannot be explained either by the above results or by the interaction theory. A different inelastic (or phase-breaking) mechanism is presumably needed. For example, inelastic scattering from two-level systems (such as these that are invoked to explain the anomalous low temperature properties of glasses) has been suggested (Lee, 1981) and shown to yield the required results under some assumptions (Stone et al., 1982). Clearly, more work is needed on the inelastic, phase breaking, processes in dirty metals.

## 4. Direct Interaction Effects On The Transport

In addition to the electron-electron interaction making a relevant contribution to the DOS and the inelastic scattering, it also influences directly the conductivity and the magnetotransport. These in principle straightforward, but quite complicated perturbation calculations in the interaction done only to lowest orders in $(k_F \ell)^{-1}$, have yielded (Altshuler and Aronov, 1979) the surprising result that to first order in $(k_F \ell)^{-1}$, the interactions cause anomalous corrections to e.g. $\sigma(T)$ which look quite similar (see below) to the weak localization correction. However, these are different processes and it looks like they should be added to the localization ones. They do however resemble the latter terms, but with $L_T$ replaced by $\sqrt{D\hbar/kT}$ (which, incidentally, would be the correct scale were Eq.(VIII-15) true more generally). There is no deep insight on why these two different contributions are so similar and whether there is some underlying connection between them. It should also be emphasized that the interaction corrections to transport are not due to the DOS anomalies. The latter were shown to cancel from $\sigma$, at least in the low frequency limit. This subject has been very adequately reviewed (Lee, 1981; Fukuyama, 1981, 1982) including the corrections to the magnetotransport, where some theoretical work is still needed on the Hall effect, (Fukuyama, 1983). We only mention that the interaction picture has predicted some qualitative effects that were confirmed experimentally and do not appear to be explainable by the localization theory alone. It turns out (Altshuler et al., 1980) that the direct terms have opposite signs to the exchange ones and that the relative weight is a function of the carrier density. In 3D the sign of the anomalous terms should be reversed by changing the doping. This indeed happens in the experiment (Rosenbaum et al., 1981), which is an important confirmation of these effects.

## IX THE DESTRUCTION OF SUPERCONDUCTIVITY IN DIRTY AND GRANULAR METALS

The M-I transition may be coupled to other order parameters. For example, the electron-electron interactions involve the spin and thus there may be coupling to antiferomagnetic (Hubbard, 1964) or other magnetic arrangements. Another consideration is the disorder which is varying, e.g., due to a changing magnetic field. This may introduce interesting behavior (Imry, 1981) and help to tune the M-I transition (von Molnar, 1983).

As far as the superconducting order parameter, it is due to a delicate balance between the weak attractive and the (screened) repulsive Coulomb interaction (Maekawa and Fukuyama, 1981). We shall not discuss this from the microscopic point of view, but present some simple primitive scaling considerations in a phenomenological way (Imry and Strongin, 1981).

The normal properties of a dirty metal on a scale L are determined, neglecting interactions, by the two coupling constants $v_L$ and $w_L$ (see Section V). The electron-electron interactions introduce one more coupling constant which could be looked at as the "Hubbard U" on scale L. Once $L \gg \Lambda$, the Hubbard picture of short range interactions between electrons on the same block becomes a reasonable approximation. A convenient representation for the effective coupling is as a "charging energy" (Anderson, 1964; Abeles, 1977) to put an extra electron on the block of side L,

$$E_{C,L} = \frac{e^2}{2C_L}, \tag{IX-1}$$

where $C_L$ is the capacitance of the block. The conditions to have a normal metal are

$$g_L \equiv v_L/w_L > g_c \text{ (a)}; \quad E_{C,L} \lesssim z\, v_L \text{ (b)} \tag{IX-2}$$

where the second condition means that the charging energy is smaller than an appropriate constant times the transfer energy, in the spirit of the Hubbard model results (Kawabata, 1977). L is any length scale larger than $L_0 = \text{Max}(\ell, \ell_{h0})$. (cf. Section VI) (We assume this is automatically larger than $\Lambda$). The macroscopic properties of the system are determined by those on the scale $L_0$ (e.g. the grain size d in the granular metal case).

The existence of superconductivity hinges on yet another coupling constant which, assuming that the superconducting coupling is already weak on scale $L \gtrsim L_0$, can be parameterized by the inter-block Josephson coupling theory, (see e.g. Deutscher et al., 1974), given at zero temperature by:

$$E_{J,L}\ (T=0) = \frac{\pi}{4}\frac{\hbar}{e^2}\frac{\Delta(o)}{R_L} = \frac{1}{4}\, g_L\ \Delta(0) \qquad \text{(IX-3)}$$

$\Delta(0)$ being the BCS superconducting gap relevant for the block (assuming it is superconducting by itself). At finite temperatures, $\Delta$ depends on T and $E_J$ is given by:

$$E_{J,L}(T) = E_{J,L}(0)\ \frac{\Delta(T)}{\Delta(0)}\ \tanh\left(\frac{\Delta(T)}{2k_BT}\right) \qquad \text{(IV-4)}$$

For superconductivity to exist even at T=0, it is necessary that the Josephson coupling will <u>not</u> be eliminated by the pair charging energy (Abeles, 1977; Kawabata, 1977, Simanek, 1979), i.e.

$$ZE_{J,L} \gtrsim E_{C,L} \qquad \text{(IX-5)}$$

which is analogous to the normal metal condition (Eq.(IX-2(b)). We shall discuss this condition further towards the end of this section. Once Eq.(IX-5) is satisfied, then the system is superconducting at T=0, and up to a $T_C$ which is of the order of z $E_{Jd}$ for $E_{c,d}=0$, and goes continuously to zero as $E_{c,d}$ increases. Below, we present a graph taken from Imry and Strongin (1981), showing $T_c/ZE_J$ as a function of $ZE_J/E_c$ in a simple mean-field approximation. Indeed $T_c \to zE_J$ for weak $E_c$ and $T_c \to 0$ for large $E_c$.

A relevant criterion here is whether the effective grain-size, d, is large or small compared with a length $L_S$ defined by (cf. Eq.(II-2) for a definition of w):

$$W_{L_s} \sim \Delta, \text{ or } L_s \sim [\Delta n(0)]^{-1/3} \qquad \text{(IX-5)}$$

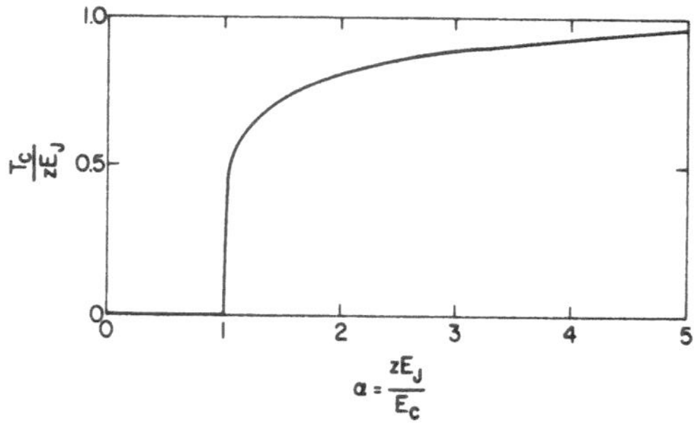


Fig.IV-1: $T_c/ZE_J$ as a function of $\alpha \equiv ZE_J/E_C$ for a granular system.

$L_s$ is on the order of 30A for typical metals. Grains that are much larger than $L_s$ behave almost like bulk superconductors while in grains much smaller than $L_s$ superconductivity is quenched (Anderson, 1959). Since the charging energy $E_c$ is of the same order of magnitude for electrons and for pairs, it is instructive to take the ratio of the intergrain coupling constants for the normal and superconducting problems, on scale L (Imry and Strongin, 1981), given by Eqs.(IX-2(b)) and Eq.(IX-5).

$$v_L/E_{J,L}=w_L/\Delta=(n(0)L^3\Delta)^{-1} = (L_s/L)^3 \qquad \text{(IX-6)}$$

where we used Eqs.(IX-2(a)), (IX-3) and (IX-6). This means that the normal metal interblock transfer energy is smaller (larger) than the superconducting Josephson coupling, for large (small) grains. This very simple observation does have strong consequences. $v_{Lo}$ and $E_{JLo}$ will determine the properties of the system, where $L_o$ is the "microscopic" scale ( $\ell$ for homogenous systems, grain size, for granular ones which are of interest to us). For large grains, $d{>>}L_s$, we find that it is <u>easier</u> to satisfy the condition for superconductivity than the one for being a normal metal, since $E_{J,L}{>>}v_L$. Thus, when the disorder is increased, for this case, there will be a range where the system will be semiconducting but will transform to a superconducting phase at low temperatures (Strongin et al., 1970; Shapiro and Deutscher, 1982). This consideration is, of course relevant only for granular or analogous systems where $d{>>}\ell$. Otherwise, $\ell{>>}L_s$ means that localization effects are weak anyway. On the other hand, for small grain systems, $d\lesssim L_s$, the condition for superconductivity is similar to or somewhat stronger than the one for a normal metal so we would expect to lose superconductivity and normal conductions around roughly the same amount of disorder.

As far as estimating the charging energy $E_c$, it turns out that the naive estimate $e^2/C$, where C is the capacitance of a metal sphere of size $L_o$, is in many cases a substantial overestimate (McLean and Stephen, 1979). What is, of course, missing is an appropriate dielectric constant in the denominator, which may be very large. In fact, (Imry and Strongin, 1981) in the metallic phase, $g_{Lo}>g_c$, we know that the Coulomb interactions are screened out over length scales $\Lambda$. Thus, for $L_o{>>}\Lambda$, which is usually the case since $\Lambda$ is typically a few A, for metals (see, however, remarks below), we expect $E_{cLo}$ to be unimportant. Similar considerations should also apply in the insulator as long as it is a "weak" one (close enough to the M-I transition). The important quantity here is not the macroscopic dielectric constant but rather $\varepsilon_q$ for $q\sim 1/L_o$. Near the M-I transition, when $\xi{>>}L_o$, both the poor metal and the weak insulator screen in roughly the same manner. These qualitative considerations are somewhat complicated by the increase in $\Lambda$, due to electron-electron interactions, near the transition, as mentioned in Section VIII-2. However, at least one need not worry about $E_c$ over most of the metallic phase.

The parameter governing the behavior is $g_{L_0}$, this is valid in both the 3D case discussed so far, where $\sigma$ or $\rho$ is the pertinent parameter, and in the 2D case. In the strict 2D limit, where one has a single layer of grains, the relevant parameter is the sheet resistance R . Our considerations suggest that the 30 kΩ limit for superconductivity in thin films is not a valid consideration for large grains or Josephson array (Voss and Webb, 1982) systems. The 2D superconductor has a special interest due to the peculiar type of long range order and vortex unbinding phase transition expected (for a recent reference see Hebard and Fiory, 1983), which is a fascinating subject that we will not discuss here.

We also did not discuss the microscopic theory of the $T_c$ depression (Maekawa & Fukuyama, 1981, Anderson et al. 1983). This is a delicate problem. The depressed DOS, the weakened Coulomb screening and perhaps the modified electron-phonon coupling (Schmid, 1974) all play their roles in determining $T_c$. In the case of extremely thin films and certainly in wires, the well-known screening anomalies (Stern, 1967; Kuper, 1964) should be relevant in depressing $T_c$, perhaps more so than the celebrated fluctuations effects.

## X CONCLUSIONS

Our aim in these lectures has been to review the physical aspects of the theory of localization and build up the ability to apply it to a variety of problems. Confronting the various predictions of the scaling theory of localization, augmented by electron-electron and electron-phonon interactions, with experiments - will be an important way to check the validity of these appealing but far from rigorous ideas. Much remains to be done on the fundamental theory (see e.g. Wegner, 1976 and the lectures by Fukuyama in these proceedings) aspects, which we have not touched in any detail here.

## ACKNOWLEDGEMENTS

The author would like to express his thanks to numerous colleagues and collaborators for instructive discussions: M. Ya Azbel, D.J. Bergman, M. Brodsky, M. Buttiker, P. Chaudhari, J. Chi, A. Fowler, A. Goldman, D. Gubser, A. Hartstein, S. Kirkpatrick, R. Laudauer, S. von Molnar, D. Prober, T.D. Schultz, F. Stern, M. Strongin, C. Tsuei, R. Voss, R. Webb, R. Wheeler and especially to Y. Gefen and Z. Ovadyahu. This research was partially supported by the U.S.-Israel Binational Science Foundation and partially conducted in the IBM Research Center, Yorktown Heights.

REFERENCES

Abeles, B., 1976, in: "Applied Solid State Sciences," R. Wolfe, ed., Academic, New York.
Abeles, B., 1977, Phys. Rev. B15:2828.
Abeles, B., and Sheng, P., p. 578 in the proceedings of LT-13 (Timmerhouse, O'Sullivan, W.J., and Hannal, E.F., eds.) vol. 3 p. 578, 1974.
Abeles, B., Sheng, P., Coutts, M.D. and Arie, Y., 1975, Adv. Phys. 23:407.
Abrahams, E., Anderson, P.W., Liccoardello, D.C., and Ramakrishnan, T.V., 1979, Phys. Rev. Lett. 42:673.
Abrahams, E. and Ramakrishnan, T.V., 1980, J. Non-Cryst. Solids, 35:15.
Abrahams, E., Anderson, P.W., Lee, P.A. and Ramakrishnan, T.V., 1981, Phys. Rev. B24:6783.
Adkins, C.J., 1977, Phil. Mag., 36:1285.
Altshuler, B.L., and Aronov, A.G., 1979, Jour. Exsp. Teor. Fiz., 77:2028 (Translated in: Sov. Phys. Jetp. 1980, 50:968).
Altshuler, B.L., Aronov, A.G., and Lee, P.A. 1980, Phys. Rev. Lett. 44:1288.
Altshuler, B.L., Khmeln'nilskii, D., Larkin, A.I., and Lee, P.A., 1980, Phys. Rev. B22:5142.
Anderson, P.W., Phys. Rev. 1958, 109:1492.
Anderson, P.W., 1964, in: "Lectures on the Many-Body Problem," E.R. Caianello, ed., Academic, New York.
Anderson, P.W., Thouless, D.J., Abrahams, E., and Fisher, D.S., 1980, Phys. Rev. B22:3519.
Anderson, P.W., 1959, J. Phys. Chem. Solids, 11:26.
Anderson, P.W., Muttalib, K.A. and Ramakrishman, T.R., Phys. Rev. 117 (1983).
Ando, T., Fowler, A.B., and Stern, F., 1982, Rev. Mod. Phys. 54:437.
Azbel, M. Ya., 1981. 1982, private communication.
Azbel, M. Ya., 1981, Phys. Rev. Lett. 47:1013.
Azbel, M. Ya., 1982, Phys. Rev., B25:849.
Azbel, M. Ya., 1982, ibid. B26:4735.
Azbel, M. Ya., and Soven, P., 1982, Phys. Rev. Lett., 49:750.
Balian, R., Maynard, R., and Toulouse G., eds., 1979, "Ill-Condensed Matter," North Hollánd, Amsterdam.
Bergmann, G., 1982, Z. Phys., B48:5 and references therein.
Bishop, D.J., Tsuei, D.C., and Dynes, R.C., 1980, Phys. Rev. Lett., 44:1153.
Borland, R.E., 1963, Proc. Roy. Soc. A274:529; Proc. Phys. Soc., 1968, Proc. Phys. Soc. London, 28:926.
Buttiker, M. Imry, Y. and Landauer, R., to be published in Phys. Lett.
Capizzi, M., Thomas, G.A., de Rosa, F., Bhatt, R.N., and Rice, T.M. 1980, Phys. Rev. Lett., 44:1019.
Castellani, C., Di Castro., C., Forgacz, G., and Tabet., E., eds., 1982, preprint.

Castellani, C., Di Castro, C., and Peliti, L., 1981, "Disordered Systems and Localization," Springer, Berlin.
Chaudhari, P., and Habermeier, H.U., 1980, Phys. Rev. Lett., 44:40.
Davies, R.A., and Pepper., M., 1982, J. Phys. C15:L371.
Davies, R.A., Pepper, M., and Kaveh, M., 1983, C16:L285.
Deutscher, G., Imry, Y., and Gunther, L., 1974, B10:4598.
Deutscher, G., and Fukuyama, H., 1982, Phys. Rev., B25:4298.
Dolan, G.J., and Osheroff, D.D., 1979, Phys. Rev. Lett., 43:721.
Domany, E., and Sarker, S., 1979, Phys. Rev., B20:4726.
Dynes, R.C., and Garno, J., 1981, Phys. Rev. Lett., 46:137.
Efetov, K.B., 1982, Zh. Exsp. Teor. Fiz, 82:872.
Finkeltein, A.M., 1983, Landau Institute, preprint.
Fisher, D.S., and Lee, P.A., 1981, Phys. Rev. B23:685.
Fowler, A.B., Hartstein, A., and Webb, R.A., 1982, 48:1961.
Friedman, L.R., and Tunstall, D.P., eds., 1978, "The Metal-Nonmetal Transition in Disordered Systems," SUSSP, Edinburgh.
Fukuyama, H., 1980, J. Phys. Soc. Japan, 1980, 48:2169; 1981, 50:3407.
Fukuyama, H., in Stern, 1981, p.489.
Fukuyama, H., in Nagaoka and Fukuyama, 1981, p.89.
Fukuyama, H., 1983, ISSP Technical Report, 1304A.
Gefen, Y., and Imry, Y., 1981, unpublished results.
Gefen, Y., and Imry, Y., 1982, submitted to PRL.
Giordano, N., Gilson, W. and Prober, D.E., 1979, Phys. Rev. Lett. 43: 725.
Giordano, N., 1980, Phys. Rev. B22:5635.
Grest, G.S., and Lee, P.A., 1983, Phys. Rev., B22:5635.
Hebard, A.F., and Fiory, A.T.,1983, Phys. Rev. Lett.
Hertel, G.H., Bishop, D.J., Spencer, E.G., Rowell, J.M. and Dynes, R.C., 1983, Phys. Rev. Lett., 50:743.
Hikami, S., Larkin, A.I., and Nagaoka, Y., 1981, Prog. Theor. Phys., 63:707.
Hubbard, J., 1964, Proc. Roy. Soc., A277:237.
Jonson, M., and Girvin, 1978, Phys. Rev. Lett., 43:1447.
Imry, Y., 1983, Phys. Rev. Lett., 50:1603.
Imry, Y., 1981, Phys. Rev. B24:1107.
Imry, Y., 1981, J. Appl. Phys., 52:1817.
Imry, Y., Gefen Y., and Bergman, D.J., 1982, Phys. Rev. B26:3436.
Imry, Y., and Ovadyahu, Z., 1982, Phys. Rev. Lett., 49:841.
Imry, Y., and Ovadyahu, Z., 1982, J. Phys., C15:L327.
Imry, Y., and Strongin, M., Phys. Rev., 1981, B24:6353, contains many prior references.
Kaveh, M., and Mott, N.F., 1981, J. Phys. C14:L179.
Kaveh, M., 1983, unpublished results.
Kawabata, A., 1977, J. Phys. Soc., Japan, 43:1491; 1980, J. Phys. Soc., Japan, 49:628; 1981, 50:2461.
Kawaguchi, Y., and Kawaji, S., 1982, Surface Sci., 113: 5051 and references therein.
Kaveh, M., Uren, M.J., Davies, R.A. and Pepper, M., 1981, J. Phys. C14: L413.

Kohn, W., 1967, Phys. Rev. Lett. 19:439.
Kuper, C.G., 1966, Phys. Rev. 150:189.
Kubo, R., 1957, J. Phys. Soc. Japan 12:570
Landau, L.D., and Lifshitz, E.M., 1976, "Quantum Mechanics" 3rd edition, Pergamen, Oxford.
Landauer, R., 1957, IBM J. Res. Dev. 1.
Landauer, R., 1970, Phil. Mag., 21:863.
Landauer, R., 1975, Z. Phys. B21:247.
Lee, P.A., 1981, Lecture Notes.
Lee, P.A., 1980, J. Noncryst. Solids, 35:21.
Licciardello, D.C. and Thouless, D.J., 1975, C8:4159; 1978, C11:925
Ma, M., 1982, Phys. Rev., B26:5.
Maekawa, S., 1982, in: Nagaoka and Fukuyama, p.103.
Maekawa, S., and Fukuyama, H., 1981, J. Phys. Soc. Japan, 50:2516.
Markiewicz, R.S., and Harris, L.A., 1981, Phys. Rev. Lett., 46:1149; 1982, Surface Sci. 113:505.
McLean, E., and Stephen, M.J., 1979, Phys. Rev. B19:5925.
McMillan, W.L., and Mochel, J., 1981, Phys. Rev. Lett., 46:556.
McMillan, W.L., 1981, Phys. Rev. B24:2739.
McKinnon, W., and Kramer, B., 1981, Phys. Rev. Lett. 47:1546.
Mooij, J.H., 1973, Phys. Stat. Sol. A17:521.
Mott, N.F., 1966, Phil. Mag. 13:989.
Mott, N.F., 1974, "Metal Insulator Transitions," Taylor and Francis, London.
Mott, N.F. and Davis, G.A.,1979, "Electronic Processes in Non-crystalline Materials," 2nd edition, Clarendon Press, Oxford.
Mott, N.F., 1969, Phil. Mag., 19:835.
Mott, N.F. 1970, Phil. Mag., 22:7.
Mott, N.F. and Twose, W.D., 1961, Adv. Phys., 10:107.
von Molnar, S., 1982, private communication.
Nagaoka, Y., and Fukuyama, H., 1982, eds. "Anderson Localization," Springer, Berlin.
Opperman, R., 1982, Solid State Commun. 44:1297.
Ovadyahu, Z., and Imry, Y., 1981, Phys. Rev., B24:7439.
Ovadyahu, Z., to be published.
Paalanen, M.A., Rosenbaum, T.F., Thomas, G.A., and Bhatt, R.N., 1982, Phys. Rev. Lett., 48:1284.
Pepper, M., 1981, in: Castellani et al., contains prior references.
Pichard, J.L., and Sarma, G., 1981, J. Phys. (Paris) 10:4.
Pollak, M., 1972, J. Non. Cryst. Solids, 11:1.
Rosenbaum, T.F., Andres, K., Thomas, G.A. and Lee, P.A., 1980, Phys. Rev. Lett. 46:568.
Rosenbaum, T.F., Andres, K., Thomas, G.A. and Bhatt, R.N., 1980, Phys. Rev. Lett., 45:1723.
Schmid, A., 1974, Phys., 271:251.
Shapira, Y., and Deutscher, G., 1981, to be published.
Shapiro, B., and Abrahams, E., 1981, Phys. Rev., B29:4889; 1981, B24:4025.
Shapiro, B., 1982, Phys. Rev., B25:4266.

Simanek, E., 1979, Solid State Commun., 31:419.
Stein, J., and Krey, U., 1979, Z. Phys., B34:287;1980, B37:18.
Stern, F., 1967, Phys. Rev. Lett., 18:546.
Stern, F., 1982, ed. "Electronic Properties of Two Dimensional Systems," North Holland, Amsterdam.
Stone, A.D., Joannopulos, J.D., and Thouless, D.J.,unpublished works.
Strongin, M., Thompson, R.S., Kammerer, O.F., and Cross, J.E., 1970, Phys. Rev. B1:1078.
Thouless, D.J., 1977, Phys. Rev. Lett., 39:1167.
Toyozawa, Y., 1961, Progr. Theor. Phys., 26:29.
Uren, M.J., Davies, R.A., and Pepper, M., 1980, J. Phys. C13:L985.
Uren, M.J. Davies, R.A., Kaveh, M. and Pepper, M., 1981, C14:L395; 1981, 5737.
Vollhardt, D., and Wolfe, P., 1980, Phys. Rev. B22:4678; 1982, Phys. Rev. Lett., 48:699.
Voss, R.F., and Webb, R.A., 1982, Phys. Rev. B25:3446.
Wegner, F., 1976, Z. Phys. 25:327; 1980, Phys. Repts. 67:151.
Wheeler, R.G., 1981, B24:4645.
Wiesmann, H., Gurvitch, M., Lutz, H., Gosh, A., Schwartz, B., Allen, P.B. and Strongin, M., 1977, Phys. Rev. Lett., 38, 782.

# MICROFABRICATION TECHNIQUES FOR STUDIES OF PERCOLATION, LOCALIZATION, AND SUPERCONDUCTIVITY, AND RECENT EXPERIMENTAL RESULTS

Daniel E. Prober

Applied Physics
Yale University
New Haven, CT 06520

## INTRODUCTION

In the past four years studies of localization and percolation in lower-dimensional systems have been conducted in a number of laboratories. These studies have advanced in a fundamental way our understanding of electron transport in dirty and inhomogeneous systems. The production of the experimental systems has relied directly on advanced microfabrication techniques. It is the purpose of this chapter to review these techniques and also selected experimental results for systems whose production exemplifies these microfabrication techniques.

The model systems which we shall discuss include metal line structures as small as 200 Å, produced with step-edge techniques. These have been used in studies of one-dimensional (1D) conduction effects. Experiments on MOSFETs (Metal-Oxide-Semiconductor-Field-Effect Transistors) have allowed study of the 2D electron gas, and with further lithographic patterning, nearly-1D MOSFETs have been produced. Finally, arrays of superconducting tunnel junctions have served as model 2D superconducting systems.

The examples of model systems listed above demonstrate the range of experiments which are now accessible to researchers at both universities and at major research laboratories. The studies have involved both engineering technologists and basic scientists, and their success demonstrates the viability of the subfield of microstructure science. As we shall see, the techniques are particularly well suited for studies of electron localization, percolation, and superconductivity, because intrinsic length scales for these phenomena can be in the range 1000 to 10,000 Å.

In this chapter we outline the basic principles, current capabilities and limits, and future possibilities of modern microfabrication techniques. There have been numerous reviews of specific microfabrication techniques, intended for an audience of scientific users[1,2] or engineering practitioners of the art.[3-5] The reader may consult these reviews and the references therein for specifics on implementing the varied techniques we shall discuss.

## LITHOGRAPHY PROCESSES

### Resist Processes and Resolution Limits

In current lithographic practice, incident radiation (photons, electrons, or ions) is used to "expose" areas of a thin polymer layer (the "resist"). In the exposed areas the resist has a modified removal rate in a suitable developer. For a positive resist the exposed areas develop faster. The exposure process involves the breaking of chemical bonds, polymer crosslinking, or initiation of chemical reactions. With electron, ion, and x-ray exposure, low-energy (<100 eV) electrons are ultimately produced. These "secondary" electrons have a large cross section for interaction with the resist. Thus, the ultimate limit on resolution is set by the path-length of these low-energy electrons, typically <100 Å.[4] The exposure resolution of modern photoresists can be as good as the resolution of the exposure system itself. For diffraction-limited exposure systems the resist resolution is thus comparable to the wavelength, ∿4000 Å.

A simple optical exposure system based on a high quality optical microscope[6] is diagrammed in Fig. 1. Here, a mask of the desired pattern is placed in the microscope field stop. A demagnified image of this pattern is projected onto the photoresist. (In Fig. 1 the light is shown being projected through a transparent substrate. This gives edge profiles[6] which are more favorable for use with the lift-off process, discussed below. This through-the-substrate exposure is convenient, but not essential.) With this system, a pattern resolution of 0.2 μm (2000 Å) can be achieved for simple line patterns. The field of view is small, <0.2 mm, for lenses with this resolution.

After the exposed resist is developed, the pattern can be transferred in one of two ways: 1. subtractive etching, in which the photoresist protects the underlying film or substrate, and 2. additive deposition, in which a film is deposited on the resist and on areas not covered by the resist. For this second case, the resist is then dissolved, and the film on top of the resist is removed (see Fig. 1). This is the lift-off process. It is widely used for research patterning of submicron structures. (Note: 1 micron ≡ 1 μm = 10,000 Å) Photoresist edge profiles which are undercut, as shown in Fig. 2, ensure clean liftoff.

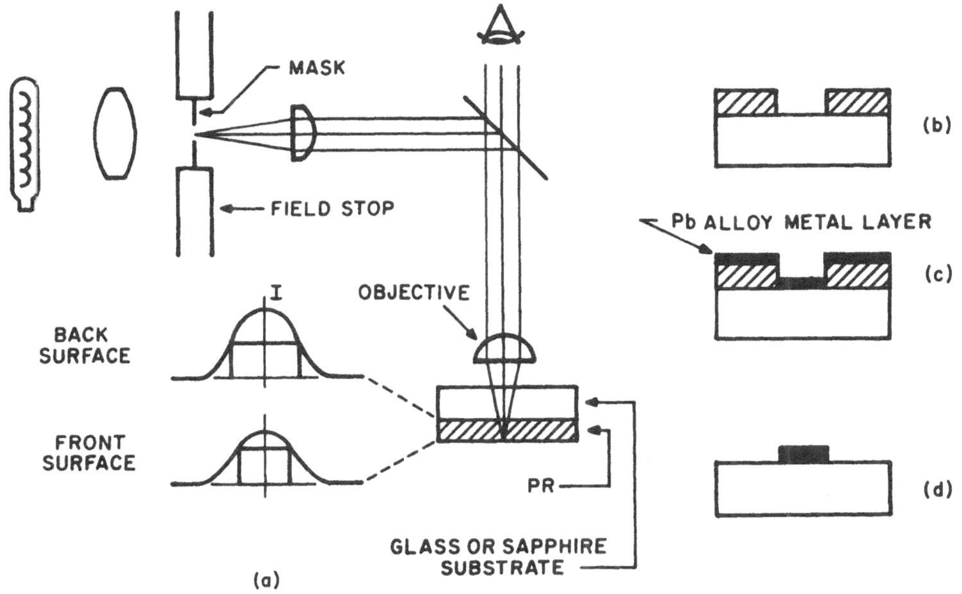


Fig. 1. Schematic of the projection microscope and lift-off procedure. (a) Projection exposure of the photoresist (PR) in a reflected-light microscope. Schematic intensity (I) of exposing radiation vs. position at two heights in PR is also shown; (b) after development; (c) following metallization; and (d) results of lift-off.

Pattern transfer by subtractive etching is required when a photoresist lift-off mask cannot withstand the deposition conditions. For example, temperature limits of $\sim$150° to 200°C apply for conventional resists, although some high temperature lift-off processes have been developed.[7] The selection of etching or lift-off may be determined by other considerations. Lift-off is well suited for producing a fine metal line, for example, but not for producing a small gap between two electrodes.

In the past 3 years it has been recognized that many of the resolution limits which were once thought to be intrinsic are not fundamental to the exposure process, but rather resulted from using single-layer resists with properties chosen to meet conflicting requirements. The resist layer must be thick to smoothly cover topography resulting from previous processes so as to avoid defects. (This consideration is of central importance in semiconductor device production, where many mask levels are required.) However, thick layers allow spreading of the incident radiation (due to scattering or diffraction), sidewards development (while development proceeds through the resist thickness), and standing-wave effects in the resist (for optical exposure).

The recent perfection of a number of multilayer resists[8] has resolved most of the conflicting requirements for the resist. Multilayer resists consist of a thin top layer, sometimes with an intermediate layer, and a thicker bottom layer for topography smoothing. (A two layer resist is shown in Fig. 2b.) The characteristics of each layer can be optimized independently. The top layer is optimized for resolution, the bottom layer for coverage. Using a multilayer resist, a standard scanning electron microscope can be used to expose 250-Å linewidth patterns.[9] In addition to the finite electron beam size, ∿100 Å, the factors determining resist resolution are the secondary electron pathlength, and the size of the coiled up polymer in the polymethylmethacrylate resist (PMMA). Limits on resist resolution are summarized in Table I.

The utility of multilayer photoresists is also seen in semiconductor device production. Hewlett-Packard has achieved a significant increase in yield for their 450,000 transistor chip when a multilayer resist process was introduced.[10] This chip serves as the processor in their 32-bit Model 9000 computer. The chip uses a lithographically aggressive design, with 1.5 μm lines and 1 μm spaces.

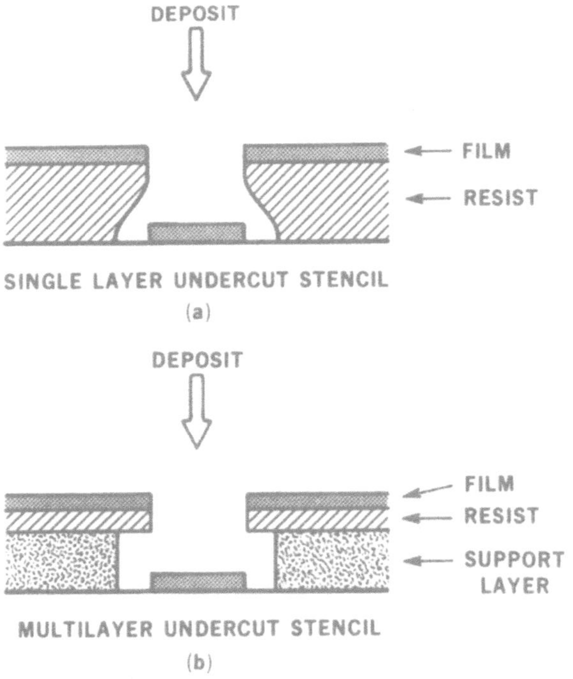


Fig. 2. Application of undercut resist profiles to lift-off.

Table I. Resolution and Other Patterning Limits

| Mechanism | Resulting Limit |
|---|---|
| Pathlength of low-energy electron | ∿100 Å |
| Polymer structure | Not established; <100 Å |
| Exposure statistics | <100 Å (PMMA) |
| Diffraction (x-ray) or scattering in resist (electron, ion exposure) | <100 Å for thin resist layer |
| Electron backscattering from thick substrate | Loss of contrast |
| Processing temperature | <150°C, most resists |

Exposure Systems

Exposure instruments for use in a research environment have been developed by a number of researchers.[2,6,9] For optical exposure, it has proven sufficient to use a high quality optical microscope.[6] A resolution of 0.2 to 0.5 μm can be achieved with care. This resolution is set by the diffraction-limited optics. Alignment, at least to 1 μm accuracy, can be accomplished with red light, which does not expose the resist. The field of view of these instruments is limited (∿0.2 mm for a typical 100X lens). Commercial optical projection systems typically achieve 1 μm resolution over a 1 cm field of view, and can align to 0.25 μm.

For electron beam exposure, a standard scanning electron microscope (SEM) can be used. A resolution of 250 Å has been achieved,[9] though careful control of the exposure dose is essential. Alignment is not trivial, since the entire substrate surface is coated with resist. Even smaller linewidths have been achieved with a scanning transmission electron microscope (STEM), which can readily achieve a beam diameter of 20 Å. On solid substrates, 100 Å lines have been produced.[11] Since the sample is inserted into the magnetic lens, its dimensions are limited to <3 mm.

Electron beam exposure may be controlled with either a flying-spot scanner,[12] or with a computer which turns the beam on or off and simultaneously drives (or is synchronized with) the X and Y beam sweeps. It is most convenient to use a beam blanker to turn the beam on and off. Simple schemes which instead "dump" the beam in an unused area of the substrate may be acceptable, but only for slow writing (due to deflection-coil response).

The resist employed in nearly all research uses of electron beam lithography is PMMA. This resist has the best resolution known. PMMA also requires a relatively large exposure dose, so that intrinsic statistical fluctuations of the exposing beam are relatively small, even for exposed areas as small as 100 Å x 100 Å.[1] Conversely, PMMA is generally considered to be too slow (insensitive) for commercial applications.

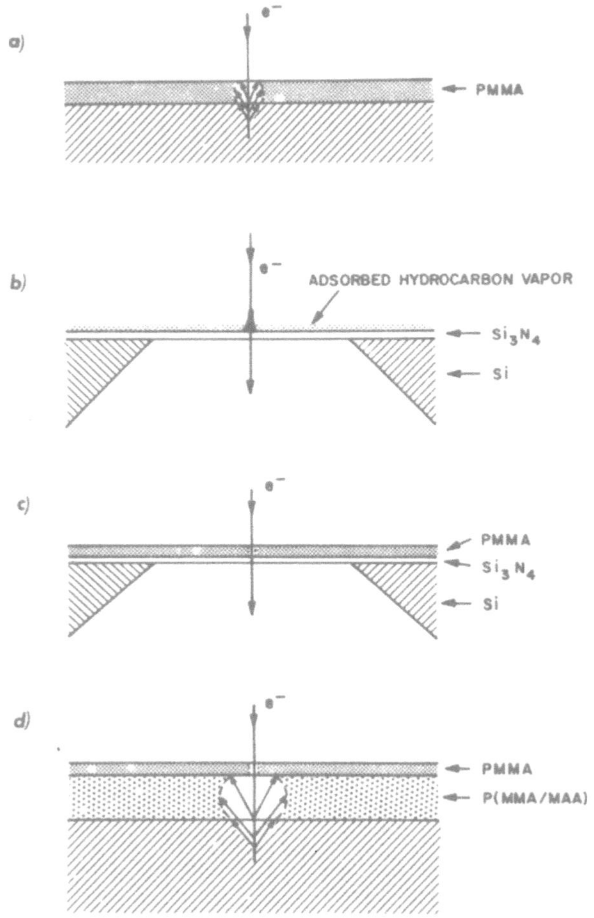


Fig. 3. Electron-beam exposure showing scattering and methods to reduce scattering. (a) thick resist; (b) contamination-resist lithography on a membrane; (c) polymer resist on a membrane; (d) high-resolution two-layer resist.

The size of the coiled-up PMMA polymer is <100 Å. Use of a resist with a smaller "grain size" might lead to even better resolution than is achieved with PMMA. IBM researchers have developed a contamination writing technique in which the electron beam converts residual organic contamination on the substrate surface into a hard carbonaceous material. This contamination resist can be used as an etch mask for Ar-ion etching. Au-Pd lines 80 Å wide have been produced with exposure in a STEM. Very careful control of the exposure intensity is required to achieve this smallest dimension. Contamination resist techniques are discussed in detail in a separate chapter.[13]

Fig. 3 shows the various configurations for electron beam exposure. Both lateral beam spreading and backscatter from the substrate are observed with a thick resist (Fig. 3a). Use of a multilayer resist, Fig. 3d, can largely eliminate these problems. Use of a thin membrane substrate, as shown in Fig. 3b and 3c, eliminates electron backscattering, and also allows inspection of the final structure in STEM or TEM instruments, which have better resolution (typically <10 to 20 Å) than usual SEMs (resolution ∼100 Å).

Table II. Pattern Resolution Achieved

| Exposure Method | Exposure Process (and Reference) | Resolution[a] | Field of View[a] |
|---|---|---|---|
| Optical | Microscope projection[6] | 2000 Å | 200 μm |
| | Commercial projection system[5] | 1 μm | 1 cm |
| Electron beam | SEM, PMMA[9] | 250 Å | 100 μm[b] |
| | STEM, PMMA[11] | 100 Å | 30 μm[b] |
| | JEOL lithography system (25 kV), PMMA | ≤200 Å[c] | 100 μm[c] |
| | STEM, contamination resist[13] | 80 Å | - |
| X-ray | $C_K$ x-ray source, PMMA[15] | 175 Å | 1 cm |

[a]Figures given are intended primarily as a general guide, but reflect actual values where available.

[b]R. E. Howard, private communication.

[c]G. J. Dolan, private communication.

Membrane substrates are particularly well suited to experiments on thin metal films, but still do not have the general applicability of the approach of Fig. 3d.

X-ray and ion-beam exposure of resist have not been widely used for laboratory studies, but are under active investigation for semiconductor device production. X-ray exposure is currently done only as a 1-to-1 contact printing process, so that mask fabrication must be done by some other high resolution process. Flanders[14] has developed a very useful exposure system and polyimide membrane mask system. A printing resolution of ∿175 Å is achieved.[15] Ion beam lithography has to date been limited in resolution by the exposing beam size (∿400 Å), but should have an intrinsic resolution of ∿100 Å. The performance characteristics of the various exposure systems is given in Table II.

## Etching and Deposition

Film patterning by etching employs either liquid or ion/plasma etchants. Liquid etching in general has the greatest material selectivity, but has the disadvantage that the liquid will also etch under the mask. (One exception is the use of anisotropic liquid etchants which etch different crystal planes of a single crystal at different rates. Such etchants for silicon are well studied.[16a]) Anodization is a liquid process which can convert many metals into their insulators. It has good dimensional resolution, and is particularly useful for patterning refractory superconductor films.[16b]

Ion or plasma etching[17] can be very directional. Material selectivity can be fair, or good to excellent, depending on whether an inert gas (e.g., Ar) or a chemically reactive gas is used. For Ar ion etching mechanical momentum transfer causes the etching. In reactive ion etching the gas forms reaction products with the material to be etched. These products are usually gases, which rapidly leave the surface and do not redeposit elsewhere. An example of such a case is the simplified reaction of $CF_4$ etching Si, where $SiF_4$, a gas, can in the end be formed. There is some amount of mechanical bombardment in all ion etching processes, so that the selectivity is rarely as good as with liquid etchants. It is only with ion etching, however, that one can etch deep slot structures or steps in polycrystalline films.

Methods of patterning which employ ion etching can achieve very high resolution, sometimes limited only by the grain size of the film to be etched. For this reason and for reasons of temperature stability, fine-grain alloy films are often desired. Examples include Au-Pd and W-Re, both used to form narrow (∿200 Å) wires. It is however possible to produce continuous and uniform films of some elements. Aluminum is an example. Cleaning the substrate by ion

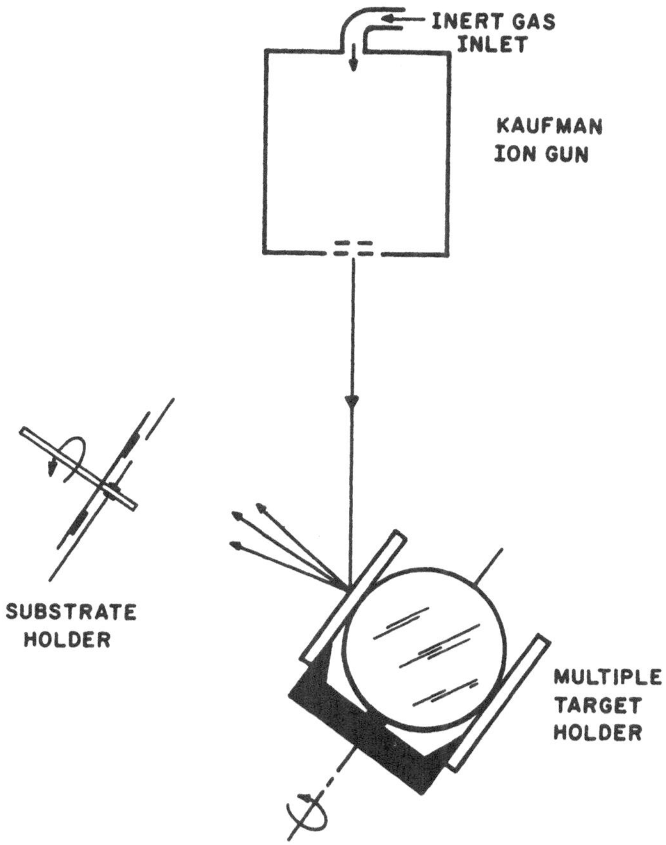


Fig. 4. Schematic of ion beam sputter deposition system, comprising a 2.5 cm Kaufman ion source, multiple target holder accommodating four 4 in. targets, and substrate holder (From Ref. 18).

etching immediately before film deposition can be critical to achieving uniform coverage and good adhesion. Such cleaning must be done "in-situ," that is, without breaking vacuum.

Sputtering can also be employed in deposition processes. Fig. 4 shows an ion beam sputter deposition system developed at Yale, which is used for depositing high quality films of refractory superconductors.[18] This sputtering system allows deposition of layers of different materials in rapid succession, to avoid interface contamination. This type of system is particularly useful in producing artificial tunneling barriers on refractory superconductors.[16b,19] In addition, certain materials, such as Ta, can be "coaxed" into growing in the desired crystal structure by deposition onto an appropriate underlayer film.[18] In general, for sputtering the directionality of deposition is not as good as with thermal evaporation. Such directional deposition is required for some of the three dimensional methods described below.

Three-Dimensional Techniques

In previous sections we have described various types of direct lithography - that is, techniques which provide a direct replica of the exposed pattern. These direct techniques are not the only, nor even the simplest, methods for producing structures on a size scale <1000 Å. A variety of innovative three-dimensional (3D) techniques which utilize edges and shadowing have been developed to fold over a height dimension into a feature dimension. These techniques can be very effective when a high aspect ratio, of height to width, is required, and they are well suited to smaller research laboratories.

Two types of edge-defined processes[20] are shown in Fig. 5.

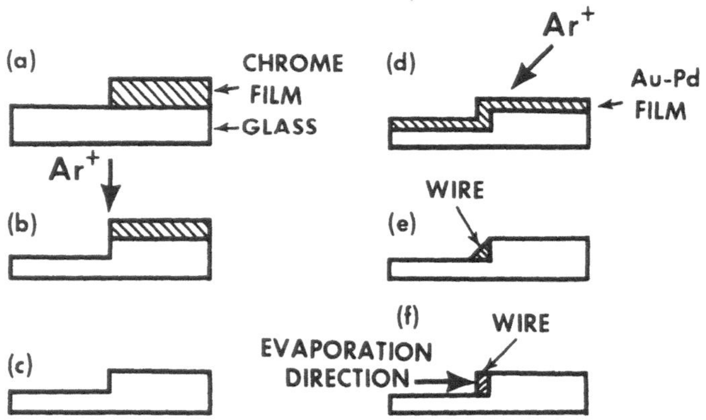


Fig. 5. Wire fabrication procedures. The substrate is shown in side view: a. Half the substrate is coated with a thin chrome film. b. The substrate is ion etched to produce a square step. c. The chrome film is removed with a chemical etch. d. The substrate is coated with the metal film and ion etched at an angle until, e. a triangular wire is formed along the step edge. (Alternate process) f. The metal film is evaporated parallel to the substrate to coat only the step edge. (From Ref. 20)

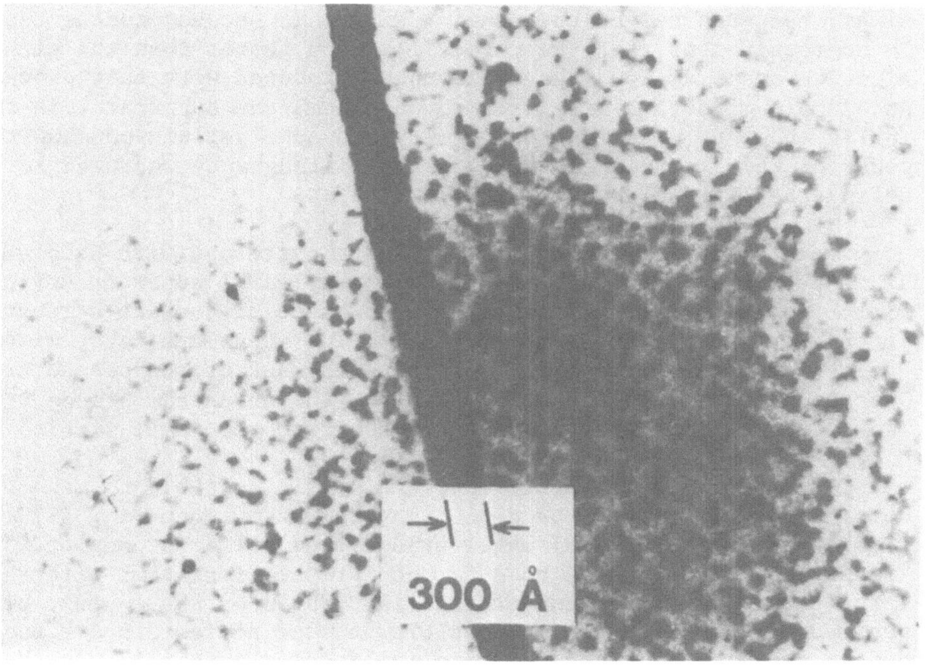


Fig. 6. Rectangular Ni wire, with 300 Å linewidth, formed on the edge of a step in $SiO_2$; transmission electron micrograph. (From Flanders and White, Ref. 14)

The first process consists of steps a thru e, and is used to form a triangular wire. The substrate step protects the metal in its "shadow" from ion etching (step d). A relatively soft material such as Au-Pd, which etches faster than glass, can be patterned in this fashion. For the angles shown, the step height sets the linewidth. Since height dimensions can be controlled to <50 Å by setting the etching time (step b), wires smaller than 200 Å can be formed. The factors which in practice limit wire dimensions are the raggedness of the photoresist lift-off mask (used for patterning the chrome etch mask), the grain size of the chrome etch mask, and finally the grain size of the Au-Pd metal film. As can be seen in Fig. 5, this process is self aligning; as a result, it has a high yield. Also, although lift-off photolithography is used, there is no diffraction limit *per se*, since only edges are employed. These edges are formed with high contrast processes. The triangular wires produced with this procedure were used in extensive studies of electron localization, described later in this chapter.

The second process for forming a narrow wire consists of steps a thru c and f in Fig. 5. Here the wire is evaporated directly on the step edge. The film thickness determines one of the wire dimensions, the step height the other. With this second process there is no requirement that the metal film etch faster than the glass. Wires as small as 200 Å have also been produced with this process. An example of such a wire, formed on a membrane substrate, is shown in Fig. 6. The excellent uniformity and edge definition show the power of this technique. Only optical lithography was used to define the lift-off mask.

More complex structures can also be patterned with step edges. It is possible to produce in this way very small superconducting microbridges, the thin film analogue of a point contact structure, in which all dimensions are <500 Å.[21] Two steps are used, with an angle-shadowed mask required for the etch mask of the second step. One step height sets the bridge length, the other the bridge width. Improved electrical performance results from the small size.

In analogy to the above approach, it is possible to make very short SNS microbridges, by depositing the superconducting (S) electrodes and the normal (N) metal bridge from different angles. High-$T_c$ SNS microbridges have been formed for the first time with this technique.[22] The superconducting film, $Nb_3Sn$ or $Nb_3Ge$, must be deposited at $\sim$800°C. This is possible because no resists are used for device patterning. The substrate (silicon on sapphire) may thus be heated during deposition. Linear arrays of SNS microbridges have been patterned with this technique. See Fig. 14 for a diagram.

The three dimensional techniques discussed above all utilize substrate patterning. Three dimensional resist structures can also be produced with multilayer resists. A useful and very elegant approach has been developed by Dolan for self- aligned fabrication of metal-oxide-metal superconducting tunnel junctions.[23] This approach is shown in Fig. 7. A tunnel junction is formed by depositing the first metal film, oxidizing it, and from the opposite direction (2) depositing the second film. The tunnel junction is thus formed without exposure to atmosphere (for lithography) or sputter cleaning between the two evaporations. As a result, one can produce small, high-current-density tunnel junctions much more easily than with conventional Josephson junction processes. Junctions produced with the Dolan process have been used in studies of macroscopic quantum tunneling[24] and as low-noise microwave mixers.[25] Arrays of such junctions are also useful for studying 2D superconducting phase transitions.

In this section we have discussed the capabilities of various lithographic techniques, and shown some examples of the results achieved. In the next section we present the major applications of microlithography in the areas of localization and superconductivity related to theoretical issues discussed in the rest of this volume.

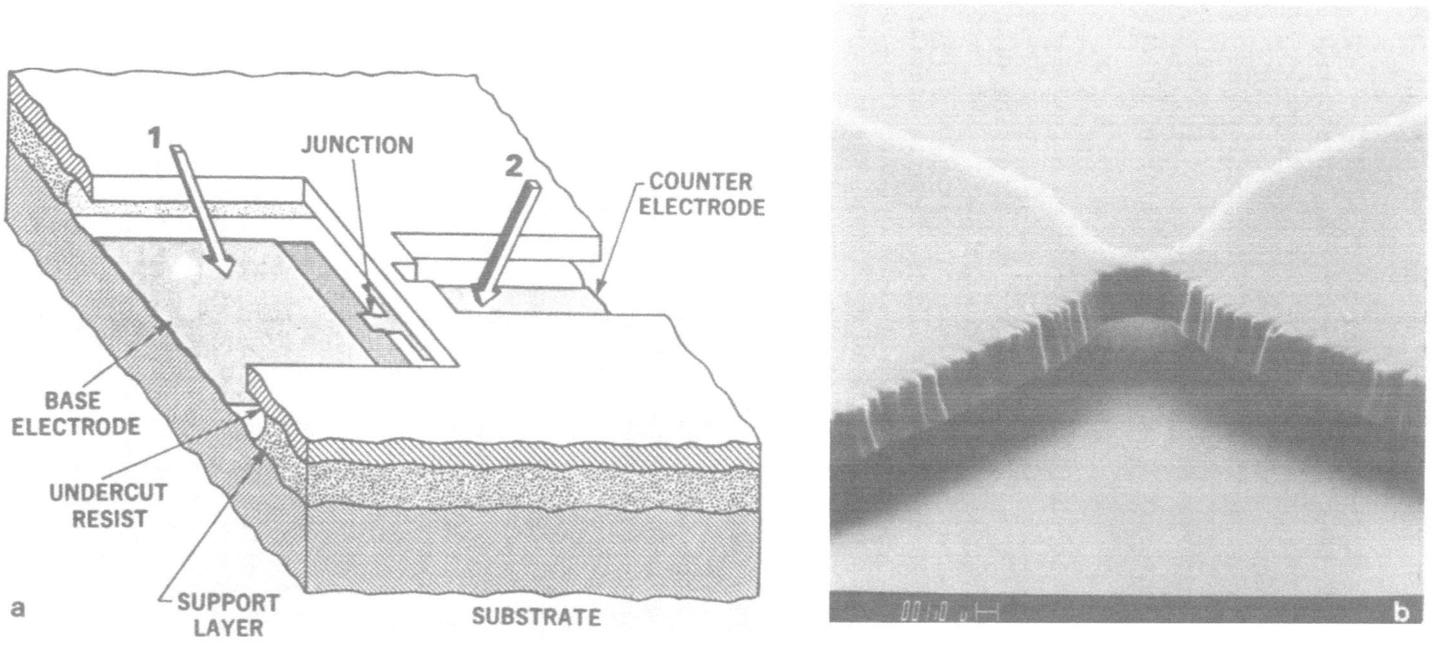


Fig. 7. Self-aligned tunnel junction fabrication, patterned with a multilayer resist stencil. The base electrode is oxidized prior to deposition of the counter electrode. a. Schematic outline; b. SEM of photoresist structure, by P. Santhanam (unpublished) for work of Ref. 25b. View is from a different orientation than Fig. a. Resist thickness is 2 μm for both the suspended layer and the support layer (not visible in Fig. b). (Fig. 7a from Ref. 1.)

SCIENTIFIC STUDIES: ELECTRON LOCALIZATION AND INTERACTION EFFECTS

In 1977, D. Thouless made a striking prediction[28] - that at $T = 0$, all wires should behave as insulators, due to electron localization effects. The only requirement is that the wires be long enough. This length requirement translates into a resistance requirement, that the high-temperature (e.g., 10K) resistance of the wire be greater than $\sim 10\ k\Omega$. Since this prediction, theoretical understanding of electron localization has advanced at a remarkable rate. Two excellent theoretical reviews from different points of view are given in this volume.[26]

In this section we review the requirements on experimental systems with regard to dimensionality and size, and discuss the experimental approaches which have been developed for producing effectively one dimensional (1D) systems. We also discuss two recent sets of experiments, on metal wires and films, and on narrow MOSFETs. We concentrate largely on 1D studies, since these require special microfabrication techniques. Studies on 2D systems require the production of very thin, continuous films or granular films. These issues are discussed elsewhere in this volume, in the chapters by Beasley, by Deutscher, and by Laibowitz.

General Considerations: Dimensional Requirements and Inelastic Times

Experiments to probe localization and electron-electron interaction effects have utilized studies of resistance, magnetoresistance, Hall effect, thermoelectric effect, and tunneling density of states. In all cases the theoretical prediction depends on sample dimensionality, and it is this aspect which can be controlled with microfabrication techniques.

At low temperatures, an experimental sample may act as a system of reduced dimensionality even though it is large compared to atomic dimensions. For localization, the relevant length scale is the Thouless length, which for finite temperatures is given as

$$L_{Thouless} = \ell_{in} = (D\tau_{in})^{1/2} \qquad (1)$$

with $\ell_{in}$ the inelastic diffusion length, D the diffusion constant ($= v_F^2\tau/3$ in 3D systems; $\tau$ is the elastic scattering time), and $\tau_{in}$ is the time between inelastic collisions. Weak (small) localization effects are expected when $\ell_{in}$ is less than the localization length.[26a] This is the usual experimental situation.

We list in Table III the essential temperature dependence of the inelastic scattering rate, $\tau_{in}^{-1}$, for electron-phonon and electron-

Table III. Temperature Dependence of Inelastic Scattering Rates

| Mechanism | System | Dimension | $\tau_{in}^{-1}$ | Ref. |
|---|---|---|---|---|
| Electron-phonon (system dimension set by $\lambda_{phonon}$) | Clean | 3D | $T^3$ | 28 |
| | | 2D | $T^2$ | 28 |
| | | 1D | $T$ | 28 |
| | Dirty ($q_{phonon}\ell<1$) | 3D | $T^4$ | 28,27a* |
| | | 3D | $T^2$ | 27b* |
| | | 2D | $T^3$ | 28 |
| | | 1D | $T^2$ | 28 |
| Electron-electron (system dimension set by $L_{int}$) | Clean | 3D | $T^2$ | 27c |
| | | 2D | $T^2 \ln T$ | 27c |
| | | 1D | $T$ | 27c |
| | Dirty ($\tau<\frac{\hbar}{kT}$) | 3D | $T^{3/2}$ | 27d |
| | | 2D | $T \ln(T_1/T)\sim T$** | 27d |
| | | 2D | $T \ln(\text{const})\sim T$ | 27e |
| | | 2D | $T$ | 27f |
| | | 1D | $T^{1/2}$ | 27d |

*See Ref. 27b for a discussion of these differences.

**Since $T_1 >> T$, the essential dependence on temperature is just $\sim$ T.

electron scattering in both clean and dirty systems. "Clean" and "dirty" are defined in the table. Mechanisms are discussed in the references.[26-28]

For electron-electron interaction effects the length scale which determines system dimensionality is the quantum diffusion length

$$L_{int} = \left(\frac{D\hbar}{kT}\right)^{1/2}. \tag{2}$$

$L_{int}$ thus depends on temperature and on the elastic scattering time (through D). The mean free path is given as $\ell = v_F\tau$.

The length scale $\ell_{in}$ or $L_{int}$ usually enters the theoretical prediction for quantum corrections to transport properties at low temperatures. For example, for 1D localization effects the resistance is increased above the Boltzmann (elastic) term by[28]

$$\frac{\Delta R^{1D}}{R} \sim \frac{\ell_{in}}{L_{36.5k\Omega}}. \tag{3}$$

$L_{36.5k}$ is the length of wire which at high temperature (~10K) has R = 36.5 kΩ. A similar formula holds for 1D interaction effects, but with $L_{int}$ instead of $\ell_{in}$. Here and in nearly all other experiments, the temperature dependence of the predicted effect is an essential feature, as $\tau_{in}$ depends on temperature, as does $L_{int}$.

It appears that a system may be of a certain dimensionality for determining the inelastic scattering process, yet of a different dimensionality with respect to localization. Thus, the scattering mechanism for a narrow wire need not be a 1D mechanism, even if 1D localization theory is applicable. Also, the total scattering rate is the sum of the rates due to all inelastic processes which contribute significantly. Thus, $\tau_{in}^{-1}$ need not have the simple temperature dependence $T^p$, with p an integer.

Experiments on 1D Metal Wires - Early Work

The theoretical predictions by Thouless of 1D localization effects inspired a number of experiments. The earliest of these 1D localization studies were at Yale.[29] In these studies, triangular wires of Au-Pd were produced with the novel step edge technique (Fig. 5a-e) discussed earlier. These and subsequent experiments by Giordano[30] showed a resistance which increased with decreasing temperature. The form of the measured increase was

$$\frac{\Delta R}{R} \sim \frac{\rho T^{-1/2}}{A}. \tag{4}$$

A is the cross sectional area, ρ the resistivity. The $A^{-1}$ dependence is as expected; it is implicit in Eq. 3. The experimental $T^{-1/2}$ dependence, however, is weaker than predicted by Eq. 3 for the expected inelastic scattering mechanisms, unless $\tau_{in}^{-1} \sim T$. In addition, the dependence on $\rho$ is not that expected from Eq. 3, which predicts that $\Delta R/R \sim \rho^{1/2}$ for $\tau_{in}$ independent of $\rho$ and $D \sim 1/\rho$. If the inelastic time had a dependence $\tau_{in} \sim \rho/T$, however, the experimental results on Au-Pd could be understood. Though no inelastic mechanism has been identified which has this dependence, magnetoresistance studies of thick, 3D In-oxide films do show such a dependence of $\tau_{in}$ on $\rho$ and T for that system.[30a]

Other experiments on 1D metal wires have been carried out by Chaudhari and co-workers at IBM[31] and by White and co-workers.[32] The IBM work used wires and films of W-Re, an amorphous superconductor. Electron beam contamination resist patterning was used. Experimental results agreed quantitatively with those for Au-Pd wires, once superconducting fluctuation effects were subtracted. Since W-Re is a superconductor with $T_c \sim 3.5$ K, measurements can also be made of the inelastic time just below $T_c$, using phase-slip resistance measurements. Results of such measurements[31b] are consistent with the inelastic times inferred from the localization measurements.[31a] However, subsequent theoretical work on localization effects above the transition temperature of a superconductor indicate that other mechanisms contribute to $\Delta R$,[35] so that a more complex analysis of the data is required for extracting $\tau_{in}$. We discuss this in the next section, in relation to experiments on Al films.

Experiments by White and co-workers on Cu, Ni, and Au-Pd wires used the procedure in Fig. 5f. Wires were produced on a thin membrane so that they could be inspected with TEM. A TEM picture of such a wire is shown in Fig. 6. In these studies it was found that low temperature resistance rise had to be carefully separated from a temperature dependent background due presumably to electron-phonon scattering. The resistance rise could be explained as being due to electron-electron interaction effects, given approximately as

$$\frac{\Delta R_{int}}{R} \sim \frac{L_{int}}{L_{36k\Omega}} . \qquad (5)$$

Furthermore, previous Au-Pd and W-Re results were shown to be consistent with the interaction mechanism, at least to within numerical factors of $\sim 2$. If the interaction mechanism is to explain the full resistance rise, then the inelastic times must be even shorter than the very short values originally inferred.[30,31] Also, the unusual dependence on $\rho$ seen in the Au-Pd results, Eq. 4, is not explained by the interaction mechanism. Since inelastic mechanisms in dirty 2D metal films were also not well understood at that time, it was clear that efforts needed to be directed towards understanding the inelastic scattering mechanism(s) itself.

Experiments on Aluminum: Films and Wires, and Other Recent Work

Recent experiments by Santhanam and Prober[33] have investigated the inelastic scattering rate in clean aluminum films and wires, specifically to understand inelastic mechanisms in a clean system where extensive transport studies have previously been done on single-crystal samples, and on superconducting properties of thin films. Al readily forms thin continuous films which adhere well to most substrates; it is also a superconductor ($T_c \sim 1.4$K for the films studied). To allow both 2D and 1D samples to be produced, a lift-off process was employed. For linewidths >1 μm, photolithography was employed. For linewidths <1 μm, x-ray lithography was used. Lines as narrow as 400 Å are readily produced with edge-defined masks.[34]

Magnetoresistance studies were carried out for a number of film strips of thicknesses 150 and 250 Å, and widths 0.6 to 40 μm. The 40 μm strips are in the 2D limit. Sheet resistances were 1-4 ohms/square, corresponding to a mean free path $\ell \sim 100$ Å. Theoretical predictions for the magnetoresistance of 2D superconducting films above $T_c$ include the effects of superconducting fluctuations,[35] localization effects,[36] and electron-electron interaction effects.[26] For the field range used, only the first two contributions are significant.

The experimental data on the wide, 2D films can be fit with the 2D theoretical expressions[35,36] at all temperatures, using a temperature-dependent value of $\tau_{in}$ and a temperature-independent value of the spin-orbit time, $\tau_{so}$. Inelastic scattering rates are found to fit the simple form:

$$\tau_{in}^{-1} = A_1 T + A_3 T^3, \tag{6}$$

where the coefficients $A_1$ and $A_3$ are determined for each sample. The $A_1$ term is due to electron-electron scattering in a dirty 2D system, and agrees with the theoretical value given by

$$A_1 = \frac{e^2 R_\square}{2\pi\hbar^2} k_B \ln (T_1/T) \tag{7}$$

with $R_\square$ the sheet resistance and $T_1 = 9 \times 10^5 (k_F \ell)^3$. The investigated films should indeed be in the dirty 2D limit for electron-electron scattering. The $A_3$ term is due to electron-phonon scattering in a 3D system. The films should indeed be in the 3D regime with respect to the phonon wavelength at the higher temperatures where the $T^3$ term is dominant. ($\lambda_{phonon} \sim 750$ Å/T for Al; $\lambda$ is thus less than the film thickness above 5K.) The theoretical value[37] of

$A_3$ is 9 x $10^6$ $sec^{-1}K^{-3}$; this value is obtained[37] by averaging over the Fermi surface. Experimental values of $A_3$ are ∿50% larger. Given the theoretical and experimental uncertainties, this can be considered to be good agreement. The theoretical value of $A_3$ at specific Fermi surface locations has been confirmed experimentally (with single crystals) to about the same accuracy.[37]

For narrower strips, a deviation from the 2D form of the magnetoresistance is seen, at temperatures which correspond roughly to the strip width W being less than $\ell_{in}(T)$. This appears to signal the 2D to 1D crossover, which is expected to occur when $\ell_{in} > W$.[26a] This crossover is a strong confirmation of the actual length scale in the localization theory. The values of $\ell_{in}$ are large - a few μm at 2K. This indicates that experiments probing localization at a size scale $\ell_{in}$ may be possible with such clean films.

Other experiments on magnetoresistance of Al films have also recently been completed.[38,39] These have all been on films with larger values of $R_\square$ . The data in all these experiments appear to be consistent with those discussed above. However in the case of Refs. 38, the methods of data analysis omitted aspects of the theory which have proven to be important.[33,39] Studies of films of high-resistivity, granular Al have been conducted by Gordon and co-workers. They find that $\tau_{in}^{-1} = A_1T + A_4T^4$ for the films with $R_\square > 50\Omega$. The $T^4$ term is ascribed to electron-phonon scattering in a dirty, 3D system. The conclusions reached by Santhanam and Prober for clean films, Eq. 6, and also for the high resistance films[39] appear to be generally applicable. The other experiments on Al films treat only 2D systems, and further discussion of the results is somewhat outside the domain of this chapter. Additional discussion is given in Refs. 33 and 39.

One other study of microfabricated 1D metal wires has been conducted recently[40] to determine the dependence of the resistance rise, ΔR, on sample length. Au-Pd triangular wires were formed, and silver contact pads were patterned by evaporating over a thin fiber which defined the gap between contact pads, and thus the wire length. The length scale found is 0.2 μm at 1.5K. This is consistent with the value of $\ell_{in}$ inferred from measurements on long wires.[29] The specific functional dependence of ΔR/R on wire length remains to be explained, however.

A final example of 1D metal wires is given by work on wires formed by filling etched particle tracks in mica with Au.[41] Wires produced were as small as 80 Å, the smallest to be studied in any experiment. The resistance rise seen is consistent in magnitude with that seen in lithographically-produced low resistivity wires,[32] although the temperature dependence of ΔR is not exactly proportional to $T^{-1/2}$ as had been found in previous studies.

In conclusion, studies of 1D conduction in fine metal wires have been fairly successful in verifying predictions of the weak localization theory. For cleaner films of Al, the inelastic scattering rate can be understood quantitatively as a sum of two terms due to electron-electron and electron-phonon scattering. Inelastic scattering processes in very dirty metal films and wires ($\rho$ > 100 $\mu\Omega$-cm) still remain to be understood.

Experiments on 1D MOSFETs

MOSFETs (Metal-Oxide-Semiconductor-Field-Effect Transistors) represent an ideal proving ground for studying 2D electron transport, and have been the primary system used for confirming 2D localization theory.[42] In addition, since the gate voltage can be varied, the number of carriers can be directly controlled, unlike the case of metals. This extra variable can be of considerable importance in emphasizing different mechanisms.

Fabrication approaches. Two dimensional MOSFETs are made with standard IC fabrication technology, typically on <100> silicon. To emphasize localization effects over interaction effects, high mobility devices are required. This demands stringent cleanliness in the fabrication process, particularly for avoidance of Na ion contamination. High energy fabrication processes, such as electron beam lithography, should also be avoided if possible, as they create charged scattering centers in the oxide, reducing the mobility. Optical photolithography does not induce such damage. The highest mobility Si devices have $\mu \sim 20{,}000$ cm$^2$/V-sec at ∿4K.[43,44]

Magnetoresistance and Hall effect studies have been conducted on 2D MOSFETs by a number of research groups.[27f,42,43,45] There appears to be consensus that the existing theory of localization can describe these results, although the underlying scaling theory has been questioned by one group.[45b] Inelastic scattering rates can be described by[27f,43b]

$$\tau_i^{-1} = A_1 T + A_2 T^2 \qquad (8)$$

with the $A_1$ and $A_2$ terms due to electron-electron scattering in dirty and clean limits respectively, or more simply by $\tau_i^{-1} = A_1 T$.[42] There is a question raised regarding the magnitude of $A_1$, and results obtained by various groups may not be in full agreement.[27f,43b]

The fabrication of 1D MOSFETs presents some special challenges which do not arise for wide MOSFETs. The essential difficulty is in producing a narrow (∿0.1 μm) gate electrode without significantly reducing the device mobility. Two main fabrication approaches have been developed. In one approach, developed by Wheeler and co-workers

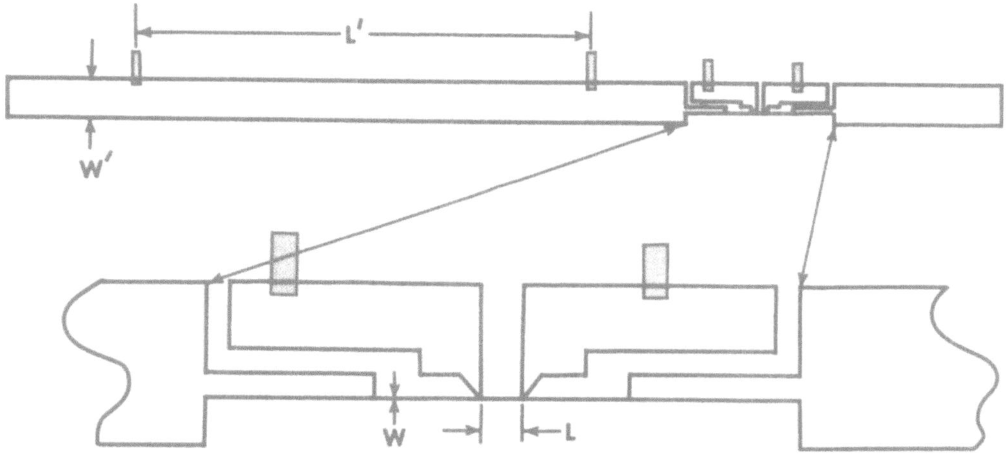


Fig. 8. Schematic representation of the gate structure of devices used in 1D MOSFET experiments at Yale. The darkened areas indicate locations of diffused-probe regions. Here W' = 100 μm, L' = 1050 μm and L = 25 μm. The data reported in Ref. 44 were from device AU310 where the mean value of W = 0.40 μm with a mean deviation of 0.02 μm.

at Yale,[44] optical projection lithography[6] is used to produce a photoresist mask for patterning a large area Al gate with wet chemical etching. With such techniques, the very high mobility achieved in the large area device is preserved. The geometry of the devices studied at Yale is shown in Fig. 8. Four-terminal measurement of resistance for both the wide and narrow sections is accomplished with doped or inversion-layer voltage probes. The oxide thickness is typically 350 Å, and the chip size is 2 mm. The minimum gate width produced is 3000 Å, with a rms width nonuniformity ≤5%.

The second major approach to 1D MOSFET fabrication has been developed at Bell Laboratories. Skocpol and co-workers[46] have produced MOSFETs as narrow as 400 Å using electron beam lithography to pattern, by lift-off, a Ni-Cr gate electrode. This gate then serves as an etch mask for reactive-ion etching of the 1000 Å $SiO_2$ layer and 2000 Å into the underlying silicon. The inversion layer thus

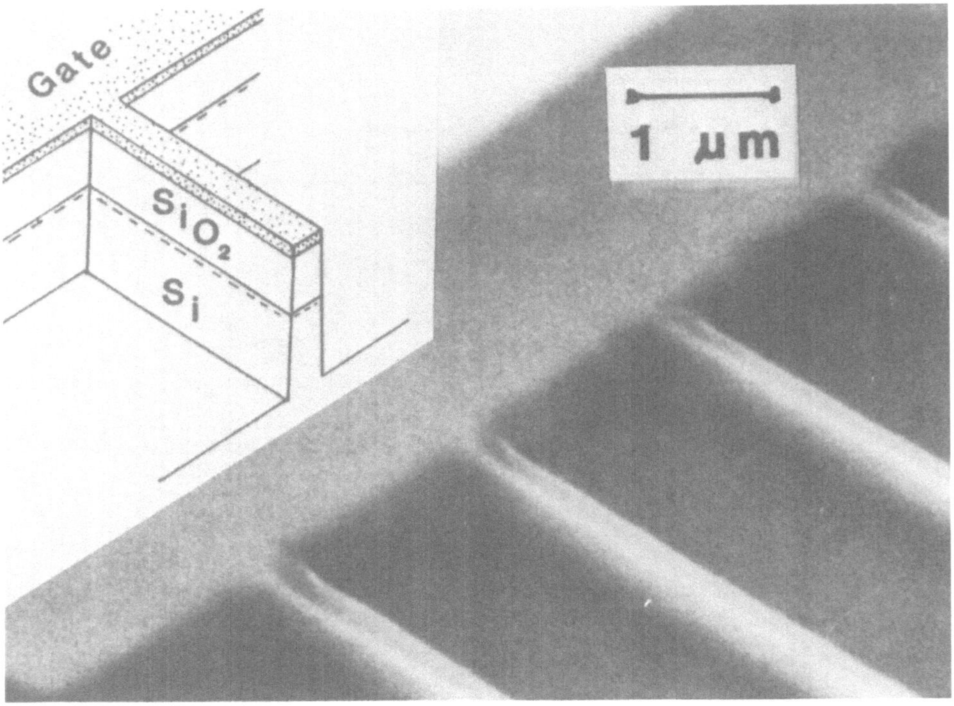


Fig. 9. Scanning electron micrograph of perspective view of several narrow MOSFET channels, and schematic diagram (inset) (From Ref. 46a)

sits in a pedestal, as shown in Fig. 9. Edge passivation is important, but fringing field effects, which can broaden the inversion layer relative to the geometrical gate width, should be negligible.

For these devices it is necessary to improve the mobility, which is significantly degraded by the lithographic processes. A 300°C thermal anneal is used after processing. The temperature limit is set by use of gold alignment marks. After annealing, mobilities up to 4000 $cm^2$/V-sec are obtained, with moderate threshold voltages. These devices are, in any case, the smallest MOSFETs produced to date. A 400 Å wide MOSFET is shown in Fig. 10.

Other approaches to the fabrication of 1D MOSFETs have been developed. The two approaches discussed above dealt with inversion layers. Fowler et al.[47] have produced very narrow accumulation layers in an FET structure where the accumulation layer is pinched down by the field of two $p^+$ regions on either side of the conducting strip. The spacing between $p^+$ regions is 1-2 μm, so that optical lithography can be used. Studies were carried out in the regime of

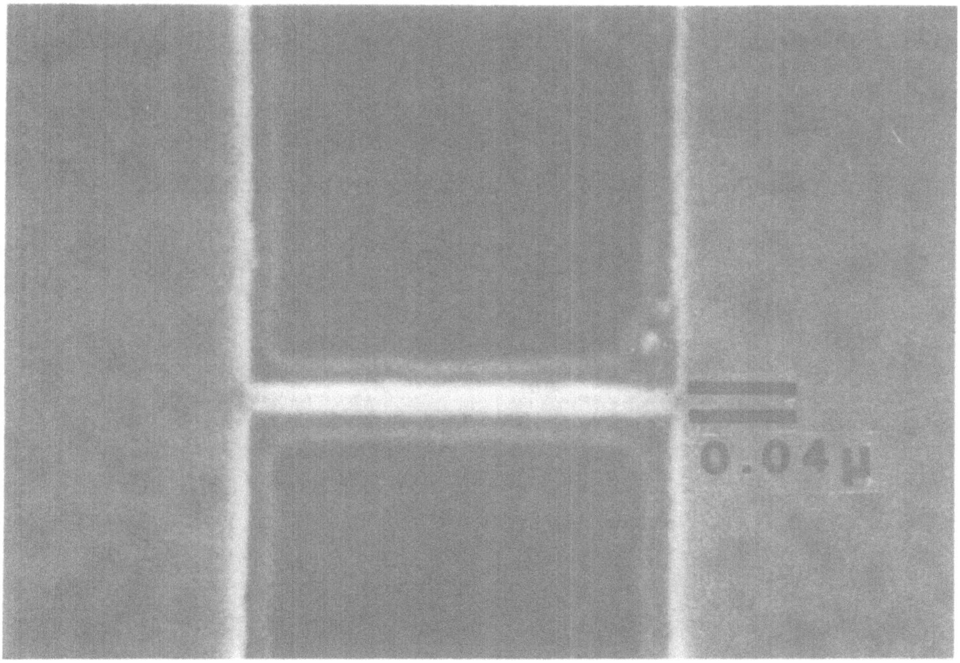


Fig. 10. SEM of narrow MOSFET; gate width is 400 Å, and gate length is 0.9 μm. (From Ref. 46b)

strong localization, where variable-range hopping is the main conduction mechanism. A transition from 2D to 1D hopping is observed as the channel width is reduced.

An edge-defined gate technique (Fig. 5f) has been developed by Kwasnick and co-workers.[48] Studies have been largely in the regime of strong localization. A topologically similar approach has been developed by Speidell[49] using the edge of a photoresist layer instead of an etched step.

Experimental Results: 1D Localization and Interaction Effects, and 1D Density-of-States. Low temperature studies of the temperature-dependent resistance and magnetoresistance have been carried out on the narrow inversion-layer devices.[44,46] Both sets of studies find that localization and interaction effects contribute in similar magnitude to the resistance rise, so that analysis of the temperature-dependent resistance does not allow a straightforward extraction of $\tau_{in}$. Magnetoresistance measurements allow a more direct determination of localization effects (and thus $\tau_{in}$), since, for the high mobility devices, the interaction effects depend only weakly on field at low fields.

Magnetoresistance measurements were carried out by Wheeler and co-workers on a number of 1D samples. The localization resistance in a finite magnetic field, H, is given by the expression[50,44]

$$\frac{\Delta R}{R} = \frac{R_{\square}}{2\pi\hbar/e^2}\,\frac{1}{W}\left(\frac{1}{D\tau_{in}} + \frac{W^2}{12\ell_H^{\,4}}\right)^{-1/2}. \tag{9a}$$

W is the channel width and $\ell_H$ is the magnetic length:[26] $\ell_H^{\,2} = (\hbar c/2eH)$. The magnetoresistance, $\delta R$, is given as

$$\delta R = \Delta R(H) - \Delta R(H = 0). \tag{9b}$$

When $\ell_{in} > W$, Eq. 9a is accurate.

For the narrow samples at low temperatures (0.5K), it is found that

$$\tau_{in}^{-1}\Big|_{narrow} \sim T^{1/2}, \tag{10}$$

whereas for the wide (2D) samples, fits to the 2D magnetoresistance theory yield a rate

$$\tau_{in}^{-1}\Big|_{wide} \sim T\ . \tag{11}$$

These dependences are in accord with the theoretical predictions for electron-electron scattering in 1D and 2D dirty systems. The experimental magnitudes are also reasonably close to the theoretical predictions. Thus, the 1D experiments seem consistent with the theories for localization and for inelastic scattering in dirty systems.

In the course of these 1D experiments, it was noted that there was a curious variation of device resistance with the surface electron density, $n_s$, which is controlled by the gate voltage, $V_g$. Although no complete theory has been developed, there are indications[51] that such a variation of resistance with $n_s$ may be due to quantization of the electron wavefunctions (and energy levels) by the narrow transverse width dimension of the gate, W. $n_s$ is proportional to the Fermi energy $E_F$, and thus is proportional to ($V_g$ - $V_{threshold}$). As $n_s$ is increased, the higher quantum levels are filled. A very high density of states occurs when the Fermi level crosses an energy level. (Conduction along the length of the channel may also be quantized, but with a much finer energy spacing, <<kT. Thus, only the width quantization should be significant.)

It is believed[51] that this particle-in-a-box width quantization affects the resistance through the localization resistance, ΔR, since ΔR depends on $\tau_{in}$ (Eqs. 1 and 3), and $\tau_{in}$ depends on the density of states at $E_F$.[44] Even though the elastic mean free path, $\ell \sim 1000$ Å, is less than the channel width, it is believed that the much longer inelastic length $\ell_{in}$ is the length over which the electron can sense quantizing boundaries. This is because over the distance $\ell_{in}$, the electron has a constant energy. At the low temperatures where these effects are observed, $\ell_{in}$ is greater than the width.

The resistance variation as a function of $V_g$ does show significant structure, with a typical spacing in gate voltage like that expected from the quantization arguments above. However, the structure is not very regular. This is due, at least in part, to width variations of the gate electrode, which cause the Fermi level to cross different quantized levels along the length of the channel.

To analyze quantitatively the lithographic requirements for width uniformity which will ensure that the Fermi level will be in a single quantum level, n, we list in Table IV the results of a calculation to determine the width difference ΔW to go from the $n^{th}$ to the (n+1) level. As seen in the table, for current device widths (3000 Å) and current $n_s$ values, a width variation of <100 Å must be achieved. This is a very demanding specification, but probably not impossible for narrower gates (500 Å) with currently available technology (See Figures 6 and 10, which indicate the type of lithographic uniformity now achieved.) Achieving such 100 Å-accuracy lithography is by no means trivial, especially if high device mobilities are required to allow smaller values of $n_s$. Producing these near-ideal 1D structures remains a significant future challenge.

Table IV. MOSFET Characteristics for 1D Density-of-States Studies

| W (Gate Width) | $n_s$ ($cm^{-2}$) | $E_F/k_B$ | n (Quantum Number at $E_F$) | $\Delta W_{n \to n+1}$ |
|---|---|---|---|---|
| 3000 Å | $4 \times 10^{12}$ | 300 K | 34 | 100 Å |
| 500 Å | $4 \times 10^{12}$ | 300 K | 6 | 100 Å |
| 500 Å | $1 \times 10^{12}$ | 75 K | 2 | 200 Å |

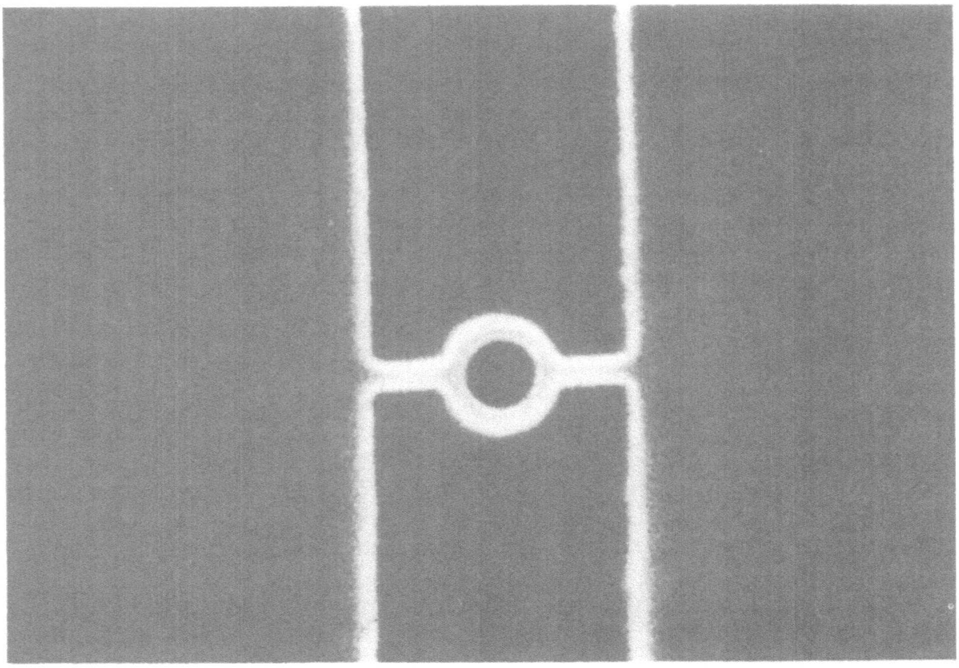

Fig. 11. SEM of 2000-Å diameter MOSFET ring, for study of normal-metal flux quantization. (From Ref. 46b)

Fluxoid Quantization in Nonsuperconducting Rings. A final issue which can be addressed with small MOSFET structures is the possibility of flux quantization in nonsuperconducting rings, first reported by Sharvin and Sharvin.[52] For such effects to be evident, the ring diameter should be less than $\sim\ell_{in}$, so that electron waves can typically remain in the same energy state and interact with each other around the ring.[26b] For metals, $\ell_{in}$ is usually short, and diameters <1 μm are required. For rings formed of an inversion layer, somewhat larger dimensions should be satisfactory. At Bell Laboratories, MOSFET rings as small as 2000 Å in diameter have been produced[46b]; an example is shown in Fig. 11. Experiments on these structures are at an initial stage.

The use of modulation-doped GaAs-AlGaAs heterostructures[53] can offer even larger values of $\ell_{in}$: a value $\ell_{in} \sim 10$ μm at 1K has recently been observed in work at Bell Laboratories.[54] Preliminary experimental evidence appears to confirm the existence of fluxoid quantization in normal rings of these heterostructures, in structures of micron size.[54] Such a confirmation provides a striking demonstration of the power of microlithography when combined with advanced thin film deposition, for producing model systems for studies of low temperature electron quantum transport.

SCIENTIFIC STUDIES: SUPERCONDUCTING JOSEPHSON ARRAYS

Many experiments in recent years have studied properties of granular and composite metal films in order to probe issues of percolation, 2D superconducting transitions, and localization. Experimental studies of percolation and superconductivity in granular films are reviewed in this volume by Deutscher. When granular or composite films are used for studies of localization and 2D superconducting transitions, a question arises as to the effect of macroscopic nonuniformity and local inhomogeneities on the intrinsic behavior. There has been some progress in answering this question, at least for model types of macroscopic inhomogeneities.[55] Still, it is clearly of interest to study model systems, such as arrays, where macroscopic inhomogeneities can, at least in principle, be eliminated.

An ideal array to consider is that of superconducting islands which are coupled by Josephson tunneling to their nearest neighbors. The Josephson coupling energy is given as

$$E_J = \Phi_o I_o (1 - \cos\theta)/2\pi, \tag{12}$$

where $\Phi_o$ is the flux quantum hc/2e, $I_o$ is the critical current, and $\theta$ is the quantum mechanical phase difference across the Josephson junction. Such a discrete system, if large, is equivalent to the 2D X-Y model,[56] and can thus serve as a system in which the Kosterlitz-Thouless vortex unbinding transition can be studied.

Josephson junction arrays have been studied by Resnick and coworkers[57], Abraham and co-workers at Harvard[58], Berchier et al.[59] and by researchers at IBM.[60] The approach of the Harvard group is conceptually the simplest. In that work, superconducting (S) islands are coupled via Josephson tunneling thru a normal (N: non-superconducting) metal. This is a SNS tunneling structure. To form the array, a superconducting film of PbBi is first evaporated through a mechanical shadow mask consisting of a thin grid with a square array of holes as small as ∿10 μm. Without breaking vacuum, the mechanical mask is quickly moved away and a uniform Cu film is evaporated on top of the PbBi squares. This procedure is intended to avoid interface contamination,which can reduce Josephson coupling. It is also simple and inexpensive. Materials were selected to have minimal interdiffusion. The grid must be held in uniform contact with the substrate to avoid penumbral shadowing or scattering of the PbBi under the mask. Also, with the mask structure employed, Josephson coupling is not simply to nearest neighbors, though this is apparently a secondary

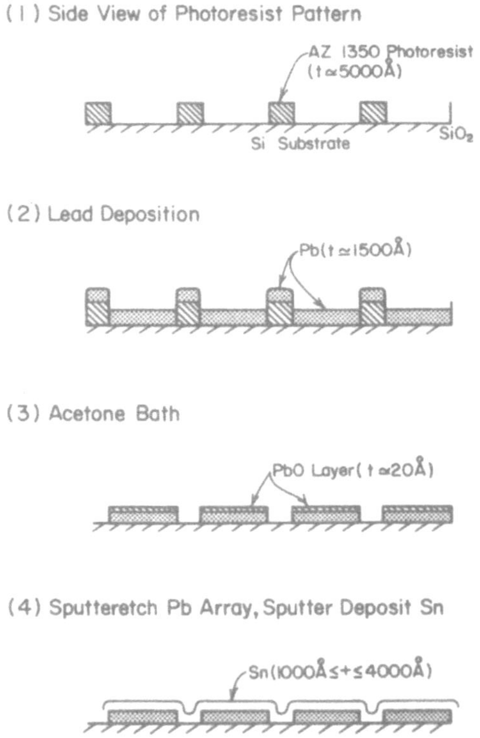


Fig. 12. Fabrication sequence for producing Pb-Sn-Pb proximity coupled Josephson arrays. (From work of Ref. 57)

issue. Vortex unbinding and periodic field response have been studied with these systems.

Patterning of SNS arrays with photolithography has been done by two groups.[57,59] In both cases, the normal metal was a second superconductor above its transition temperature. We shall describe the work of the Ohio State-Cincinnati group[57] as an example. This group formed a Pb-Sn-Pb tunnel junction array. Berchier and Sanchez have discussed the specifics of their fabrication procedure elsewhere.[59]

The procedure used to fabricate the Pb-Sn array is shown in Fig. 12. A Pb film is first evaporated onto a photoresist-patterned substrate, and a Pb dot pattern is produced by lift-off. Then, in a sputtering system, the substrate is Ar ion etched. Without breaking vacuum a Sn film is sputter deposited. The in-situ sputter cleaning is required to remove surface contamination. Ion beam sputter cleaning could serve the same purpose if the second film was to be deposited by evaporation. Such in-situ cleaning is necessary when an atomically clean surface or good metallurgical contact is required. The vortex unbinding transition was studied with the Pb-Sn arrays.

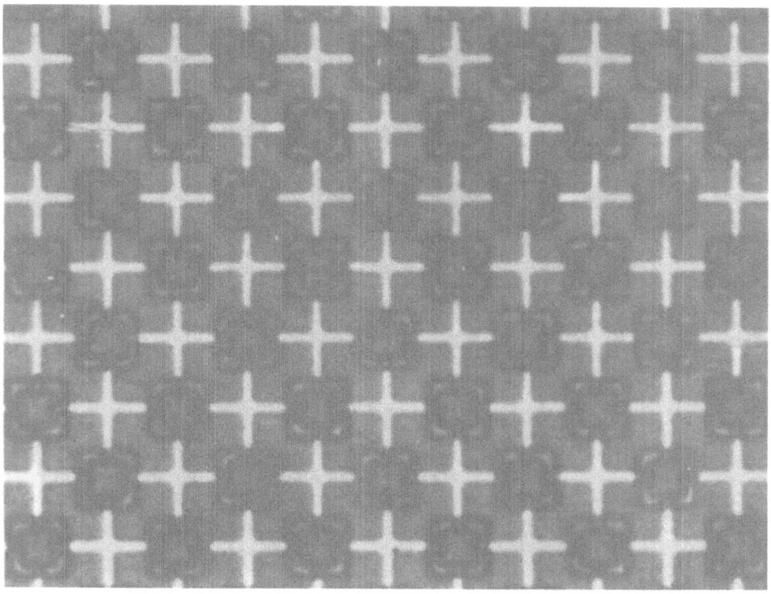

Fig. 13. Micrograph of a section of Pb-alloy Josephson junction array. Junction dimension is 2.5 μm. The total array consisted of 7200 junctions. (From Ref. 60a; see also Ref. 61.)

The last example of 2D Josephson junction arrays is work done at IBM. Standard processing was employed to form the oxide-barrier tunnel junctions.[61] Coupling was only to nearest neighbors for these arrays. An array of 7200 Pb-alloy junctions[60a] was used to study the effects of inhomogeneity when one fourth of the junctions were removed at random by laser "zapping" of the connections. A picture of a section of the uniform array is shown in Fig. 13. The tunnel junctions are located at the overlaps of the square rings and the crosses.

An array of 20,000 Nb-Nb tunnel junctions of $1\text{-}(\mu m)^2$ area was used to study the vortex unbinding transition and the periodic modulation of the zero-bias resistance with magnetic field.[60b] For a similar array, critical currents of individual junctions varied by ±50%. It is not clear if the resulting variation in $E_J$ (Eq. 12) is a problem, although experimental results were explainable in terms of behavior expected for a uniform array. The IBM Josephson process[61] can produce much more uniform Pb-alloy and Nb-Pb junctions for dimensions $\geq 2.5$ µm.

Linear arrays of SNS microbridges have recently been produced at Stanford by deLozanne and co-workers using the step edge techniques described in a previous section.[22] The geometry of a single microbridge is shown in Fig. 14. The use of a step edge to define

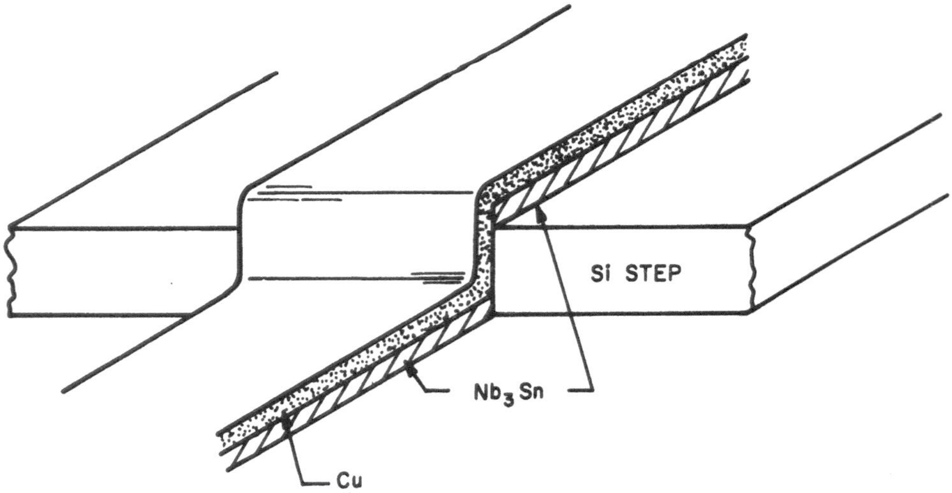


Fig. 14. Procedure for production of high-$T_c$ SNS microbridge. Depositions of the normal metal (Cu or Au) and the superconductor ($Nb_3Sn$, $Nb_3Ge$) are from different angles. (From Ref. 22.)

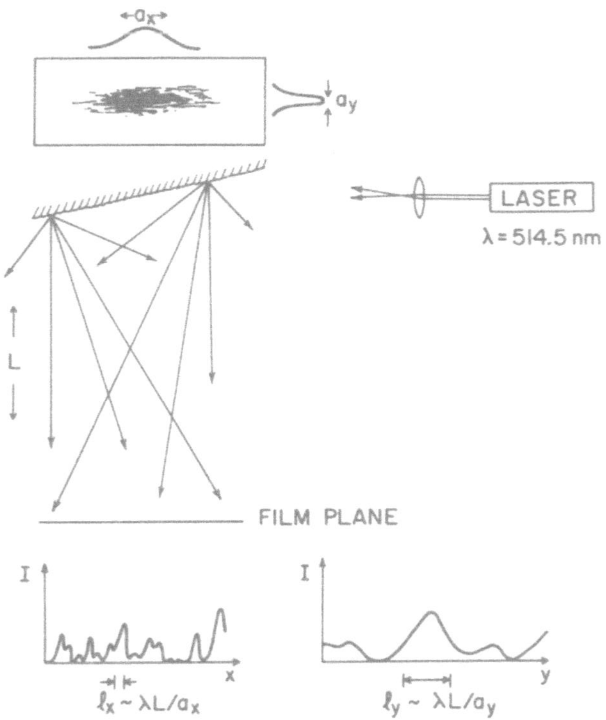


Fig. 15. Schematic of the optical system used to create the speckle patterns. The y direction is out of the plane of the figure. I is the exposure intensity at the film plane. This film is used as the exposure mask for photolithography.

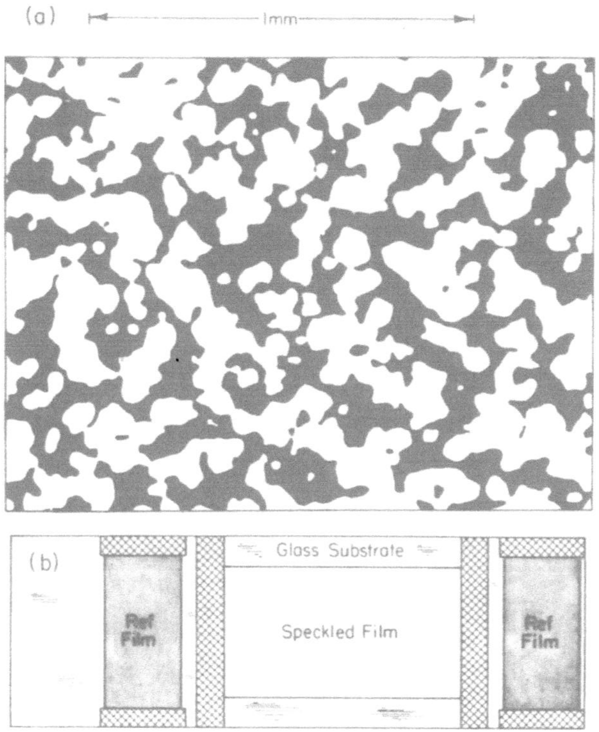


Fig. 16. (a) Photograph of a small section from an isotropic sample. The black areas are metal. (b) Sample geometry. The cross-hatched regions are the thick aluminum or NiCr contacts. (From Ref. 64.)

microbridge length leads to good uniformity of the critical currents for bridges in an array.

Other examples of array-like structures, though not of Josephson junctions, include Al films patterned with a triangular hole lattice[62] and films patterned by holographic resist exposure with a grating-like grid.[63] Both types of samples were used to study flux flow and pinning. Partially continuous films have been patterned by photoresist lift-off, where the exposure mask was produced by a laser speckle pattern, as shown in Figs. 15 and 16.[64] These partly continuous films were used in studies of percolation, in particular to test anisotropy effects. A final example is the construction of a metal dot pattern by angle evaporation onto an insulating substrate which had been etched with two crossed grating patterns. These samples were used to study surface-enhanced Raman scattering.[65]

Arrays have thus made effective contributions in a variety of scientific areas. One of the main difficulties in producing large arrays is production of the very complex masks required. Here again, the capabilities of modern integrated circuit facilities may be utilized. The large range of materials and fabrication approaches which can be employed should in the future allow many new scientific issues to be addressed.

## CONCLUSIONS

In this chapter we have reviewed trends in modern microlithography. It is now possible to pattern on the 100 Å scale using either direct (electron beam) lithography or edge defined approaches. Etching and deposition now commonly employ directional ion processes, such as reactive ion etching and ion beam sputtering. These varied techniques provide a powerful set of microfabrication methods. These techniques have already led to real advances in our understanding of electron quantum transport in systems of reduced dimensionality, and the 2D vortex unbinding transition in superconductors. In the next five years, the ability to pattern at these unprecedented size scales will open up many new scientific challenges and opportunities.

### Acknowledgements

Instructive discussions with numerous colleagues and collaborators are gratefully acknowledged. Particular thanks go to D. W. Face, R. E. Howard, Y. Imry, P. Santhanam, W. J. Skocpol, R. G. Wheeler and S. Wind for discussions relevant to this article. Research work at Yale University has been supported by the National Science Foundation and the Office of Naval Research.

REFERENCES

1. R. E. Howard and D. E. Prober, "Nanometer Scale Fabrication Techniques," in: VLSI Electronics: Microstructure Science, Vol. V, N. E. Einspruch, ed., Academic, New York (1982).
2. J. E. Lukens, AIP Conf. Proc. 44:198 (1978); R. E. Howard, P. F. Liao, W. J. Skocpol, L. D. Jackel, and H. G. Craighead, Science 221:117 (1983).
3. H. I. Smith, Proc. IEEE 62:1361 (1974).
4. A. N. Broers, IEEE Trans. Electron Devices ED-28:1268 (1981).
5. R. Newman (ed.), "Fine Line Lithography," North Holland Publ., Amsterdam (1980); I. Brodie and J. J. Muray, "Physics of Microfabrication," Plenum Press, New York (1982); also, Refs. 1-3, 6-12 in Ref. 1.
6. M. D. Feuer and D. E. Prober, IEEE Trans. Electron Devices ED-28: 1375 (1981); chromatic aberration of microscope lenses is discussed by M. J. Brady and A. Davidson, Rev. Sci. Instrum. 54: Oct. (1983), to be published.
7. P. Grabbe, E. L. Hu, and R. E. Howard, J. Vac. Sci. Technol. 21:33 (1982), and references therein.
8. J. M. Moran, Solid State Technol. 24(4):195 (1981); M. Hatzakis, Solid State Technol. 24(8):74 (1981); B. J. Lin, E. Bassous, V. W. Chao, and K. E. Petrillo, J. Vac. Sci. Technol. 19:1313 (1981).
9. L. D. Jackel, R.E. Howard, E. L. Hu, P. Grabbe, and D. M. Tennant, Appl. Phys. Lett. 39:268 (1981); D. M. Tennant, L. D. Jackel, R. E. Howard, E. L. Hu, P. Grabbe, R. J. Capik, and B. S. Schneider, J. Vac. Sci. Technol. 19:1304 (1981).
10. P. S. Burggraaf, Semicond. Intl. (6)55 June (1983); K. Bartlett, G. Hillis, M. Chen, R. Trutna and M. Watts, SPIE Proc. 394, no. 05 (1983).
11. H. G. Craighead, R. E. Howard, L. D. Jackel, and P. M. Mankiewich, Appl. Phys. Lett. 42:38 (1983).
12. P. Grabbe, Rev. Sci. Instrum. 51:992 (1980).
13. R. B. Laibowitz, this volume; also A. N. Broers, W. W. Molzen, J. J. Cuomo, and N. D. Wittels, Appl. Phys. Lett. 29:596 (1976) and W. W. Molzen, A. N. Broers, J. J. Cuomo, J. M. E. Harper, and R. B. Laibowitz, J. Vac. Sci. Technol. 16:269 (1979).
14. D. C. Flanders and A. E. White, J. Vac. Sci. Technol. 19:892 (1981); D. C. Flanders, J. Vac. Sci. Technol. 16:1615 (1979), and D. C. Flanders, Ph.D. Thesis, MIT (1978).
15. D. C. Flanders, Appl. Phys. Lett. 36:93 (1980).
16a. K. E. Bean, IEEE Trans. Electron Devices ED-25:1185 (1978).
16b. H. Kroger, L. N. Smith, and D. W. Jillie, Appl. Phys. Lett. 39:280 (1981).
17. H. W. Lehmann and R. Widmer, J. Vac. Sci. Technol. 15:319 (1978); L. M. Ephrath, IEEE Trans. Electron Devices ED-28: 1315 (1981).

18. D. W. Face, S. T. Ruggiero, and D. E. Prober, J. Vac. Sci. Technol. A1:326 (1983); S. T. Ruggiero, D. W. Face, and D. E. Prober, IEEE Trans. Magn. MAG-19:960 (1983); a general review of ion beam techniques for material processing is given by J. M. E. Harper, J. J. Cuomo, and H. R. Kaufman, Ann. Rev. Mat. Sci. 13:413 (1983).
19. M. Gurvitch, M. A. Washington, and H. A. Huggins, Appl. Phys. Lett. 42:472 (1983); S. T. Ruggiero, E. Track, and D. E. Prober, to be published. See also Ref. 16b.
20. D. E. Prober, M. D. Feuer, and N. Giordano, Appl. Phys. Lett. 37:94 (1980).
21. M. D. Feuer and D. E. Prober, Appl. Phys. Lett. 36:226 (1980).
22. A. de Lozanne, M . S. DiIorio, and M. R. Beasley, Appl. Phys. Lett. 42:541 (1983), and private communication
23. G. J. Dolan, Appl. Phys. Lett. 31:337 (1977); E. L. Hu, L. D. Jackel, and R. E. Howard, IEEE Trans. Electron Devices ED-28:1382 (1981).
24. L. D. Jackel, J. P. Gordon, E. L. Hu, R. E. Howard, L. A. Fetter, D. M. Tennant, R. W. Epworth, and J. Kurkijärvi, Phys. Rev. Lett. 47:697 (1981).
25. a. G. J. Dolan, T. G. Phillips, and D. P. Woody, Appl. Phys. Lett. 34:347 (1979) and b. A. D. Smith, R. A. Batchelor, W. R. McGrath, P. L. Richards, H. van Kempen, D. E. Prober, and P. Santhanam, Appl. Phys. Lett. 39:655 (1981).
26. a. Y. Imry, this volume, and J. Appl. Phys. 55:1812 (1981); b. H. Fukuyama, this volume, and Surf. Sci. 113:489 (1982).
27. a. A. Schmid, Z. Phys. 259:421 (1973); b. G. Bergmann, Z. Phys. B48:5 (1982); c. G. Giuliani and J. J. Quinn, Phys. Rev. B26:4421 (1982) - we identify $|p-p_F|$ with the thermal energy kT; d. E. Abrahams, P. W. Anderson, P. A. Lee, and T. V. Ramakrishnan, Phys. Rev. B24:6783 (1981); e. B. L. Altshuler, A. G. Arovov, D. E. Khmel'nitsky, J. Phys. C 15:7367 (1982); f. R. A. Davies and M. Pepper, J. Phys. C. 16:L353 (1983).
28. D. J. Thouless, Phys. Rev. Lett. 39:1167 (1977) and Solid State Commun. 34:683 (1980).
29. a. N. Giordano, W. Gilson, and D. E. Prober, Phys. Rev. Lett. 43:725 (1979); b. N. Giordano, Phys. Rev. B22:5635 (1980).
30. N. Giordano, in Physics in One Dimension, J. Bernasconi and T. Schneider, eds., Springer-Verlag, New York (1981), p.310, and Ref. 29b.
30a. Z. Ovadyahu, to be published.
31. P. Chaudhari and H.-U. Habermeier, Phys. Rev. Lett. 44:40 (1980), and Solid State Commun. 34:687 (1980); b. P. Chaudhari, A. N. Broers, C. C. Chi, R. Laibowitz, E. Spiller, and J. Viggiano, Phys. Rev. Lett. 45:930 (1980).
32. A. E. White, M. Tinkham, W. J. Skocpol, and D. C. Flanders, Phys. Rev. Lett. 48:1752 (1982).
33. P. Santhanam and D. E. Prober, to be published.

34. S. Wind, unpublished.
35. A. I. Larkin, Pis'ma Zh. Eksp. Teor. Fiz. 31:239 (1980), [JETP Lett. 31:219 (1980)].
36. S. Hikami, A. I. Larkin, and Y. Nagaoka, Prog. Theor. Phys. 63:707 (1980).
37. W. E. Lawrence and A. B. Meador, Phys. Rev. B18:1154 (1978).
38. a. Y. Bruynseraede, M. Gijs, C. Van Haesendonck, and G. Deutscher, Phys. Rev. Lett. 50:277 (1983); a recent reanalysis of this data yields the result of Eq. 6. b. M. E. Gershenson, V. N. Gubankov, and Yu. E. Zhuralev, Solid State Commun. 45:87 (1983).
39. J. M. Gordon, C. J. Lobb, and M. Tinkham, Phys. Rev. B, to be published.
40. J. T. Masden and N. Giordano, Phys. Rev. Lett. 49:819 (1982).
41. W. D. Williams and N. Giordano, Bull. Am. Phys. Soc. 28:486 (1983), and private communication.
42. R. C. Dynes, Physica 109-110B:1857 (1982); D. J. Bishop, R. C. Dynes and C. C. Tsuei, Phys. Rev. B26:773 (1982).
43a. R. G. Wheeler, Phys. Rev. B24:4645 (1981)b. K. K. Choi, Phys. Rev. B, to be published.
44. R. G. Wheeler, K. K. Choi, A. Goel, R. Wisnieff, and D. E. Prober, Phys. Rev. Lett. 49:1674 (1982).
45. a. Y. Kawaguchi and S. Kawaji, Surf. Sci. 113:505 (1982); b. R. A. Davies, M. Pepper, and M. Kaveh, J. Phys. C 16:L285 (1983).
46. a. W. J. Skocpol, L. D. Jackel, E. L. Hu, R. E. Howard, and L. A. Fetter, Phys. Rev. Lett. 49:951 (1982) and Physica 117-118B:667 (1983); b. L. D. Jackel, Bull. Am. Phys. Soc. 28:401 (1983); W. J. Skocpol, Bull. APS 28:276 (1983).
47. A. B. Fowler, A. Hartstein, and R. A. Webb, Phys. Rev. Lett. 48:196 (1982).
48. R. F. Kwasnick, M. A. Kastner, and J. Melngailis, Bull. Am. Phys. Soc. 28:322 (1983).
49. J. L. Speidell, J. Vac. Sci. Technol. 19:693 (1981).
50. B. L. Altshuler and A. G. Aronov, JETP Lett. 33:499 (1981).
51. R. G. Wheeler, Bull. Am. Phys. Soc. 28:276 (1983); also, D. E. Prober, Bull. Am. Phys. Soc. 28:401 (1983).
52. B.L. Al'tshuler, A. G. Aronov, B. Z. Spivak, D. Yu. Sharvin, and Yu. V. Sharvin, JETP Lett. 35:588 (1983).
53. R. Dingle, H. Störmer, A. Gossard and W. Wiegmann, Appl. Phys. Lett. 33:665 (1978).
54. G. E. Blonder and R. C. Dynes, private communication and to be published.
55. Y. Gefen, D. J. Thouless, and Y. Imry, Phys. Rev. B, to be published.
56. J. E. Mooij and P. Minnhagen, this volume; Y. Imry, AIP Conf. Proc. 58:141 (1980).
57. D. J. Resnick, J. C. Garland, J. T. Boyd, S. Shoemaker, and R. S. Newrock, Phys. Rev. Lett. 47:1542 (1981).

58. D. W. Abraham, C. J. Lobb, M. Tinkham, and T. M. Klapwijk, Phys. Rev. B26:5268 (1982).
59. J. L. Berchier and D. Sanchez, Rev. de Physique Appl. 14: 757 (1979), and references therein.
60. a. A. Davidson and C. C. Tsuei, Physica 108B:1243 (1981); b. R. F. Voss and R. A. Webb, Phys. Rev. B25:3446 (1982).
61. J. H. Greiner et al., IBM J. Res. Dev. 24:195 (1980); R. F. Broom et al., IEEE Trans. Electron Dev. ED-27:1998 (1980).
62. A. T. Fiory, A. F. Hebard and S. Somekh, Appl. Phys. Lett. 32:73 (1978).
63. O. Daldini, P. Martinoli, J. L. Olsen, and G. Berner, Phys. Rev. Lett. 32:218 (1974).
64. L. N. Smith and C. J. Lobb, Phys. Rev. B20:3653 (1979).
65. P. F. Liao, J. G. Bergman, D. S. Chemla, A. Wokaun, J. Melngailis, A. M. Hawryluk, and N. P. Economou, Chem. Phys. Lett. 82:355 (1981).

# FABRICATION OF SUB-0.1 μm FINE METAL LINES USING HIGH RESOLUTION ELECTRON BEAM TECHNIQUES WITH CONTAMINATION RESIST

R. B. Laibowitz and C. P. Umbach

IBM Research Center
PO Box 218
Yorktown Heights, NY 10598

## INTRODUCTION

Along with the interest in decreasing the size of the devices used in microelectronics, there is also a great deal of interest in determining the physical properties of these small devices and, in particular, the properties of the fine metal lines used in them. These fine metal lines are already known to exhibit many interesting physical properties in their own right such as Josephson effects and electron localization[1]. Many new techniques and processes are necessary in order to fabricate and test fine metal lines with thicknesses of about 10 to 30 nm and with linewidths less than 0.1 μm. Experiments to develop such techniques are in themselves very useful in extending our knowledge of the limits of present day lithography and in discovering the physical processes that may limit the density and complexity of the structures used in future large scale integrated circuits. The main fabrication processes described in this paper utilize high resolution electron beam techniques, thin (window)substrates and contamination resist in addition to a variety of thin film fabrication processes. In addition to forming single fine lines, more complex patterns such as multi-voltage probe samples, bridge SQUIDS and circular loops can be fabricated using these techniques. Many other patterns are also possible, demonstrating the versatility of electron beam writing. It is worth emphasizing that, in addition to its use in the fabrication of the fine lines, electron microscopy, both transmission (TEM) and scanning (SEM), is also used extensively in the examination and evaluation of the fine metal lines. Other techniques[2] for fine line fabrication such edge shadowing are discussed in the references and elsewhere in these proceedings.

We have made fine wires from a variety of metals, both normal and superconducting, whose structure varied from amorphous to fine grain to large grain to single crystal. For many of the thin metal films that we have used in the fabrication of the fine lines, it has been found that the grain size can play a large role in the mechanical and the electrical properties of the fine line. Some of the metals that we have studied include polycrystalline Nb with an average grain size of about 15 nm, Au whose structure could be varied from small grain to single crystal and amorphous W-Re alloys. Metals such as Cu, Al and AuPd alloys have also been used.

In the sections below we shall first discuss the use of electron beams in the fabrication process and then the role of ion milling in forming the fine line pattern. A separate section covers the metallizations used for the fine lines. In general this paper mainly attempts to describe the fabrication processes and the structure of the fine lines. Detailed descriptions of the electrical conductivity of narrow lines are left to the references[1,2].

## LITHOGRAPHIC FABRICATION PROCESSES

The source of electrons used in the fabrication of our fine metal lines is shown in Fig. 1. It is basically a scanning transmission electron microscope (STEM) which allows us to use the same source of electrons in both the fabrication and the examination of

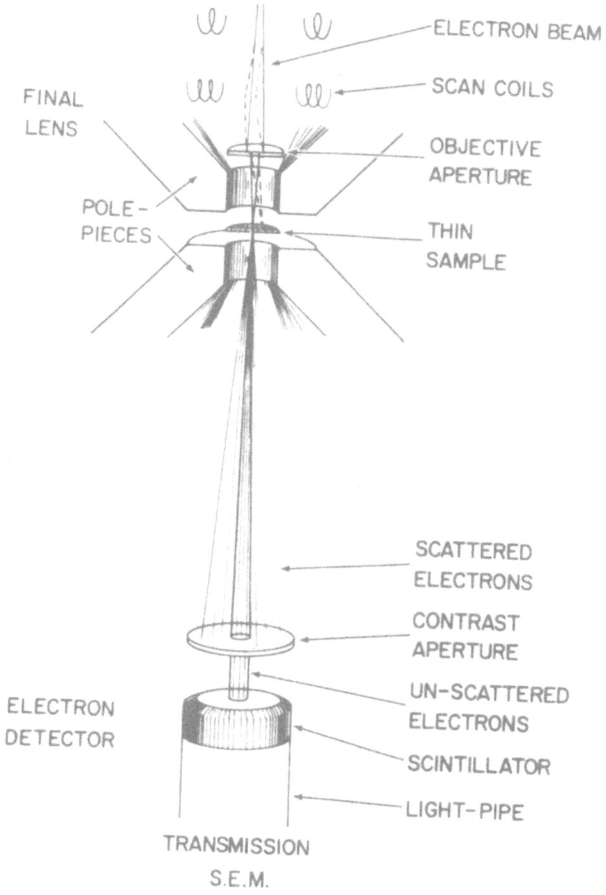


Figure 1. A schematic representation of a scanning transmision electron microscope showing the placement of a thin sample between the pole pieces of the electromagnet. A scheme for detecting the transmitted electrons is also shown.

the samples. As shown, the sample is placed at the focal point in the beam column; the size of the focal spot is only a few tenths of a nm in our STEM. This is to be compared with spot sizes of several hundred Å which is typical for a scanning electron microscope (SEM). Standard TEM was also used to study the fine lines and, in particular, to obtain diffraction patterns. SEM studies were also performed where possible as they provided unique views of the fine lines and the connecting pads.

In order to use transmitted electrons and transmission microscopy in general, thin samples and substrates are required as indicated in Fig.1 and in Fig. 2 (a) These figures show schematically a thin membrane covering a hole etched in a slice of Si. The fine lines are formed on the top surface of these window substrates. The Si thickness is about 200 um while the membrane thickness is generally about 0.2 um. Membranes as thin as 20 nm have been used successfully. The membrane is an amorphous material and consists of either $SiO_2$ or $Si_3N_4$. grown or deposited at elevated temperatures ( about 1000 C). The 'back side' of the wafer is patterned using a photoresist mask with the desired window opening and then etched preferentially to remove the Si. Since the etch rate of Si is much faster than the membrane material, there is ample time to remove the sample from the etching solution after the Si has been removed but before the membrane is etched through. While windows as large as several mm could be fabricated, the windows used for these fine line studies were generally around 40 μm on a side. The window size was also limited by the requirement that the final chip fit into a three mm diameter sample holder which is typical for many transmission electron microscopes. Two windows were made on each of these chips and fine line patterns were made on each window.

While the thin window substrates were vital for examining the fine lines in a transmission electron microscope, they can also be important in the reducing the width of the lines. This effect is illustrated in Fig. 2 (b) where a side view of a window sample on which a thin metal film has been deposited is shown. As illustrated, the thin substrate leads to a considerable reduction in the

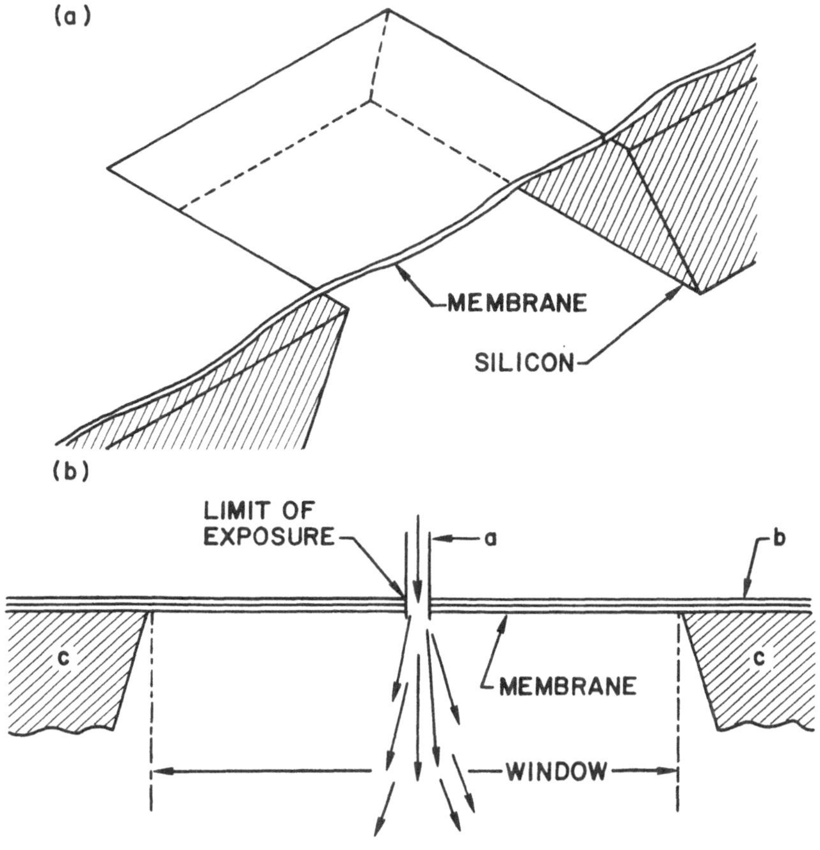


Figure 2 (a). A schematic view of a window substrate showing a thin membrane covering a hole in a Si wafer. Typical membrane material consisted of thin amorphous films of $Si_3N_4$ about 200 nm thick. The wafer thickness was about 250 μm.

Figure 2 (b). A side view of a window substrate fabricated in a Si wafer, c, with a thin metal film, b, deposited on the top surface. Also shown is a beam of electrons, a, which is transmitted through the thin membrane and the thin metal without creating any backscattered electrons. Such backscattering can lead to an increase in the exposure limits and an increase in the minimum linewidth.

extent of exposure caused by backscattered electrons. In using the electron beam to expose an area of the substrate, it is desirous to use only the primary, highly focussed beam and to eliminate any exposure due to backscattered electrons. The elimination of backscattered electrons was found to be important in obtaining linewidths below about 70 nm using contamination resist. Recent work, however, has shown that lines with smaller widths can be fabricated using other e-beam resist techniques on thick substrates[3].

An important consideration in the fabrication procedure is to provide for electrical contact to the fine line. This is generally accomplished using a series of contact pads patterned in a single layer. Several procedures have been established to contact the fine lines and provide good electrical contact; the choice of a particular procedure will in general depend on the desired pattern and on the metallurgy being used. For example, the fine line material may be deposited as the first layer on the window substrate and the pad pattern established as the second layer. However, the patterning for the pads is done using electron-beam lithography and a PMMA resist which requires a 160 C bake before exposure. Not all fine line metals remain unchanged during this baking process. A second procedure is to establish the pad pattern as the first layer on the window and then deposit the fine line material as an overlay. In this way the fine line material is not exposed to the resist processing steps. However, this second procedure requires good edge definition of the pads as electrical contact to the fine lines is difficult to make across ragged pad edges. In both techniques, good electrical contact between the pads, the leads and the fine lines is required if small resistance changes in a sample are to be measured in a reliable manner. One technique for accomplishing this is to plasma etch the pads, in-situ, prior to the deposition of the fine line material.

Two scanning electron micrographs of Nb pads on a $Si_3N_4$ windows are shown in Fig. 3. Reasonably good edges are obtained in this sample. The pad material continues onto the solid portion of the

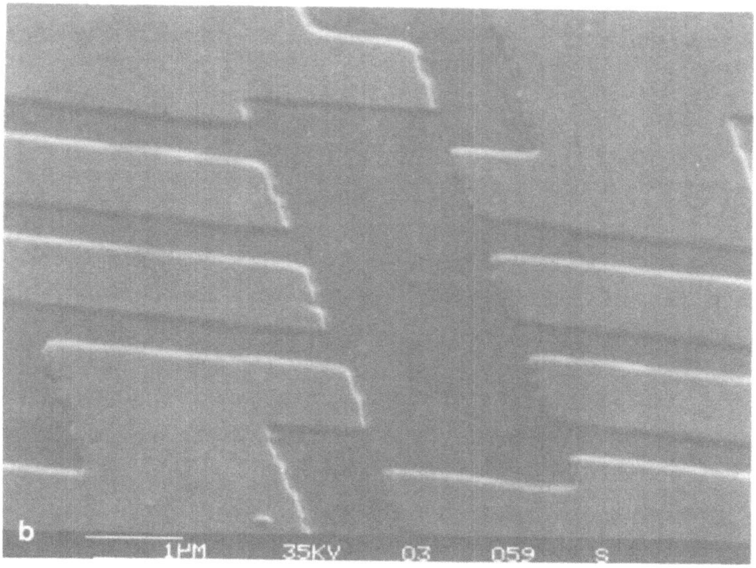


Figure 3. Two examples of pad patterns that have been fabricated on the thin membranes of the window substrates. The SEM upper picture shows four Nb pads with a separation of about 1 μm while the lower shows a more complex sample with about a 1.5 μm separation in the main channel. Fine lines are formed between these pads which are formed using 'standard' electron beam lithographic techniques to define the geometry.

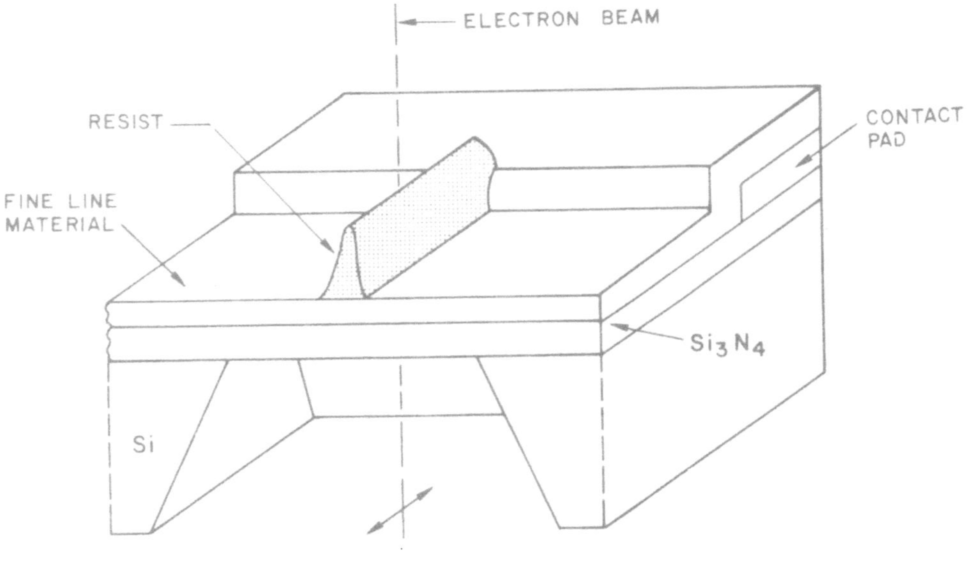


Figure 4. A pictorial representation of an important step in the fabrication process showing the contamination resist in place on the fine line material. At this stage the sample is ready for the etching procedure, e. g. ion milling.

subtrate where contact to the outside world can be made by a variety of standard techniques. As shown in Fig. 3 a typical pad spacing for the experiments in this paper was about 1 µm with pad separations as small as 0.1µm possible. Much smaller pad separations are possible with the use of contamination lithography techniques but this was not generally done as the fabrication of large pad patterns with these procedures is somewhat tedious. Much larger pad separations were of course easily acheivable.

In order to fabricate a fine line pattern on a substrate as shown for example in Fig. 2 (b), use was made of a resist consisting of a carbonaceous layer that can build up on a surface upon exposure by an electron beam. The source for this resist is believed to be the contamination layer that forms on surfaces left in oily environments. After the fine line material is deposited over a substrate and the surface is contaminated with this organic film, an electron beam is rastered over the surface in the desired pattern. The contamination resist then builds up to a thickness of 100 nm and more. The sample is shown schematically in Fig. 4 after this electron beam exposure has been performed. This contamination resist will then serve as a protective layer in the subequent processing to remove the unprotected metal while the metal under the resist forms the fine line. The amount of contamination on a given surface is limited and after lengthy electron-beam exposure the background contamination layer must be replaced by removing the sample from from the relatively good vacuum of the microscope and recontaminating it.

One significant advantage that electron-beam fine line fabrication techniques appears to have over that of edge shadowing is the ability to 'write' relatively complex patterns on a small scale. This is demonstrated in the four terminal resistor geometry of Fig. 5. Additional examples of the capabilities of this proceesing technique are shown in Figs. 6 and 7. Fig. 6 a shows a small loop or ring of Au which required the drawing of four straight line

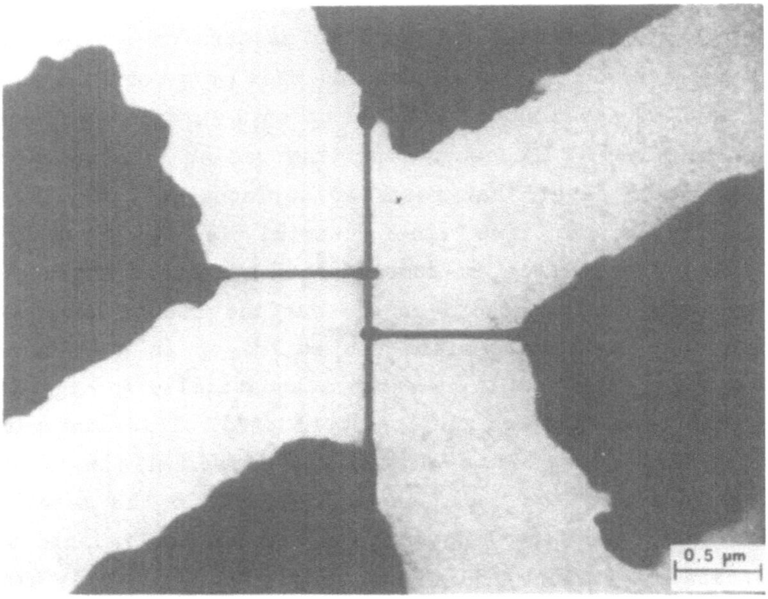


Figure 5. A transmission micrograph of Nb fine lines in a four terminal resistor pattern. The Nb fine lines in this sample had a linewidth of about 60 nm.

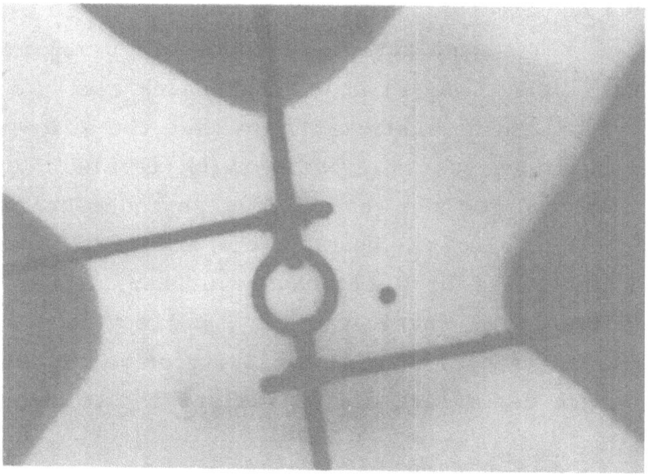

Figure 6 (a). This transmission micrograph shows a small loop of gold withconnecting leads fabricated as part of an experiment to search for quantum effects in normal metals at very low temperatures[4]. Five separate lines have to be formed for this structure. The loop diameter is about 0.35 μm.

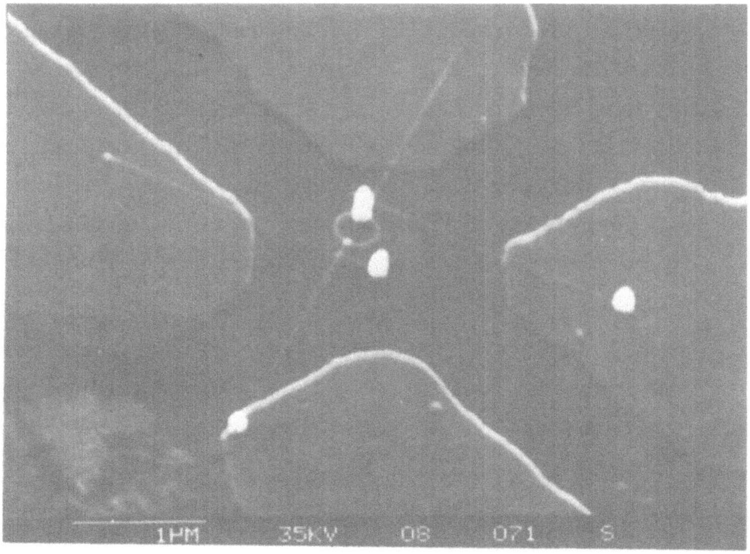


Figure 6 (b) An SEM view of a similar sample as in (a) which also shows the cones of contamination where the rastered electron beam has paused. The height of such contamination resist can easily be greater than 100 nm.

segments and the circle itself. This type of sample was developed for studies of flux quantization in loops of normal metals[4] at low temperatures. Fig. 6 (b) shows a scanning electron micrograph of a similar loop and demonstrates also that the contamination resist can be built up to relatively thick levels. The cones of contamination shown are the result of a periodic overexposure of the surface by the electron beam at the end of the raster sweep. Cone heights in excess of 0.12 μcron have been observed. Figure 7 shows a long fine line fabricated for localization studies with a series of voltage probes. The multiplicity of voltage leads allows studies of length dependence of the resistivity to made quite conveniently.

After the contamination resist is in place as a protective layer, the unwanted material must be removed or etched away. In our present work this was accomplished by placing the sample in an ion milling apparatus in which 500 v Ar atoms sputter etch the metal films. Of course some of the contamination is also removed during this process but as long as the milling is completed while some resist remains the properties of the underlying metal remain unaffected by the Ar impingement. Typical milling times are about 100 sec.; however this time can vary considerably depending on the ion current density and the background pressure in the system. In addition, the milling time is also sensitive to the choice of material, e. g. the sputter yield of Au films using 600V Ar atoms is expected to be about 3 Au atoms per incident Ar ion while the yield[5] for Nb is only about 0.5. Refractory oxides such as $SiO_2$ and $Al_2O_3$ generally have much lower yields[5]. The low sputter yield of these oxides makes it necessary to keep the background pressure in the milling system as low as possible to minimize the formation of oxides during the actual ion milling. Otherwise the protective contamination layer will be milled away before the metal film. In-situ monitoring of the sample resistance during the milling process has been used to characterize and establish the actual milling times.

METALLURGY

Fine lines have been made from a variety of different materials, including amorphous W-Re and $Nb_3Ge$ and polycrystalline Nb, Cu, Al, AuPd and Au. The films were vapor deposited either by thermal or electron beam evaporation or by sputtering. Film thicknesses ranged between 10 and 60 nm. Background pressure during the evaporations depended on the particular deposition system and were generally between $10^{-6}$ and $10^{-9}$ torr.

Fine lines have also been made from single crystal Au films. Making these novel samples involved first growing a 20 -60 nm thick single crystal Au film by vapor phase epitaxy on a one μcron thick Ag film which had been just evaporated onto a freshly cleaved NaCl substrate. This is illustrated in Figure 8. The Au was deposited at a rate of 1-2 A/sec, at a substrate temperature of 90 C in a background pressure of about $10^{-6}$ torr. The relatively thick Ag film served to smooth out irregularities in the NaCl surface and permitted a smoother Au film to be formed. After the Au film had been fabricated, the NaCl substrate was dissolved away in water. The remaining Au/Ag composite film was then floated in an aqueous nitric acid bath to remove the Ag. Once the Ag was removed, the Au film was rinsed in a water bath and then picked out of the bath on a window substrate. Electrical contact to Au film was made via metal pads already in place on the window chip. Lines of contamination were then drawn on the Au in a high resolution STEM. Subsequent ion milling removed all Au not covered by contamination and left behind a fine line metal pattern of single crystal Au. Fig. 9c shows a square loop pattern drawn with contamination on single crystal Au.

The electral properties of the fine lines could be varied in a number of ways such as changing the fine line material, the cross sectional area of the line or the background pressure during the evaporation. Sample resistivity ranged from greater than 200 μΩ-cm for amorhpous W-Re to 30 μΩ-cm for Au evaporated in a poor

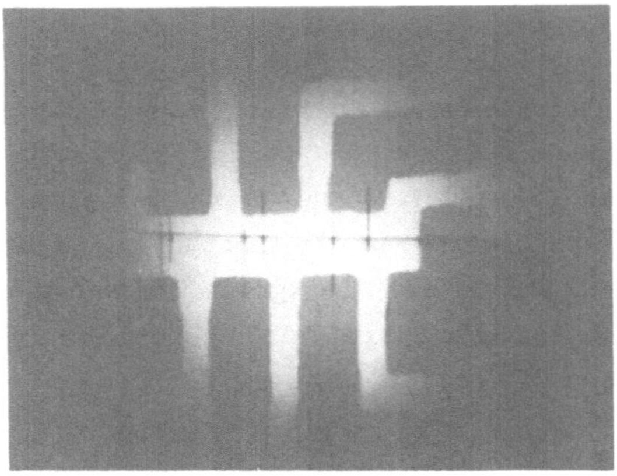

Figure 7. A multi voltage probe fine line sample showing the complexity possible with the contamination processes. Complex structures often require recontamination of the surface of the samples.

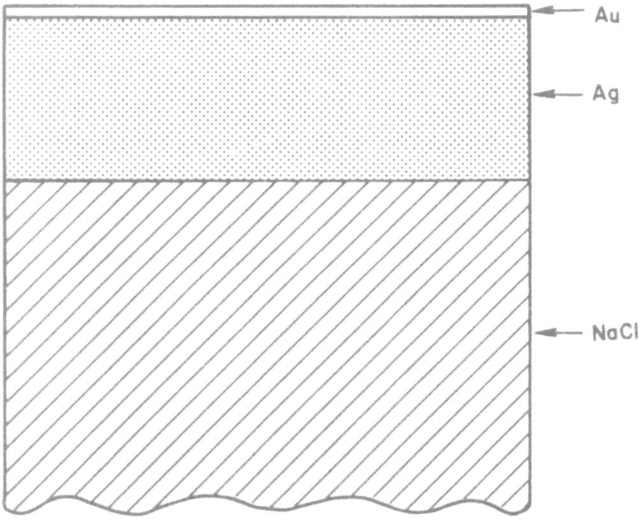


Figure 8. A cross sectional view of the sample configuration necessary to grow a single crystal film of Au. The substrate for the growth of the Au film is the single crystal Ag film grown on the crystalline slab of NaCl. The Au film is separated from the lower substrate as described in the text and floated onto a window substrate for use as a fine line material.

vacuum to 5 $\mu\Omega$-cm for Au deposited in a clean vacuum system or in single crystal Au samples. In samples with dimensions of several tens of nm, size effects also become important in limiting the conductivity of the fine lines[6]. Figure 9 a,b,and c shows the morphology of high resistivity, small grain Au films formed in a poor vacuum; low resistivity, larger grain Au evaporated in a clean vacuum, and single crystal Au respectively.

Grain size appears to play a role in determining the mechanical properties of fine lines. Lines which consist of a few large grains may be more susceptible to going open circuit than lines of many smaller grains. For example, while we have been able to make and measure lines made from fine grain Au, we have had more difficulity measuring lines of large grain Au. Figure 6(a) shows a STEM micrograph of a loop pattern made of large grain Au. Electrical measurements of this sample showed that there was no continuity between any of the pads even though there are no obvious breaks in the lines. Figure 10 shows a negative electron micrograph taken with a 200 KeV TEM of the same sample after it was measured electrically. Large grains of Au underneath the lines of contamination are visible, however, once again, there are no obvious visible breaks in the lines. It may be, however, that microcracks were formed between grains some time during the processing and subsequent cooling to low temperatures. In fine grain material, it appears to be less likely for such a failure mechanism to take place. Of course, amorphous materials do not have these grain boundary problems. Additional study on the question of suitable metallurgies for fine line fabrication is required. A great deal of work still remains to be done in the study of the properties of ultra-small structures and the processes for making them. This paper has outlined one particular set of techniques for making fine lines, that of high resolution electron beam lithography using contamination resist. These techniques hold great promise as scientific tools for the fabrication ultra-small. devices and fine lines.

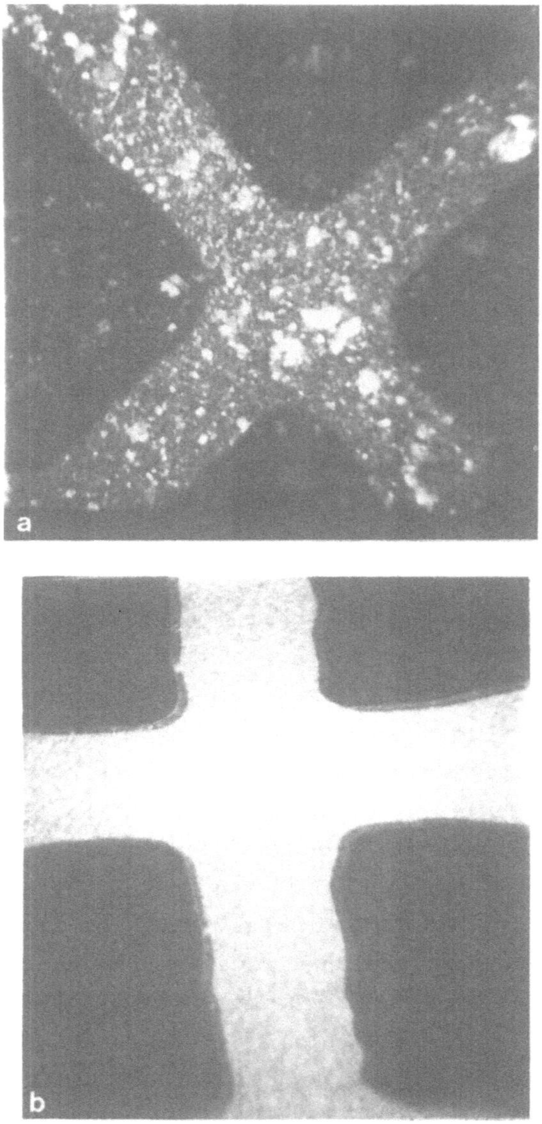
a
b

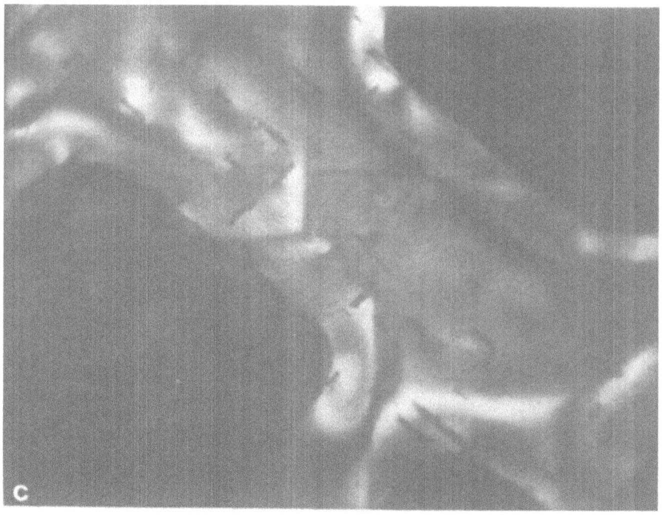


Figure 9. This figure demonstrates that Au films of various grain size can be formed on the window substrates used for fine line fabrication. The pad separation in all cases is about 1 μm. (a) shows fine grained Au with a grain size of about 20 nm while (b) shows large grained Au with about 100 nm grain size. In (c) we show a micrograph of the single crystal film containing a variety of defects. Also shown in this figure is a fine line structure of a square loop after contamination but prior to the ion milling.

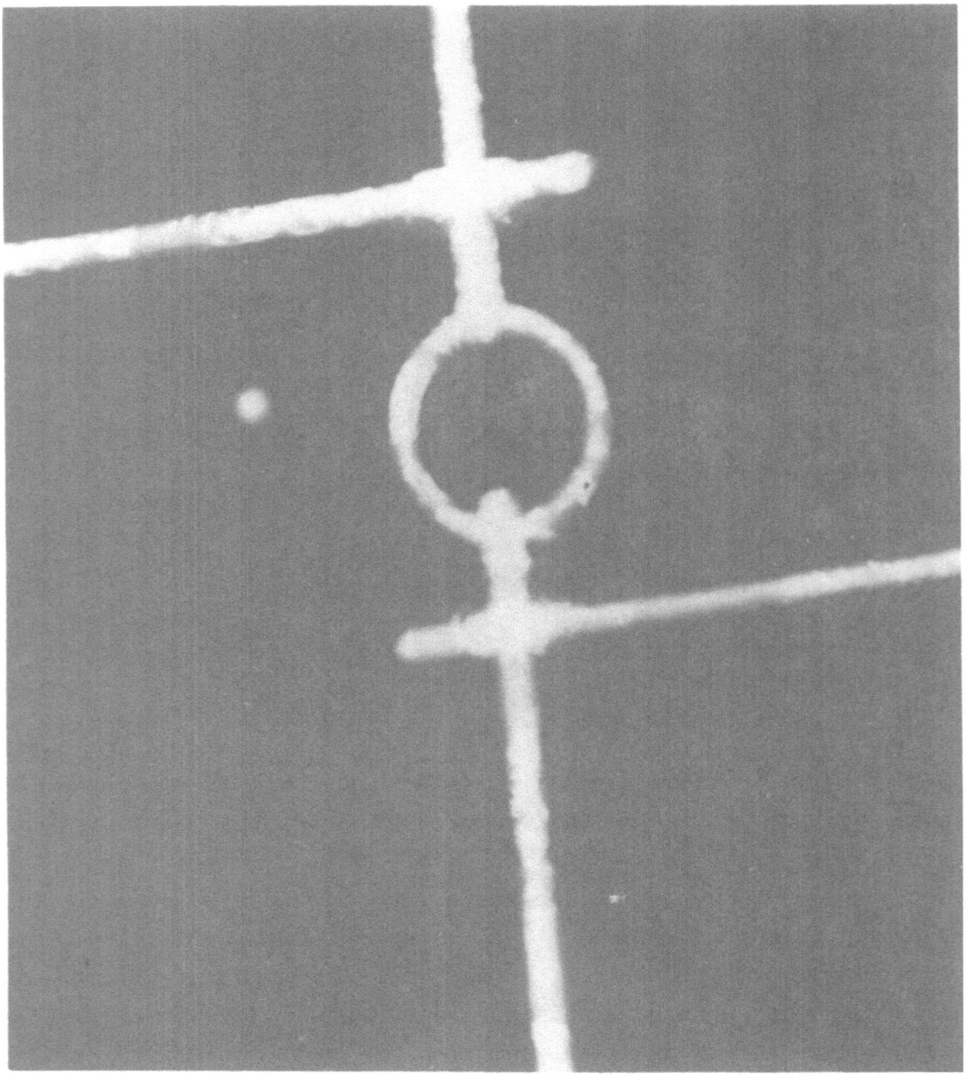

Figure 10. This figure shows a transmission micrograph of a loop pattern made from a large grain Au film similar to the sample of Figure 6 (a) after it was cycled to low temperatures and measured electrically. The metal film is no longer uniform in appearance and there was no electrical conductivity between the pads along the Au lines.

ACKNOWLEDGEMENTS

The authors wish to acknowledge helpful discussions with P. Chaudhari, C. C. Chi, R. F. Voss, S. P. McAlister, E. I. Alessandrini, and A. N. Broers, the expertise and collaboration of J. M. Viggiano, W. W. Molzen and J. L. Speidell and the technical assistance of many of their coworkers in paricular K. H. Nichols, J. A. Lacey and J. Kuran.

REFERENCES

1. see, for example, R. B. Laibowitz, A. N. Broers, J.T.C. Yeh and S. M. Viggiano, "Josephson Effect in Nb Nanobridges", Applied Physics Letters, **35**, 891 (1979); R. B. Laibowitz and A. N. Broers, "Fabrication and Physical Properties of Ultra-Small Structures" Treatise on Materials Science and Technology Vol. **5** 285, Academic Press 1982. J. T. Masden and N. Giordano; Length-Dependent Resistance of Thin Wires," Phys. Rev. Lett. **49**, (1982); A. E. White, M. Tinkham, W. J. Skocpol and D. C. Flanders, "Evidence for Interaction Effects in the Low-Temperature Resistance Rise in Ultrathin Metallic Wires," Phys. Rev. Lett. **48**, 1752 (1982); and references therein.

2. N. Giordano, W. Gilson, and D. E. Prober, "Experimental Study of Anderson Localization in Thin Wires," Phys. Rev. Lett. **43**, 725 (1979); R. E. Howard and D. E. Prober, "Nanometer Scale Fabrication Technique," VSLI Electronic Microstructure Science Vol. **5** 285, Academic Press 1982.

3. H. G. Craighead, R. E. Howard, L. D. Jackel and P. M. Maniewich, "10-nm Linewidth Electron Beam Lithography on GaAs," Appl. Phys. Lett. **42** 38 (1983).

4. D. Yu. Sharvin and Yu. V. Sharvin, "Magnetic Flux Quantization in a Cylindrical Film of a Normal Metal," JETP Lett. **34**, 274 (1981); B. L. Al'tshulu, A. G. Aronov, B. Z. Spivak, D. Yu. Sharvin and Yu. V. Sharvin, "Observation of the Aaronov-Bohm Effect in Hollow Metal Cylinders," JETP Lett. **35**, 589 (1982).

5. Handbook of Thin Film Technology edited by L. I. Maissel and R. Glang, McGraw-Hill Book Co., New York, New York (1970), p. 4-40.

6. E. H. Sondheimer, "The Mean Free Path of Electrons in Metals," Adv. in Physics **1**, 1 (1952).

# TWO-DIMENSIONAL COULOMB GAS AND CONNECTIONS TO SUPERCONDUCTING FILMS

Petter Minnhagen

Department of Theoretical Physics
Lund University
Sölvegatan 14A, S-223 62 Lund, Sweden

## INTRODUCTION

The present set of lectures deals with the thermodynamic properties of a two-dimensional Coulomb gas and its connection to "dirty" superconducting films. The general background is the following: "Dirty" superconducting films are a class of effectively two-dimensional superconductors for which vortex-fluctuations may be expected to play an important role in the physical description close to the transition between the superconducting and normal state. This expectation stems from the theoretical development by Berezinskii[1], Kosterlitz and Thouless[2,3] who found that topological excitations like vortices are crucial for destroying the quasi-long-range order in two-dimensions. Subsequently these ideas were developed in case of two-dimensional superfluids and successfully explained experiments on helium films[4-6]. At this point several authors realized that the situation should be very analogous for "dirty" superconducting films[7-10]. This was the starting point of a considerable and ongoing research activity.

There already exist several excellent review articles. Among those covering the present subject are the review articles by Halperin[11] and by Nelson[12] which review the theoretical development and the articles by Hebard and Fiory[13] and by Mooij[14] which review the experimental situation for superconducting films.

The perspective of the present set of lectures will be somewhat limited. It will be focused on the Coulomb gas description of vortex-fluctuations. The lectures consist roughly of two parts. The first deals with the properties of the Coulomb gas in itself. The second deals with the connection to superconducting films and with some consequences that the description leads to.

## 2. DEFINITION OF 2-D COULOMB GAS

A Coulomb gas (C.G.) consists of positive and negative charges of equal magnitude. The charges interact via the Coulomb interaction which is defined by Poisson's equation

$$\bar{\nabla}^2 V(\bar{r}) = -2\pi\,\delta(\bar{r}) \tag{2.1}$$

(the unit of charge is taken to be 1). In 2-dimensions this interaction depends logarithmically on distance i.e.

$$V(\bar{r}) \sim \ln(r) \tag{2.2}$$

The "pure" C.G. is somewhat unphysical in that charges of opposite sign would collapse into each other for small enough temperatures[15] unless the interaction is modified for small distances. In order to make the model well-defined for all temperatures we will redefine it slightly. In the modified version, which we will consider, the charges have a finite extension of the order of $\xi$ and the logarithmic interaction is cut-off after a length of the order of $\lambda_c$ [16]. Consequently, the spatial extension of a charge will be given by a function $f_\xi(\bar{r})$ where $\lim_{\xi\to 0} f_\xi(\bar{r}) = \delta(\bar{r})$. The precise form of $f_\xi$ will be of little importance in the following. To be explicit we take $f_\xi$ as the Fourier transform of $e^{-k\xi/2}$. Likewise it will be of little importance precisely how the interaction is cut-off at large distances. To be explicit we will introduce $\lambda_c$ by modifying $\bar{\nabla}^2$ to $\bar{\nabla}^2 - \lambda_c^{-2}$ in Poisson's equation (2.1)[16].

The interaction energy between two charges i and j separated by the distance $r_{ij}$ and with equal (opposite) charge is, in this modified C.G., given by $\overset{+}{(-)}\, U(r_{ij})$ where $U(r)$ has the limiting forms

$$U(r) \approx \begin{cases} e^{-r/\lambda_c} & \lambda_c \ll r \\ -\ln(r/\lambda_c) & \xi \ll r \ll \lambda_c \end{cases} \tag{2.3a}$$

and

$$U(0) \approx \ell n(\lambda_c/\xi) \quad \text{for} \quad \lambda_c/\xi \gg 1 \tag{2.3b}$$

We will define a statistical mechanics for this model. In this statistical mechanics we will take into account the interaction energy between the charges as well as the energy needed to create the charges. The creation energy of a single charge is taken to be

$$E_{\pm} = 1/2\ U(0) + |\mu_{\pm}| \tag{2.4}$$

where $\pm$ refer to the sign of charge. The first term is just the electro-static self-energy and the second term, as will be apparent, can be thought of as a chemical potential $(\mu_{\pm} = -|\mu_{\pm}|)$.

The statistical mechanics is defined through the Boltzmann factor $e^{-H_N/T}$ and a phase-space division $\sim \xi^2$. $H_N$ is the energy of a configuration of $N$ charges. It is given by

$$H_N = \frac{1}{2} \sum_{i \neq j} s_i s_j\, U(r_{ij}) + \frac{N}{2} U(0) + N_+|\mu_+| + N_-|\mu_-| \tag{2.5}$$

Here $s_i = \pm 1$ denotes the sign charge $i$ and $N_{(\pm)}$ is the number of positive (negative) charges in the configuration. The partition function $Z$ is given by

$$Z = \mathrm{Tr}\{e^{-H_N/T}\} = \sum_{N=0}^{\infty} \prod_i \int \frac{d\bar{r}_i}{\xi^2}\, e^{-H_N/T}\, (N_+!\,N_-!)^{-1} \tag{2.6}$$

Equation (2.6) completes the definition of the model.

We will be interested in the properties of the model as a function of three variables. These we choose as temperature $T$, a fugacity $z = \exp\{-(|\mu_+|+|\mu_-|)/2T\}$, and the difference in creation energy for positive and negative charges $\Delta\mu = |\mu_+|-|\mu_-|$. Our interest will be in the large $\lambda_c/\xi$-limit. In this limit the interaction between the charges is predominantly logarithmic.

Using the identity

$$\frac{1}{2} \sum_{i \neq j} s_i s_j = \frac{1}{2}(N_+ - N_-)^2 - \frac{1}{2} N \tag{2.7}$$

we can express the configuration energy as

$$H_N = \frac{1}{2} \sum_{i \neq j} s_i s_j (U(r_{ij}) - U(0)) +$$

$$+ \frac{1}{2}(N_+ - N_-)^2 U(0) - NT\, \ell nz + \frac{N_+ - N_-}{2} \Delta\mu \tag{2.8}$$

Now, since $U(r_{ij}) - U(0) \approx - \ell n(r/\xi)$ for $\xi \ll r \ll \lambda_c$ whereas $U(0) \approx \ell n(\lambda_c/\xi)$, non-neutral configurations (i.e. configurations with $N_+ \neq N_-$) will have a huge energy in the large $\lambda_c/\xi$ -limit caused by the term proportional to $U(0)$ compared to the neutral ones (i.e. the ones with $N_+ = N_-$). As a consequence the neutral configurations will completely dominate the partition function in the large $\lambda_c/\xi$-limit. The model reduces to the neutral C.G. for which the energy of a configuration is given by

$$H_N \approx - \frac{1}{2} \sum_{i \neq j} s_i s_j\, \ell n(r_{ij}/\xi) - NT\, \ell nz \tag{2.9}$$

Another limiting case is obtained if $\Delta\mu \sim \lambda_c^2$ [16]. In this case the C.G. will not reduce to the neutralC.G. in the $\lambda_c/\xi \to \infty$ limit. Both limiting cases will, as we will see, have connections to superconducting films[16].

## 3. NEUTRAL C.G.

We first consider the neutral C.G. in the $\lambda_c \to \infty$ limit. In this limit the neutral C.G. undergoes a Kosterlitz-Thouless (K-T) transition[2] at a certain temperature $T_c$ for any fixed $z$ (the value of $T_c$ is a function of $z$ ). A sketch of this transition is shown in fig. 1. On the high-temperature side of the transition line correlations fall off exponentially for large distances whereas on the low temperature side the correlations fall off as powerlaws[1-3]. One may interpret the physical origin of this change of behaviour in the following way: On the high-temperature side some of the charges are free whereas on the low-temperature side all charges form neutral dipole pairs[2].

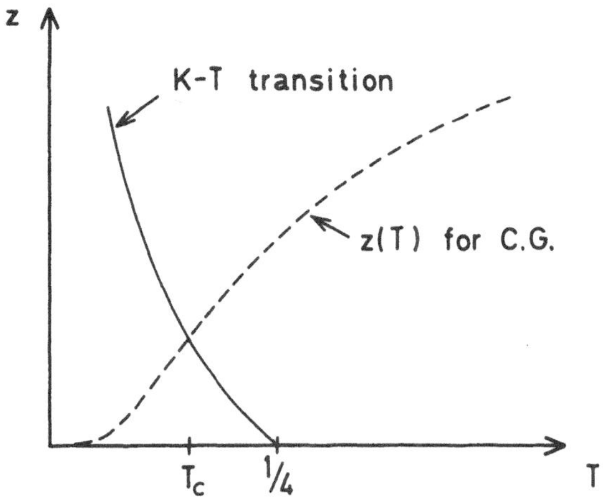


Fig. 1 Sketch of the K-T transition in the {z,T}-plane. Full-drawn line is the K-T transition line $4T\varepsilon = 1$ . Dashed line is the z(T)-trajectory of a specific C.G.

Imagine that an infinitesimal test-charge is inserted into the C.G. The free charges will screen-out the potential caused by the test charge. The dipole-pairs will diminish its strength due to polarization. On the level of a Poisson-Boltzmann equation one gets

$$\bar{\nabla}^2 t\, V_{test} = -\frac{2\pi t}{\varepsilon} f_\xi(\bar{r}) - \frac{2\pi\, n_F^+}{\varepsilon}\left(e^{-t\, V_{test}/T} - 1\right)$$

$$+ \frac{2\pi\, n_F^-}{\varepsilon}\left(e^{t\, V_{test}/T} - 1\right) \tag{3.1}$$

$$= -\frac{2\pi t}{\varepsilon} f_\xi(\bar{r}) + \frac{2\pi\, n_F}{\varepsilon T}\, t\, V_{test}$$

Here $V_{test}$ is the test potential, t the infinitesimal test-charge, $n_F$ is the density of free charges ($n_F^{(\pm)}$ refers to free positive (negative) charges). The dielectric constant $\varepsilon$ takes into account the "charge renormalization" caused by the polarization[16]. In Fourier space this means that

$$V_{test} = \frac{1}{\varepsilon}\frac{2\pi\ e^{-k\xi/2}}{k^2 + \lambda^{-2}} \tag{3.2a}$$

where

$$\lambda^{-2} = \frac{2\pi\ n_F}{\varepsilon T} \tag{3.2b}$$

It is reasonable to assume that the leading k-dependence of $V_{test}$ for small k can always be expressed on the form of eq. (3.2b) (compare the following section). It is then natural to define $n_F$ by eq.(3.2b). It is also instructive to keep track of the large distance cut-off, $\lambda_c$, which leads to the modification[16]

$$\lambda^{-2} - \lambda_c^{-2}/\varepsilon = \frac{2\pi\ n_F}{\varepsilon T} \tag{3.2c}$$

In r-space eq. (3.2a) gives an exponential decay of $V_{test}$ proportional to $e^{-r/\lambda}$. In the same way the potential outside a C.G. charge is screened out. Since it is screened out it follows that its large distance behaviour is identical to that of $V_{test}$. As a consequence the leading k-behaviour of $U_{eff}$ (= the effective interaction energy between two C.G.-charges of equal sign) for small k is given by

$$U_{eff}(k) = \frac{1}{\varepsilon}\frac{2\pi\ e^{-k\xi}}{k^2 + \lambda^{-2}} \tag{3.3}$$

What happens at the Kosterlitz-Thouless transition is that $\lambda^{-1}$ vanishes and remains zero below the transition in the limit $\lambda_c \to \infty$. One can get an idea of why this happens from the following argument: The energy needed to create a single charge is in the C.G. model given by

$$H_1 = - T \ln z + \frac{1}{2} U(0) \tag{3.4}$$

The energy needed to create $N_F$ free particles with net charge zero in presence of bound pairs is then given by

$$H_{N_F} = - N_F T \ln z + \frac{N_F}{2} U_{eff}(0) \tag{3.5a}$$

which means that the energy per free particle is given by

$$\{H_1\}_{eff} = - T \ln z + \frac{1}{2} U_{eff}(0) \tag{3.5b}$$

In equilibrium one may expect that the density of free particles $n_F$ is proportional to the corresponding effective Boltzmann factor i.e.

$$n_F \sim \xi^{-2} \exp[- \frac{\{H_1\}_{eff}}{T}] \tag{3.6}$$

From eq. (3.3) one finds that the leading contribution to $U_{eff}(0)$ for large $\lambda/\xi$ is $\frac{1}{\varepsilon} \ln(\lambda/\xi)$ and hence

$$n_F \sim \frac{z}{\xi^2} (\xi/\lambda)^{1/2\varepsilon T} \qquad \text{for} \qquad \xi/\lambda \ll 1 \tag{3.7}$$

Finally, since $n_F$ is related to $\lambda$ by eq. (3.2c) one obtains

$$\lambda^{-2} - \lambda_c^{-2}/\varepsilon = z\, g(z,T) \xi^{-2} (\xi/\lambda)^{1/2\varepsilon T} \tag{3.8}$$

where $g(z,T)$ is some positive function. In the limit $\lambda_c \to \infty$ eq. (3.8) has the solution (provided $zg < 1$)

$$\lambda^{-2} = 0 \qquad \text{for} \qquad T < 1/4\varepsilon \tag{3.9a}$$

$$(\xi/\lambda)^2 = (gz)^{\frac{1}{1-1/(4\varepsilon T)}} \qquad \text{for} \qquad T > 1/4\varepsilon \tag{3.9b}$$

So one concludes that a C.G. undergoes a K-T transition at a temperature $T_c$ defined by $T_c = 1/4\varepsilon(T_c)$ . Note that $\lim_{z \to 0} \varepsilon = 1$

because the dielectric constant is 1 in the absence of charges and that $\varepsilon$ increases for increasing $z$ because the number of dipoles increases. This accounts for the qualitative features of the K-T transition line sketched in fig. 1. A specific C.G. corresponds to a trajectory $z(T)$ in the $\{z,T\}$-plane, e.g. the dashed line in fig. 1. Note that a C.G. always has $T_c \leq 1/4$ .

Fig. 2 illustrates how the density of free charges $n_F$ vanishes as the K-T transition is approached (as given by eq.(3.8) with $zg$ = constant). Full-drawn line is for finite $\lambda_c$ and the dashed is the $\lambda_c \to \infty$ limit. The deviation comes when $\lambda_c^{-2} \approx \frac{2\pi\, n_F}{\varepsilon T}$ (arrows in the figure). So the $\lambda_c \to \infty$ limit gives a good approximation for temperatures not too close to $T_c$ .

On the other hand, close enough to $T_c$ one expects for any specific C.G. that $z(T)\, g(T)$ is well approximated with $z(T_c)\, g(T_c)$ . So the leading temperature dependence for the screening-length $\lambda$ close to $T_c$ should come from the exponent

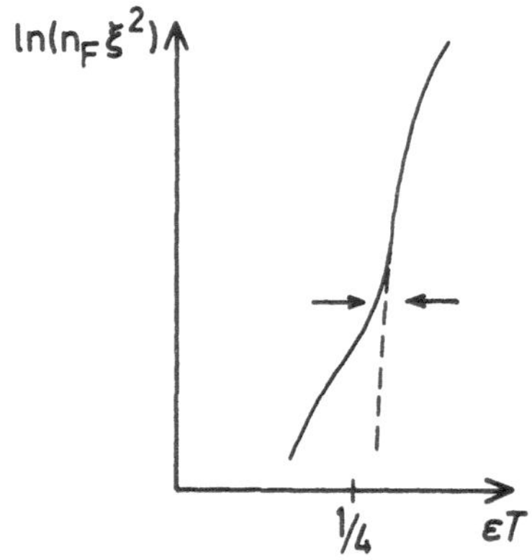


Fig. 2 Density of free charges $n_F$ close to the K-T transition at $\varepsilon T = 1/4$ . Full line is for finite $\lambda_c$ and the dashed one is the $\lambda_c \to \infty$ limit. The arrows indicate the $\lambda_c^{-2} = \frac{2\pi\, n_F}{\varepsilon T}$ . (The curves are solutions to eq. (3.8) with $zg$ = constant).

in eq. (3.9b). To extract this dependence one must know how $\varepsilon(z,T)$ varies close to $T_c$ (for constant $z$).

The following heuristic argument gives an idea[17]: It is reasonable to assume that $\varepsilon(T)$ can be Taylor expanded around any $T$ except $T_c$ and that the K-T transition is reflected in some non-analytic behaviour at $\bar{T}_c$. In fact, if we choose the "renormalized" temperature $\bar{T} \equiv \varepsilon T$ as the expansion parameter then the effect of the dipoles is already taken into account so that $\varepsilon$ as a function of $\bar{T}$ might be expected to be more well-behaved. Expanding around a temperature $\bar{T}_o$ gives

$$\varepsilon(\bar{T}) = \varepsilon(\bar{T}_o) + \varepsilon'(\bar{T}_o)(\bar{T}-\bar{T}_o) + \frac{\varepsilon''(\bar{T}_o)}{2}(\bar{T}-\bar{T}_o)^2 + O\{(\bar{T}-\bar{T}_o)^3\} \tag{3.10a}$$

On the other hand directly from the definition of $\bar{T}$ one gets

$$\varepsilon(\bar{T}) = \varepsilon(\bar{T}_o) + \frac{1}{T_o}(\bar{T}-\bar{T}_o) - \frac{\varepsilon(\bar{T}_o)}{T_o}(T-T_o) + O\{(\bar{T}-\bar{T}_o)(T-T_o)\} \tag{3.10b}$$

These two expansions are compatible only if $\bar{T}-\bar{T}_o \sim T-T_o$ or if $(\bar{T}-\bar{T}_o)^2 \sim T-T_o$. The second possibility demands that $\varepsilon'(\bar{T}_o) = \frac{1}{T_o}$, $\varepsilon''(\bar{T}_o^+) < 0$ and $\varepsilon''(\bar{T}_o^-) > 0$. Obviously the first possibility is consistent with $\varepsilon(T)$ having a Taylor expansion around $T_o$ while the second possibility is not. It is then natural to associate the second possibility with the behaviour at the K-T-transition. As a consequence one gets close to $T_c$

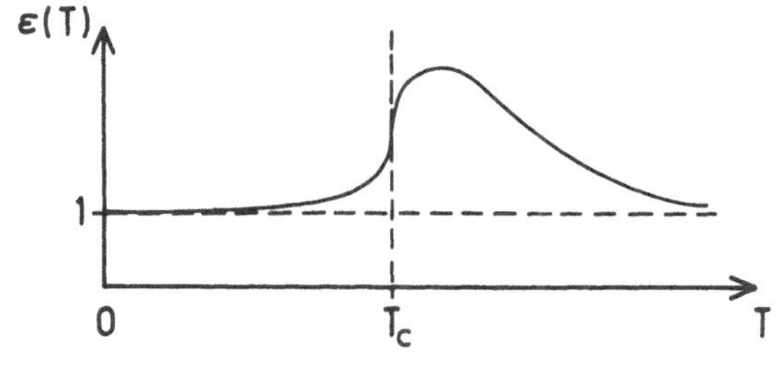


Fig. 3 Sketch of $\varepsilon(T)$ for a specific C.G. The behaviour close to $T_c$ is given by eq. (3.11).

$$\varepsilon(T) = \varepsilon(T_c)(1 \overset{+}{(-)} |c_{\overset{+}{(-)}}| \sqrt{|1 - T/T_c|} \qquad T - T_c \overset{>}{(<)} 0 \tag{3.11}$$

where $c_{\pm}$ are constants. The behaviour of $\varepsilon(T)$ for a specific C.G. is sketched in fig. 3. It now follows that the exponent of eq. (3.9b) can close to $T_c$ be approximated by (compare eq.(3.11))

$$\frac{1}{1-1/4T\varepsilon} \sim \frac{1}{\sqrt{T-T_c}} \qquad T > T_c. \tag{3.12a}$$

and hence the leading temperature dependence of the screening-length, $\lambda$, is close to $T_c$ given by[3]

$$\lambda \sim \exp\left\{\frac{|\text{const}|}{\sqrt{T - T_c}}\right\} \tag{3.12b}$$

The dielectric response function $\varepsilon(k)$ is defined by $V_{test}(k) = \frac{2\pi}{k^2\, \varepsilon(k)}$. In the $k \to 0$ limit one finds that (compare eq.(3.2a))

$$\frac{1}{\varepsilon_0} \equiv \lim_{k\to 0} \frac{1}{\varepsilon(k)} = \lim_{k\to 0} \frac{1}{\varepsilon} \frac{1}{1+(\lambda k)^{-2}} = \begin{cases} \frac{1}{\varepsilon} & \text{for} \quad T < T_c \\ 0 & \text{for} \quad T > T_c \end{cases} \tag{3.13}$$

This means that the quantity $1/T\varepsilon_0$ at $T_c$ jumps from the value 4 to zero (because $T_c\ \varepsilon(T_c) = \frac{1}{4}$). This is the famous universal jump for a 2-D superfluid when phrased in the C.G.-language[4,18]. The jump is illustrated in fig. 4.

We have now obtained some basic properties of the neutral C.G. in a rather intuitive way. One might then ask if there is not a more systematic way of obtaining the same properties. One such approach is briefly indicated in the next section.

## 4. FIELD THEORY ANALOGY

It turns out that the partition function Z for the neutral C.G. can be expressed as a functional integral over a real field $\phi(\bar{r})$[19,20] i.e.

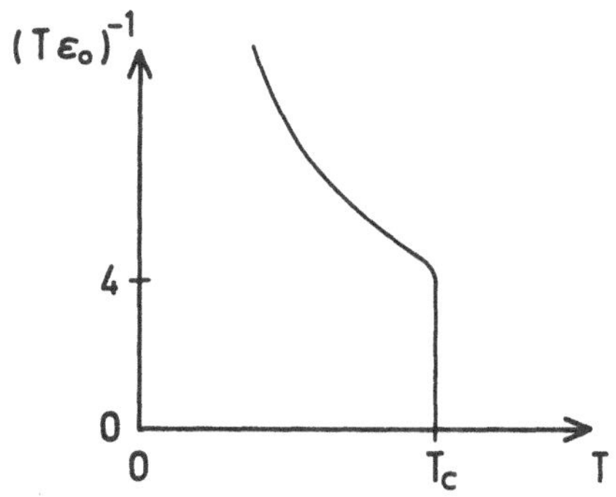


Fig. 4 Sketch of $(T\ \varepsilon_o)^{-1}$ for a specific C.G. The jump from 4 to 0 at $T_c$ is <u>universal</u> (i.e. independent of the specific z(T)-trajectory).

$$Z = \mathrm{Tr}\,\{e^{-H_N/T}\} = \langle e^{-\int d\bar{r}\, H_{sG}(\bar{r})} \rangle \qquad (4.1a)$$

where $\langle O \rangle$ is defined as

$$\langle O \rangle \equiv \frac{\int d\phi\ O}{\int d\phi\ \exp\{-\int d\bar{r}\left(\frac{(\bar{\nabla}\phi)^2}{2} + \lambda_c^{-2}\ \phi^2/2\right)\}} \qquad (4.1b)$$

$H_{sG}$ is given by

$$H_{sG}(\bar{r}) = \frac{(\bar{\nabla}\phi)^2}{2} + \frac{\lambda_c^{-2}}{2}\ \phi^2 - \frac{2z}{\xi^2}\cos\{\sqrt{\frac{2\pi}{T}}\ \tilde{\phi}(\bar{r})\}$$

and $\tilde{\phi}(\bar{r}) \equiv \int d\bar{r}'\ f_\xi(\bar{r}' - \bar{r})\ \phi(\bar{r}')$ .

From a practical point of view this means that the statistical mechanics of the C.G. has been turned into a field theory where one can plunge ahead with standard diagrammatics and renormalization methods. The field theory we landed in is called the massive sine-Gordon theory in two Euclidean dimensions. $V_{test}$, which gave the information on the dielectric constant $\varepsilon$ and the screening-length $\lambda$, corresponds in the field theory language to the one-particle Green's function[16,21] i.e.

$$G(\bar{r}) = \frac{\langle e^{-\int d\bar{r}\, H_{sG}(\bar{r})}\, \phi(\bar{r})\, \tilde{\phi}(0)\rangle}{\langle e^{-\int d\bar{r}\, H_{sG}(\bar{r})}\rangle} \tag{4.2a}$$

$$V_{test} = 2\pi\, G(\bar{r}) \tag{4.2b}$$

Note that the $T \to \infty$ limit for $V_{test}$, which is the Debey-Hückel limit, corresponds to a Klein-Gordon limit in the field theory language[21].

The Green's function is related to the one-particle irreducible self-energy $\Sigma$ by

$$G(k) = \frac{e^{-k\xi/2}}{k^2 - \Sigma(k)} \tag{4.3}$$

It is reasonable to assume that $\Sigma(k)$ can be expanded in powers of $k^2$. By comparing the leading behaviour of $V_{test}$ as $k \to 0$ with $G(k)$ one finds that

$$\varepsilon = 1 - \frac{\partial}{\partial k^2}\Sigma(k)\Big|_{k=o} \tag{4.4a}$$

$$\lambda^{-2} = \frac{-\Sigma(0)}{1 - \frac{\partial}{\partial k^2}\Sigma(k)\Big|_{k=o}} \tag{4.4b}$$

This means that we have related the dielectric constant $\varepsilon$ and the screening-length $\lambda$ to the diagrammatics of a sine-Gordon field theory.

So it is possible to attack the C.G. by standard field-theory methods. The results of the previous section have been verified by this route in the limit of small $z$ [21-23]. For $T > 1/4$ the $z \to 0$ limit is particularly straight-forward since the field theory in this case only involves a mass-renormalization[21]. The screening length is in this limit given by[21]

$$(\xi/\lambda)^2 = (g(T)z)^{1/(1-1/4T)} \tag{4.5}$$

i.e. in this limit $\varepsilon = 1$ and $g$ is a function only of $T$. As the K-T-transition is approached a wave-function renormalization is also needed. The wave-function renormalization constant $Z_\phi$ is in fact in the C.G.-language just $1/\varepsilon$. This turns out to be a safe but rather technical way of obtaining $\varepsilon$. An alternative, more intuitive way of obtaining $\varepsilon$ is given in the next section.

Standard thermodynamics give a relation between the free energy density $f = -\frac{T}{\Omega}\,\ell n(Z)$ and the density of charges $n$

$$n = -\frac{z}{T}\frac{\partial}{\partial z} f(T,z) \tag{4.6}$$

In the Debey-Hückel limit (i.e. $T \to \infty$) the screening-length is given by

$$\lambda^{-2} = \frac{2\pi n}{T} \tag{4.7}$$

which means that in the limit of high temperatures there are no bound pairs left. One notes that eqs (4.5-4.7) are compatible only if

$$f \sim \frac{z^{1/(1-1/4T)}}{\xi^2} \tag{4.8}$$

On the other hand, expandning the partition function $Z$ directly in powers of $z$ gives the leading z-dependence of $Z$ and $f$ proportional to $z^2$ [24]. Note that $\frac{1}{1-1/4T} = 2$ for $T = \frac{1}{2}$. It follows that[21,25]

$$\left.\begin{aligned} f &\sim \frac{z^{1/(1-1/4T)}}{\xi^2} \\ \lambda^{-2} &\sim n \end{aligned}\right\} \quad \text{for} \quad T > \frac{1}{2} \tag{4.9a}$$

and

$$\left.\begin{aligned} f &\sim \frac{z^2}{\xi^2} \\ \lambda^{-2} &\sim \xi^{-2}(n\xi^2)^{1/(2-1/2T)} \end{aligned}\right\} \quad \text{for} \quad T < \frac{1}{2} \tag{4.9b}$$

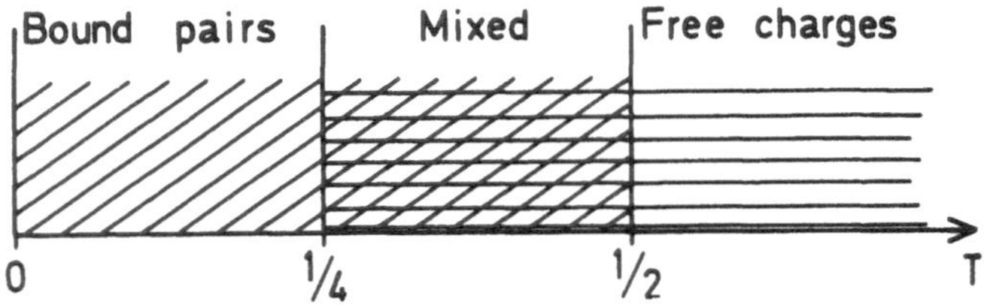


Fig. 5 Phase-diagram in the $z \to 0$ limit. The three regions are: $T > 1/2$, only free charges; $1/4 < T < 1/2$, both free charges and bound pairs; $T < 1/4$, only bound pairs.

Also note that $\frac{1}{2-1/2T} > 1$ for $T < \frac{1}{2}$. This means that in the limit $z \to 0$ the charge-binding starts at $T = 1/2$ [15,21,25]. In the same limit $T_c = 1/4$. This is illustrated in Fig. 5.

## 5. KOSTERLITZ EQUATIONS

In order to find the position of the K-T-line in the $\{z,T\}$-plane one must calculate $\varepsilon(z,T)$. The following reasoning, due to A.P. Young[26,27] is an extension of Kosterlitz and Thouless original approach[2]: Below $T_c$ the charges form dipole pairs. Focus on a dipole pair with charge-separation $r$. In the absence of other charges the binding force would have been $F = 1/r$. However, dipoles of smaller separation will polarize it, so the effective force is modified to $F_{eff} = \frac{1}{\varepsilon_{KT}(r)r}$. $\varepsilon_{KT}(r)$ is a distance dependent dielectric constant introduced by Kosterlitz and Thouless[2] and $\lim_{r\to\infty} \varepsilon_{KT}(r) = \varepsilon$. Treat the smaller dipoles as a dielectric medium i.e.

$$\varepsilon_{KT}(r) = 1 + 2\pi\, \chi(r) \tag{5.1}$$

where $\chi$ is the electric susceptibility caused by dipoles with separations less than $r$. $\chi$ is then given by

$$\chi(r) = \frac{1}{\Omega} \int^{r} n(r')\, \alpha(r')dr' \tag{5.2}$$

where $\Omega$ is the volume, $n(r')$ is the density of dipoles with separation $r'$ and $\alpha(r')$ is their polarizability

$$\alpha(r') = \frac{(r')^2}{2T} \tag{5.3}$$

It is reasonable to assume that for $z \ll 1$ the density $n(r')$ is given by the corresponding effective Boltzmann factor

$$n(r') = \frac{\Omega 2\pi r'}{\xi^4} e^{-\{H_2\}_{eff}/T} = \frac{\Omega 2\pi r' z^2}{\xi^4} e^{\frac{1}{T}\{U(r') - U(0)\}_{eff}} \tag{5.4}$$

where $-\{U(r) - U(0)\}_{eff}$ is the energy needed to separate the charges and $\frac{\Omega 2\pi r' dr'}{\xi^4}$ is the appropriate phase-space factor. Self-consistency now requires that

$$\frac{\partial}{\partial r'} \{U(r') - U(0)\}_{eff} = - F_{eff}(r') = - \frac{1}{r' \varepsilon_{KT}(r')} \tag{5.5}$$

This self-consistent set of equations is conveniently expressed in differential form using the variable $\ell = \ell n(r/\xi)$:

$$\frac{d}{d\ell} \{T\, \varepsilon_{KT}(\ell)\} = 2\pi^2 z^2(\ell) \tag{5.6a}$$

$$\frac{d}{d\ell} \{z(\ell)\} = 2z(\ell) \{1 - \frac{1}{4\varepsilon_{KT}(\ell)T}\} \tag{5.6b}$$

where $z(\ell)$ is given by

$$z(\ell) = z\, e^{2\ell}\, e^{\frac{1}{2T}\{U(r) - U(o)\}_{eff}} \tag{5.6c}$$

These equations were first derived by Kosterlitz using a renormalization group method[3]. $\varepsilon$ is obtained from these equations by integrating to $\ell = \infty$. It is reasonable to assume that the effect of dipoles with very small separation is negligible so that $\varepsilon(\ell = 0) \approx 1$ and $z(\ell = 0) \approx z$. The flow diagram for the differential equations are illustrated in fig. 6.

The Kosterlitz equations are easily combined into

$$\frac{d}{d\ell} \left\{ (\frac{1}{4T\varepsilon_{KT}(\ell)} - 1)^2 - \frac{z^2(\ell)\pi^2}{4T^2 \varepsilon_{KT}^2(\ell)} \right\} = 0 + \mathcal{O}(z^4(\ell)) \tag{5.7}$$

which means that[3]

$$(\frac{1}{4T\varepsilon_{KT}(\ell)} - 1)^2 - \frac{z^2(\ell)\pi^2}{4T^2 \varepsilon_{KT}^2(\ell)} = (\frac{1}{4T} - 1)^2 - \frac{z^2 \pi^2}{T^2 4} \tag{5.8}$$

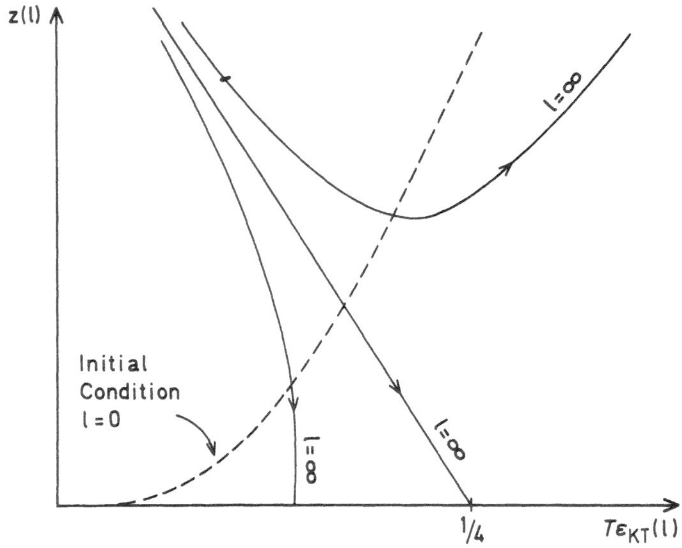


Fig. 6 Flow-diagram for Kosterlitz equations. Dashed line is a $z(T)$-trajectory giving initial conditions for $\ell = 0$. Full lines indicate the flow in the $\{z(\ell),\ T\,\varepsilon_{KT}(\ell)\}$-plane when integrating towards $\ell = \infty$ from an initial condition. The initial condition which flows into $T\,\varepsilon_{kT}(\infty) = T\varepsilon = 1/4$ is on the K-T line.

if the initial condition $(z(0),\ T\,\varepsilon_{KT}(0)) = (z,T)$ at $\ell = 0$ is used. Since $z(\infty) = 0$ for $T < T_c$ one finds from eq. (5.8) that[3]

$$T_c = \frac{1}{4} - \frac{\pi z}{2} + \mathcal{O}(z^2) \tag{5.9}$$

and that

$$\varepsilon(T) = \varepsilon(T_c)(1 - c_- \sqrt{1 - T/T_c}) \tag{5.10a}$$

with

$$c_- = 2\sqrt{2\pi z} + \mathcal{O}(z) \tag{5.10b}$$

It is also possible to obtain information on the screening-length $\lambda$ by integrating the Kosterlitz equations. It turns out that the leading behaviour for small $z$ close to $T_c$ is[28]

$$\lambda = c\,\exp\left\{\frac{1}{4}\sqrt{\frac{\pi}{z}}\ \frac{1}{\sqrt{T/T_c - 1}}\right\} \tag{5.11a}$$

where

$$c \sim e^{-1/(4\pi z)} \qquad (5.11b)$$

(The proportionality constant is not known.)

It should be kept in mind that the results derived from Kosterlitz equations are valid in the limit of small $z$ .

## 6. NON-NEUTRAL COULOMB GAS

The neutral Coulomb gas has the property that the energy needed to create a positive and negative charge is equal and that as a consequence the average charge-density $\langle n\rangle$ is zero. Let us consider a case when the creation energy for positive and negative charges is unequal and the average charge-density $\langle n\rangle \neq 0$ is uniform (independent of position). In this case we obtain the energy needed to create an additional positive charge in the following way[16]: First introduce the additional charge uniformly distributed over the 2-dimensional volume.

This requires an overall charging energy given by

$$\int d^2r\, U(\bar{r})\, \langle n\rangle = 2\pi\, \lambda_c^2\, \langle n\rangle \qquad (6.1)$$

The next step is to assemble the charge into the distribution $f_\xi(r)$ . This requires a similar self-energy as in the neutral case (see eq. (3.5b)

$$-\,T\,\ell nz + \frac{\Delta\mu}{2} + \frac{1}{2}\,U_{eff}(0) \qquad (6.2)$$

It follows that the total energy needed to create a positive (negative) charge is

$$\{H_{1\,\overset{+}{(-)}}\}_{eff} = -\,T\ell nz \overset{+}{(-)} \frac{\Delta\mu}{2} + \frac{1}{2}\,U_{eff}(0) \overset{+}{(-)} 2\pi\lambda_c^2\,\langle n\rangle \qquad (6.3)$$

As in the neutral case we argue that in equilibrium the density of positive (negative) free charges $n_F^{\overset{+}{(-)}}$ is proportional to the corresponding Boltzmann factor

$$n_F^{\overset{+}{(-)}} = g_{\overset{+}{(-)}}\,\frac{e^{-\{H_{1\,\overset{+}{(-)}}\}_{eff}/T}}{\xi^2} \qquad (6.4)$$

The density of free charges $n_F^{\overset{+}{(-)}}$ is related to the screening-length by (see eq. (3.2c)

$$\lambda^{-2} - \lambda_c^{-2}/\varepsilon = \frac{2\pi(n_F^+ + n_F^-)}{\varepsilon T} = \frac{2\pi\ n_F}{\varepsilon T} \tag{6.5}$$

and to the average charge density by

$$\langle n\rangle = n_F^+ - n_F^- \tag{6.6}$$

If $g_+$ is approximately equal to $g_-$ (i.e. $g_\pm = g$ ) then these equations may for $\lambda/\lambda_c \ll 1$ be combined into[16,29]

$$\langle n\rangle = n_F\left[1 - (zg)^2\left(\frac{2\pi\ \xi^2\ n_F}{\varepsilon T}\right)^{\frac{1}{2\varepsilon T} - 2}\right]^{1/2} \tag{6.7}$$

The equation describes a cross-over from a low temperature regime where $n_F \approx \langle n\rangle$ to a high temperature regime where the dominant contribution to $n_F$ is generated by thermal pair-breaking. Fig. 7 is a sketch of the solution to eq. (6.7) (for $zg$ = const.). The dashed curve is the corresponding result of the neutral Coulomb gas (compare e.g. eq. (3.9b)). Note that eqs (6.3-6.6) can also be combined to

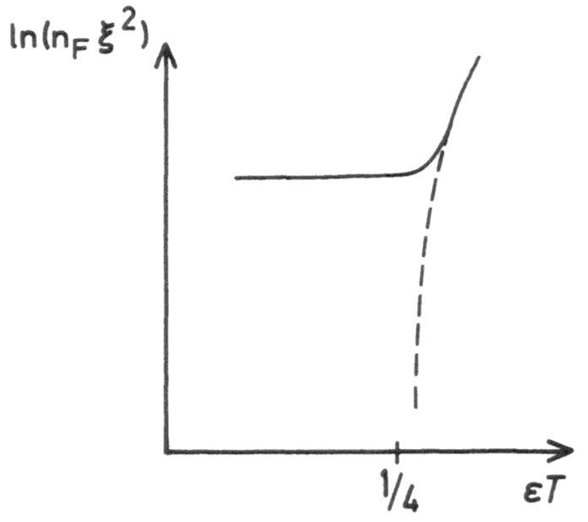


Fig. 7 Density of free charges $n_F$ for a non-neutral C.G. close to the K-T transition at $\varepsilon T = 1/4$ . Full line shows the cross-over from $n_F \gg \langle n\rangle$ for $\varepsilon T > 1/4$ to $n_F \approx n$ for $\varepsilon T < 1/4$ (as given by eq. (6.7) with $zg$ = const.). The dashed line is the corresponding result for a neutral C.G.

$$\langle n\rangle = n_F \tanh\left[\frac{\Delta\mu}{2T} - \frac{2\pi\lambda_c^2}{T}\langle n\rangle\right] \tag{6.8}$$

which means that in the limit $\lambda_c \to \infty$ one must have[16]

$$\frac{\Delta\mu}{2} \approx 2\pi\lambda_c^2 \langle n\rangle \tag{6.9}$$

Another property of interest is how $n_F$ varies as a function of $\Delta\mu$. This behaviour is sketched in fig. 8 for one temperature above $T_c$ and one below. For temperatures below $T_c$ there appears to be a "quasi"-critical $\Delta\mu_c$ given by[16,29]

$$\frac{|\Delta\mu_c|}{2} = \frac{1-4T\varepsilon}{2\varepsilon}\ \ln\left(\frac{\lambda_c\sqrt{\varepsilon}}{\xi}\right) + T\ \ln\left(\frac{2}{gz}\right) \tag{6.10}$$

$|\Delta\mu_c|$ corresponds in the 2-D superconductor to the lower "quasi"-critical magnetic field $H_{c1}$ discussed by Doniach and Huberman[8]. Note that for $T=0$ $|\Delta\mu_c|/2$ is the self-energy for a single charge (i.e. $\frac{1}{2}$ U(0)).

The non-neutral C.G. may for lower temperatures also form a Wigner-lattice[30]. This Wigner-lattice may possibly melt into a hexatic phase[31,32]. A possible phase-diagram may look something like the one sketched in fig. 9.

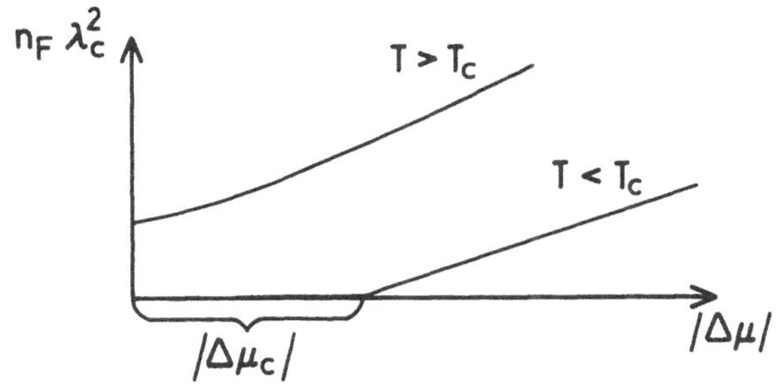


Fig. 8 Density of free charges $n_F$ as a function of the difference in chemical potential $\Delta\mu$. For $T < T_c$ there is a "quasi" critical $\Delta\mu_c$.

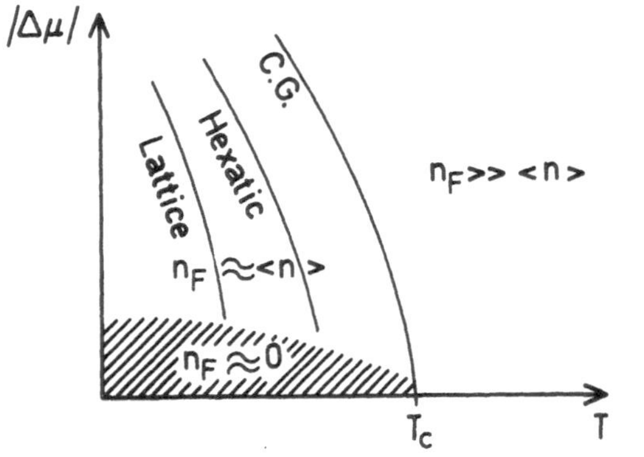


Fig. 9 Possible phase-diagram for a C.G. in the plane. There are three regions with respect to $n_F$ : $n_F \gg \langle n\rangle$ for $T > T_c$; $n_F \approx 0$ for $T < T_c$ and $|\Delta\mu| < |\Delta\mu_c|$; $n_F \approx \langle n\rangle$ for $T < T_c$ and $|\Delta\mu| > |\Delta\mu_c|$ . The last region may be further divided into a Wigner-lattice at low temperatures and a C.G. at higher temperatures. In between there may possibly be a hexatic phase.

7. COULOMB GAS IN ELECTRIC FIELD

Suppose that an external electric field $\bar{E}$ is applied on the C.G. The charges of a bound pair will then be pulled apart by a force $F_{ext} = \frac{E}{\varepsilon}\cos\theta$ where $\theta$ is the angle between the axis of the pair and the field. The attractive force for a pair of separation $r$ is given by (compare eq. (5.5))

$$F_{eff}(r) = -\frac{\partial}{\partial r}\{U(r) - U(0)\}_{eff} = \frac{1}{r\varepsilon} \tag{7.1}$$

where the last equality holds for large $r$ . A pair breaks up if $F_{ext} \geq F_{eff}$ . This means that in the presence of an external electric field there are free charges present even below $T_c$. In zero electric field there are no free charges present below $T_c$.

In a steady state situation one may obtain the field dependence of $n_F$ by the following heuristic reasoning[11]: The minimum energy needed for breaking a pair is

$$\frac{1}{\varepsilon}\,\ell n\left(\frac{1}{E\xi}\right) \quad \text{for} \quad E\xi \ll 1 \tag{7.2}$$

which is just the energy needed to stretch the pair to the breaking point, $F_{ext} = F_{eff}$ . In thermal equilibrium one expects that the rate of pair-breaking is proportional to the corresponding Boltzmann factor

$$\frac{\partial n_F}{\partial t} \sim e^{-\frac{1}{\varepsilon T}\,\ell n\left(\frac{1}{E\xi}\right)} \tag{7.3}$$

This pair-breaking must in a steady state situation be balanced by an equal recombination rate which, if it can be thought of as a simple collision process, is proportional to $n_F^2$ . So one expects that

$$\frac{\partial n_F}{\partial t} \sim n_F^2 \tag{7.4}$$

and consequently

$$n_F \sim (E\xi)^{1/2\varepsilon T} \tag{7.5}$$

Note that if $E^{-1}$ could be thought of as an effective screening length then eq. (7.5) is analogous to eq. (3.7).

## 8. CONNECTION TO 2-D SUPERCONDUCTORS

The connection between 2-D superconductors and the C.G. is that vortices play a role of C.G. charges[33,34]. We will first make this connection explicit for the following case: The superconductor is assumed to be well described by a phenomenological Ginzburg-Landau (G.L.) theory in the absence of vortices. The vortices are added on top of the G.L. description. In a G.L. description[35] the free-energy difference between the superconducting and normal state is expressed by a complex order parameter $\psi = |\psi| e^{i\theta}$

$$f_s - f_n = \alpha\left(\frac{T-T_{co}}{T_{co}}\right)|\psi|^2 + \frac{\beta}{2}|\psi|^4 + \frac{1}{2m^*}\left|\left(\frac{\hbar}{i}\bar{\nabla} - \frac{e^*\bar{A}}{c}\right)\psi\right|^2 + \frac{h^2}{8\pi} \tag{8.1}$$

Here $f_s - f_n$ is the difference in free energy density, $\alpha$ and $\beta$ are positive constants, $\bar{A}$ is the vector-potential and $\bar{h}$ is the corresponding magnetic field $\bar{h} = \bar{\nabla} \times \bar{A}$ , $m^*$ $(e^*)$ is the mass (charge) of a Cooper-pair ($m^* = 2m$, $e^* = 2e$) , $T_{co}$ is the Ginzburg-

Landau temperature. Thermodynamic stability requires that the appropriate free-energy (it turns out to be Gibbs free energy) is stable with respect to variations of $\psi$ and $\bar{A}$. In the G.L.-description the transition between the superconducting and normal state occurs at a phenomenological temperature $T_{co}$. The theory gives the temperature dependence of the free energy close to $T_{co}$. The results may be phrased in the following way for a 3-dimensional superconductor: The density of superconducting electrons is given by

$$n_s^o = 2|\psi|^2 = n_s^o(T=0)\,\frac{T_{co}-T}{T_{co}} \tag{8.2a}$$

the scale of magnetic penetration is given by the London penetration depth

$$\lambda_L = \lambda_L(T=0)\left(\frac{T_{co}}{T_{co}-T}\right)^{1/2} \tag{8.2b}$$

the scale of spatial variations of $n_s^o$ is given by the Ginzburg-Landau coherence length $\xi$

$$\xi = \xi(T=o)\left(\frac{T_{co}}{T_{co}-T}\right)^{1/2} \tag{8.2c}$$

the difference in free energy density, $f_s - f_n$, is given by a critical field $H_c$

$$f_s - f_n = -\frac{H_c^2}{8\pi} \quad , \qquad H_c = \frac{\phi_o}{2\pi\sqrt{2}\,\xi\,\lambda_L} \tag{8.3}$$

where $\phi_o \equiv \frac{hc}{2c}$ is the flux-quantum.

One may hope that the G.L. description in a phenomenological way takes into account fluctuations up to length-scales of the order of $\xi$ sufficiently close to $T_{co}$.

Let us now suppose that the superconductor has a thickness $d$ which is smaller than $\xi$. This means that we can to good approximation ignore spatial variations of $\psi$ over the thickness. The superconductor is effectively 2-dimensional. We will also assume that $\lambda_L/\xi \gg 1$. A lab-realization of the 2-D superconductor we have in mind is e.g. a "dirty" type II thin superconducting film[7]. For a 2-D superconductor we will let $n_s^o$ denote the areal density of superconducting electrons $n_s^o = 2d|\psi|^2$. It turns out that the scale of magnetic penetration is different for a 2-d superconductor as compared with a 3-D one[33,34]. This is because the empty space outside the superconducting plane helps the magnetic field to penetrate into the superconductor. The penetration length for the 2-D case is given by[34]

$$\Lambda = \frac{2\lambda_L^2}{d} \tag{8.4}$$

For a "dirty" superconducting film this may be of the order of cm, so it is typically larger than the sample size[7].

Next we incorporate vortices into the description. We will only consider the lowest energy vortices which correspond to vorticity $s = \pm 1$, since it turns out that the others are less important. In terms of the order parameter $\psi$ a vortex at a point $\bar{r}_i$ means that

$$\psi(\bar{r}_i) = 0 \tag{8.5a}$$

$$2d|\psi(\bar{r})|^2 \approx \begin{cases} 0 & |\bar{r}_i - r| < \xi \\ n_s^o & |\bar{r}_i - \bar{r}| > \xi \end{cases} \tag{8.5b}$$

and

$$\oint \bar{\nabla}\,\theta(\bar{r})\cdot d\bar{\ell} = \pm\, 2\pi \tag{8.5c}$$

where the line integral is around a contour enclosing $\bar{r}_i$. Note that the line-integral has to be a multiple of $2\pi$ due to the single-valuedness of $\psi$. A sketch of a vortex core is given in fig. 10. The current associated with a vortex is given by the London equation[34]

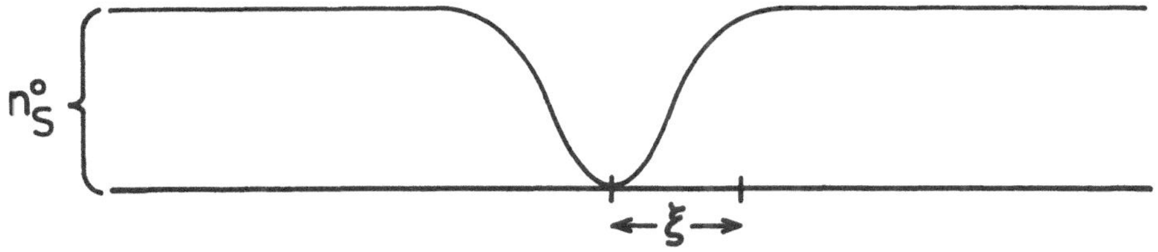


Fig. 10 Sketch of a vortex core. $n_s^o$ is zero at the center of the core. The linear dimension is given by the Ginzburg-Landau coherence length $\xi$.

$$\bar{j}_s(\bar{r},z) = \bar{j}_s(\bar{r})\,\delta(z) \tag{8.6a}$$

$$\bar{j}_s(\bar{r}) = \frac{e^2 n_s^o}{mc}\{\frac{\phi_o}{2\pi}\bar{\nabla}\,\theta(\bar{r}) - \bar{A}(\bar{r})\} \tag{8.6b}$$

where $\bar{r} = (x,y)$ is a vector in the plane of the superconductor and the z-axis is normal to it. It follows that a vortex is associated with a flux-quantum passing through the superconducting plane[34].

We need to know the total energy associated with a vortex. This we may think of as consisting of two parts; the vortex core part associated with the variation of the magnitude of $\psi$ around the vortex center and the current and magnetic field contribution outside the vortex core. The vortex core energy $E_c$ can be estimated from the volume of the core-region $\sim \pi\,\xi^2 d$ and the free energy density difference between the normal and superconducting state $\sim \frac{H_c^2}{8\pi}$. This gives[34]

$$E_c \sim \pi\,\xi^2\, d\,\frac{H_c^2}{8\pi} \tag{8.7a}$$

or

$$E_c = c_1\left(\frac{\phi_o}{2\pi}\right)^2 \frac{1}{\Lambda} \tag{8.7b}$$

The constant of proportionality $c_1$ may in principle be calculated. The current and magnetic field is obtained by solving eq. (8.6). Let us assume that the 2-D superconductor has $N$ vortices distributed according to

$$n(\bar{r}) = \sum_i s_i\,\delta(\bar{r} - \bar{r}_i) \tag{8.8}$$

where $s_i = \pm 1$ is the vorticity and $\sum_i s_i = 0$. In this case the current and magnetic field contribution to the energy can be expressed as

$$\left(\frac{\phi}{2\pi}\right)^2 \frac{1}{\Lambda}\frac{1}{2}\int d^2r\, d^2r'\, n(\bar{r})\,\tilde{U}(\bar{r} - \bar{r}')\,n(\bar{r}') \tag{8.9a}$$

$\tilde{U}(r)$ has the limiting forms[34] (specializing to the case when the linear dimension of the superconductor $R$ is smaller than $\Lambda$)

$$\tilde{U}(r) \approx -\ln(r/R) \qquad \xi \ll r \ll R$$
$$\tilde{U}(0) \approx \ln(R/\xi) \tag{8.9b}$$

So the interaction between two vortices, $\tilde{U}$ , has the same limiting forms as the interaction between two C.G. charges, U , where R plays the role of the cut-off $\lambda_c$ . This means that vortex-fluctuations on the superconductor should for $R/\xi \gg 1$ be well described by the $\lambda_c/\xi \to \infty$ limit of the thermodynamics of the C.G. Table I gives a translation between the superconductor and the C.G. A perpendicular magnetic field B gives an additional energy[8,16] $-(N_+ - N_-)mB$ where m is the average magnetic moment. This translates over to the difference in chemical potential for the non-neutral C.G. The Lorentz force acting on a vortex in the presence of a supercurrent translates into an electrical field for the C.G.[36]. Note that the $T \to \infty$ limit for the C.G. corresponds to the $T \to T_{co}$ limit for the superconductor.

Another situation when a 2-D superconductor can be connected with the C.G. is the following[37-39]: Let us assume that the superconductor consists of a square array of superconducting grains (compare fig. 11). A Hamiltonian describing the situation is given by

$$H = J \sum_{\langle ij \rangle} \{1 - \cos(\theta_i - \theta_j)\} \tag{8.10}$$

where J is a (temperature dependent) coupling constant, $\theta_i$ is the phase of the order-parameter at grain i , and the summation is over all pairs of nearest-neighbour grains. This is (upto a constant) identical to the classical XY-model[40]. It turns out that the phase-coherence is destroyed by a Kosterlitz-Thouless vortex-unbinding transition[2]. One can approximately obtain the vortex-Hamiltonian which governs the transition in the following way[2]: First approximate H with a quadratic Hamiltonian

$$H \approx J_R \frac{1}{2} \sum_{\langle ij \rangle} (\theta_i - \theta_j)^2 \tag{8.11}$$

Note that if we try to optimize eq.(8.11) so as to approximate H as well as possible using e.g. the self-consistent harmonic approximation[38,41], $J_R$ will be different from J[38]. Next we take a continuum limit and define a vortex of vorticity s as before i.e. $\oint d\bar{l}\cdot\bar{\nabla}\theta(\bar{r})=2\pi s$. The energy needed to create a neutral configuration of N vortices can then be estimated.[2]

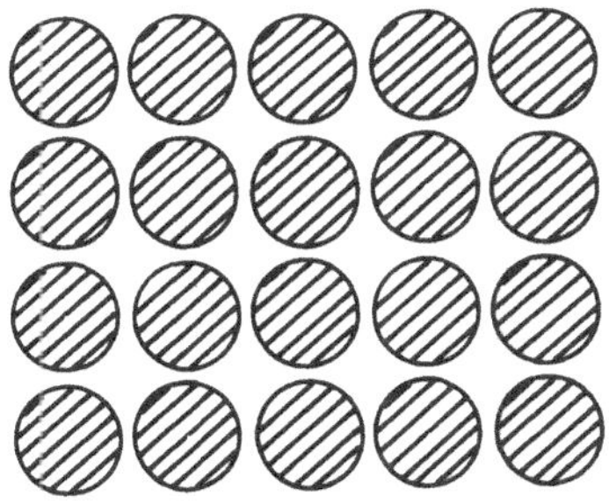

Fig. 11 Square array of superconducting grains

Table I. Translation between 2-D Coulomb gas and 2-D superconductor

| C.G. | Superconductor |
|---|---|
| Charge | Vortex |
| Charge dimension $\xi$ | Vortex core dimension<br>Ginzburg-Landau coherence length<br>$\xi = \xi(T = 0)\left[\frac{T_{co}-T}{T_{co}}\right]^{-1/2}$<br>$T_{co}$ = Ginzburg-Landau temperature |
| Sign of charge $s = \pm 1$ | Vorticity |
| Temperature $T$ | $T \cdot \frac{4\pi^2\Lambda}{\phi_o^2} = \frac{4\pi^2\ \Lambda(T=0)}{\phi_o^2}\ \frac{T_{co}T}{T_{co}-T}$ |
| Chemical potential $\mu$ | $- c_1$ |
| Difference in chemical potential $\Delta\mu$ | $2mB \cdot \frac{4\pi^2\Lambda}{\phi_o^2}$<br>m = average magnitude of magnetic moment<br>B = applied perpendicular magnetic field |
| Electric field $E$ | $\frac{I\ \phi_o}{c} \cdot \frac{4\pi^2\Lambda}{\phi_o^2}$<br>I = supercurrent (areal density) |

$$H_N = - 2\pi\ J_R \frac{1}{2} \sum_{i \neq j} \ell n(r_{ij}/a) + \frac{\pi^2}{2}\ J_R\ N \tag{8.12}$$

where $a$ is a lattice constant. It follows that the vortex-fluctuations close to $T_c$ of a weak-link superconductor should again be well-described by the C.G. The C.G. temperature $T$ and the chemical potential correspond to $T/(2\pi\ J_R)$ and $-\frac{\pi}{4}$, respectively, for the weak-link superconductor. Note that the fugacity $z = e^{-|\mu|/T}$ close to $T_c$ is small ($\approx .05$) so that the Kosterlitz equations should give a good approximation. This is in contrast to the C.G. stemming from the Ginzburg-Landau description where $z$ close to $T_c$ is approximately $.5$ which means that the Kosterlitz equations can not be used to describe this case. Also note that it is possible that the Ginzburg-Landau C.G. is also a good approximation of the fluctuations for a weak-link superconductor provided that the coherence length $\xi$ is much larger than the lattice spacing $a$.

## 9. COULOMB GAS SCALING

The thermodynamics of the Coulomb gas should, as we have seen, describe the vortex-fluctuations on a 2-D superconductor. One may then ask in what way the predictions of the C.G. are reflected in the experiments on the superconductor. One link to experiments is through the Bardeen-Stephen model for dissipation[42] which relates the density of the free vortices, $n_F$, to the resistance

$$\frac{R}{R_N} = 2\pi\,\xi^2\,n_F \tag{9.1}$$

Here $R$ is the measured resistance, $R_N$ is the normal state resistance, and $\xi$ is the Ginzburg-Landau coherence length. Let us first assume that there is no perpendicular magnetic field i.e. $B = 0$. The limit of vanishing current $I$ corresponds to vanishing electric field $E$ in the C.G. language (compare table I). Below $T_c$ there are then no free vortices. However, above $T_c$ there are thermally created free vortices. Consequently, if the C.G. description applies, the resistive tail (which is indeed observed experimentally[14]) should be due to thermally created free vortices. One way of making this link firmer is to observe that within the Ginzburg-Landau-C.G. description all superconductors are described by the same C.G. (see table I). It follows that any dimensionless thermodynamic quantity when plotted against the C.G. temperature is universal (= sample independent)[43,44]. This we refer to as Coulomb gas scaling. Now, since $R/R_N$ is proportional to $\xi^2 n_F$, which is a dimensionless thermodynamic C.G. quantity, it follows that $R/R_N$ when plotted against the C.G. temperature should be universal. Since $T_c$ in the C.G. language is defined by $4T_c\varepsilon_c = 1$ one finds that the C.G. temperature, when translated to superconductor temperatures, is proportional to[43]

$$X = \frac{T}{T_{co}-T}\cdot\frac{T_{co}-T_c}{T_c} \tag{9.2}$$

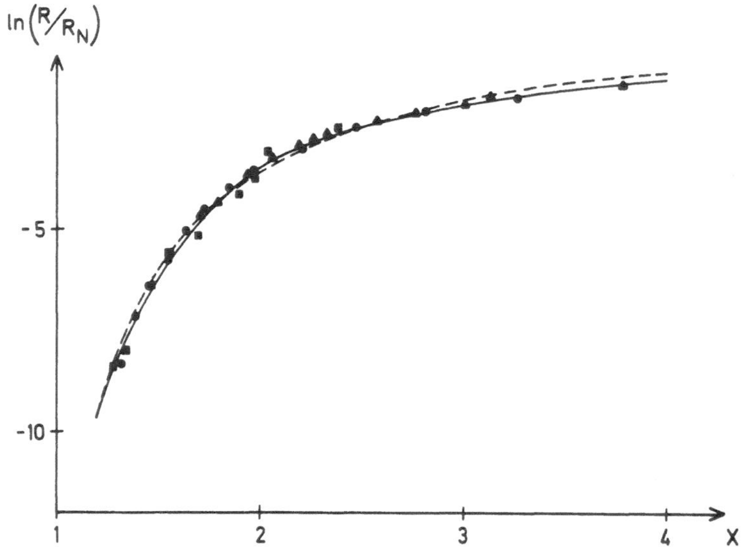


Fig. 12 C.G.-scaling for resistance. Resistance ratio $R/R_N$ plotted against the scaling-variable X for five different samples (from ref. 44, the data for the five samples are given by full line, dashed line, dots, squares, and triangles).

The C.G. scaling for the resistance appears to be born out experimentally. This is illustrated in fig. 12 which shows data for five (microscopically rather different) 2D-superconductors[44].

Close to $T_c$ one expects from the C.G. description that[10] (compare eqs (3.2b), (3.12b), (5.11), (9.1), and (9.2))

$$\frac{R}{R_N} \sim \xi^2 n_F \sim \xi^2/\lambda^2 \sim e^{-\frac{|\text{const}|}{\sqrt{X-1}}} \qquad X - 1 \ll 1 \qquad (9.3)$$

It turns out that the data in fig. 12 is not well approximated by eq. (9.3). However, this is hardly surprising for two reasons: First of all the data is not in the region where eq. (9.3) is expected to be valid[43,44]. Secondly, the square-root dependence in eq. (9.3) hinges on the largest length-scales which may in practice be masked by finite -size, -current, and -magnetic field effects[44].

Below $T_c$ there should be no free vortices in the limit of vanishing current I. However, since I corresponds to E in the C.G. gas, it follows that for any finite I free vortices will be generated according to eq. (7.5)[45]. Translated to the superconductor variables this is[10] (compare table I)

$$R \sim \xi^2 n_F \sim (E\xi)^{1/2\varepsilon T} \sim (I/I_o)^{\frac{2\varepsilon_c}{\varepsilon(X)X}} \tag{9.4a}$$

where

$$I_o = \frac{\phi_o c}{4\pi^2 \Lambda \xi} \tag{9.4b}$$

and $\varepsilon_c$ is the dielectric constant at $T_c$ i.e. $\varepsilon_c = \varepsilon(X=1)$. In terms of the voltage V this means that

$$V \sim (I/I_o)^{a(X)} \tag{9.5a}$$

where

$$a(X) = 1 + \frac{2\varepsilon_c}{\varepsilon(X)X} \tag{9.5b}$$

This type of non-linear I-V characteristics is observed experimentally[46-48]. From these measurements it is possible to extract $\varepsilon(X)$. This is illustrated in fig. 13 for two Hg-Xe alloy samples[47,49]. Note that $\varepsilon(X)$ for the two samples fall on one curve. This is in accordance with C.G.-scaling[49]. Also note that there is no trace of the square-root dependence of $\varepsilon(X)$ expected close to $T_c$ (see eqs (3.11) and (5.10a)). However, as pointed out above, the square-root dependence may in practice be masked by finite -size, -current, and -magnetic-field effects.

Another situation of interest is finite perpendicular field B and the limit of vanishing current. This takes us to the non-neutral C.G. In this case a dimensionless thermodynamic quantity will be a function of two scaling variables[44] X and $B/B_o$ where $B_o = \frac{\phi_o}{2\pi\xi^2}$. It follows that $R/R_N$ as a function of $(X, B/B_o)$ should be universal. The cross-over behaviour close to $T_c$ as described by eq. (6.7) can be expressed as

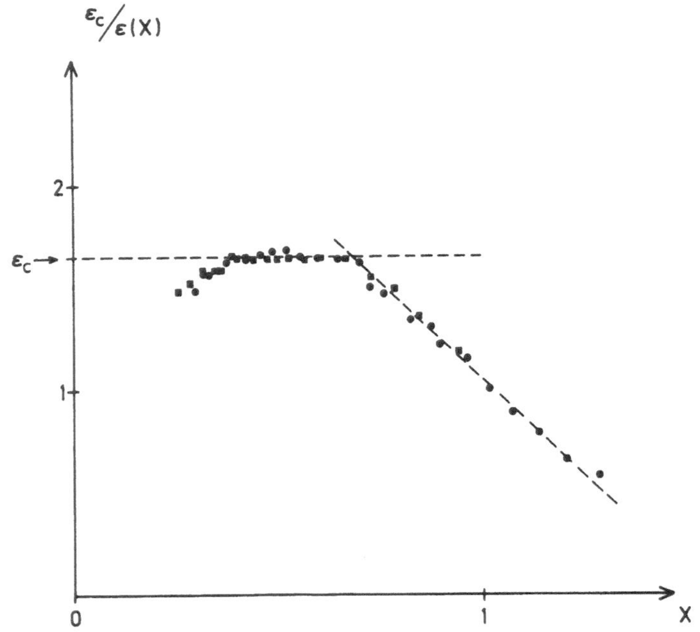


Fig. 13 $\varepsilon(X)$ for two Hg-Xe alloy samples (from ref. 49). Dashed lines are guides for the eye. Crossing point between horizontal dashed line and vertical axis gives $\varepsilon_c$ .

$$\frac{B}{B_o} = \frac{R}{R_N}\left\{1 - \kappa^2\left(\frac{R}{R_N}\right)^{\frac{2\varepsilon_c}{\varepsilon X} - 2}\right\}^{1/2} \tag{9.6}$$

where $\kappa(X,B/B_o)$ and $\varepsilon(X,B/B_o)$ are, at present, two unknown functions[16].

Finally we note that in general a dimensionless thermodynamic quantity will be a function of three scaling variables $(X, B/B_o, I/I_o)$ .

## 10. ESTIMATE OF $\varepsilon_c$

The dielectric constant at $T_c$, $\varepsilon_c$, may in principle be extracted from I-V characteristics of the 2D-superconductor. Within the Ginzburg-Landau-C.G. description this should be a universal number. The analysis illustrated in fig. 13 gives $\varepsilon_c \approx 1.65$. One would then like to have a theoretical estimate of the same quantity. As pointed out above, the Kosterlitz equations cannot be used for this purpose. However, it is possible to estimate $\varepsilon_c$ by a somewhat roundabout reasoning: Let us assume that there exists a 2D-superconductor which at $T_c$ is well-described both by the XY-model and the Ginzburg-Landau-C.G. From the Ginzburg-Landau-C.G. we have

$$\varepsilon_c = \frac{\phi_o^2}{16\pi^2 \Lambda(T_c) T_c} \tag{10.1}$$

and from the dirty limit of BCS-theory[35]

$$\Lambda = \frac{\phi_o^2 e^2 R_N}{2\pi^4 \hbar} \left[\Delta(T) \tanh\left(\frac{\Delta(T)}{2T}\right)\right]^{-1} \tag{10.2}$$

where $R_N$ is the normal-state-sheet-resistance and $\Delta(T)$ is the BCS energy gap. On the other hand, the zero-voltage-maximum current, $i_c$, through an ideal junction is given by[50]

$$i_c(T) r_N = \frac{\pi\Delta(T)}{2e} \tanh\left(\frac{\Delta(T)}{2T}\right) \tag{10.3}$$

where $r_N$ is the normal state tunnelling resistance per unit area. Assuming that all resistance is by tunnelling through the weak links gives $R_N = r_N$. It follows that

$$\Lambda\, i_c(T) = \frac{\phi_o^2}{8\pi^3} \frac{2e}{\hbar} \tag{10.4}$$

Finally, $i_c$ is related to the coupling constant $J$ by[51]

$$\frac{\hbar\, i_c}{2e} = J \tag{10.5}$$

This chain of relations gives

$$\varepsilon_c = \frac{\pi J}{2T_c} \tag{10.6}$$

At this point one may take advantage of existing Monte-Carlo calculations for the XY-model[52]. The Monte-Carlo estimates give

$$T_c \approx (.90 \pm .05)\ J \tag{10.7}$$

This in its turn gives

$$\varepsilon_c \approx 1.75 \pm .10 \tag{10.8}$$

Note that this theoretical estimate is in fair agreement with the experimental value mentioned above $(\varepsilon_c \approx 1.65)$.

## 11. DYNAMICS

As discussed above (see section 3) the C.G. quantity $1/T\varepsilon_0$ jumps from 4 to 0 at $T_c$. We will now connect this to the famous universal jump of the superfluid density[4]. The superfluid density in case of superconducting films corresponds to the areal density of superconducting electrons $n_s$. This latter quantity may be defined as follows: Let an infinitesimal macroscopic supercurrent $\bar{J}$ flow across the superconductor. The areal density of superconducting electrons, $n_s$, can then be defined by the corresponding increase of free energy[18]

$$\Delta F = \frac{1}{2m\, n_s\, \Omega} \{\frac{mJ}{e}\}^2 \tag{11.1}$$

The current density at point r can be written as a sum of two parts[11]

$$\bar{j}(\bar{r}) = \bar{j}_{\parallel}(\bar{r}) + \bar{j}_v(\bar{r}) \tag{11.2}$$

where $j_{\parallel}$ has zero curl and is given by

$$\text{where } j_{\parallel}(\bar{r}) = \frac{\hbar e n_s^o}{2m} \bar{\nabla}\, \theta(\bar{r}) \tag{11.3}$$

and $\bar{j}_v$ is the contribution from the vortices. This latter part has zero divergence and can be expressed as[16]

$$\bar{j}_v(\bar{r}) = \frac{\hbar\, n_s^o\, e}{2m} (\bar{\nabla} \times \hat{z} \int d^2r'\ U(\bar{r} - \bar{r}')\ n(r')) \tag{11.4}$$

where $U$ is the C.G. potential and $n(r)$ is the C.G. charge density. Using the above definition of $n_s$ it is straight-forward to show that[18]

$$\frac{n_s}{n_s^o} = \lim_{k\to o} \frac{1}{\varepsilon} \frac{k^2+\lambda_c^{-2}}{k^2+\lambda^{-2}} = \lim_{k\to o} \frac{1}{\varepsilon(k)} \equiv \frac{1}{\varepsilon_o} \tag{11.5}$$

The prediction of a universal jump comes from the Kosterlitz-Thouless transition and so requires $\lambda_c \to \infty$ before $k \to 0$. For a superconductor this is, of course, an approximation. The kinetic inductance is related to $n_s$ by

$$L_K = \frac{m}{e^2 n_s} \tag{11.6}$$

In terms of this quantity the universal jump is given by

$$(L_K\, T_c)^{-1} = \frac{e^2 n_s}{m\, T_c} = \frac{e^2 n_s^o}{m\, T_c\, \varepsilon} = \begin{cases} \dfrac{8e^2 k_B}{\pi\, \hbar^2} \approx 81\cdot10^6 (HK)^{-1} & T \to T_c^- \\ 0 & T \to T_c^+ \end{cases} \tag{11.7}$$

This universal jump prediction for superconductors is illustrated in fig. 14 by a set of data from measurements by Hebard and Fiory[53]. The dashed line in the figure is $\{8\, e^2\, k_B/(\pi\hbar^2)\}T$. However, this

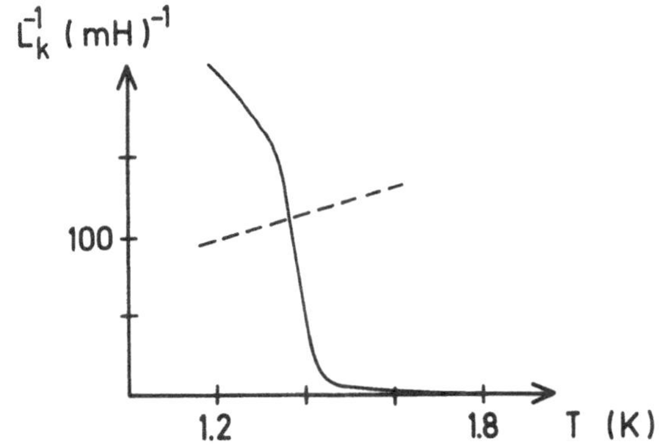


Fig. 14 Kinetic inductance for a granular Al-film. (1 MHz-data from ref. 53). Dashed line is $\{8e^2 k_B/(\pi\hbar^2)\}T$. This line gives the prediction of the static theory for the jump at $T_c$.

type of measurements are made at finite frequences, so the static theory is not quite adequate.

The extension of the C.G. description to finite frequences has been developed by Ambegaorkar et al.[5,45,54] and the translation to superconducting films was made by Halperin and Nelson[10]. It turns out for the static case, that if the supercurrent density is $\bar{I} = \int \frac{d^2r}{\Omega} \langle \bar{j}_{\parallel}(\bar{r}) \rangle$ in the absence of vortices, then the inclusion of vortices modifies the supercurrent to $\bar{I}/\varepsilon_o$[36]. Let us assume that the effect of vortices can be included by a dielectric constant also for finite frequences. This means that if the current in the absence of vortices is $\bar{I} = \bar{I}_o e^{i\omega t}$, then inclusion of vortices modifies it to $\bar{I}/\varepsilon(\omega)$, where $\varepsilon(\omega)$ is a frequency dependent dielectric function[5,45]. Next assume that a free vortex, $i$, moves according to[55]

$$\frac{d\bar{r}_i}{dt} = s_i \, \mu(\hat{z} \times \bar{F}) \tag{11.8}$$

where $\mu$ is the mobility, $F$ is the average magnitude of the force $F = \frac{\phi_o}{c} I/\varepsilon(\omega)$, and $s_i$ is the vorticity. On the other hand, the average current can be expressed as (see eq. (11.2))

$$I/\varepsilon(\omega) = I + \int \frac{d^2r}{\Omega} \langle j_v(\bar{r}) \rangle \tag{11.9}$$

and the last term can be rewritten as

$$\begin{aligned} \int \frac{d^2r}{\Omega} \langle \bar{j}_v(\bar{r}) \rangle &= \frac{\hbar n_s^o e}{2m} \langle (\bar{\nabla} \times \hat{z} \int d^2r' \, U(\bar{r} - \bar{r}') \, n(r') \rangle \\ &= 2\pi \frac{\hbar n_s^o e}{2m} \langle \sum_i s_i \, \bar{r}_i \rangle \times \hat{z} \end{aligned} \tag{11.10}$$

The sum on the left-hand side may be approximately split up into a sum over bound pairs and a sum over free vortices

$$\langle \sum_i s_i \, \bar{r}_i \rangle \times \hat{z} = \Big( \sum_{\substack{\text{bound} \\ \text{pairs} \\ \langle ij \rangle}} \langle \bar{r}_{ij} \rangle + \sum_{\substack{\text{free} \\ \text{vortices} \\ i}} s_i \langle \bar{r}_i \rangle \Big) \times \hat{z} \tag{11.11}$$

It turns out that the first term is proportional to the polarization due to bound pairs[5]. The corresponding contribution to the average current can then be expressed in terms of a "bound pair" dielectric constant, $\varepsilon_b(\omega)$

$$I_{\substack{\text{bound}\\ \text{pairs}}} = (1 - \varepsilon_b(\omega))\, I/\varepsilon(\omega) \tag{11.12}$$

The time-derivative of the sum over free vortices is given by

$$\sum_{\substack{\text{free}\\ \text{vortices}\\ i}} s_i \left\langle \frac{d\bar{r}_i}{dt} \right\rangle \times \hat{z} = \mu \bar{F} n_F \tag{11.13}$$

The time-derivative of the total average current $I/\varepsilon(\omega)$ is then given by

$$\frac{i\omega I}{\varepsilon(\omega)} = i\omega I + \frac{i\omega(1-\varepsilon_b(\omega))}{\varepsilon(\omega)} I + \frac{2\pi\hbar n_s^o e}{2m} \frac{\phi_o}{c} \frac{\mu n_F I}{\varepsilon(\omega)} \tag{11.14}$$

Solving for $\varepsilon(\omega)$ gives[5]

$$\varepsilon(\omega) = \varepsilon_b(\omega) + i/\omega\tau \tag{11.15a}$$

where[10]

$$\tau = \frac{L_K^o}{\mu n_F} \frac{c^2}{\phi_o^2} \quad , \quad L_K^o \equiv \frac{m}{e^2 n_s^o} \tag{11.15b}$$

The voltage drop per length of the superconductor, $E$, is given by[10]

$$E = \frac{\hbar}{2e} \frac{d}{dt} \langle \bar{\nabla}\theta \rangle = \frac{mi\omega}{e^2 n_s^o} I \tag{11.16}$$

The complex impedance $Z$ is defined as the ratio between $E$ and the average current density $I/\varepsilon(\omega)$, i.e.

$$Z = \frac{E}{(I/\varepsilon(\omega))} = i\omega\, L_K^o\, \varepsilon(\omega) \tag{11.17}$$

So measurements of the complex impedance is directly related to the C.G.-dielectric constant $\varepsilon(\omega)$. As a special case note that somewhat below $T_c$ one has $n_F = B/\phi_o$ and approximately $\varepsilon_b \approx 1$. This means that, for a fixed $T$ in this region, $Z$ should only be a function of $B/\omega$. This expectation is in fair agreement with measurements by Hebard and Fiory[56].

## 12. FINAL REMARKS

We have in the present lectures discussed the thermodynamic properties of a two-dimensional Coulomb-gas and the connection to vortex-fluctuations for superconducting films. Questions that immediately come to the mind are: "How well are vortex-fluctuations described by the C.G.-model?" and "How important are vortex-fluctuations for superconducting films?" Some perspective on these questions will be provided by Professor Mooij's lectures[57]. However, I do not think that clear-cut answers can be given yet. But then something remains for the future!

## REFERENCES

1. V.L. Berezinskii
Zh. Eksp. Theor. Fiz. 61 (1971) 1144 (Sov. Phys. JETP 34 (1972) 610
2. J.M. Kosterlitz and D.J. Thouless
J. Phys. C6 (1973) 1181
3. J.M. Kosterlitz
J. Phys. C7 (1974) 1046
4. D.R. Nelson and J.M. Kosterlitz
Phys. Rev. Lett. 39 (1977) 1201
5. V. Ambegaokar, B.I. Halperin, D.R. Nelson, and E.D. Siggia
Phys. Rev. Lett. 40 (1978) 783
6. D.J. Bishop and J.D. Reppy
Phys. Rev. Lett. 40 (1978) 1727
7. M.R. Beasley, J.E. Mooij, and T.P. Orlando
Phys. Rev. Lett. 42 (1979) 1165
8. S. Doniach and B.A. Huberman
Phys. Rev. Lett. 42 (1979) 1169
9. L.A. Turkevich
J. Phys. C12 (1979) L 385
10. B.I. Halperin and D.R. Nelson
J. Low Temp. Phys. 36 (1979) 599
11. B.I. Halperin
in Physics of Low-Dimensional Systems, Proc. Kyoto Summer Inst., Sept. 1979, ed. Y. Nagaoka and S. Hikami (Publication Office, Progr. Theor. Phys., Kyoto 1979) p. 53
12. D.R. Nelson
in Fundamental Problems in Statistical Mechanics V, ed. E.G.D. Cohen (North-Holland, 1980) p. 53
13. A.F. Hebard and A.T. Fiory,
in Proceedings of the 16th International Conference on Low Temperature Physics, ed. by. W.G. Clark (Physica, 109 & 110 B+C, (1982) 1637)
14. J.E. Mooij in NATO Advanced Study Institute on Advances in Superconductivity (Plenum to be published)
15. E.H. Hauge and P.C. Hemmer
Phys. Norv. 5 (1971) 209

16. P. Minnhagen
Phys. Rev. B23 (1981) 5745
17. P. Minnhagen
NORDITA-preprint 80/25 (1980)
18. P. Minnhagen and G.G. Warren
Phys. Rev. B24 (1981) 2526
19. A.M. Polyakov
Nucl. Phys. B20 (1977) 429
20. S. Samuel
Phys. Rev. D18 (1978) 1916
21. P. Minnhagen, A. Rosengren, and G. Grinstein
Phys. Rev. B18 (1978) 1356
22. P.B. Wiegmann
J. Phys. C11 (1978) 1583
23. D.J. Amit, Y.Y. Goldschmidt, and G. Grinstein
J. Phys. A13 (1980) 585
24. J. Zittartz
Z. Phys. B23 (1976) 277
25. H.U. Everts and W. Koch
Z. Phys. B28 (1977) 117
26. A.P. Young
J. Phys. C11 (1978) L 453
27. A.P. Young
in NATO Advanced Study Institute on Ordering in Strongly Fluctuating Condensed Matter Systems (Plenum, 1980) p. 271
28. A.P. Young and T. Bohr
J. Phys. C14 (1981) 2713
29. P. Minnhagen
Solid State Commun. 36 (1980) 805
30. A.A. Abrikosov
Sov. Phys. JETP 5 (1957) 1174
31. B.A. Huberman and S. Doniach
Phys. Rev. Lett. 43 (1979) 950
32. D.S. Fisher
Phys. Rev. B22 (1980) 1190
33. J. Pearl
Appl. Phys. Letters 5 (1964) 65
34. J. Pearl
in Low Temperature Physics -L79, ed. by J.G. Daunt, D.O. Edwards, F.J. Milford, and M. Yagub (Plenum, New York, 1965) p. 566
35. See. e.g. M. Tinkham
Introduction to Superconductivity (McGraw-Hill, New York, 1975) and references therein
36. R.J. Myerson
Phys. Rev. B18 (1978) 3204
37. Y.E. Lozovik and S.G. Akapov
Solid State Commun. 35 (1980) 693
38. Y.E. Lozovik and S.G. Akapov
J. Phys. C14 (1981) L 31

39. C.J. Lobb, D.W. Abraham, and M. Tinkham
Phys. Rev. B27 (1982) 150
40. See e.g. M. Suzuki
in Physics of Low-Dimensional Systems, Proc. Kyoto Summer Inst. Sept. 1979, ed. Y. Nagaoka and S. Hikami, p. 39 (Publication Office, Progr. Theor. Phys. Kyoto, 1979)
41. B. Mühlschlegel and H. Zittartz
Z. Physik 175 (1963) 553
42. J. Bardeen and M. J. Stephen
Phys. Rev. 140A (1965) 1197
43. P. Minnhagen
Phys. Rev. B24 (1981) 6758
44. P. Minnhagen
Phys. Rev. B27 (1983) 2807
45. V. Ambegaokar, B.I. Halperin, D.R. Nelson, and E.D. Siggia
Phys. Rev. B21 (1980) 1806
46. K. Epstein, A.M. Goldman, and A.M. Kadin
Phys. Rev. Lett. 47 (1981) 534
47. K. Epstein, A.M. Goldman, and A.M. Kadin
Phys. Rev. B26 (1982) 3950
48. J. Resnick, J.C. Garland, J.T. Boyd, S. Shoemaker, and R.S. Newrock
Phys. Rev. Lett. 47 (1981) 1542
49. P. Minnhagen
Phys. Rev. B, in press
50. V. Ambegaokar and A. Baratoff
Phys. Rev. Lett. 10 (1963) 486; 11 (1963) 104 (E)
51. B.D. Josephson
Phys. Lett. 1 (1962) 251
52. J. Tobochnik and G.V. Chester
Phys. Rev. B20 (1979) 3761
53. A.F. Hebard and A.T. Fiory
Phys. Rev. Lett. 44 (1980) 291
54. V. Ambegaokar and S. Teitel
Phys. Rev. B19 (1979) 1667
55. Y.B. Kim and M.J. Stephen
in Superconductivity, ed. R.D. Park (Marcel Dekker, New York, 1969) Chapter 19
56. A.T. Fiory and A.F. Hebard
Phys. Rev. B25 (1982) 2073
57. J.E. Mooij in this volume, p. 325.

TWO-DIMENSIONAL TRANSITION IN SUPERCONDUCTING FILMS

AND JUNCTION ARRAYS

J.E. Mooij

Department of Applied Physics
Delft University of Technology
Delft, The Netherlands

## 1. INTRODUCTION

In two-dimensional systems a special phase transition is possible which has become known as the Kosterlitz-Thouless transition. The original authors[1] characterized the transition as the breakdown of topological long range order. Below a critical temperature $T_c$ topological excitations, vortices, occur only as bound vortex-antivortex pairs. Above the critical temperature single vortices and anti-vortices are present in thermodynamic equilibrium. A renormalization procedure was developed to account for the influence of bound pairs with small separation on pairs with larger separation. A Kosterlitz-Thouless phase transition occurs only if the energy of a bound vortex-antivortex pair depends logarithmically on the separation r. As the number of free vortices grows extremely slowly with temperature above $T_c$, the transition is not very pronounced from an experimental point of view.

In superconducting thin films the vortex interaction energy is logarithmic in r only for separations below the penetration depth for perpendicular fields Λ. For this reason it was originally supposed that Kosterlitz-Thouless transitions would not occur in two-dimensional superconductors. A few years ago[2-5], it was realized that Λ is very large in films with a high sheet resistance and that Kosterlitz-Thouless-like transitions should be present to the same extent as in other finite systems. A large number of papers, reporting on experimental results in superconducting thin films has been published since. In addition, two-dimensional arrays of Josephson junctions and superconducting weak links have been fabricated.

Experimental studies of interacting vortices in thin superfluid helium films have preceded those in superconducting films. The connection between the general theory of Kosterlitz-Thouless (KT) transitions and experiments has been worked out by Ambegaokar, Halperin, Nelson and Siggia[6,7] and Ambegaokar and Teitel.[8] Halperin and Nelson[4] have adapted the results to superconducting films.

An attempt is made here to review the present experimental status in superconducting systems, from a limited point of view. Many experiments have made clear that single and paired vortices are present in the temperature region below the BCS transition temperature $T_{co}$. As discussed by Minnhagen in these proceedings[9] their influence can be described with the model of a two-dimensional Coulomb gas. In this general picture, a Kosterlitz-Thouless transition is one of several possible patterns. The discussion of experimental results in thin films is focused here on the quantitative aspects in direct relation to the KT transition, in the critical temperature region around $T_c$. In this discussion scales play a central role. Below $T_c$, there is a characteristic value for the separation of bound pairs that is called $\xi_-$. Above $T_c$, the average distance between free vortices is $\xi_+$. According to theory, both $\xi_-$ and $\xi_+$ diverge near $T_c$, proportional to the inverse square root of $|T-T_c|$. This square-root-cusp behaviour is a key feature of KT transitions. It is of much importance for the interpretation of experiments whether the characteristic sample size W and the penetration depth $\Lambda$ are smaller or larger than $\xi_+$ or $\xi_-$. In addition, methods of measuring introduce their own characteristic scale that has to be compared with both W or $\Lambda$ and $\xi_+$ or $\xi_-$. Only by few experimenters attention has been paid to these matters of scale and to the quantitative aspects of the renormalized theory. Two groups, Fiory and Hebard[10,11] and Epstein, Goldman and Kadin[12] are mentioned in this respect.

With arrays, fewer and less-detailed experimental results are available. The minimum scale, which is put by the coherence length in thin films, is connected to the lattice spacing in arrays. As a consequence, the effective size of arrays is considerably smaller than of films. No thorough experimental investigation of quantitative aspects has been performed yet. Theoretical aspects of the behaviour of superconducting arrays have been worked out recently by Lobb, Abraham and Tinkham.[13]

In thin films as well as in arrays, very interesting effects are found in an external perpendicular magnetic field. However, as these effects are only indirectly related to a KT transition of the superconducting system, the discussion in this review is limited to zero or near-zero field values.

Hebard and Fiory[14] and Mooij[15] have given previous reviews of experimental techniques and results.

Although the aim of this chapter is to give a review of experiments and although Minnhagen in his chapter gives a general discussion of the theory, to arrive at a detailed, quantitative discussion of experimental results it is first necessary to develop the detailed, quantitative theoretical picture of KT transitions in films and arrays of finite size, as far as that is available at present. In section 2 the relevant results of the Kosterlitz-Thouless theory are collected, in two different degrees of approximation. Section 3 gives the applications to superconducting films and the theoretical predictions for experimentally measurable quantities, based largely on the work of Halperin and Nelson. Section 4 discusses the experimental methods and results and compares these results with the theory. In section 5 the consequences of the theory are developed for arrays. Section 6 contains the experimental results on the arrays that have so far been fabricated. Although little can be said with definiteness at this time, section 7 consists of some final remarks under the heading 'conclusions'.

## 2. KOSTERLITZ-THOULESS TRANSITIONS

### 2.1. Introduction

In two-dimensional systems in which the order is described with one continuously variable parameter, so-called Kosterlitz-Thouless transitions are possible. In theoretical studies, very often the two-dimensional X-Y model is considered as the characteristic model system. It contains spins in a regular 2D array, that can rotate in the plane. The angle that spin i makes with a reference direction is called $\varphi_i$. In the array nearest neighbours i,j interact with interaction energy $U_{ij} = -J\cos(\varphi_i - \varphi_j)$, where J is a positive constant. At T = 0 all spins are parallel, at finite temperatures 'spin waves' occur, which are large length variations of $\varphi$. As neighbouring spins are still almost parallel, spin waves do not require much energy and waves with long wavelength are already present at low temperatures.

At higher temperatures topological excitations, vortices, start occurring in pairs. Going around one vortex the phase changes by $2\pi$. Going in the same sense around a vortex of opposite sign, an antivortex, the phase changes by $-2\pi$. Around a vortex-antivortex pair no phase change is present. The free energy of one single vortex-antivortex pair depends logarithmically on the separation of the vortex centers r:

$$U(r) = A\ln(r/a_o) + 2\mu_c \tag{2.1}$$

where $a_o$ is the smallest separation possible, one lattice constant in the discrete array. In the X-Y model, A is equal to $2\pi J$. The core potential $\mu_c$ is half the free energy of a pair at smallest separation. As the temperature increases, the number of bound pairs present increases as well as their average separation. In an infinite system no single free vortices occur until T is equal to $T_c$, the Kosterlitz-Thouless critical temperature. Above $T_c$ single vortices and bound pairs are both found. The interaction between vortices over a distance r is modified by bound pairs with separation smaller than r. A renormalization procedure is required to calculate properties.

In superconducting thin films and in arrays of Josephson junctions, vortices and antivortices show the same logarithmic interaction. However, the interaction constant is temperature dependent, approaching zero when the temperature goes to $T_{co}$, the BCS critical temperature. Although the universal behaviour, as following from the renormalized theory, is not changed by this additional temperature dependence, for the calculation of practical properties it is extremely important.

With increasing temperature, apart from the occurrence of vortices, the density and the average wave vector of the spin waves increases. Their presence may modify the interaction between vortices, in a scale-dependent way. Ohta and Jasnow[16] have treated the influence of spin waves on some aspects of KT transitions. Key features are retained, in some cases with modified parameters. In the analyses of superconducting systems that are discussed in the following sections, the spin waves are ignored completely. It is highly desirable that a thorough study be made of the influence of spin waves, which are not directly observed, on the aspects of the vortex system that are investigated in experiments.

A review of phase transitions in two-dimensional systems has been written by Halperin[17]. The notation there is not the same as in this chapter.

## 2.2. Renormalized theory

A scaling parameter $\ell$ is defined as:

$$\ell = \ln (r/a_o) \tag{2.2}$$

Functions of r are now usually considered as functions of the scale $\ell$. The potential for a vortex-antivortex pair at separation r, in the presence of other vortices, is:

$$U(\ell) = \int_o^{\ell} \frac{A(T)}{\varepsilon(\ell')} d\ell' + 2\mu_c(T) \tag{2.3}$$

A(T) is the interaction constant without other vortex pairs. The quantity $\varepsilon(\ell)$ depends on the density of bound pairs with separation smaller than $r = a_o \exp(\ell)$, which modify the interaction. As no pairs exist with separation smaller than $a_o$, $\varepsilon(0)=1$.

Two quantities are introduced: the reduced stiffness constant

$$K(\ell) = \frac{1}{2\pi\varepsilon(\ell)} \frac{A(T)}{k_B T} \tag{2.4}$$

and the pair excitation probability or activity:

$$y(\ell) = \exp\{2\ell - U(\ell)/2k_B T\} \tag{2.5}$$

It turns out that the most interesting behaviour of the system occurs in a region of values of K near $2/\pi$. It is therefore convenient to use a transformation

$$x(\ell) = \frac{2}{\pi} \frac{1}{K(\ell)} - 1 \tag{2.6}$$

In terms of x and y the Kosterlitz scaling equations, valid for $y << 1$, are:

$$\begin{aligned} \frac{dx}{d\ell} &= 8\pi^2 y^2 \\ \frac{dy}{d\ell} &= 2\,\frac{xy}{1+x} \end{aligned} \tag{2.7}$$

Together, these equations provide a relation between x and y:

$$x - \ln(1+x) - 2\pi^2 y^2 + C = 0 \tag{2.8}$$

C is an integration constant that is independent of $\ell$. Eq.(2.8) defines the so called trajectories in the plot of y versus x. Along the trajectories the scale $\ell$ varies. If for a particular system x and y are known at scale zero, the trajectories show how x and y change with increasing scale. In Fig.1 trajectories are shown, arrows indicate increasing $\ell$.

At this stage, it is not assumed that x is small. Later the approximation $x<<1$, leading to $x-\ln(1+x) \approx \frac{1}{2}x^2$, will be used. In practical systems, this is often not justified.

For $C<0$, y goes to zero for very large $\ell$. The value of x where the trajectory crosses the $y=0$ axis is called $x_\infty$ and follows from

$$x_\infty - \ln(1+x_\infty) = -C \tag{2.9}$$

If $|x_\infty|<<1$, which is true for sufficiently small $|C|$, $x_\infty \approx -|2C|^{\frac{1}{2}}$. For $C>0$ the trajectories lead to large y and x. The two regions are separated by the critical trajectory, found for $C=0$. It approaches

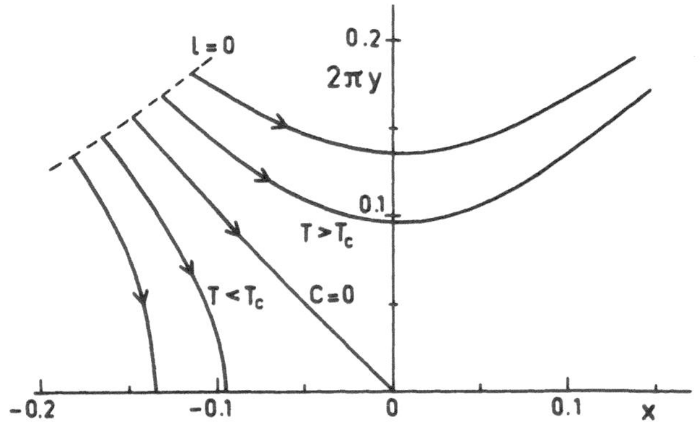


Fig. 1. Trajectories that show the dependence of the activity on the renormalized stiffness constant at fixed temperature. Along the trajectories $\ell$ increases in the direction of the arrow. The trajectory indicated with C = 0 is the critical trajectory at $T_c$. The dashed line indicates a particular choice for the non-universal starting conditions $y_0(T)$-$x_0(T)$.

the point x = 0, y = 0 as a straight line $2\pi y + x = 0$. The trajectories are universal, but the scale $\ell$ corresponding to a point on one of them is dependent on non-universal properties of the system, in particular the core potential. We take the temperature as constant on a trajectory. For C<0 T is below $T_c$, for C>0 T is above $T_c$.

## 2.3. Starting conditions, critical temperature

The relation between T and C has to be determined from the starting conditions. The values of x and y at scale zero, $x_0 \equiv x(0)$ and $y_0 \equiv y(0)$, are:

$$x_0 = 4k_BT/A(T) - 1 \tag{2.10}$$

$$y_0 = \exp\{-\mu_c(T)/(k_BT)\} \tag{2.11}$$

These starting conditions define the $\ell = 0$ curve in the y-x plot. The critical temperature $T_c$ follows from the intersection with the critical isotherm, i.e. the trajectory with C = 0:

$$\frac{4k_BT_c}{A(T_c)} - 1 - \ln\left[\frac{4k_BT_c}{A(T_c)}\right] - 2\pi^2\exp\left[-\frac{2\mu_c(T_c)}{k_BT_c}\right] = 0 \qquad (2.12)$$

At $T_c$, x goes to zero for large $\ell$. Consequently, if $\varepsilon_c$ is defined as the value of $\varepsilon(\ell)$ at $T_c$ and at infinite $\ell$, one finds from Eqs.(2.4) and (2.6):

$$k_BT_c = \frac{1}{4\varepsilon_c} A(T_c) \qquad (2.13)$$

This equation determines $T_c$. To calculate it, the core potential that determines $\varepsilon_c$ must be known as a function of temperature. The fact that at $T_c$, the renormalized interaction parameter for infinite scale, $A(T_c)/\varepsilon_c$, is equal to $4\,k_BT_c$ is a universal result as shown by Nelson and Kosterlitz.[18] It is also valid in the presence of spin waves.[16] Because in helium films the interaction constant is proportional to the superfluid density, and because above $T_c$ the interaction at infinite scale is zero, Eq.(2.13) is known as the 'jump in the superfluid density'.

For practical systems, the core potential can often be written as:

$$\mu_c(T) = \gamma A(T) - k_BT \ln N_0 \qquad (2.14)$$

where in the first term $\gamma$ is independent of temperature. The second term contains the entropy $k_B \ln N_0$ with $N_0$ a measure of the number of relevant independent configurations. If $\gamma$ and $N_0$ are known, $\varepsilon_c$ follows from Eqs.(2.12) when (2.13) and (2.14) are substituted:

$$\frac{1}{\varepsilon_c} - 1 + \ln\varepsilon_c - 2\pi^2N_0^2 \exp(-8\gamma\varepsilon_c) = 0 \qquad (2.15)$$

In Fig.2 values of $\varepsilon_c$ are given as a function of $\gamma$ for three values of $N_0$. Unfortunately no predictions for $N_0$ have been given from theory. It is considered unlikely that $N_0$ is smaller than 1. A simple relation exists between $\varepsilon(\ell)$, $\varepsilon_c$ and $x(\ell)$:

$$\varepsilon(\ell) = \varepsilon_c\{1 + x(\ell)\} \qquad (2.16)$$

### 2.4. Temperature dependence

A system with a temperature dependent interaction constant can best be compared with other systems by introducing an effective temperature

$$T' = T\, A(T_c)/A(T) \qquad (2.17)$$

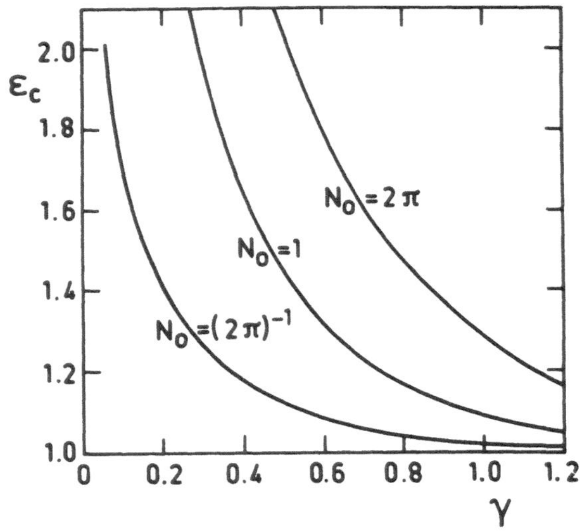


Fig. 2. Values of $\varepsilon_c$ as a function of $\gamma$ for different values of of $N_o$, as calculated from Eq.(2.15).

and a dimensionless temperature parameter:

$$\tau' = T'/T_c - 1 \tag{2.18}$$

The temperature scale is fixed by the starting conditions. Along the $\ell = 0$ curve, $x_o$ and $y_o$ are functions of $\tau'$:

$$x_o(\tau') = \varepsilon_c^{-1}(1 + \tau') - 1 \tag{2.19}$$

$$y_o(\tau') = N_o \exp\{-4\gamma\varepsilon_c/(1+\tau')\} \tag{2.20}$$

and the temperature dependence of C follows from Eq.(2.8):

$$C(\tau') = - x_o(\tau') + \ln\{1+x_o(\tau')\} + 2\pi^2 y_o^2(\tau') \tag{2.21}$$

To find the leading temperature dependence near $T_c$, C is expanded with respect to $\tau'$. Realizing $C(0) = 0$, one can write:

$$C(\tau') \approx B\,\tau' \qquad \tau' \ll 1 \tag{2.22}$$

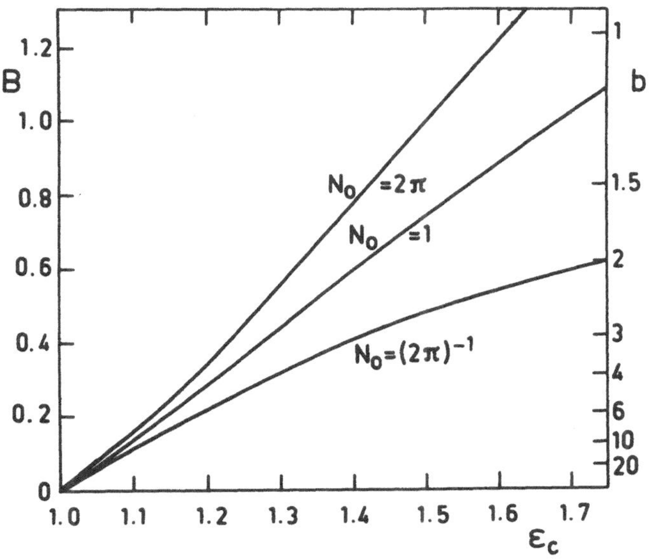


Fig. 3. Expansion parameter B, as defined in Eq.(2.22) against $\varepsilon_c$. B follows from Eq.(2.25). On the right-hand scale the corresponding values of b, Eq.(2.26), are indicated.

with B following from:

$$B = C'(0) = -x_o(0)x_o'(0)/\{1+x_o(0)\} + 4\pi^2 y_o(0)y_o'(0) \tag{2.23}$$

From Eq.(2.9) we see that near $T_c$, where $x_\infty$ is small, $x_\infty^2 \approx 2C$. This means that the reduced stiffness constant at infinite scale $K(\infty) = \{\frac{1}{2}\pi(1+x_\infty)\}^{-1}$ just below $T_c$ is:

$$\frac{\pi}{2} K(\infty) \approx 1 + (2B)^{\frac{1}{2}} |\tau'|^{\frac{1}{2}} \tag{2.24}$$

This is the well-known square-root-cusp behaviour, typical for these two-dimensional systems near $T_c$. The value of B follows from Eq.(2.23). With Eqs.(2.19) and (2.20) it is found that:

$$B = 1-\varepsilon_c^{-1} + 16\pi^2\gamma\varepsilon_c N_o^2 \exp(-4\gamma\varepsilon_c) \tag{2.25}$$

In Fig.3 values of B are plotted as a function of $\varepsilon_c$ for the same values of $N_o$ as in Fig.2. As shown in that figure, there is a direct connection between $\gamma$ and $\varepsilon_c$ for given $N_o$. A plot against $\varepsilon_c$ is preferred in Fig.3 to allow comparison between two quantities that can be derived from experimental results.

In the literature a quantity b is often used, which has the following relation to B:

$$b = \pi^2/(8B) \tag{2.26}$$

The values of b have been indicated on the right of Fig.3.

## 2.5. Characteristic scales

Both below and above $T_c$ there are characteristic, temperature-dependent scales that have to be kept in mind when considering experimental results. Below $T_c$, a length $\xi_-$ and a corresponding scale $l_-$ are defined with:

$$\ell_- = \ln(\xi_-/a_o) = \frac{1}{4}(|x_\infty|^{-1}-1) \tag{2.27}$$

$\xi_-$ is a measure of the average separation of bound pairs that are present at that temperature. Near $T_c$, $|x_\infty|$ is small and approximately equal to $(2B|\tau'|)^{\frac{1}{2}}$. As a consequence:

$$\ell_- \approx (32B|\tau'|)^{-\frac{1}{2}} = (2\pi)^{-1}(b/\tau')^{\frac{1}{2}} \qquad |\tau'|<<1 \tag{2.28}$$

Above $T_c$ the characteristic length $\xi_+$ with corresponding scale $\ell_+ = \ln(\xi_+/a_o^c)$ is connected with the average distance between free vortices. It is defined indirectly with:

$$y(\ell_+,\tau') = y_o(\tau') \tag{2.29}$$

As $y(\ell)$ can only be determined from integration, starting at $\ell=0$, it is even for small $\tau'$ not possible to give a simple expression for $\xi_+$ or $l_+$ without making the additional assumption that $|x_o|$ is small (see section 2.6). However, it is clear that both $\ell_-$ and $\ell_+$ diverge near $T_c$. In practice this means that in any experiment on a necessarily finite sample, there is a finite temperature region immediately above and below $T_c$ where the sample size is smaller than $\xi_+$ or $\xi_-$.

## 2.6. Approximations for x << 1

If the core energy is high, $\varepsilon_c-1 << 1$, it may be assumed that not only $y << 1$, as required for the Kosterlitz scaling relations to be valid, but also $x << 1$. In all relations we can now replace $x - \ln(1+x)$ by $\frac{1}{2}x^2$. The trajectories are given by:

$$x^2 - 4\pi^2 y^2 + 2C = 0 \tag{2.30}$$

The simpler equations have analytical solutions. Using the notation $c \equiv |2C|^{\frac{1}{2}}$, these solutions are:

$$\begin{aligned}
T < T_c:\quad & x(\ell) = -c\,[\mathrm{tgh}\{2c\ell + \mathrm{tgh}^{-1}(-c/x_o)\}]^{-1}\\
& 2\pi y(\ell) = c\,[\sinh\{2c\ell + \mathrm{tgh}^{-1}(-c/x_o)\}]^{-1}\\
T = T_c:\quad & x(\ell) = 2\pi y(\ell)\,(x_o^{-1} - 2\ell)^{-1}\\
T > T_c:\quad & x(\ell) = -c\,[\mathrm{tg}\{2c\ell + \mathrm{tg}^{-1}(-c/x_o)\}]^{-1}\\
& 2\pi y(\ell) = c\,[\sin\{2c\ell + \mathrm{tg}^{-1}(-c/x_o)\}]^{-1}
\end{aligned} \tag{2.31}$$

These analytical expressions make it easy to calculate $y(\ell)$ and $x(\ell)$ for given starting conditions. It is also possible to find an expression for $\ell_+$, as defined in Eq.(2.29) without resorting to numerical integration. In the expression for $y(\ell)$ above $T_c$, the argument of the sin function is equal at $\ell = 0$ and $\ell = \ell_+$ if:

$$2c\ell_+ = \pi - 2\mathrm{tg}^{-1}(-c/x_o)$$

For small values of $\tau'$, one has $c = (2B\tau')^{\frac{1}{2}} = \frac{1}{2}\pi(b/\tau')^{-\frac{1}{2}}$ and:

$$\ell_+ \approx \left[1-(2/\pi)\mathrm{tg}^{-1}\{-\tfrac{1}{2}\pi(b/\tau')^{-\frac{1}{2}}/x_o\}\right] (b/\tau')^{\frac{1}{2}} \qquad |x_o| \ll 1,\ \tau' \ll 1 \tag{2.32}$$

If $\tau'$ is even small enough that the $\mathrm{tg}^{-1}$ function can be neglected, one finds:

$$\ell_+ \approx (b/\tau')^{\frac{1}{2}} \qquad \tfrac{1}{2}\pi(\tau'/b)^{\frac{1}{2}} \ll |x_o| \ll 1 \tag{2.33}$$

As $|x_o|$ has to be small, this expression is only relevant for temperatures extremely close to $T_c$. Note that $\ell_-$, as given in Eq.(2.28) is smaller by a factor of $2\pi$ near $T_c$, at the same value of $|\tau'|$. The expression for $\ell_-$ is valid in a much wider range of parameter values.

The approximation $|x| \ll 1$ also yields a different relation between the temperature parameters. As C is now equal to $-\frac{1}{2}x^2 + 2\pi^2 y$, the expansion parameter B becomes:

$$B = -x_o(0)x_o'(0) + 4\pi^2 y_o(0)y_o'(0) \tag{2.34}$$

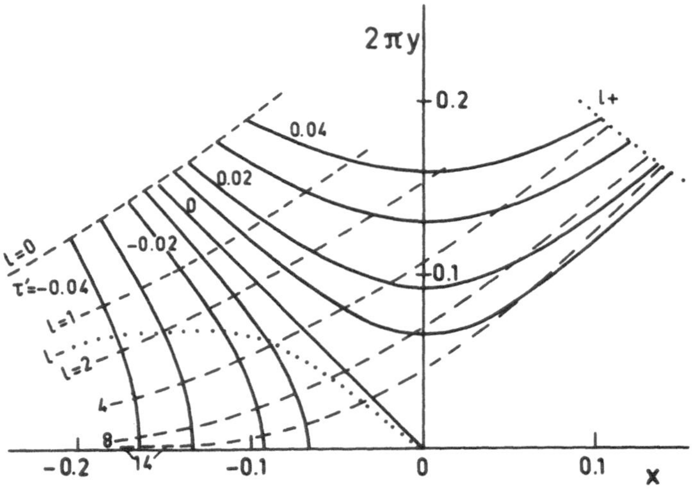


Fig. 4. Values of y and x as functions of $\tau'$ and $\ell$ for the model system of section 2.7 with $\gamma = \pi/4$ and $N_0 = 1$. Trajectories with fixed $\tau'$ and varying $\ell$ are indicated as drawn lines. Lines of constant scale are dashed. The characteristic scales $\ell_-$, below $T_c$, and $\ell_+$, above $T_c$, are indicated with dotted lines.

which has to be compared with Eq.(2.23). In terms of $\gamma$, $N_o$ and $\varepsilon_c$ this corresponds to:

$$B = \varepsilon_c^{-1}(1-\varepsilon_c^{-1}) + 16\pi^2 \gamma\varepsilon_c N_o^2 \exp(-8\gamma\varepsilon_c). \tag{2.35}$$

The relation between $\varepsilon_c$, $\gamma$ and $N_o$ is also slightly different in this approximation:

$$(1-\varepsilon_c^{-1})^2 - 4\pi^2 N_o^2 \exp(-8\gamma\varepsilon_c) = 0 \tag{2.36}$$

There is no reason to use (2.35) and (2.36) as they are not simpler than the original expressions (2.23) and (2.15). In practice it is often doubtful whether $|x_o|$, equal to $1-\varepsilon_c^{-1}$ at $T_c$, is really small enough for the approximations of this section to be valid.

## 2.7. Numerical example

As an example some values are given for one particular system, defined by starting conditions that correspond to $N_o = 1$ and $\gamma = \pi/4$. These values have been mentioned as valid for the X-Y model.[1] For this system one finds, without using the approximation $|x| \ll 1$, that $x_o(0) = -0.149$, $2\pi y_o(0) = 0.159$, $\varepsilon_c = 1.175$, $B = 0.240$ and $b = 5.15$.

When the values are derived from the equations of section 2.6, using $|x| \ll 1$, one finds $-x_o(0) = 2\pi y_o(0) = 0.154$, $\varepsilon_c = 1.181$, $B = 0.217$ and $b = 5.67$. For this system, where $|x_o|$ is reasonably small, the difference with the exact results are relatively small. The differences increase fast with increasing $|x_o|$, for systems with $\varepsilon_c$ higher than 1.2 the approximation $|x| \ll 1$ is not valid.

To give some indication for the dependence of x and y on temperature and scale, Fig.4 gives trajectories and lines of equal scale for the model system, calculated using the approximation $|x| \ll 1$. In the figure $\ell_+$ and $\ell_-$ are indicated. Near $T_c$, they both diverge. The value of $\ell_-^+$ varies from 1.89 at $\tau' = -0.04$ to 3.79 at $\tau' = -0.01$.

# 3. SUPERCONDUCTING THIN FILMS

## 3.1. Vortex interaction

In dirty, thin films with thickness d smaller than the bulk penetration depth $\lambda$, the effective penetration depth for perpendicular films is:

$$\Lambda = \frac{2\lambda^2}{d} \tag{3.1}$$

If the sheet resistance is $R_n$, it can be expressed as:

$$\Lambda = \frac{\Phi_o^2}{2\pi^4} \frac{R_n}{\hbar/e^2} \left[\Delta(T) \tanh\{\tfrac{1}{2}\Delta(T)/k_B T\}\right]^{-1} \tag{3.2}$$

where $\Phi_o = hc/2e$ is the flux quantum and $\Delta$ the energy gap. For films with a sheet resistance of 1 kΩ or more, which will turn out to render the relevant samples, $\Lambda$ is of the order of millimeters, even at low temperatures.

The current density in a vortex is proportional to $r_1^{-1}$ when the distance from the vortex centre $r_1$ is smaller than $\Lambda$, but proportional to $r_1^{-2}$ for $r_1 \gg \Lambda$. As a consequence the vortex-antivortex interaction depends logarithmically on the separation r only for $r \ll \Lambda$. As $\Lambda$ is usually larger than the sample size, the latter dimension plays a more limiting role. In the logarithmic regime, $r \ll \Lambda$, the vortex-antivortex interaction constant is:

$$A(T) = \frac{\Phi_o^2}{4\pi^2\Lambda(T)} = \tfrac{1}{2}\pi^2 \frac{\hbar/e^2}{R_n} \Delta(T) \tanh\{\tfrac{1}{2}\Delta(T)/k_BT\} \qquad (3.3)$$

The critical temperature follows from $k_BT_c = A(T_c)/(4\varepsilon_c)$. The expression for $T_c$ in superconducting films, assuming $\varepsilon_c = 1$ was first derived by Beasley, Mooij and Orlando (BMO).[2] $T_c$ is only significantly below $T_{co}$ in films of which the sheet resistance is of the order of 1 kΩ or higher. The value of $\Lambda$ at $T_c$ follows directly from Eqs.(3.3) and (2.13):

$$\Lambda(T_c) = \Phi_o^2/\{4\pi^2A(T_c)\} = \Phi_o^2/(16\pi^2\varepsilon_c k_BT_c) \qquad (3.4)$$

BMO have pointed out that $\Lambda(T_c)$ is equal to about 2 cm devided by the value of $T_c$ in K.

In practice the interesting temperature region is close to the BCS critical temperature $T_{co}$. In this Ginzburg-Landau regime where $\Delta(T)/k_BT \ll 1$, the interaction parameter A(T) is:

$$A(T) = 23.1 \frac{\hbar/e^2}{R_n} k_B(T_{co}-T) \qquad 1 - T/T_{co} \ll 1 \qquad (3.5)$$

and Eq.(2.13) yields the modified BMO expression:

$$\frac{T_c}{T_{co}} = \left[1 + 0.173\,\varepsilon_c \frac{R_n}{\hbar/e^2}\right]^{-1} \qquad (3.6)$$

3.2. Core energy

The size of a vortex core is close to the Ginzburg-Landau coherence length $\xi$. In a London-type model of a vortex, the core is approximated by a normal cylinder with radius $\xi$. The associated loss of condensation energy is the volume $\pi\xi^2d$ times $H_c^2/8\pi$ where $H_c$ is the thermodynamic critical field. One finds, using the simple relation $\Phi_o = 2\sqrt{2}\pi\xi\lambda H_c$, that this part of the core energy is equal to $A(T)/8$. In a more sophisticated approach, the energy associated with the gradient of the order parameter must be added and a higher value might be expected.

However, it is not sufficient to know the core energy of a single isolated vortex, in order to know the vortex potential $\mu_c$, as used in chapter 2. We need the free energy of a pair with smallest separation $a_o$, that behaves according to the expressions used to derive the recursion relations. In the renormalization procedure, the presence of small pairs has a screening effect on pairs with large separation. Pairs that have no polarizability in the 2D Coulomb gas analogy are irrelevant. To determine $a_o$ for vortex pairs in superconducting films, one needs to calculate the 'polarizability' for small separation as well as the free energy. Clem[19] has shown that, apart from the energy associated with condensation and with the gradient of the order parameter, the vortex energy is proportional to $\ln(r/\xi)$, even for very small values of the separation r. If the smallest polarizable pair has a separation $p\xi$, a contribution $\frac{1}{2}A(T)\ln p$ has to be added to the core energy.

No theoretical estimate is available at this time for the core potential of vortex pairs in superconducting films. As the energy terms can be expected to be proportional to A(T), it is reasonable to assume that Eq.(2.14) holds, with unknown $\gamma$ and $N_o$. One may expect that for all films that are homogeneous on a scale down to $\xi$, the same values of $\gamma$, $N_o$ and consequently $\varepsilon_c$ should be found in experiment. Inhomogeneities on this scale, as may be caused by grain boundaries or percolative structures, break this universality.

In the following, $a_o$ is assumed to be $\xi$ in the relations between scale and separation.

3.3. General aspects

In the discussion on thin films, it is assumed that the Ginzburg-Landau approximations are valid. When Eq.(3.5) is used, the temperature parameter $\tau'$, as defined in Eq.(2.18) is:

$$\tau' = (T - T_c)/T_{co} - T) \qquad (3.7)$$

Much of what is discussed in this section 3 originates from the work of Halperin and Nelson[4] (HN), who adapted the calculations on helium films of Ambegaokar, Halperin, Nelson and Siggia[6-8] (AHNS) to superconducting films. HN take the interaction A as constant in the critical region around $T_c$. They define $\tau_c$ as:

$$\tau_c = (T_{co} - T_c)/T_c \qquad (3.8)$$

and $\tau$ as:

$$\tau = (T - T_c)/T_c \qquad (3.9)$$

For $|\tau| << \tau_c$, to a good approximation:

$$\tau' \approx \tau/\tau_c = (T - T_c)/(T_{co} - T_c) \qquad (3.10)$$

There is a direct relation between A(T) and the two-dimensional density of superconducting electrons $n_s$:

$$A(T) = \tfrac{1}{2}\pi\hbar^2 n_s/m \tag{3.11}$$

where m is the electron mass. Sometimes the effective density of superconducting electrons is put equal to $A(T)/\varepsilon(\ell)$. The effective superfluid density for infinite scale just below $T_c$ follows from:

$$\tfrac{1}{2}\pi\hbar^2 (n_s)_{eff}/m = A(T_c)/\varepsilon_c = 4k_B T_c \tag{3.12}$$

Just above $T_c$, the effective density for infinite scale is zero. This is called the jump in the superfluid density.[18] In this chapter A(T) is used instead of $n_s$.

3.4. Resistivity

The dc resistance in films is connected with the presence of unpaired vortices. They can be 'free' vortices, present in thermodynamic equilibrium due to unbinding of vortex pairs, or induced by an external field. Pinning is supposed to be absent. Considering that the core condensation energy A(T)/8 is smaller that $k_B T$ near $T_c$, this seems a reasonable assumption unless the film is very inhomogeneous. The areal density of vortices is:

$$n_v = n_f + n_H \tag{3.13}$$

where $n_f$ and $n_H$ are the areal densities of free and field-induced vortices respectively. If H is the perpendicular component of the external field, $n_H \approx H/\Phi_o$.

The sheet resistance of the film, normalized to the normal state sheet resistance $R_n$ is:

$$R_s/R_n = 2\pi\xi^2 n_v \tag{3.14}$$

When the assumption is made that the mean distance between single vortices $r_m$ is connected to $n_v$ as in a triangular lattice, $R_s/R_n = 7.3\ (\xi/r_m)^2$. In Fig.5 a plot is made of $R_s/R_n$ versus $r_m$ for $\xi$ = 30 nm. In the figure, $H_{eq}$ on the right hand scale indicates the equivalent field that gives rise to the same vortex density. To observe a resistance around $10^{-7}$ times the normal state resistance due to free vortices, the external field must be smaller than $10^{-4}$ Oersted. At this resistance level, the mean distance $r_m$ corresponds to a scale $\ln(r_m/\xi)$ of about 9.

In an infinite sample in zero field, neglecting the finiteness of $\Lambda$, no free vortices are present below $T_c$. Above $T_c$, the vortex density according to HN is:

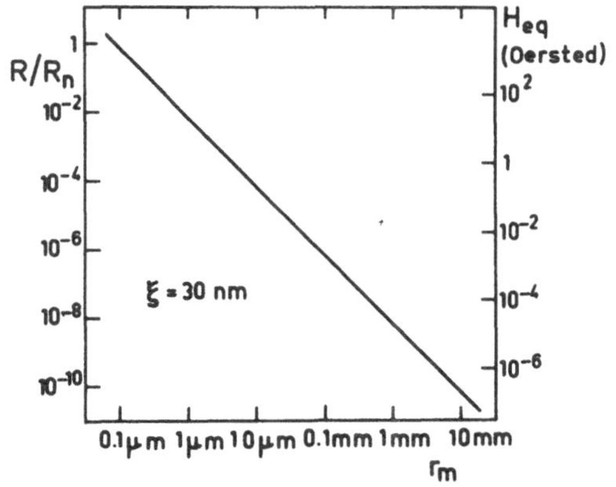


Fig. 5. Relation between the normalized resistance and the mean distance between vortices. On the right the value of the perpendicular external magnetic field that gives rise to the same resistance is indicated.

$$n_f = C_1/(2\pi\xi_+^2) \tag{3.15}$$

where $C_1$ is an unspecified constant of order unity and $\xi_+$ is defined by Eq.(2.29). If $\varepsilon_c-1$ is small enough that the approximations of section 2.6 for $x \ll 1$ are valid, $\xi_+$ is given by Eq.(2.32). If in addition $\tau'$ is extremely small, the relation (2.33) may be used leading to:

$$R_s/R_n = C_1 \exp(-2\ell_+) \approx C_1 \exp\{-2(b/\tau')^{\frac{1}{2}}\} \tag{3.16}$$

This relation, which predicts a linear relation between ln $R_s$ and $(T-T_c)^{\frac{1}{2}}$, is often quoted as the expected behaviour of the resistance in the critical region. However, Eq.(3.16) is only valid when $\frac{1}{2}\pi(\tau'/b)^{\frac{1}{2}}$ is much smaller than $|x_o|$, when $|x_o|$ is small enough that the approximate solutions of section 2.6 apply and moreover $\xi_+$ is smaller than W. Together:

$$\tfrac{1}{2}\pi\ell_W^{-1} < \tfrac{1}{2}\pi(\tau'/b)^{\frac{1}{2}} \ll |x_o| \ll 1 \tag{3.17}$$

As $\ell_W$ has a maximum value of about 14, it is impossible to satisfy all requirements simultaneously.

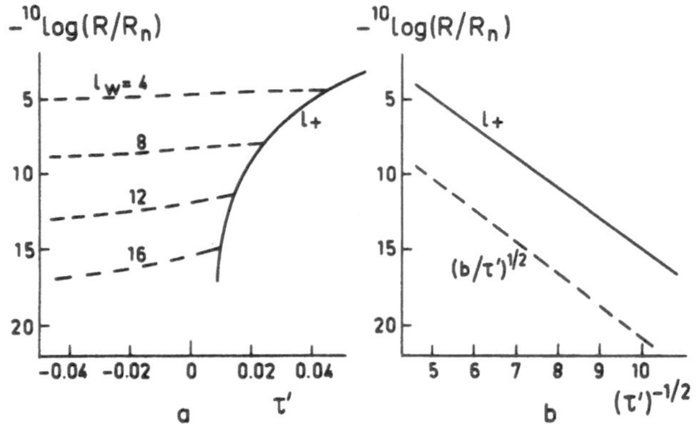


Fig. 6. Temperature dependence of resistance. The drawn line indicated with $\ell_+$ gives the result of Eq.(3.19) for the model system of section 2.7. In a the influence of finite size according to Eq.(3.20) is shown. In b the dashed line shows the result of Eq.(3.16).

It is possible to obtain a better approximation with wider applicability from Eq.(2.32). When the $tg^{-1}$ function is not neglected but approximated by its argument, one has:

$$\ell_+ \approx (b/\tau')^{\frac{1}{2}} - |x_0|^{-1}$$

According to Eq.(2.19), for small $\tau'$ it is found that $|x_0|^{-1} \approx \varepsilon_c(\varepsilon_c-1)^{-1}$. Consequently:

$$\ell_+ \approx (b/\tau')^{\frac{1}{2}} - \varepsilon_c(\varepsilon_c-1)^{-1} \qquad \tau', |x_0| << 1 \qquad (3.18)$$

For the resistance this renders:

$$R_s/R_n \approx C_1 \exp\{2\varepsilon_c(\varepsilon_c-1)^{-1}\} \exp\{-2(b/\tau')^{\frac{1}{2}}\} \qquad (3.19)$$

For $\varepsilon_c$ around 1.2, Eq.(3.19) predicts a resistance that is $10^5$ higher than as predicted by Eq.(3.16). The temperature dependence is the same.

Another factor of importance is the sample size. In a finite sample free vortices are present even below $T_c$ and there is no true phase transition. In a superconducting film $\Lambda$ provides an upper limit to the useful sample size. For high resistivity films, $\xi$ is of the order of 10 to 100 nm and the maximum useful scale about 14. The minimum of sample size and $\Lambda$ is indicated with W, the corresponding scale with $\ell_W$. HN give the free vortex density in a finite sample as $nf = W^{-2}\ y(\ell_W)$, which leads to:

$$R_s/R_n = 2\pi y(\ell_W)\ \exp(-2\ell_W) \qquad W < \xi_+ \tag{3.20}$$

In Fig.6 the resistance, as calculated with Eq.(3.19), is shown, as indicated with $\ell_+$. In Fig.6a the plot is linear against $\tau'$ and the influence of finite $\ell_W$ is indicated for several values of $\ell_W$. Fig.6b contains the resistance against $(\tau')^{-\frac{1}{2}}$. With a dashed line the result of Eq.(3.16) is given.

The discussion is limited to the free vortices here. If an external field is present, the effects as discussed will only be seen if $n_f > n_H$.

## 3.5. Current-induced vortex unbinding

From a transport current through the film, vortices experience a force which is perpendicular to the current. The force has opposite directions for the two partners of a vortex-antivortex pair, which favours a specific orientation of the pair. In this direction with lowest energy, unbinding is easier than without the transport current. A reference sheet current density is defined as:

$$J_o = A(T)\ h/(2e\xi) \tag{3.21}$$

$J_o$ is equal to 2.60 times the Ginzburg-Landau critical current. The derivative with respect to $\ell$ of the pair potential at separation $\xi\exp\ell$, for an angle $\theta$ between the pair orientation and the current is:

$$\frac{\partial U(\ell,\theta)}{\partial \ell} = \frac{A(T)}{\varepsilon(\ell)}\left[1 - (J_s/J_o)\exp\ell\ \sin\theta\right] \tag{3.22}$$

In the direction $\theta = \pi/2$ the potential has a minimum. There a saddle point is found for a scale:

$$\ell_j = \ln(J_o/J_s) \tag{3.23}$$

with corresponding separation $r_j = \xi(J_o/J_s)$. The saddle point energy is:

$$U_{sp} = 2\mu_c + \frac{A(T)}{\varepsilon(\ell_j)}(\ell_j - 1) \tag{3.24}$$

Results for the voltage V in the nonlinear regime have been

given by HN. Current-induced vortex unbinding can best be studied below $T_c$ where there are few free vortices without the current. The expression for $V(J_s)$ is given in the two limits where $\ell_j$ is either much smaller or much larger than the characteristic scale $\ell_-$.

$$V = 2J_oR_n\{a(T)-3\}\ (J_s/J_o)^{a(T)} \tag{3.25}$$

with

$$a(T) = 3 + \tfrac{1}{2}\ell_-^{-1} \qquad \ell_j >> \ell_- \tag{3.26a}$$

$$a(T) = 3 + \tfrac{1}{2}\ell_j^{-1} \qquad \ell_j << \ell_- \tag{3.26b}$$

For larger currents and consequently smaller $\ell_j$ it is found, making use of the fact that $J_s/J_o = \exp(-\ell_j)$, that:

$$V = 0.61\ J_oR_n(J_s/J_o)^3\{\ln(J_o/J_s)\}^{-1} \qquad \ell_j << \ell_- \tag{3.27}$$

For small current values, on the other hand, using Eq.(2.28) one finds near $T_c$:

$$a(T) \approx 3 + \pi(b/\tau')^{-\frac{1}{2}} \tag{3.28}$$

and one might hope to see the square-root cusp behaviour. However, near $T_c$ the characteristic length $\xi_-$ is large and the condition $\ell_j >> \ell_-$ is difficult to satisfy.

A different evaluation, also based on the work of AHNS, has been given by Kadin, Epstein and Goldman.[12] The unbinding process can be seen as thermally activated excitation across the saddle point energy (3.24). The density of free vortices due to this process is proportional to the square root of the excitation rate.[9] The resultant value of the exponent a in the relation $V = I^a$ is now found to be a simple function of the Boltzmann factor $U_{sp}/k_BT$. It can be written as:

$$a(\tau') = 1 + \pi K(\ell_j, \tau') \tag{3.29}$$

where $K(\ell)$ is the reduced stiffness constant defined in Eq.(2.4), equal to $(2/\pi)\{x(\ell)+ 1\}^{-1}$. The assumption $\ell_j >> 1$ has been used. Eq.(3.29) shows that measuring $a(\tau')$ is a very direct way of probing the renormalized vortex interaction. As $\ell_j$ depends on the current level, the measurements at small currents are the most interesting. Kadin et al. have calculated the behaviour of K as a function of $\tau$ for different values of $\ell$. Their result is reproduced in Fig.7. The dashed line for $\ell = \infty$ shows the jump from $\pi K = 2$ to zero. One should realize that at all times one needs $\ell_j < \ell_+$ and $\ell_j < \ell_W$.

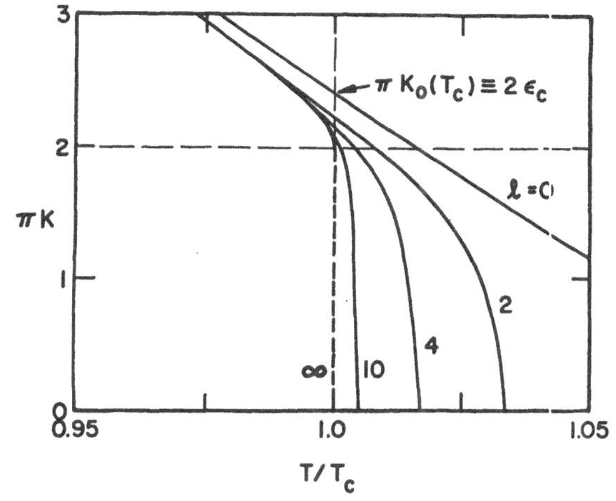


Fig. 7. Reduced stiffness constant K, directly related to the exponent a with Eq.(3.29), as a function of temperature for different values of $\ell_j$. Reproduced from Kadin et al.[12]

## 3.6. Dynamic behaviour

In the absence of vortices, a superconducting film shows a purely inductive response to time-varying fields. The geometric inductance and the capacitance are small, important is the kinetic inductance per square:

$$L_{KO} = 2\pi\Lambda/c^2 = \Phi_o^2/\{2\pi c^2 A(T)\} \quad (3.30)$$

The influence of bound pairs and free vortices can be accounted for by a complex 'dielectric function' $\varepsilon_v(\omega)$ which relates the complex impedance of the film per square Z to $L_{KO}$:

$$Z = i\omega\, L_{KO}\, \varepsilon_v(\omega) \quad (3.31)$$

The dielectric function can be separated in a part representing the bound pairs $\varepsilon_{vb}$ and a part representing the free vortices which is imaginary and therefore dissipative:

$$\varepsilon_v(\omega) = \varepsilon_{vb}(\omega) + 2\pi\xi^2 R_n n_v/(i\omega L_{KO}) \quad (3.32)$$

$n_v$ is the density of vortices, equal to the sum of 'free' vortices $n_f$ and field-induced vortices $n_H$, if present.

The response of bound pairs is strongly dependent on the separation. A pair with large r cannot adjust to the field if the period is small. An important parameter is the diffusion constant:

$$D = \mu\, k_B T \tag{3.33}$$

where $\mu = 2e^2\xi^2 R_n/(\pi\hbar^2)$ is the mobility. Pairs with separation smaller than a special value $r_\omega$ respond to a field with frequency $\omega$, larger pairs do not. The frequency-dependent characteristic separation $r_\omega$ is given by:

$$r_\omega = (14D/\omega)^{\frac{1}{2}} \tag{3.34}$$

The corresponding scale $\ln(r_\omega/\xi)$ is called $\ell_\omega$. In a film with a sheet resistance of 4 kΩ and a coherence length of 10 nm at a temperature of 2 K, D is equal to 0.16 $cm^2$/s and $r_\omega$ is equal to 6.0 μm for 1 MHz frequency, $\ell_\omega$ is 6.4.

The contribution of the bound pairs is given as an integral over pair separations:

$$\varepsilon_{vb}(\omega) = 1 + \int_0^{\ell_m} dr \left[\frac{d\varepsilon(\ell)}{d\ell}\, g(\ell,\omega)\right] \tag{3.35}$$

In this expression $\varepsilon(\ell)$ is the static scale dependent 'dielectric constant', introduced in Eq.(2.3). $\ell_m$ is the maximum scale that is available for bound pairs which is:

$$\begin{aligned} \ell_m &= \ell_W & T &< T_c \\ \ell_m &= \min(\ell_W, \ell_+) & T &> T_c \end{aligned} \tag{3.36}$$

The response function $g(\ell,\omega)$ is equal to one for local equilibrium (very low frequencies). In general, according to AHNS , two scales enter in $g(\ell,\omega)$: $\ell_\omega$ and $\ell_-$. For the case that $\ell_\omega \gg \ell_-$, the response function can be approximated by:

$$g(r,\omega) \approx (1 - i\omega r^2/r_\omega^2)^{-1} \tag{3.37}$$

Making use of the fact that $\varepsilon(\ell)$ varies slowly with $\ell$, Eq.(3.35) is then found to give:

$$\begin{aligned} \mathrm{Re}\left[\varepsilon_{vb}(\omega)\right] &= \varepsilon(\ell_\omega) \\ \mathrm{Im}\left[\varepsilon_{vb}(\omega)\right] &= \frac{\pi}{4}\left[\frac{d\varepsilon(\ell)}{d\ell}\right]_{\ell=\ell_\omega} \end{aligned} \tag{3.38}$$

With Eq.(2.16), $\varepsilon(\ell) = \varepsilon_c\{1 + x(\ell)\}$, and the recursion relation Eq.(2.7a) the factor $d\varepsilon/d\ell$ can be expressed in the activity $y(\ell)$. The complex impedance is found to be:

$$\begin{aligned} \mathrm{Re}Z &= 2\pi^3\varepsilon_c\omega L_{KO}\, y^2(\ell_\omega) + 2\pi\xi^2 R_n n_v \\ \mathrm{Im}Z &= \varepsilon_c\omega L_{KO}\{1 + x(\ell_\omega)\} \end{aligned} \tag{3.39}$$

under the conditions:

$$\begin{aligned} &0 < \ell_\omega < \ell_m \\ &\ell_\omega \ll \ell_- \text{ for } T < T_c \end{aligned} \tag{3.40}$$

Clearly, the complex impedance at frequency $\omega$ is a probe of the properties at a scale $\ell_\omega$. If $\ell_\omega$ falls outside the range of the integral in Eq.(3.35), the previous results do not apply. If $r_\omega$ is smaller than $\xi$, clearly properties are probed on a too microscopic scale to be relevant. However, $r_\omega > L_m$ is a very interesting situation. Assume the sample size W and the frequency are small enough that $r_\omega \gg W$. Now $g(\ell,\omega)$ is equal to 1, and $\varepsilon_{vb}(\omega)$ equal to $\varepsilon(\ell_m)$ This gives:

$$\begin{aligned} \mathrm{Re}Z &= 2\pi\xi^2 R_n n_v \\ \mathrm{Im}Z &= \varepsilon_c\omega L_{KO}\{1 + x(\ell_m)\} \end{aligned} \qquad r_\omega \gg W \tag{3.41}$$

One should realize that all expressions are temperature dependent in a complicated way. In (3.39) and (3.41) $x, y, n_v$ as well as $L_{KO}$, $\xi$, $\ell_\omega$ and $\ell_m$ vary with temperature.

## 4. EXPERIMENTAL RESULTS ON FILMS

### 4.1. Determination of background parameters

For studies of the two-dimensional transition films must exhibit a high sheet resistance, preferably more than 1 kΩ. The films used are either granular, percolative or amorphous. A high degree of homgeneity is required, in the first place on a macroscopic scale because variation of properties across the sample leads to smearing of the transition. Homogeneity on a microscopic scale, i.e. the scale of $\xi$, is required to prevent pinning and to allow use of the theoretical expressions. Inhomogeneity on the scale of $\xi$ will not prevent the system from showing the type of phenomena discussed. However, as the core energy will be very different in relation to the interaction constant A(T), compared with a purely homogeneous system, and strongly dependent on the type and scale of the inhomogeneities, non-universal behaviour must be expected. This will be true with respect to Kosterlitz-Thouless transition phenomena as

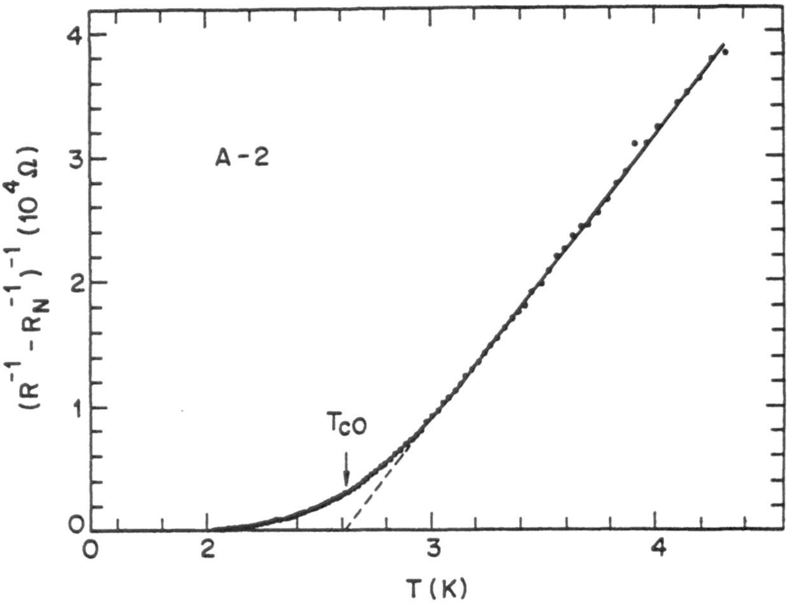


Fig. 8. Determination of $T_{co}$ from fluctuation-enhanced conductivity at high temperatures, Fiory et al.[11]

well as to the more general Coulomb gas approach, as discussed by Minnhagen. It is certainly not an easy task to make homogeneous films with high sheet resistance. It is also difficult to analyze the degree to which one has succeeded in this respect.

For the analysis of the behaviour around $T_c$, knowledge of the basic superconducting properties in that temperature range is required. $T_{co}$ can be determined in different ways. The resistive transition is continuous from the low temperature side, dominated by fluctuations of the phase, to the high temperature side, dominated by fluctuations of the modulus of the order parameter. The transition has no special feature at $T_{co}$. It is possible to find $T_{co}$ from a fit of the high-temperature tail to fluctuation theory. For high resistance films the Aslamazov-Larkin theory without pair breaking corrections is adequate. This theory gives in two dimensions for the fluctuation-enhanced conductivity:

$$\sigma = \sigma_n + \frac{e^2}{16\hbar d}\,\frac{T}{T - T_{co}}$$

or by multiplying with d and rearranging:

$$(R_s^{-1} - R_n^{-1})^{-1} = R_o(1 - T_{co}/T) \approx R_o(T/T_{co} - 1) \qquad (4.1)$$

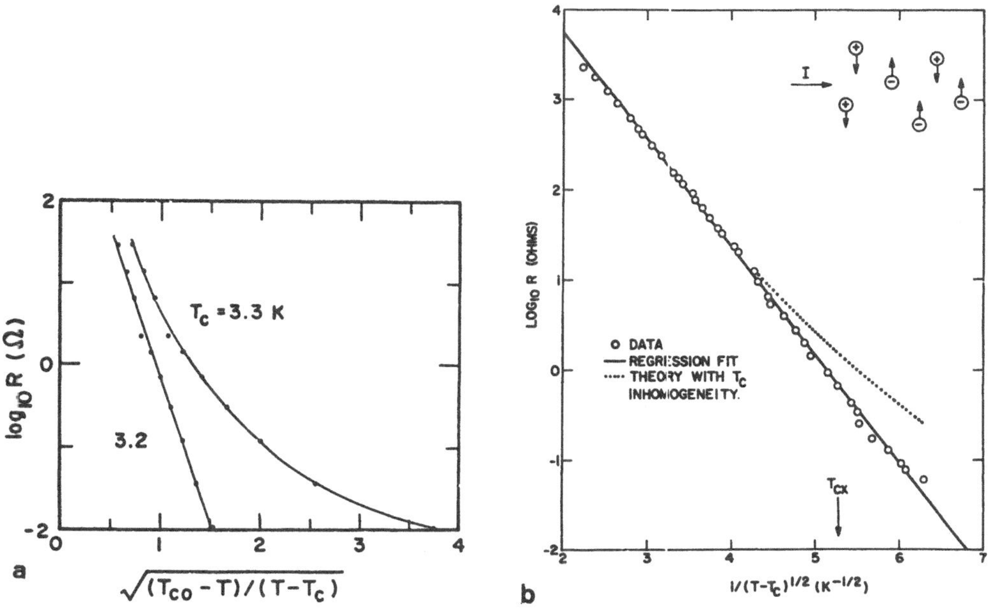


Fig. 9. Resistance measurements. a. Results of Kadin et al.[12] on a film of Hg-Xe. b. Resulsts of Hebard and Fiory[10] on a film of indium/indium oxide.

$R_o = 16\hbar/e^2$ is equal to 66 kΩ·per square. Fiory et al.[11] have used this method to determine $T_{co}$, in Fig.8 an example is shown.

Both current-induced vortex unbinding and the kinetic inductance provide additional methods to measure $T_{co}$. Outside the critical region a(T), the exponent in $V = I^a$, and $L_K^{-1}$, the inverse kinetic inductance, are proportional to $T_{co}-T$, as long as the Ginzburg-Landau approximations are valid. Extrapolation of low temperature data to higher temperatures gives the value of $T_{co}$.

The kinetic inductance also provides Λ and A(T), as Eq.(3.30) shows. The coherence length ξ can be determined from the low current resistivity in a known external field. Determination of the normal state resistance in the temperature region around $T_c$ is not a trivial matter. For high resistance films there are several mechanism that may lead to temperature dependence of $R_n$.

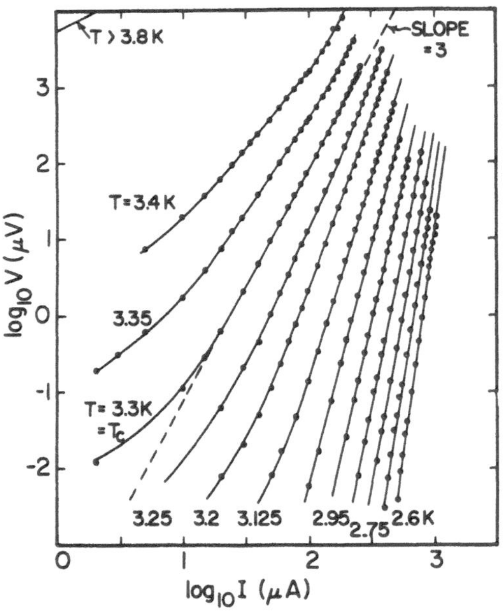


Fig. 10. Nonlinear resistance. Measurements of Epstein et al.[21]

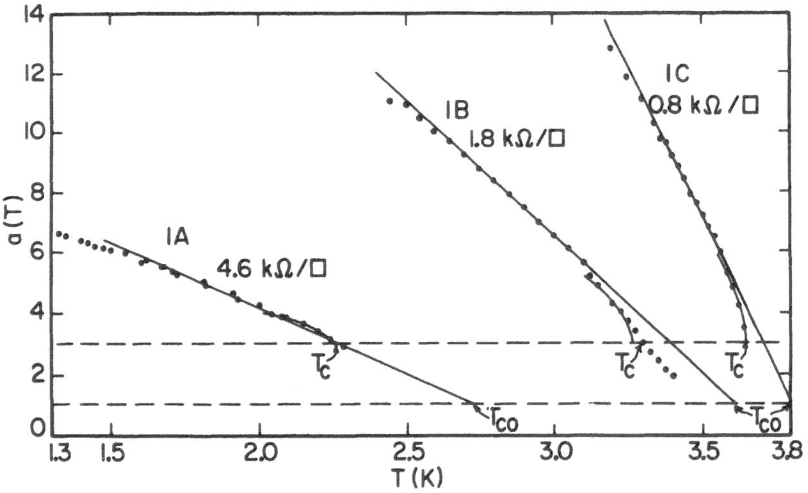


Fig. 11. Results of Epstein et al.[21] for the exponent of a(T) in $V \propto I^a$, as measured on three different films.

## 4.2. Resistive transition

Experimental values of the linear resistance above $T_c$ are usually compared with the theoretical expression (3.16). It can, using Eq.(3.7), be rewritten as:

$$\ln R_s/R_n = \ln C_1 - 2\{b(T_{co}-T)/(T-T_c)\}^{\frac{1}{2}} \quad (4.2)$$

Very close to $T_c$, when $T-T_c \ll T_{co}-T_c$, Eq.(3.10) can be used and one obtains:

$$\ln R_s/R_n = \ln C_1 - 2\{b(T_{co}-T_c)/(T-T_c)\}^{\frac{1}{2}} \quad (4.3)$$

Both equations are only relevant if the extreme conditions (3.17) would be satisfied. Nevertheless, a reasonable fit is not too difficult to obtain, due to the many unknown parameters. In Fig.9 two sets of data are shown. Fig.9a is taken from Kadin et al.[12] and shows results for a film of Hg - Xe, plotted for comparison with Eq.(4.2). The value of $T_{co}$ was known from other measurements, these same measurements suggested 3.3 K for $T_c$. However, as Fig.9a shows, with $T_c$ = 3.3 K no fit to Eq.(4.2) is possible while $T_c$ = 3.2 K yields the expected straight line. The value of b that corresponds to the slope observed is 16.6, which is a rather high value.

The data of Hebard and Fiory[10] in Fig.9b are compared with Eq.(4.3). They refer to a film of indium/indium oxide with a sheet resistance of about 20 kΩ. The temperature range extends from near $T_c$ to near $T_{co}$ and $T-T_c$ is not small with respect to $T_{co}-T_c$ everywhere. Nevertheless, a remarkably straight line is obtained. The value of b that is found is 6.3. A fit to Eq.(4.2) would yield a different value.

In section 3.4 it has been pointed out that Eq.(3.16) cannot be valid for practical films. However, Eq.(3.19) with a wider range of relevance shows the same functional dependence of $R_s/R_n$ on τ'. In Eq.(4.3) a constant term has to be added to $\ln C_1$. A straight line in the plot of $\ln(R_s/R_n)$ versus $(\tau')^{-\frac{1}{2}}$ is to be expected and the value of b obtained from the slope should be correct.

## 4.3. Nonlinear resistance

The resistance measurements of the previous section must be performed with a small enough measuring current that no current-induced unbinding may occur. For larger currents the I-V characteristic is nonlinear. Studies of the nonlinear resistance have been performed in particular by Epstein, Goldman and Kadin[20-22,12] and recently also in detail by Hebard and Fiory.[10,11]

In Fig.10 a set of I-V curves is given, as published by Epstein

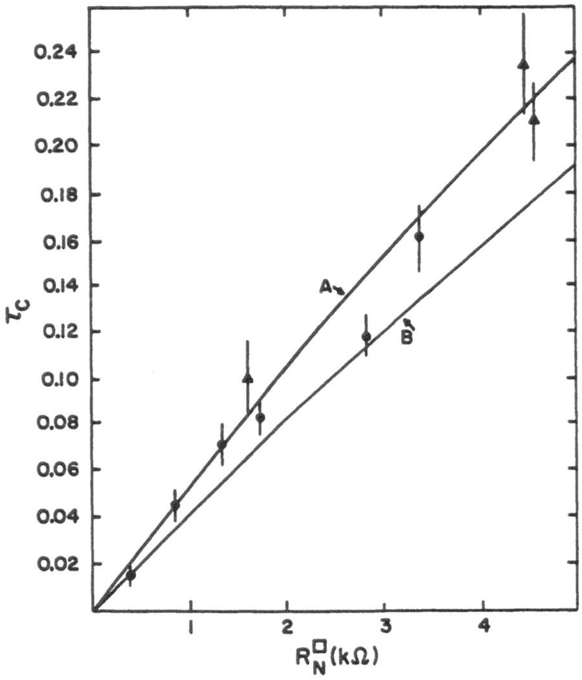


Fig. 12. Relation between $T_c$, $T_{co}$ and $R_n$ as determined by Kadin et al.[12] $\tau_c = (T_{co}-T_c)/T_c$. The drawn lines are theoretical predictions with $\varepsilon_c = 1.2$ (curve A) and $\varepsilon_c = 1$ (curve B).

et al.[21] The lines in the logV-logI plot are straight over a reasonably long range, from their slope the exponent a in $V \propto I^a$ can directly be determined. Fig.11 shows the results for a(T) for three different samples with different sheet resistance. This figure also illustrates how $T_{co}$ and $T_c$ are obtained. The low temperature values lie on a more or less straight line that is extrapolated towards higher temperatures. Eq.(3.29) indicates that without vortices, when K is proportional to A(T) and consequently to $T_{co}$-T, a linear dependence of a on T is expected and that the value at $T_{co}$ would be 1. In the region immediately below $T_c$, a(T) is expected to deviate from the linear behaviour, following Eq(3.28). Where a(T) is equal to 3, the temperature is equal to $T_c$. As discussed in section 3.5, Eq.(3.28) is only valid as long as $r_j >> \xi_-$. Near $T_c$, $\xi_-$ is large and one will probably switch from Eq.(3.26a) to Eq.(3.26b). The approach of Kadin et al.[12], using Eq.(3.29) rather than Eq.(3.26) or (3.28), is probably required. It is unlikely that a theoretical expression for a(T) can be obtained for finite currents other than by numerical integration.

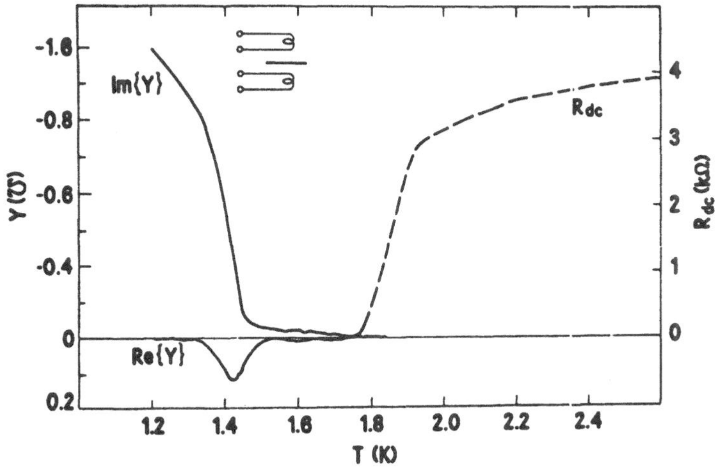


Fig. 13. Imaginary and real parts of the complex impedance of a granular aluminum film at 17.5 MHz, as measured by Hebard and Fiory.[23] The dashed line gives the DC resistance.

Kadin et al.[12] have determined $T_c$, $T_{co}$ and $R_n$ for a number of Hg-Xe samples. The results are represented in Fig.12. It shows $\tau_c = (T_{co}-T_c)/T_c$ versus $R_n$. The drawn line B indicates the theoretical prediction of Beasley et al.[2], for $\varepsilon_c = 1$. Eq.(3.6) is the Ginzburg-Landau limit of that prediction. Curve A is the prediction for $\varepsilon_c = 1.2$ and as shown is in very good agreement with the experimental data points.

Hebard and Fiory[10] have derived $T_c$ from their measurements of the non-linear resistance by observing a distinct qualitative change of the logV-logI plot. Below $T_c$ the lines are straight, above $T_c$ curvature sets in. This value of $T_c$ has been compared with values, obtained with different methods, and is found to agree.

## 4.4. Measurements of AC impedance

Dynamic measurements on films in the critical temperature region have been performed by Hebard and Fiory, first on films of granular aluminum[23], recently on films of indium/indium oxide.[11] In section 3.6 theoretical aspects have been discussed. Measurements at higher frequencies are performed by placing the sample between generator and detector coils and measuring the shielding properties of the film. At low frequencies measurements with contacts as in DC measurements are also possible.

Fig.13 shows the complex admittance $Y = Z^{-1}$ of a granular aluminum film with a sheet resistance of 4.1 kΩ. Structure is

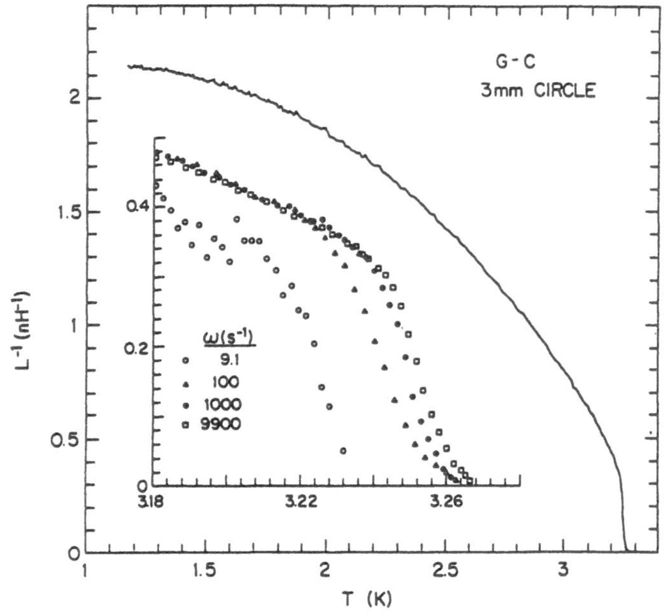


Fig. 14. Inverse kinetic inductance of a film of indium/indium oxide according to Fiory et al.[11] The drawn line is the result of measurements at $\omega = 10^3\ s^{-1}$. In the inset the transition near $T_c$ is shown for four different frequencies.

visible in both Im(Y) and Re(Y) at temperatures clearly below $T_{co}$. The latter quantity can be estimated from the DC resistance, also shown in the figure. In these measurements the expression for ReZ and ImZ of Eq.(3.39) can be expected to apply. ReZ and ReY = ReZ/$\{$ReZ)$^2$+(ImZ)$^2\}$ are zero without bound pairs and free vortices. As soon as a significant number of bound pairs of scale $\ell_\omega$ are present, ReZ and ReY start growing. ReY decreases again when ImZ becomes very large and the peak in ReY of Fig.13 is obtained. As for ImY, without vortices it is directly proportional to $L_{KO}^{-1}$ and A(T). In the Ginzburg-Landau regime it should vary linearly with $T_{co}$-T. In the presence of bound pairs of scale $\ell_\omega$, ImY decreases faster with increasing temperature. In fact, as ImZ is proportional to $A(T)^{-1}\varepsilon(\ell_\omega)$ according to Eqs.(2.16) and (3.30), as long as ImZ >> ReZ one has:

$$\text{ImY} \propto \frac{A(T)}{\varepsilon(\ell_\omega)} \propto k_B T\ K(\ell_\omega) \qquad (4.4)$$

where K is the temperature and scale dependent reduced stiffness constant of Eq.(2.4). The curve for ImY in Fig.13 should be compared with the plot of K versus T at different values of $\ell$ of Fig.7. The correspondence is clear.

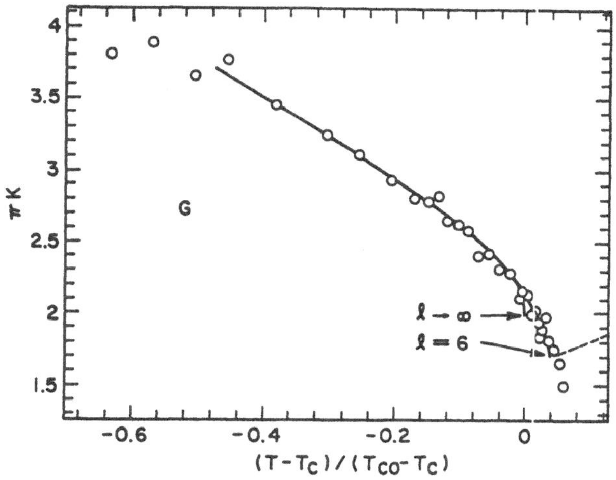


Fig. 15. Reduced stiffness constant, derived from the kinetic inductance, of a sample of indium/indium oxide according to Fiory et al.[11] The drawn lines indicate theory fits for scale $\ell = \infty$ and $\ell = 6$. The dashed line indicates $\ell_+$.

In Fig.14 the inverse kinetic inductance, equivalent to ImY, is shown as measured by Fiory et al.[11] on a film of indium/indium oxide with a sheet resistance of about 1.2 kΩ. The critical region has been enlarged in the inset, results at four different measuring frequencies are shown. The lower the frequency, the higher the corresponding value of $\ell_\omega$. As Fig.7 illustrates, the transition to small K occurs at a lower temperature for larger $\ell$. This is exactly what is observed.

When $\ell_\omega >> \ell_W$, or the characteristic diffusion length for frequency $\omega$ larger than the sample size, the frequency dependence should disappear. The results of Eq.(3.41) must be used now. Fiory et al.[11] have performed a series of measurements on relatively narrow samples where this regime is applicable. Fig.15 shows the results for one sample, where the imaginary component of the impedance has been worked out for a comparison with theory. Values of $T_c$ and $T_{co}$, obtained from these and other measurements on this sample, have been used to determine the temperature scale which is about equal to $\tau'$. The drawn lines are theory predictions for $\ell = \infty$ and $\ell = 6$. Here the scale is determined by the sample size and not by $\ell_\omega$.

## 4.5. Analysis of Fiory, Hebard and Glaberson

In the previous sections, experimental results obtained by Hebard and Fiory[10] and Fiory, Hebard and Glaberson[11] on films of

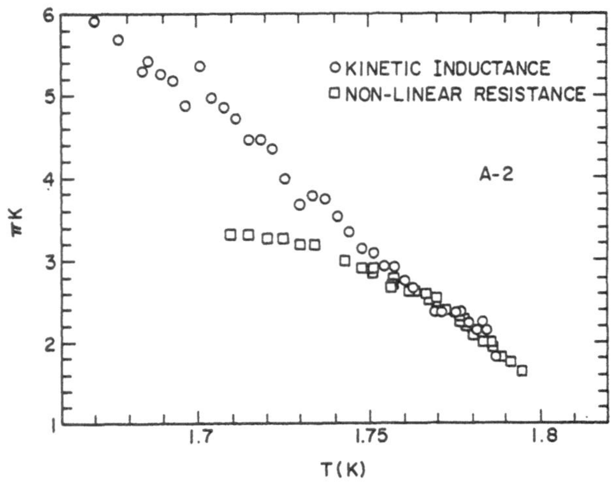


Fig. 16. Reduced stiffness constant as derived from the kinetic inductance (circles) and the nonlinear resistance (squares), from Fiory et al.[11]

indium/indium oxide have been mentioned. These authors have analyzed their data in a complete and detailed way that is far beyond any other work in this field. It is impossible to do justice to their analysis in this review and a study of the original publications is recommended. Three different samples have been investigated, on all three the kinetic inductance, the nonlinear resistance and the low-current linear resistance in zero and finite field have been determined. From the data $\xi$, $L_{KO}$, $R_n$ and $T_{co}$ have been derived which are required for the analysis of the critical behaviour.

An example of the data is shown in Fig.16. Here results for the reduced stiffness constant K in one sample, obtained from measurements of kinetic inductance as well as nonlinear resistance have been brought together. In particular at the higher temperatures near $T_c$ (1.78 K for this sample) very good agreement is obtained.

Fiory et al. compare their results with the theory including the Kosterlitz recursion relations. They do not use the approximations described in section 2.6, but use numerical integration as a function of $\ell$ at each temperature. As discussed in section 2, it is in all cases first necessary to know the properties at zero scale which are represented as $y_o$ and $x_o$. The zero-scale properties are determined by the core potential, which in this paper is represented as $\mu_c = \gamma A(T) - k_B T \ln N_o$, see Eq.(2.14). Fiory et al. use starting conditions that are equivalent to $N_o = 1$ and a variable $\gamma$. From these starting conditions the properties at different scales and temperatures are calculated by numerical integration of the recur-

Table I. Results obtained from $InO_x$ samples of Fiory et al.[11] The values of $T_c$BMO and b theor. have been calculated from their data.

| $R_n$ (KΩ) | $T_{co}$ (K) | $T_c$ (K) | $\gamma$ | $\varepsilon_c$ | $T_c$BMO (K) | b exp. | b theor. |
|---|---|---|---|---|---|---|---|
| 1.25 | 3.404 | 3.234 | 0.73 | 1.21 | 3.20 | 7 | 4 |
| 3.74 | 2.29 | 1.903 | 0.73 | 1.2 | 1.93 | 9 | 4 |
| 1.78 | 2.62 | 1.782 | 0.35 | 1.80 | 2.31 | 2 | 1 |

sion relations and compared with the experimental data. From the comparison, the value of $\gamma$ is deduced (in the notation of Fiory et al. a quantity C is used, which is equal to $2\pi\gamma$). Also from the best fit, the value of $\varepsilon_c$ is found. As should be the case, the value of $\varepsilon_c$ is the same as follows from Eq.(2.15) with $N_o = 1$ and $\gamma$ as obtained from the fit. No new information is contained here. The maximum scale in the samples is about 9.

In Table I some of the deduced values on the three samples are given. In a column the value of $T_c$ as calculated with the modified BMO relation (3.6) is represented, calculated with the values of $T_{co}$, $R_n$ and $\varepsilon_c$ as given by Fiory et al. For the first sample the agreement with the experimental $T_c$ is fine, in the second not too bad, but in the third the experimental $T_c$ is much lower. This seems to indicate nonideal properties.

In the table values of the temperature expansion parameter b are also given. They have been derived from the critical behaviour of the linear and the nonlinear resistance. Fiory et al. do not make a connection to theoretical values of b. The predicted value of b follows from Eqs.(2.25) and (2.26) as illustrated in Fig.3, using $N_o = 1$ and the value of $\varepsilon_c$ or $\gamma$ found from the experimental data. This 'theoretical' value of b is also given in the table. The agreement is not too good. It seems that better agreement might be found between $\gamma$ and $\varepsilon_c$ on one hand and b on the other if a smaller value of $N_o$ would be used. Of course, using a smaller value of $N_o$ alters the whole analysis of Fiory et al. All numerical integrations would have to be redone from different starting conditions and different values of $\gamma$ as well as $\varepsilon_c$ would result.

### 4.6. Other experiments

As announced in the introduction, a very limited choice has been made. No data in finite fields have been given. From all measurements of R(T) in various materials only few have been mentioned.

Noise in high resistance granular aluminum and tin films has been studied near $T_c$ by Voss, Knoedler and Horn.[24] Bancel and Gray[25] have investigated the time-dependent response of the vortex plasma to short laser pulses. For both experiments it seems that a new analysis of the data might be useful, based on improved understanding of experimental aspects of the theory as has been developed in the last years.

Very interesting new experiments have been performed by Garland, van Harlingen, Rudman and Lee. They measure the resistance and other properties of a vortex plasma in the presence of a thermal gradient perpendicular to the current direction. No publication has come available yet.

## 5. ARRAYS OF JUNCTIONS

### 5.1. Introduction

Two-dimensional arrays of Josephson junctions provide very interesting possibilities from two different points of view: a controlled simulation of some aspects of granular films and as a system for the study of two-dimensional phase transitions. The latter aspect will be discussed here. Experimental studies of arrays of junctions are in a sense intermediate between computer 'experiments' and measurements on thin films. Large junction arrays can be made with modern lithographic techniques, larger than the arrays that computers can handle. If the parameters are not as well known and as homogeneous over the sample as in software-defined arrays, in comparison with high-resistivity films significant advantages may be obtained.

Many aspects of the theory of transitions in arrays of junctions have been worked out by Lobb, Abraham and Tinkham (LAT).[13]

A junction array consists of superconducting islands, arranged two-dimensionally, that are weakly coupled to their neighbours. The coupling energy depends on the difference in phase of the order parameter and on the critical current of the junction. There is direct correspondence with the two-dimensional X-Y model mentioned in section 2.1. As in that case, the interaction energy between neighbouring islands with phases $\varphi_i$ and $\varphi_j$, is $U_{ij} = -J\cos(\varphi_i-\varphi_j)$ where J is the coupling energy of the connecting junction:

$$J = (\hbar/2e)\ I_o(T) \qquad (5.1)$$

$I_o$ is the critical current of the junction. The interaction constant is equal to $2\pi J$:

$$A(T) = (h/2e)\ I_o(T) \qquad (5.2)$$

According to Ambegaokar and Baratoff the critical current of an ideal Josephson junction is:

$$I_o(T) = \frac{\pi}{2eR_J}\ \Delta(T)\ \tanh\{\tfrac{1}{2}\Delta(T)/k_BT\} \qquad (5.3)$$

$R_J$ is the junction resistance. In practice the product $I_oR_J$ is very often below its theoretical value. If Eq.(5.3) holds, an interaction constant A(T) is obtained that is the same as for the homogeneous film, Eq.(3.3). The normal state film sheet resistance $R_n$ is replaced by $R_J$ in the expression for A(T). For a square array the normal state array sheet resistance is equal to $R_J$, providing exact equivalence.

In general, the critical current does not follow Eq.(5.3) and junction arrays show significantly different behaviour from that of homogeneous thin films.

## 5.2. Properties

The smallest pair separation $a_o$ in an array is equal to the lattice spacing, that we indicate with s. If an array has N × N islands, the largest distance in the array is W = Ns, the largest scale available is

$$\ell_W = \ln N \qquad (5.4)$$

In an square array containing $2\times10^6$ junctions, N is equal to $10^3$ and $\ell_W = 6.9$. The sample size is a more serious limitation in arrays than for thin films. As in thin films, the logarithmic interaction between vortices is cut-off at the penetration depth for perpendicular films Λ, which near $T_c$ is of the order of 1 cm as in films. If lithography limits s to a minimum of 5 μm, for example, and if Λ is 1 cm, the maximum useful value of N is 2000, the maximum useful scale 7.6.

The transition temperature for the arrays follows from Eq. (2.13). As for films, $\varepsilon_c$ is not known. For a specific lattice type, providing the coupling energy has the same dependence on the phase difference, $\varepsilon_c$ should be the same irrespective of the temperature dependence of $I_o$. Also $\varepsilon_c$ should be the same as for the corresponding computer arrays. For a square X-Y model, Kosterlitz and Thouless[1] give as value for the core potential $\mu_c = \frac{1}{2}\pi^2 J$, which corresponds to $N_o = 1$ and $\gamma = \pi/4$ in Eq.(2.14). These two values have been used in the numerical example of section 2.7 and for Fig.4. For this sytem $\varepsilon_c$ is equal to 1.175 as calculated from Eq.(2.15). Minnhagen[5] quotes a value of $\varepsilon_c$ near 1.75 obtained from Monte Carlo calculations for the X-Y model. Keeping $\varepsilon_c$ as an unknown, but possibly universal, parameter, the critical temperature follows from:

$$k_B T_c = \frac{1}{4\varepsilon_c} \frac{h}{2e} I_o(T_c) \tag{5.5}$$

To find $T_c$, the function $I_o(T)$ has to be known. The numerical factors are such that[7]

$$I_o(T_c)/T_c \approx 30 \text{ nA/K} \tag{5.6}$$

5.3. Junctions and weak links

Various types of weak links and junctions can be used for arrays. So far experiments have been performed on arrays with tunnel junctions and with superconductor-normal metal-superconductor (SNS) weak links. For both types the characteristic voltage is introduced:

$$V_o(T) = I_o(T)R_J \tag{5.7}$$

$V_o$ is typical for a specific fabrication procedure. For ideal tunnel junctions at low temperatures, Eq.(5.3) gives $eV_o = 2.76\ k_B T_{co}$, which is of the order of 1 meV. In practice $V_o$ is usually lower than the theoretical value.

The critical current of SNS bridges depends exponentially on the ratio of the bridge length $L_B$ and the coherence length in the normal metal $\xi_N$. The latter quantity is proportional to $T^{-\frac{1}{2}}$, so that the temperature dependence of $I_o$ is very different from that of tunnel junctions as long as $L_B$ is not much smaller than $\xi_N$. For arrays, the advantage of SNS bridges lies in the possibility of obtaining low values of $I_o$ in a well-controlled way, by using relatively long bridges. The temperature dependence of $\exp(-L_B/\xi_N)$ has to be taken into account. LAT give the approximate relation:

$$I_o(T) \propto (1 - T/T_{co})^2 \exp\{-L_B/\xi_N(T)\} \tag{5.8}$$

for SNS bridges near the critical temperature of the islands $T_{co}$.

Strictly speaking, the coupling energy is not directly determined by the critical current, but by the current-phase relation. The energy associated with a phase difference $\Phi$ across the junction is called $U(\Phi)$, $I_s(\Phi)$ is the dependence of the junction supercurrent on $\Phi$.

Both U(0) and $I_s(0)$ are zero. To calculate $U(\Phi)$, one takes the integral of the power needed to change the phase difference from zero at a time $t = t_1$ to $\Phi$ at $t_2$:

$$U(\Phi) = \int_{t_1}^{t_2} I_s(t)\, V(t)\, dt$$

With the Josephson relation $V = (\hbar/2e)(d\Phi/dt)$ this leads to:

$$U(\Phi) = (\hbar/2e) \int_0^{\Phi} I_s(\Phi')d\Phi' \qquad (5.9)$$

For $I_s(\Phi) = I_o \sin\Phi$, the expression for the supercurrent in a Josephson junction, one obtains:

$$U(\Phi) = (\hbar/2e)\, I_o(1 - \cos\Phi) \qquad (5.10)$$

The important two-dimensional properties are connected with small phase differences. Expansion of $U(\Phi)$, Eq.(5.9), yields:

$$U(\Phi) \approx \frac{\hbar}{2e} \left(\frac{dI_s}{d\Phi}\right)_{\Phi=0} \tfrac{1}{2}\Phi^2 \qquad \Phi \ll 1 \qquad (5.11)$$

In a junction with a non-sinusoidal current-phase relation, $dI_s/d\Phi$ at $\Phi = 0$ is not equal to $I_o$. For an array with such junctions the interaction constant is:

$$A(T) = (h/2e) \left(\frac{dI_s}{d\Phi}\right)_{\Phi=0} \qquad (5.12)$$

Near the core, the phase differences can be large. To calculate the core energy, one needs to know $I_s(\Phi)$ also for larger values of $\Phi$. For non-sinusoidal $I_s(\Phi)$, the core energy is different from the value for pure Josephson junctions. This may provide a way of obtaining systems with higher core energy and smaller $\varepsilon_c$.

## 5.4. Resistivity

When vortices move across an array of junctions, a voltage results as for thin homogeneous films. LAT have calculated the resistance of a junction array, their treatment is reproduced here. The array has a length L and a width W. The phase difference over the array in the direction of the length is $\theta$. When vortices move across the sample, the average change in time of $\theta$ is:

$$< \frac{d\theta}{dt} > = 2\pi L\ v_d\ n_v$$

$v_d$ is the average drift velocity of the vortices which is equal to $s/\tau_J$ when $\tau_J$ is the time it takes for the vortex core to move across one junction. $n_v$ is the vortex density. The Josephson relation for the array $<V> = (\hbar/2e)\ <d\theta/dt>$ gives:

$$<V> = (\hbar/2e)\ 2\pi L\ (s/\tau_J)\ n_v$$

The sheet resistance of the array is:

$$R_s = \frac{<V>}{I}\frac{W}{L} = \frac{\hbar}{2e}\,2\pi n_v\,\frac{W}{I}\frac{s}{\tau_J} \qquad (5.13)$$

I is the total current through the array. The crossing time $\tau_J$ is coupled to the current through one junction $I_J$, which on the average is $(s/W)I$. The barrier height against a phase change of $2\pi$ in one single junction is twice the coupling energy. In an array LAT show the barrier to be lower by a factor 5. From the correspondence with the crossing time in a single junction they deduce:

$$\tau_J = \frac{h}{2e}\,\frac{1}{I_J R_J}\,I_{BO}^2\left(\frac{A(T)}{10\pi k_B T}\right) \qquad (5.14)$$

Here the notation $I_{BO}(x)$ is used for the modified Bessel function of zero order with the argument x. For small x:

$$I_{BO}(x) \approx 1 + \frac{1}{4}x^2\ \ldots\ldots$$

The argument of the Bessel function in Eq.(5.14) can be written as:

$$\frac{A(T)}{10\pi k_B T} = \frac{1}{10\pi}\,4\varepsilon_c\,\frac{1}{1+\tau'}$$

If $\varepsilon_c$ is not too large and if the temperature is not too far below $T_c$, the argument is small and the Bessel function can be replaced with 1. It is found that:

$$R_s = n_v\, s^2\, R_J \qquad (5.15)$$

When the vortices present are free vortices, not due to current-induced pair unbinding, their density is connected with $\xi_+$ according to Eq.(3.15) and the small current resistance is:

$$R_s = (C_1/2\pi)(s/\xi_+)^2\, R_J = (C_1/2\pi)\,\exp(-2\ell_+)R_J \qquad (5.16)$$

## 5.5. Temperature dependence

For arrays of tunnel junctions and weak links with sinusoidal current-phase relation, the critical temperature follows from Eq.(5.5). More generally, using Eq.(5.12) it is found that:

$$k_B T_c = \frac{1}{4\varepsilon_c} \frac{\hbar}{2e} \left(\frac{dI_s}{d\Phi}\right)_{\Phi=0,\,T=T_c} \tag{5.17}$$

For junctions with sinusoidal current-phase relation Eqs.(2.17), (2.18) and (5.2) lead to:

$$T' = T \;\; I_o(T_c)/I_o(T) \tag{5.18}$$

and

$$\tau' = \frac{T/I_o(T) - T_c/I_o(T_c)}{T_c/I_o(T_c)} \tag{5.19}$$

In the anlysis of experimental results on real arrays, $T_c$ is not known beforehand. Contrary to the usual case in thin films, a linear Ginzburg-Landau like approximation for $A(T) \propto I_o(T)$ near $T_{co}$ cannot be used in most cases. It is necessary to introduce an effective temperature proportional to $T/I_o(T)$. LAT suggest the dimensionless quantity ($T'$ in their notation)

$$\tilde{T} = \frac{k_B T}{(\hbar/2e) I_o(T)} \tag{5.20}$$

The real temperature T, as a function of which experimental results come available, has to be transformed to $\tilde{T}$ before analysis. For this purpose $I_o(T)$ must be known. In terms of $\tilde{T}$, $\tau'$ is:

$$\tau' = \frac{\tilde{T} - \tilde{T}_c}{\tilde{T}_c} \tag{5.21}$$

## 5.6. Arrays of linear strips

Connections between islands in an array can be made with long narrow strips of homogeneous material. The current-phase relation of such a strip which is longer than the coherence length is:

$$I_s = (S/L)(\hbar n_s e/2m)\Phi \tag{5.22}$$

Where S and L are the cross-section and the length of the strip and $n_s$ is the density of superconducting electrons. With Eq.(5.12) this yields:

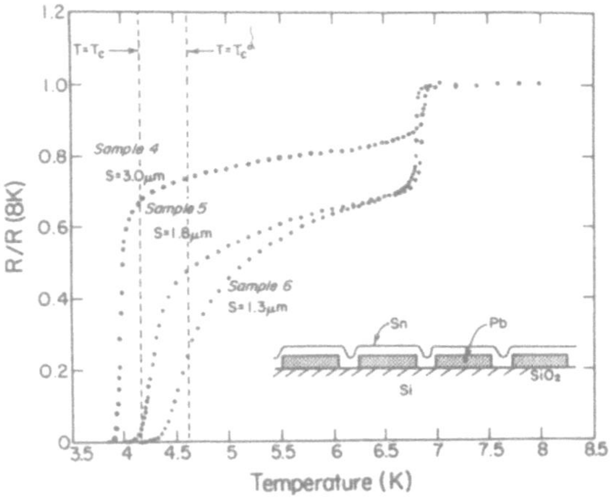


Fig. 17. Resistance of three arrays of lead islands with a connecting film of tin, as a function of temperature according to Resnick et al.[27]

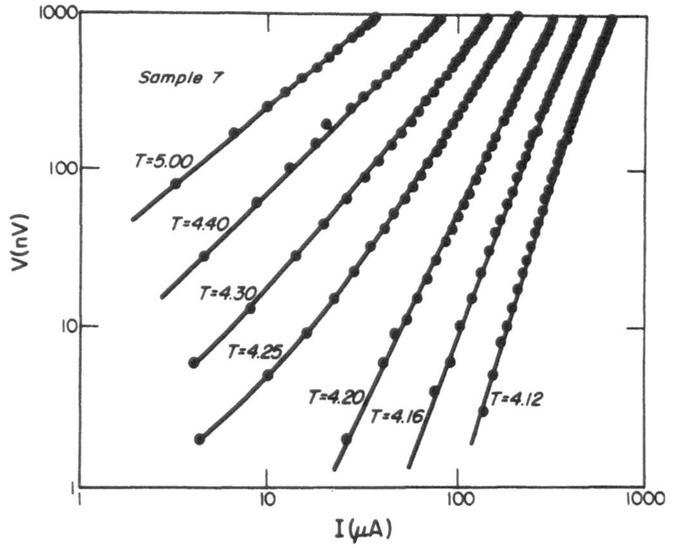


Fig. 18. Nonlinear resistance of a lead/tin array at various temperatures, from Resnick et al.[27]

$$A(T) = (h/2e)(S/L)(\hbar n_s e/2m)$$

which for a dirty superconductor, using expressions of the free electron model is found to be:

$$A(T) = \tfrac{1}{2}\pi^2 \frac{\hbar/e^2}{R_s} \Delta(T) \tanh\{\tfrac{1}{2}\Delta(T)/k_B T\} \tag{5.23}$$

where $R_s$ is the nromal state resistance of one strip. In a square array, the array sheet resistance is equal to $R_s$ and (5.23) is exactly the same as the expression for A(T) in a thin film, Eq. (3.3). The main difference between a homogeneous film and an array of islands connected with linear strips is the smallest distance $a_0$, which in films is equal to $\xi$, in the array s. In the array s is known which is a significant advantage, on the other hand the maximum scale is restricted by the fabrication techniques. The effective resistance in the array can be much higher, enhanced by the length/width ratio of the strips. Arrays of strips have not yet been studied experimentally.

## 6 EXPERIMENTAL RESULTS ON ARRAYS

### 6.1. Arrays of SNS bridges

Arrays of superconductor-normal metal-superconductor weak links are fabricated as arrays of superconducting islands on top of a continuous film of normal metal. This geometry facilitates fabrication with a small lattice spacing, homogeneous across the substrate. Because of the exponential length dependence of the coupling energy, the coupling to nearest neighbours is dominating. The individual links are not very well defined, the normal-state resistance of the effective junction with unknown width is not known.

The first to fabricate such arrays were Sanchez and Berchier.[26] They used indium islands on a film of gold/indium with a transition temperature of about 1.8 K. The largest arrays were 40 × 40 islands. The resistive transition of the arrays was measured.

Resnick, Garland, Boyd, Shoemaker and Newrock[27] fabricated a triangular array of lead disks with a diameter of about 13 µm, covered with a thin film of tin. The smallest distance between neighbouring lead islands was about 1.3 µm. In the temperature region between the transition temperature of lead, 7.2 K, and of tin, 3.7 K, one is dealing with an SNS system. The size of the array is 5 × 20 mm$^2$, which indicates an effective width of about 400 weak links, or a maximum scale $\ell_W$ of about 6. Fig.17 shows the resistance of three samples as a function of temperature. It shows the same characteristic behaviour on cooling down: a small drop of the

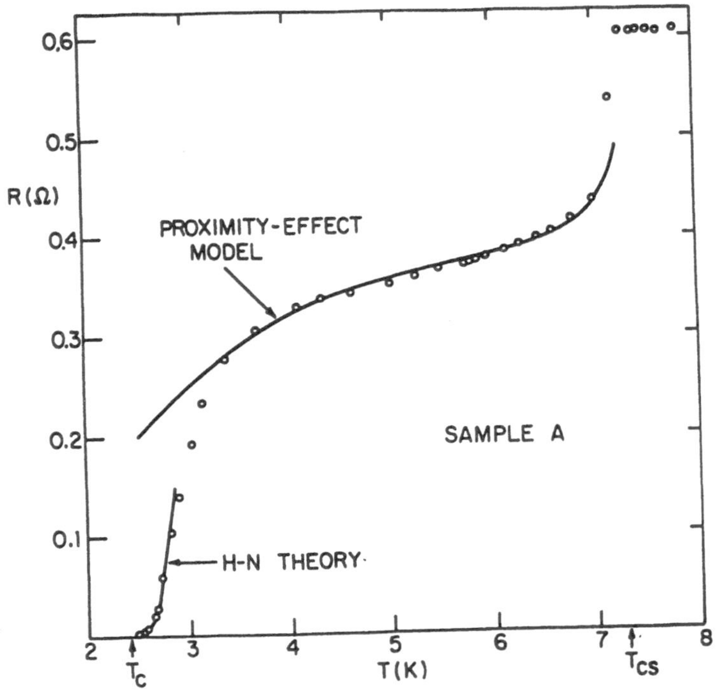


Fig. 19. Resistive transition of a 1000 × 1000 array of lead-bismuth islands on copper. The circles indicate experimental points. The drawn lines are fits to theory. Abrahams et al.[28]

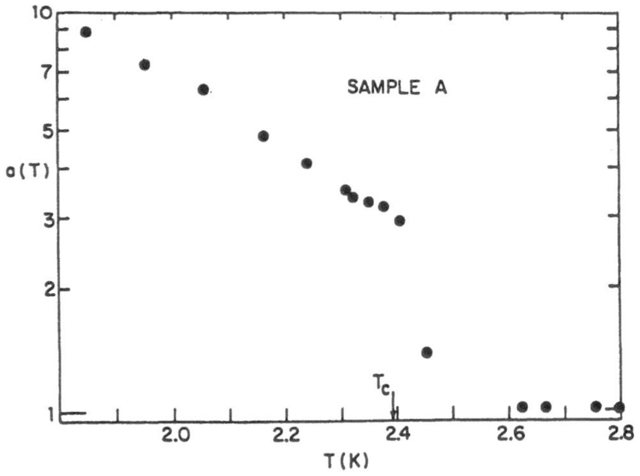


Fig. 20. Exponent a(T) from $V \propto I^a$ for array of lead-bismuth/copper as measured by Abraham et al.[28]

resistance at the transition of the lead islands, a plateau, followed by a steep decrease. In the plateau region, there is no phase coherence, the steep region marks the two-dimensional transition.

Resnick et al. also measured the nonlinear resistance at higher current levels. An example is shown in Fig.18. Clearly the behaviour is very similar to that of thin films. The authors have derived $T_c$ and an effective $T_{co}$ for their samples from both the linear resistance and the nonlinear V-I characteristics. Agreement is found. No details of the analysis are given.

Abraham, Lobb, Tinkham and Klapwijk[28] have studied square arrays of lead-bismuth islands on a copper film. The center-to-center spacing was 25 µm, the copper length between the islands about 6 µm. The arrays contained about $10^6$ islands, which gives an effective width of 1000 links and a maximum scale of about 7. Contacts were placed along two opposing edges of the array. It would seem that such rigid boundary conditions, imposed on an array that is square in its external shape, should have considerable influence. In Fig.19 the resistive transition of such an array is shown. The same qualitative behaviour is seen as in Fig.17.

Abraham et al. were able to fit the temperature dependence of R in the plateau region to a simple model. By the proximity effect, the region with stable phase coherence is larger than the lead-bismuth area. With decreasing temperature, the proximity effect strengthens and the islands become effectively larger. The upper drawn line is a two parameter fit to their simple model, which seems very adequate. At lower temperatures, phase coherence is established from island to island and the model ceases to be applicable.

The exponent a(T) of the nonlinear resistance, defined with $W \propto I^a$, has been determined by Abraham et al. Fig.20 shows the temperature dependence for the same sample as in Fig.19. In the same temperature region where the linear resistance grows fast with increasing temperature, a(T) decreases fast. Unfortunately few data points are given for the actual transition, which on comparison with Fig.7 is certainly as steep as can be expected for a scale of 7.

## 6.2. Arrays of Josephson junctions

Arrays of Josephson junctions have been studied by Voss and Webb.[29] With the technology developed for digital applications, a square array of 100 × 100 islands of niobium was fabricated, connected with tunnel junctions. The junction area was about 1 $\mu m^2$, the lattice spacing about 10 µm. The low temperature critical current of each junction was about 0.4 nA. From Eq.(5.5) it follows that $T_c$ for such an array should be around 12 mK. The experimental data indicate a transition at a temperature of that order of magnitude.

Tunnel junctions have a hysteretic I-V characteristic and hysteresis is also found for the array over a considerable temperature range. This makes the interpretation of the data more difficult.

### 6.3. Discussion

So far, the measurements on arrays and the analysis of the results have been rather limited. Effects of finite size, which are more serious than in thin films, have not been considered yet in the interpretation. As the size is exactly known, a quantitative analysis may be possible in a more accurate way than in films.

In arrays, the behaviour in an external magnetic field is very interesting. When a basic cell of the array contains a rational fraction of a flux quantum, extrema are expected and observed in the resistance. As this is a separate effect and only indirectly connected with the two-dimensional transition of the array, it is not discussed here.

## 7. CONCLUSIONS

The question should be addressed, whether or not a Kosterlitz-Thouless phase transition occurs and has been observed in thin superconducting films. A pure transition is only possible in an infinite system. Still, if the finite system shows the phenomena as predicted from the Kosterlitz recursion relations, one may consider the transition to be of the Kosterlitz-Thouless class.

In thin films, bound pairs and free vortices are clearly present below the BCS transition temperature. The experimental results are in general in good qualitative agreement with expectations from KT theory. This is not enough to distinguish between a general Coulomb-gas-like behaviour or, more specifically, as a Kosterlitz-Thouless system. A KT transition has two main features: the jump of the 'superfluid density' or $K^{-1}(\infty)$ at $T_c$ and the square-root temperature dependence near $T_c$. The data, in particular those of the current-induced pair unbinding and the kinetic inductance, show such a fast decrease as well as may be expected in systems of the given size. The square-root cusp as a function of temperature has not been shown convincingly. In many cases where its occurrence was claimed, the Halperin-Nelson expressions were not applicable. None of the data are in obvious disagreement with the expectations either.

If homogeneous, thin, high-normal-resistance superconducting films form a class of Kosterlitz-Thouless systems, one would expect quantities such as $\varepsilon_c$ and b to have the same values in all of them. Unfortunately no prediction exists from theory. Both $\varepsilon_c$ and b are connected to the parameters of the core potential, $\gamma$ for the energy and $N_o$ for the entropy. If theory could at least predict one of those, a much more meaningful comparison between theory and experiment could be made. Too few experimental values of $\varepsilon_c$ and b have come available and the results are too contradictory to draw any conclusions about their values.

It is necessary to perform more detailed studies of thin films, of the type performed recently by Fiory, Hebard and Glaberson, taking into account the full consequences of the renormalized interactions, the finite dimensions and the specific scale associated with a specific measuring method. By including $N_o$ as an unknown variable in the analysis, a better agreement might be possible for values from different films.

Arrays of junctions and weak links are promising systems for studies of Kosterlitz-Thouless transitions. The background parameters may be better defined than in thin films. As the maximum scale is small, a detailed analysis is impossible without taking the finite size into consideration.

ACKNOWLEDGEMENTS

I have profited from discussions with P. Minnhagen, A.M. Kadin, C.J. Lobb, H.J. Hilhorst and J.G.F. Kablau. I thank many authors for sending preprints prior to publication, in particular A.F. Hebard, A.T. Fiory and W.I. Glaberson who made their beautiful new results available before the summer school.

REFERENCES

1. J.M. Kosterlitz and D.J. Thouless, J.Phys.C6, 1181(1973).
2. M.R. Beasley, J.E. Mooij and T.P. Orlando, Phys.Rev.Letters 42, 1165(1979).
3. S. Doniach and B.A. Huberman, Phys.Rev.Letters 42, 1169(1979).
4. B.I. Halperin and D.R. Nelson, J.Low Temp.Phys.36, 599(1979).
5. L.A. Turkevich, J.Phys.C12, L385(1979).
6. V. Ambegaokar, B.I. Halperin, D.R. Nelson and E.D. Siggia, Phys.Rev.Letters 40, 783(1978).
7. V. Ambegaokar, B.I. Halperin, D.R. Nelson and E.D. Siggia, Phys.Rev.B21, 1806(1980).
8. V. Ambegaokar and S. Teitel, Phys.Rev.B19, 1667(1979).
9. P. Minnhagen, these proceedings.
10. A.F. Hebard and A.T. Fiory, Phys.Rev.Letters 50, 1603(1983).
11. A.T. Fiory, A.F. Hebard and W.I. Glaberson, Phys.Rev.B28, 5075 (1983).
12. A.M. Kadin, K. Epstein and A.M. Goldman, Phys.Rev.B27, 6691(1983).
13. C.J. Lobb, D.W. Abraham and M. Tinkham, Phys.Rev.B27, 150(1983).

14. A.F. Hebard and A.T. Fiory, Proceedings LT16, Physica 109 & 110 B+C, 1637(1982).
15. J.E. Mooij, in Nato Advanced Study Institute on Advances in Superconductivity, Erice 1982, B. Deaver and J. Ruvalds, editors, Plenum 1983.
16. T. Ohta and D. Jasnow, Phys.Rev.B20, 139(1979).
17. B.I. Halperin, in Proceedings of the Kyoto Summer Institute on Low Dimensional Systems, 1979.
18. D.R. Nelson and J.M. Kosterlitz, Phys.Rev.Letters 39, 1201(1977).
19. J.R. Clem, Inhomogeneous Superconductors-1979, AIP Conference proceedings nr.58, 1980, editors D.U. Gubser, T.L. Francavilla, S.A. Wolf and J.R. Leibowitz, page 245.
20. K. Epstein, A.M. Goldman and A.M. Kadin, Phys.Rev.Letters 47, 534(1981).
21. K. Epstein, A.M. Goldman and A.M. Kadin, Proceedings LT16, Physica 109 & 110 B+C, 2087(1982).
22. K. Epstein, A.M. Goldman and A.M. Kadin, Phys.Rev.B26, 3950(1982).
23. A.F. Hebard and A.T. Fiory, Phys.Rev.Letters 44, 291(1980).
24. R.F. Voss, C.M. Knoedler and P.M. Horn, Phys.Rev.Lettters 45, 1523(1980).
25. P.A. Bancel and K.E. Gray, Phys.Rev.Letters 46, 148(1981).
26. D.H. Sanchez and J.-L. Berchier, J.Low Temp.Phys.43, 65(1981).
27. D.J. Resnick, J.C. Garland, J.T. Boyd, S. Shoemaker and R.S. Newrock Phys.Rev.Letters 47, 1542(1981).
28. D.W. Abraham, C.J. Lobb, M. Tinkham and T.M. Klapwijk, Phys.Rev.B26, 5268(1982).
29. R.F. Voss and R.A. Webb, Phys.Rev.B25, 3446(1982).

# STATIC AND DYNAMIC PROPERTIES OF COMMENSURATE AND INCOMMENSURATE PHASES OF A TWO-DIMENSIONAL LATTICE OF SUPERCONDUCTING VORTICES

P. Martinoli, H. Beck, M. Nsabimana and G.A. Racine

Institut de Physique, Université de Neuchâtel
CH - 2000 Neuchâtel, Switzerland

## INTRODUCTION

Phase transitions in two-dimensional (2D) systems have received considerable attention recently. In several experiments the 2D crystal under consideration is exposed to the force field created by a periodic substrate. Among other situations this is the case of a 2D lattice of superconducting vortices interacting with a periodic pinning potential. As pointed out by Martinoli and coworkers[1-3] some years ago, thin superconducting films, whose thickness is periodically modulated in one dimension, provide such a system. In this lecture we discuss the static and dynamic behaviour of this model system in which the 2D vortex lattice can be driven through a variety of phases[4,5] simply by changing the conditions of flux-line density and/or temperature. In particular, we show how measurements of the critical currents and of the complex rf impedance of thickness-modulated layers can be used to probe the transition of the 2D vortex lattice from a "locked" commensurate (C) phase in registry with the substrate periodicity to a "floating" incommensurate (I) solid phase exhibiting 2D topological order[6,7] or to a fluid-like phase.

## THE PHASE DIAGRAM

The phase diagram of 2D crystals interacting with a periodic force field has been studied by a number of authors[4,5]. It is determined by considering, in addition to phonons, two types of topological excitations: domain walls (also called discommensurations, kinks, or solitons) which trigger the instability of a C-phase with respect to an I-phase (CI-transition) and dislocations which drive melting of the floating-solid phase (I-phase) into a liquid-like phase through the dislocation unbinding mechanism proposed by Kosterlitz and Thouless[6,7].

Dealing with situations where the periodic substrate is, as in our case, anisotropic, a recent theory by Pokrovsky and Talapov[8] (PT) is particularly relevant for the understanding of our experiments, where the 2D vortex lattice experiences the 1D periodic pinning potential created by the thickness modulation. In the following we review some of the basic concepts and results of the PT-theory. The only important difference between our treatment and the one by PT is that in establishing the CI-phase boundary we use, instead of their renormalization-group technique, an alternative approach based on the more transparent *S*elf-*C*onsistent *H*armonic *A*pproximation (SCHA) obtaining a similar result[9,10]. The modifications of the phase diagram resulting from the presence of thermally excited dislocations, which are not included in the PT-model, are briefly discussed at the end of this section.

## The CI-Transition at Zero Temperature

We consider a 2D triangular lattice of superconducting vortices, with lattice parameter $a$, in static interaction with a 1D harmonic potential of amplitude $\Delta$ and wave vector $\vec{q}$ ($q = 2\pi/\lambda_g$). We focus our attention on situations where $\vec{q}$ is very close to one of the vectors, $\vec{g}$, of the reciprocal vortex lattice, the condition $\vec{q} = \vec{g}$ defining a configuration of perfect matching between the (undistorted) lattice and the sinusoidal pinning potential. It is assumed that the flux-line lattice is incompressible and, further, that the pinning is weak when compared to the lattice stiffness, i.e. $\Delta < \mu$, where $\mu$ is the shear modulus of the vortex lattice[11]. Under these conditions only long-wavelength shear deformations turn out to be relevant and, as a consequence, the vortex lattice can be treated as an elastic continuum. Then, the energy $\mathcal{E}$ of the system can be written as the sum of an elastic contribution due to the pinning-induced lattice distortions and of a potential energy contribution due to the periodic pinning field :

$$\mathcal{E} = \int \left[\frac{\mu}{2}\left(\frac{\partial u}{\partial y} + \frac{\partial v}{\partial x}\right)^2 + \Delta(1 - \cos q\phi)\right] dx\, dy \, . \tag{1}$$

In writing this expression we have jumped ahead to the conclusion of PT asserting that, at $T = 0$, the ground state of the system is characterized by a quasi 1D deformation field $\vec{w}$, whose components $(u,v)$ in an $(x\text{-}y)$-reference frame with x pointing in a direction forming an angle of $45^0$ with $\vec{q}$ are given by :

$$u = \delta y , \qquad v = \delta x - \sqrt{2}\,\phi(x) , \tag{2}$$

where $\delta = 1 - (g/q)$ measures the degree of mismatch. It clearly emerges from these expressions that $\vec{w}$ results from the superposition of two uniform shear deformations (one along x and the other along y) and of a 1D transverse field $\phi(x)$ which, for an incompressible lattice, is found to propagate along x[3,8]. In an (x'-y')-coordinate system rotated by $45^0$ with respect to (x-y) the uniform part of $\vec{w}$ is an area conserving deformation consisting of a uniform compression ($\delta > 0$) or expansion ($\delta < 0$) $-\delta x'$ along x' (parallel to $\vec{q}$) combined with a uniform expansion ($\delta > 0$) or compression ($\delta < 0$) $\delta y'$ along y'. This uniform deformation is such that the potential energy contribution to $\mathcal{E}$ vanishes : the vortices are forced to lie at the bottom of the potential wells of the cosine-potential.

To determine $\phi(x)$, we simply minimize the functional $\mathcal{E}[\phi(x)]$ with respect to $\phi(x)$, thereby obtaining the following sine-Gordon equation[12] for the "phase" field $\Phi(x) = q\phi(x)$ :

$$\sin\Phi - \ell^2 \frac{\partial^2\Phi}{\partial x^2} = 0 , \tag{3}$$

where $\ell^2 = 2\mu/\Delta q^2$. Its solution in terms of elliptic functions

$$\Phi(x) = \pi + 2\mathrm{am}(x/k\ell) \tag{4}$$

is a stair-shaped function representing a regular sequence of kinks, whose period L is related to k by

$$L = 2k\ell K(k) , \tag{5}$$

where K(k) is a complete elliptic integral of the first kind. Using Eqs. (4) and (5), $\mathcal{E}$ can be expressed as a function of the variational parameter k. Then, minimization of $\mathcal{E}(k)$ with respect to k leads to :

$$|\delta| = (2/\pi)(\Delta/\mu)^{1/2}[E(k)/k] , \tag{6}$$

where E(k) is a complete elliptic integral of the second kind. There are solutions of Eq. (6) satisfying the condition $0 \le k \le 1$ only if $|\delta|$ is larger than a critical mismatch $\delta_c$ given by :

$$\delta_c = (2/\pi)(\Delta/\mu)^{1/2} \tag{7}$$

For $|\delta| > \delta_c$ $\Phi(x)$ is of the form (4). As a consequence, for $|\delta| > \delta_c$ the vortex lattice is characterized by the formation of a superstructure, of period L, consisting in a regular 1D sequence of domain walls propagating at $45^{0}$ with respect to $\vec{q}$. This is the incommensurate I-phase shown in Fig. 1. In constructing this figure we have assumed that the starting matching configuration is that defined by $\vec{q} = \vec{g}_{10}$, $\vec{g}_{10}$ being one of the six nearest-neighbour reciprocal lattice vectors ($g_{10} = 4\pi/a\sqrt{3}$). Since $\delta > 0$, large portions of the lattice shown in Fig. 1 appear to be uniformly compressed along x' and expanded along y' and are essentially commensurate to the underlying 1D periodic substrate. These regions are separated from each other by discommensurations where the phase field $\Phi(x)$, which is essentially constant in the nearly commensurate regions between successive kinks, changes by $2\pi$ over distances of the order of $\sim k\ell$. The period L of the superstructure diverges logarithmically as $|\delta|$ approaches $\delta_c$ [$k \to 1$ in Eq. (5)].

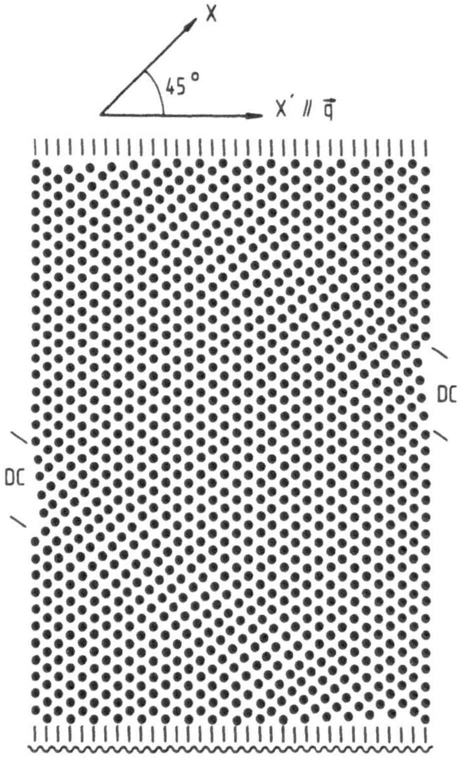


*Fig. 1. Incommensurate I-phase for $\delta = 0.13$ ($B/B_{10} = 0.76$). Discommensurations (DC) form a periodic 1D sequence propagating along the x-direction.*

For $|\delta| < \delta_c$ there are no solutions of Eq. (6) and, consequently, $\Phi(x)$ is no longer given by Eq. (4). In this case $\mathcal{E}$ has its minimum value when the potential energy contribution due to the periodic pinning field vanishes in Eq. (1), i.e. when $\Phi(x) = 0$ everywhere. This, of course, corresponds to the commensurate C-phase shown schematically in Fig. 2 for $\delta = 0$ (matching configuration $\vec{q} = \vec{g}_{10}$) and for vortex densities corresponding to deviations from perfect registry but still such that $|\delta| < \delta_c$.

The areal free energy density $F_\square$ of the 2D vortex lattice can be written in the form :

$$F_\square = 2\mu\delta^2 - 2\Delta\{[\delta/\delta_c E(k)]^2 - 1\}\ \theta(|\delta| - \delta_c)\ , \qquad (8)$$

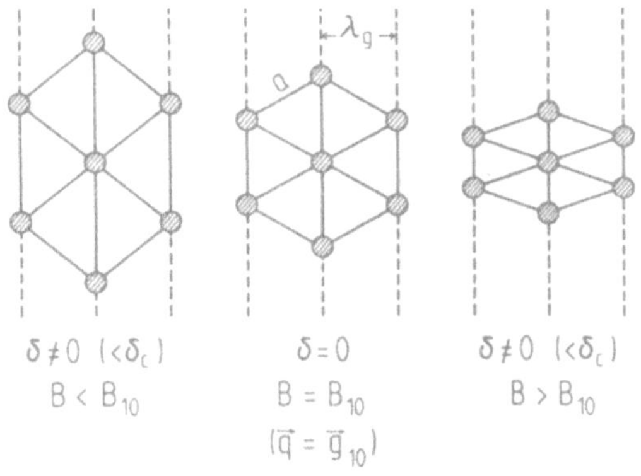


*Fig. 2. The fundamental commensurate $C_{10}$-phase in three different states of deformation.*

where $\theta(z)$ is the Heaviside step-function. The first term on the right-hand side of Eq. (8) is the elastic energy density associated with the uniform deformation appearing in both the C- and the I-phase, whereas the second one is due to the phase field $\Phi(x)$ and therefore contributes to $F_\square$ only in the I-phase.

Phase Transitions at Finite Temperatures

To study the CI-transition at finite temperatures, we first consider the case of perfect matching ($\delta = 0$), which is particularly simple. For $\vec{q} = \vec{g}$ the vortices execute a Brownian motion around the equilibrium positions they would assume at the bottom of the potential wells at T = 0. Accordingly, the Langevin equation of motion for a vortex at the lattice site $\underline{l}$ can be written as :

$$\eta\dot{\vec{u}}_{\underline{l}} = -\sum_{\underline{l}'}{}' \tilde{G}(\underline{l}-\underline{l}')\vec{u}_{\underline{l}'} - \vec{q}\Delta'\sin(\vec{q}\cdot\vec{u}_{\underline{l}}) + \vec{f}_{\underline{l}}(t) , \tag{9}$$

where the four terms represent, successively, the viscous damping force, the lattice restoring force, the sinusoidal pinning force and the fluctuating Langevin force acting on the vortex at $\underline{l}$. $\eta^{-1} = R_\Box/B\phi_0$, where $R_\Box$ is the sheet flux-flow resistance of the superconducting film, is the mobility[13] of a free vortex, $\tilde{G}(\underline{l}-\underline{l}')$ the elastic matrix and $\Delta'$ is related to $\Delta$ by $\Delta = n_\Box\Delta'$, where $n_\Box = B/\phi_0$ is the areal vortex density. $\vec{f}_{\underline{l}}(t)$ is assumed to have a white noise spectrum defined by the correlation function :

$$\langle f_{\underline{l}\alpha}(t) f_{\underline{l}'\beta}(t')\rangle = 2\eta k_B T\,\delta_{\alpha\beta}\,\delta_{\underline{l}\underline{l}'}\,\delta(t-t') , \tag{10}$$

stating that the Langevin force is uncorrelated in direction, space and time. To find the mean square fluctuation $\langle u^2\rangle$ of the vortices, which is the quantity of interest here, it is convenient to expand $\vec{u}_{\underline{l}}(t)$ in normal modes of the vortex lattice[10]. Then, linearizing Eq. (9) within the framework of SCHA and considering, as before, only transverse (t) modes of the lattice the following expression for the t-component of the normal mode amplitude $\vec{u}_{\underline{k}}(\omega)$ is obtained :

$$u_{\underline{k}t}(\omega) = \frac{n_\Box f_{\underline{k}t}(\omega)}{D_{\underline{k}t} + \Delta_R(\vec{q}\cdot\hat{e}_{kt})^2 - i\eta\omega} , \tag{11}$$

where $D_{\underline{k}t}$ is the matrix element of the (diagonal) dynamical matrix associated with t-modes, $\hat{e}_{\underline{k}t}$ the polarization vector for t-deformations and $f_{\underline{k}t}(\omega)$ the t-Fourier component of the Langevin force. In our SCHA-approach the effective strength, $\Delta_R$, of the pinning field experienced by the vortices is given by :

$$\Delta_R = \Delta\, e^{-\frac{1}{2}q^2\langle u_{tx}^2\rangle} , \tag{12}$$

where $\langle u_{tx}^2\rangle$ is the mean square t-fluctuation along the x-direction parallel to $\vec{q}$. Therefore, in our treatment the renormalization effect of the thermal fluctuations, which provides the essential mechanism for the phase transition, enters through a Debye-Waller factor which weakens the periodic pinning force acting on the vortices. To calculate the mean square t-fluctuation $\langle u_t^2 \rangle$ we assume a Debye model, for which $D_{\underline{k}t} = \mu k^2$, and replace the required sum over $\underline{k}$ by an integral over a smooth density of states. Then, in the weak pinning limit $\Delta \ll \mu$ considered here we deduce from Eqs. (10) and (11) :

$$\langle u_t^2 \rangle = \frac{k_B T}{4\pi\mu} \ln(\mu/\Delta_R) . \qquad (13)$$

This result shows quite clearly that in a C-phase the vortex lattice is a 2D solid with conventional long-range order as long as $\Delta_R$ remains finite. As expected for 2D systems, $\langle u_t^2 \rangle$ diverges logarithmically as $\Delta_R$ vanishes. Since, by equipartition, $\langle u_{tx}^2 \rangle \approx (1/2)\langle u_t^2 \rangle$ in the limit $\Delta \ll \mu$, from Eqs. (12) and (13) one deduces :

$$\Delta_R/\Delta = (\Delta/\mu)^{T/(T_{LU} - T)} , \qquad (14)$$

where $T_{LU}$ is implicitly given by :

$$k_B T_{LU} = (4/\pi)\mu(T_{LU})\lambda_g^2 . \qquad (15)$$

It clearly emerges from Eq. (14) that at the "*L*ocking-*U*nlocking" temperature $T_{LU}$ the vortex lattice undergoes a transition from a perfectly matched ($\delta = 0$) "locked" C-phase ($\Delta_R \neq 0$) to an "unlocked" phase ($\Delta_R = 0$) whose precise nature (floating solid or liquid) will be discussed in a moment. It should be noticed that the expression for $T_{LU}$ deduced with our SCHA-scheme is the same as that obtained by PT using a renormalization-group technique.

For a moderately dense lattice of vortices in dirty superconducting films $\mu$ can be written in the form[11] :

$$\mu(T) = (1/2)n_{\Box}(\phi_0/4\pi)^2 \Lambda^{-1}(T) , \qquad (16)$$

where $\Lambda = 2\lambda^2/d$ is the effective penetration depth for 2D superconducting layers[10,14]. Since $\mu$ is a function of $n_{\Box} = B/\phi_0$, Eq. (15) shows that $T_{LU}$ depends upon the matching configuration under consideration. For a triangular lattice such configurations are defined by[3] :

$$B_{mn} = (\sqrt{3}/2)(\phi_0/\lambda_g^2)(m^2 + n^2 + mn)^{-1} , \qquad (17)$$

where m and n are integers. Expressing $\Lambda(T)$ in terms of the normal state sheet resistance $R_{n\square}$ of the film, the transition temperature $T_{LU}$ for the fundamental $C_{10}$-phase (m = 1, n = 0 corresponding to $\vec{q} = \vec{g}_{10}$) deduced from Eqs. (15) and (16) can be written in the form :

$$T_{LU}/T_c \approx 1 - 0.31(R_{n\square}/R_u) \quad \text{for} \quad R_{n\square} << R_u , \tag{18}$$

where $T_c$ is the BCS-transition temperature and $R_u$ the universal sheet resistance $\hbar/e^2$. LU-transition temperatures for $C_{mn}$-phases defined by higher values of m and n lie below that given by Eq. (18).

The case of finite mismatch ($\delta \neq 0$) is more delicate. It is clear, however, that, as a consequence of the fluctuating Brownian motion of the vortices, the critical degree of mismatch $\delta_c(T)$ tolerated by a C-phase becomes smaller and smaller as the temperature rises and finally vanishes at $T = T_{LU}$. PT determined the phase boundary $\delta_c(T)$ using a renormalization-group technique, in which the only renormalizable parameter is the pinning potential amplitude $\Delta$[8]. More recently, Puga et al.[15,16] have generalized the PT-treatment by considering the effect of renormalization also on $\mu$ and $\delta$. Although several aspects of the CI-phase transition emerging from their calculation turn out to be different from those following from the much simpler SCHA-scheme, the shape of the phase boundary $\delta_c(T)$ resulting from their approach is very similar to that predicted by SCHA. In the latter approximation $\delta_c(T)$ simply follows from Eq. (7) by replacing $\Delta$ with its renormalized value $\Delta_R$ given by Eq. (14). The resulting phase diagram is shown in Fig. 3, where, instead of $\delta_c(T)$, we have plotted the related quantity $B_c(T) = B_{mn}[1 \pm \delta_c(T)]^2$. Temperatures are conveniently measured in units of $T_M$, the melting temperature of the 2D vortex lattice[17,18], which, as shown by Eq. (16) and the following equation, is independent of B at moderate vortex densities :

$$k_B T_M = (1/4\pi)\mu(T_M)a^2 . \tag{19}$$

So far, only domain wall excitations, which drive the CI-phase transition, have been included in our analysis. To obtain the complete phase diagram, however, several authors[7,10,19-25] have emphasized that it is necessary to add the effect of thermally excited dislocation pairs[6,7] in order to assess the stability of the I-phase against melting into a fluid-like phase. Recently, Haldane et al.[26] have given a rather unified description of the phase diagram showing that it strongly depends on the order of commensurability p, which for our triangular lattice is related to m and n by $p = (2/\sqrt{3})(m^2 + n^2 + mn)^{1/2}$. For $p < \sqrt{8}$ they find that,

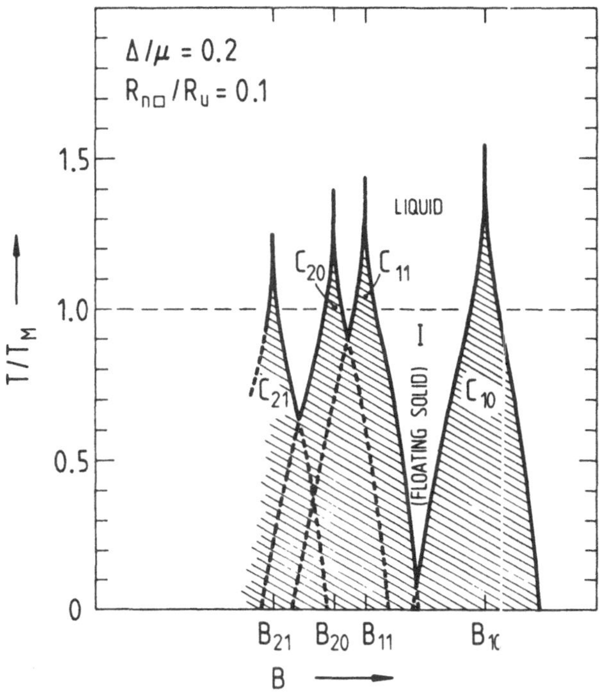


*Fig. 3. Phase diagram in the (B,T)-plane of a 2D vortex lattice in a 1D periodic potential.*

at finite temperatures, there is always a liquid phase separating the C-phase from the I-phase, whereas for $p > \sqrt{8}$ a direct CI-phase transition is always possible. Although important for a deeper understanding of the physics of 2D system, these features play only a marginal role in the analysis of the experiments reported later on in this lecture. Thus, for simplicity, in constructing the phase diagram of Fig. 3 we have assumed that, in the I-phase, the vortex lattice is a 2D floating solid undergoing a melting transition driven by the unbinding of dislocation dipoles at $T = T_M$. It follows that, if the CI-transition occurs for $T > T_M$, it is actually a transition from a C-phase to a fluid-like phase. This is the case for the lower order C-phases of Fig. 3, where $T_{LU}$ is larger than $T_M$. A straightforward calculation based on Eqs. (15), (17) and (19) shows, however, that there is a particular commensurate phase, the $C_{22}$-phase corresponding to $p = 4$, for which $T_{LU}$ becomes equal to $T_M$. For C-phases of higher order ($p > 4$) $T_{LU}$ is always lower than $T_M$

and, consequently, a floating solid phase always separates the C-phase from the liquid phase. This agrees with the findings of other authors[19,20,22,23,26].

CRITICAL CURRENTS

To test some of the theoretical ideas put forward in the previous section, critical current ($I_c$) measurements were performed on thickness modulated granular Al-films as a function of magnetic field and temperature. A combined holographic-photolithographic technique[1] was used to fabricate grating-like film profiles with $\lambda_g \lesssim 1$ μm and with a relative thickness modulation, $\Delta d/d$, less than 20 - 25%. The most relevant superconducting and normal state parameters of the two Al-films studied in this work are summarized in Table 1.

Since a registered C-phase is pinned by the periodic film structure, a finite current, flowing parallel to the 1D grooves, is required to depin the vortex lattice and, subsequently, to sustain vortex motion in the dissipative flux-flow régime. An I-phase, on the other hand, is not pinned by the periodic substrate, its energy being independent of the relative position of the discommensurations with respect to the pinning potential. Therefore, the critical current for entering the flux-flow régime vanishes in this case.

In Fig. 4 $I_c(B)$-curves for the film Al*1* are shown for different reduced temperatures $t = T/T_c$. Using Eq. (17), one can easily verify that the peak at $B \approx 28$ Gauss corresponds to the fundamental $C_{10}$-configuration ($p = 2/\sqrt{3}$) shown in Fig. 2, while the small structure at $B \approx 7$ Gauss can be assigned to the $C_{11}$- and $C_{20}$-phases which, on account of their strong overlap (see Fig. 3), are hard to resolve from each other. The shape of the $I_c(B)$-peak associated with a given C-phase has been calculated by Burkov and Pokrovsky[27]. In

Table 1. Parameters of the Al-Films

| Film | $d$[Å] | $\Delta d/d^{(a)}$ | $\lambda_g[\mu m]$ | $R_{n\square}[\Omega]$ | $T_c$[K] | $(\xi_0 l)^{1/2}$[Å]$^{(b)}$ | $\lambda_L(0)\left(\frac{\xi_0}{l}\right)^{1/2}$[Å]$^{(b)}$ |
|---|---|---|---|---|---|---|---|
| Al1 | 200 | ~0.2 | 0.79 | 15 | 1.89 | 365 | 4300 |
| Al2 | 200 | ~0.2 | 0.77 | 35 | 2.16 | 223 | 6140 |

a Determined by combined optical and electrical methods.
b Calculated using $\rho l = 4 \times 10^{-12}\,\Omega\,cm^2$ and $\lambda_L(0) = 157$ Å for Al. $\xi_0$ was scaled from the bulk Al value (1, 6 μm) according to our $T_c$.

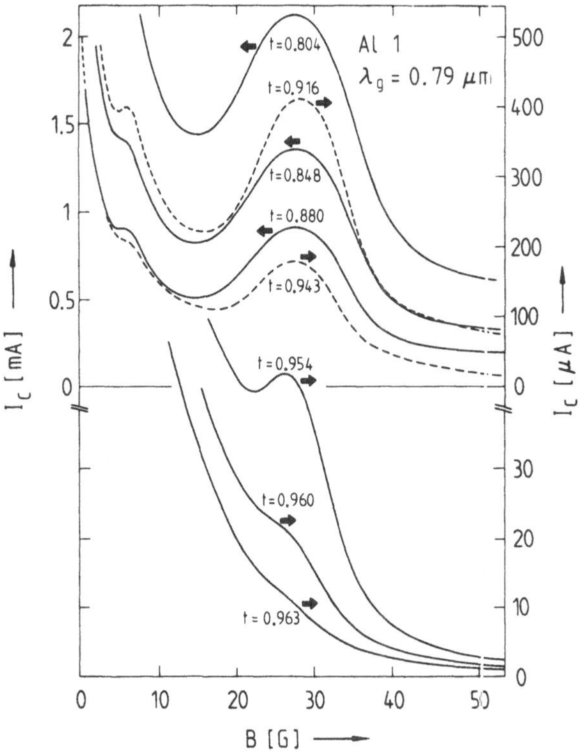


*Fig. 4. Critical current vs. magnetic field curves of a thickness modulated film (Al1) at different reduced temperatures $t = T/T_c$.*

a sample of finite size it is determined by the stability of the C-phase against the nucleation of a domain wall at the boundary of the film. Therefore, a detailed comparison of the shape of the main peak in Fig. 4 with the theoretical prediction of this model could, in principle, allow a determination of the $C_{10}$I-phase boundary. For two reasons, however, this appears to be, in practice, a difficult problem. The first and more important one is that in our films $\Delta$ is found to be of the order of $\mu$, typically $\Delta/\mu \approx 0.9$. Under this condition one expects considerable overlap of the $C_{10}$-phase with the $C_{11}$-phase (in Fig. 3 the overlap of the various C-phases is enhanced as $\Delta/\mu$ increases). This is certainly at the origin of the relatively high shoulder on the low field side of the fundamental $I_c$-peak in Fig. 4. The second reason is that in real films one is dealing with unavoidable pinning effects due to randomly distributed inhomogeneities, which result in a finite contribution to $I_c$ even in the I-phase. Clearly, both overlapping and random pinning effects render the analysis of the peak shape quite difficult. We shall therefore concentrate on a much more accessible experimental quantity : the temperature dependent strength $I_{cM}(T)$ of a critical current peak.

For perfect matching the equilibrium position of a vortex is determined[3,28,29] by balancing the Lorentz driving force $\vec{F}_L = d\phi_0(\vec{j} \times \hat{z})$ against the pinning force experienced by the vortex in the effective cosine-potential $\Delta'_R$ $(1 - \cos q\phi)$. This results in the following expression for the transport current density :

$$j = (q\Delta'_R/\phi_0 d)\sin\Phi \ . \qquad (20)$$

The critical current density, $j_{cM}$, is reached for $\Phi = \pi/2$, a condition corresponding to vortices located halfway between the bottom and the top of the potential wells. Thus, using Eq. (14) $j_{cM}$ can be written as :

$$j_{cM} = (q\Delta'/\phi_0 d)(\Delta/\mu)^{T/(T_{LU} - T)} \ . \qquad (21)$$

In order to compare our $I_c$-data with Eq. (21) we need a model for $\Delta = n_{\square}\Delta'$, the characteristic energy scale of the pinning mechanism operating in our thickness modulated films. In the thin film limit $(d << \lambda)$ the potential energy $\varepsilon(\vec{r})$ of a vortex located at $\vec{r}$ can be expressed by the convolution[30] :

$$\varepsilon(\vec{r}) = \int f(\vec{r}' - \vec{r})d(\vec{r}')d^2r' \ , \qquad (22)$$

where $d(\vec{r}') = d + \Delta d \cos qx$ is the periodically varying film thickness and $f(\vec{r}' - \vec{r})$ the free energy density distribution within the flux line. Using Clem's model[31] for $f(\vec{r}' - \vec{r})$, $\varepsilon(x)$ can be easily evaluated[10] from Eq. (22) in the limits $q\xi < 1$ ($\xi$ is the GL-coherence length) and $q\Lambda >> 1$ of interest here. From the expression for $\varepsilon(x)$ one immediately identifies $\Delta$ as :

$$\Delta \approx 2n_{\square}(\Delta d/d)(\phi_0/4\pi)^2\Lambda^{-1}(T) \ . \qquad (23)$$

This result shows that $\Delta$ has the same temperature dependence as $\mu$ (Eq. 16), a considerable simplification in the analysis of the $I_c$-data. By combining Eqs. (21) and (23), $I_{cM}$ can finally be written in the form :

$$\frac{I_{cM}(T)}{I_{cM}(0)} = \frac{\Lambda(0)}{\Lambda(T)} \left(\frac{\Delta}{\mu}\right)^{T/(T_{LU}-T)} , \qquad (24)$$

where $\Delta/\mu \approx 4(\Delta d/d)$. This expression shows very clearly how, with rising temperature, thermal fluctuations further reduce $I_{cM}(T)$ with respect to the BCS-value which corresponds to the limit $T_{LU} \to \infty$. After substraction of the background due to random pinning, which was deduced from a flat but otherwise identical reference film, the critical currents $I_{cM}(T)$ of Al*1* and Al*2* were fitted to Eq. (24) using $I_{cM}(0)$, $\Delta/\mu$ and $T_{LU}/T_c$ as fitting parameters. The result of this analysis is shown in Fig. 5 where, for comparison, theoretical curves calculated by neglecting the effect of thermal fluctuations are also shown. Good agreement with Eq. (24) is found for a reasonable choice of the parameters. $T_{LU}/T_c$ scales with $R_{n\square}$

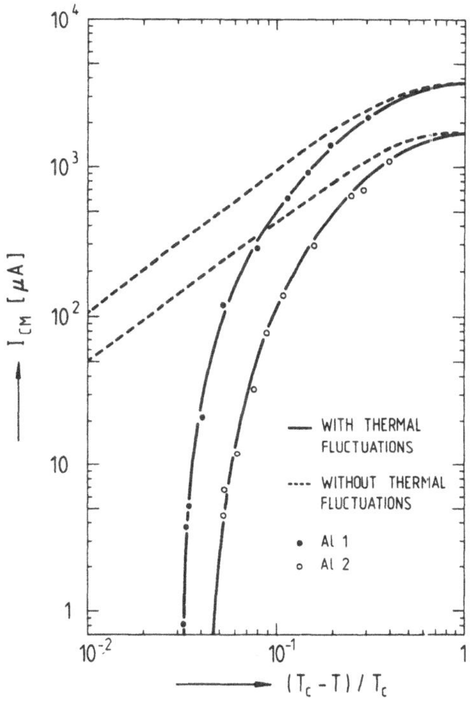


*Fig. 5. Temperature dependence of the critical currents of thickness-modulated films in the perfectly registered ($\delta = 0$) $C_{10}$-phase. The experimental data are fitted to Eq. (24) using : $\Delta/\mu = 0.95$, $T_{LU}/T_c = 0.978$ for Al1 and $\Delta/\mu = 0.90$, $T_{LU}/T_c = 0.972$ for Al2.*

approximately as predicted by Eq. (18) where, however, the numerical coefficient 0.31 is found to be about an order of magnitude too small to account for values deduced from the fit of Fig. 5. As for $\Delta/\mu$, there is good agreement between the values obtained from the fit and those estimated with $\Delta/\mu \approx 4(\Delta d/d)$ using the experimental data of $\Delta d/d$ listed in Table 1.

DYNAMICS

The motion of superconducting vortices in a periodic field shows interesting quantum features reminiscent of ac-Josephson phenomena in arrays of superconducting weak links. When the dc driving current $I_{dc}$ exceeds $I_c$, a particular flux-flow régime characterizes a C-phase. Dynamic coupling of the vortex lattice with the periodic substrate results, in this case, in a highly coherent velocity oscillation of the vortices which, in turn, generates a weak but detectable macroscopic voltage oscillation[2,3], typically in the radiofrequency (rf) range. Indirect evidence for the collective oscillation of the vortices in a C-phase is obtained by exposing thickness-modulated films to rf-radiation. Pinning-induced coupling at rf-frequencies between the oscillating motion of the vortex lattice and the rf-field gives rise to quantum interference transitions in the I-V-curves at discrete values, $E_n = n\nu\lambda_g B$, of the flux-flow dc electric field $E_{dc}$[3,9,32] ($\nu = \Omega/2\pi$ is the frequency of the rf-radiation). Derivatives of the I-V-characteristics for Al*l* exposed to 100 MHz-radiation are shown in Fig. 6. As in the critical current case (Fig. 4), from the evolution of the n = 1 interference transition with increasing temperature one is led to the conclusion that dynamic coupling of the matched vortex lattice to the periodic substrate is totally absent for $T > T_{LU} \approx 1.85$ K. This provides additional evidence for the occurrence of the locking-unlocking phase transition discussed in the previous section. More detailed information about the dynamics of the 2D vortex lattice in the various phases it can assume on the periodic substrate can be obtained from a study of the complex rf-impedance of thickness-modulated layers. In the rest of this lecture we shall therefore focus our attention on some theoretical and experimental aspects of the dynamic response of the vortex medium to a small oscillating driving field.

Low Temperature Vortex Mobility

At low temperatures ($T \ll T_{LU}$) thermal fluctuations of the vortices can be neglected. Assuming a driving Lorentz force of the form $\vec{F}_L = d\phi_0(\vec{j}_{rf} \times \hat{z})\exp(-i\Omega t)$, the equation of motion for the position $\vec{r}_{\underline{l}}$ of the vortex associated with the lattice site $\underline{l}$ can

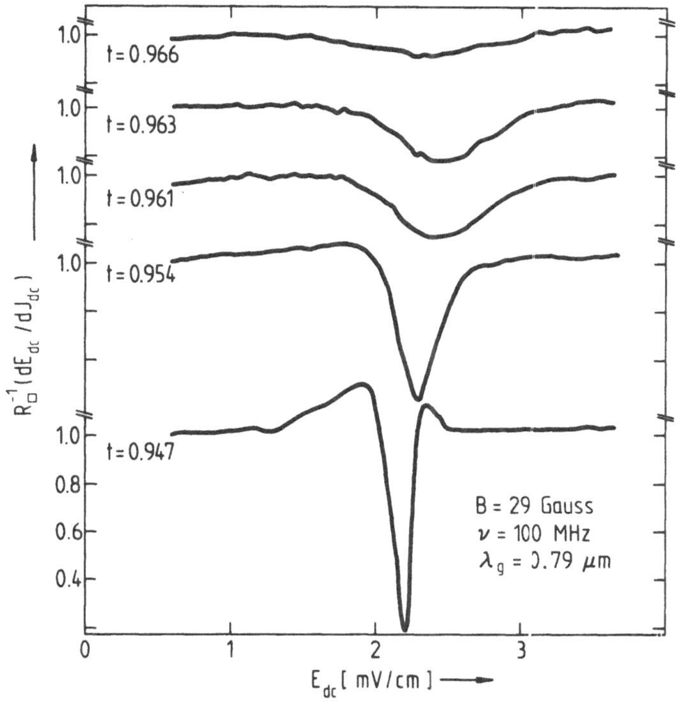


*Fig. 6. Derivatives of the rf-excited current-voltage characteristics of Al1 showing the temperature dependence of the $n = 1$ interference transition.*

then be written as[3] :

$$\eta \dot{\vec{r}}_{\underline{l}} = \vec{F}_L + \Delta' \vec{q} \sin \vec{q} \cdot (\vec{r}_{\underline{l}} - \vec{r}_o) - \sum_{\underline{l}'}{}' \frac{\partial W_{\underline{l}\underline{l}'}}{\partial r_{\underline{l}\underline{l}'}} \hat{r}_{\underline{l}\underline{l}'} , \qquad (25)$$

where $W_{\underline{l}\underline{l}'}$ is the interaction energy between the two vortices at $\vec{r}_{\underline{l}}$ and $\vec{r}_{\underline{l}'}$, $r_{\underline{l}\underline{l}'} = |\vec{r}_{\underline{l}} - \vec{r}_{\underline{l}'}|$ and $\vec{r}_o$ defines the relative position of the vortex lattice with respect to the 1D thickness modulation. Thus, the last term in Eq. (25) is the restoring force set up by the lattice in response to the periodic pinning force. To determine the linear response of the vortex medium to $\vec{F}_L$, one should study the oscillating motion of the vortices about their static equilibrium configuration characterized by the deformation field $\vec{W} = (u,v)$ described by Eq. (2). In the continuum limit this problem can be solved exactly in terms of elliptic functions[33]. To avoid the rather complex algebra of this treatment, in this lecture we shall rely on a simpler perturbative approach which contains, however, all the essential physical features of the vortex dynamics. We assume that $\Delta$ is small and, consequently, write the solution of Eq. (25) in the form[3] :

$$\vec{r}_{\underline{l}} = \vec{\underline{l}} + i(\vec{v}_{rf}/\Omega)\, e^{-i\Omega t} + \vec{u}_{\underline{l}} \,, \tag{26}$$

where $\vec{v}_{rf} = (d\phi_0/\eta)(\vec{j}_{rf} \times \hat{z})$. The first two terms in Eq. (26) represent the solution of Eq. (25) in the flat-film case ($\Delta = 0$), whereas $\vec{u}_{\underline{l}}$ is the (small) additional dynamic displacement caused by the (weak) periodic pinning force. Substituting $\vec{r}_{\underline{l}}$, as given by Eq. (26), into Eq. (25) and expanding up to first order in the small quantities $\vec{q} \cdot \vec{u}_{\underline{l}} \ll 1$ and $(\vec{q} \cdot \vec{v}_{rf})/\Omega \ll 1$ one obtains :

$$\eta\dot{\vec{u}}_{\underline{l}} = - \sum_{\underline{l}'}{}' \tilde{G}(\underline{l} - \underline{l}')\vec{u}_{\underline{l}'} + \frac{\Delta'}{2i} \sum_{\vec{q}} \vec{q}(1+i\vec{q} \cdot \vec{u}_{\underline{l}})\left(1 - \frac{\vec{q} \cdot \vec{v}_{rf}}{\Omega}\, e^{-i\Omega t}\right) e^{i\vec{q}\cdot(\vec{l} - \vec{r}_0)} \,, \tag{27}$$

where the sum in the last term is over $\vec{q}$ and $-\vec{q}$ (two terms). To solve Eq. (27) it is again convenient to expand $\vec{u}_{\underline{l}}$ in (transverse) normal modes of the lattice. The resulting expression contains terms proportional to the normal mode amplitudes $\vec{u}_t(\vec{k},\omega)$, $\vec{u}_t(\vec{k} \pm \vec{q},\omega)$ and $\vec{u}_t(\vec{k} \pm \vec{q}, \omega \pm \Omega)$. Since we are interested in situations where $\vec{q}$ is close to one of the reciprocal lattice vectors $\vec{g}$ ($\vec{q} \approx \vec{g}$), one can set $\vec{u}_t(\vec{k} \pm \vec{q},\omega) \simeq \vec{u}_t(\vec{k},\omega)$. Moreover, one can neglect, to a first approximation, terms involving $\vec{u}_t(\vec{k} \pm \vec{q}, \omega \pm \Omega)$ which, arising from the product $(\vec{q} \cdot \vec{u}_{\underline{l}})(\vec{q} \cdot \vec{v}_{rf})/\Omega$ in Eq. (27), are small compared to those proportional to $\vec{u}_t(\vec{k},\omega)$. Then, the equation of motion can be solved for $\vec{u}_t(\vec{k},\omega)$ and the resulting expression is used to calculate the contributions, proportional to $\vec{u}_t(\vec{k} \pm \vec{q}, \omega \pm \Omega)$, neglected in the first order approximation. Finally, a new solution, now accurate to second order in $\Delta$, is worked out from which the vortex velocity $\vec{v}_{\underline{l}}(t) = \dot{\vec{r}}_{\underline{l}}$ is easily deduced. At this point, in order to describe the anisotropic dynamic response of the 2D vortex lattice in the 1D periodic potential, we introduce a vortex mobility tensor $\tilde{\mu}(\Omega)$ defined by $\vec{v}(\Omega) = \tilde{\mu}(\Omega)\vec{F}_L(\Omega)$, where $\vec{v}(\Omega)$ is the vector amplitude of the oscillating flux-flow velocity of the vortex medium. $\vec{v}(\Omega)$ follows by averaging $\vec{v}_{\underline{l}}(t)$ over all vortices of the lattice. An alternative way to describe the lattice response is in terms of a vortex impedance tensor $\tilde{Z}_{v\square}(\Omega)$ defined by $\vec{E}(\Omega) = \tilde{Z}_{v\square}(\Omega)\vec{J}(\Omega)$, where $\vec{E}(\Omega) = -\vec{v}(\Omega) \times \vec{B}$ and $\vec{J}(\Omega) = d\vec{j}(\Omega)$ are, respectively, the electric field and sheet (□) current amplitudes. Since in a coordinate system with the x-axis pointing in the $\vec{q}$-direction both $\tilde{\mu}(\Omega)$ and $\tilde{Z}_{v\square}(\Omega)$ are found to be diagonal, their components are simply related to each other by $Z_{v\square xx} = n_\square \phi_0^2 \mu_{yy}$ and $Z_{v\square yy} = n_\square \phi_0^2 \mu_{xx}$. Within our perturbative approach we find $\mu_{yy} = \eta^{-1}$, as expected, and :

$$\mu_{xx} = \eta^{-1}\left(1 - \frac{1}{2}\frac{1}{1-i\Omega\tau}\delta_{\vec{q},\vec{g}} - \frac{1}{2}\frac{\Delta/\mu}{(1-i\Omega\tau)[(b-1)^2+(\Delta/\mu)]}\right) \quad (28)$$

where $b = B/B_{mn}$ and $\tau = \eta/\Delta' q^2$ is the characteristic time for vortex relaxation in the potential wells. In writing the last term of Eq. (28) we have expressed the restoring force constant $\mu k^2$, where $\vec{k} = \vec{q} - \vec{g}$ is the wave vector of the pinning-induced shear deformation propagating at $45^0$ with respect to $\vec{q}$, in terms of b (or, alternatively, of $\delta$) using the results of Ref. 3. Real and imaginary parts of $\mu_{xx}$ (or $Z_{v\square yy}$), as deduced from Eq. (28), are shown in Fig. 7 (dashed curves) as a function of b. There is a discontinuous jump in both components at $b = 1$ ($\delta = 0$). This reflects the onset of the CI-phase transition which, in our approximation, occurs for $\delta_c = 0$. In the exact calculation[33] both $Re[Z_{v\square}]$ and $Im[Z_{v\square}]$ are constant within a C-phase and equal to the corresponding values given by Eq.

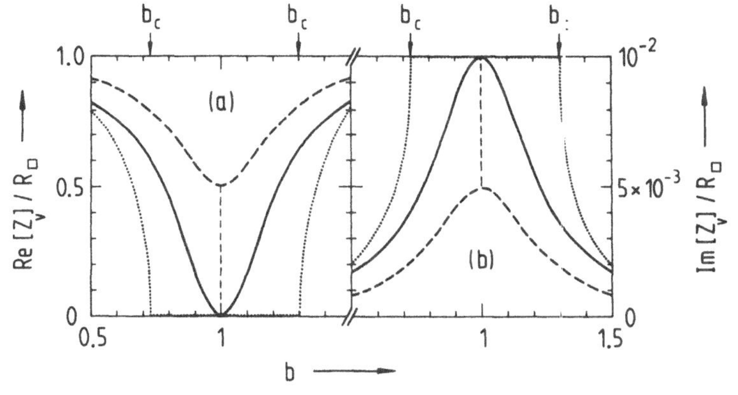


*Fig. 7. Real (a) and imaginary (b) part of the normalized vortex impedance as a function of $b = B/B_{mn}$. The driving current flows in the y-direction perpendicular to $\vec{q}$. $\Delta/\mu = 0.05$, $\Omega\tau = 10^{-2}$. The significance of the various curves is explained in the text.*

(28) for $b=1$ ($\vec{q} = \vec{g}$) up to the critical mismatch $\delta_c$ (dotted curves in Fig. 7). One way to correct for the jump at $b=1$ and to improve our approximation is to write $\mu_{xx}$ in the following form :

$$\mu_{xx} = \eta^{-1}\left(1 - \frac{\Delta/\mu}{(1 - i\Omega\tau)[(b-1)^2 + \Delta/\mu]}\right) \tag{29}$$

As shown in Fig. 7 (full curves) this expression provides a reasonable description of the actual lattice response. For this reason we shall rely on Eq. (29) for the interpretation of the sheet conductance measurements reported in the last section of this lecture.

There is a simple physical interpretation for the general behaviour of $Re[Z_{v\square}]$ and $Im[Z_{v\square}]$ shown by Fig. 7. In a C-phase dissipation arises from the excitation, in the viscous vortex medium, of a collective mode in which the vortices oscillate in phase around their equilibrium positions at the bottom of the potential wells. In the I-phase additional dissipation results from the excitation of the soliton superstructure. Thus, $Re[Z_{v\square}]$ must have its minimum value in a C-phase. $Im[Z_{v\square}]$, on the other hand, measures the delay in lattice response caused by the periodic pinning structure. As the effect of pinning is strongest in a C-phase, $Im[Z_{v\square}]$ is obviously largest in such a phase.

## Sheet Conductance Measurements

To measure the dynamic response of the vortices in our thickness modulated layers, we rely on the two-coil experimental technique developed by Fiory and Hebard[34]. The superconducting film is mounted in a transverse plane between two coaxial closely-spaced coils, one (drive coil) providing the rf-excitation of the vortex medium, the other (receive coil) to detect its rf-response. For an isotropic 2D superconductor, as it is the case for a flat ($\Delta = 0$) superconducting film, Fiory and Hebard have shown that, in the limit of weak screening, the rf-voltage amplitude $V_R(\Omega)$ at the receive coil due solely to the rf-response currents flowing in the sample is proportional to $\Omega^2 G_\square I_D(\Omega)$, where $G_\square = Z_\square^{-1}$ is the (complex) sheet conductance of the film and $I_D(\Omega)$ the (constant) amplitude of the rf-current in the drive coil. A straightforward extension of this calculation to our anisotropic samples shows that, in the same weak-screening limit, $V_R(\Omega)$ can be written in the form :

$$V_R(\Omega) = \Omega^2 I_D(\Omega)(K_1 G_{\square xx} + K_2 G_{\square yy}) , \tag{30}$$

where $K_1$ and $K_2$ are constants depending on the geometry of the sample and of the two-coil configuration and the x-axis is still pointing in the $\vec{q}$-direction. According to Fiory and Hebard[35 36], $\tilde{Z}_\square = \tilde{G}_\square^{-1}$ can be expressed as a series connection of the vortex impedance $\tilde{Z}_{v\square}$ with the kinetic inductance $L_k = (1/2)\mu_0\Lambda$ associated with the superfluid background. Thus, using the results of the previous section, $Z_{\square xx} = G^{-1}_{\square xx} = (n_\square\phi_0^2/\eta) + i\Omega L_k$ and $Z_{\square yy} = G^{-1}_{\square yy} = n_\square\phi_0^2\mu_{xx} + i\Omega L_k$, where $\mu_{xx}$ is given by Eq. (29). Using typical parameters for our Al-films (Table 1), it can be easily verified that, except very close to $T_c$ or at extremely low magnetic fields, the kinetic inductance term is always much smaller than the vortex impedance and can therefore be neglected in the analysis of most of our experimental data. Furthermore, in the vicinity of a registered phase ($B \approx B_{mn}$), the lattice configuration of interest here, $Z_{\square xx} \approx n_\square\phi_0^2/\eta$ varies weakly with B, while, on the contrary, $Z_{\square yy} \approx n_\square\phi_0^2\mu_{xx}$ shows pronounced structures (Fig. 7). Thus, if one uses a modulation technique, in which B is weakly modulated

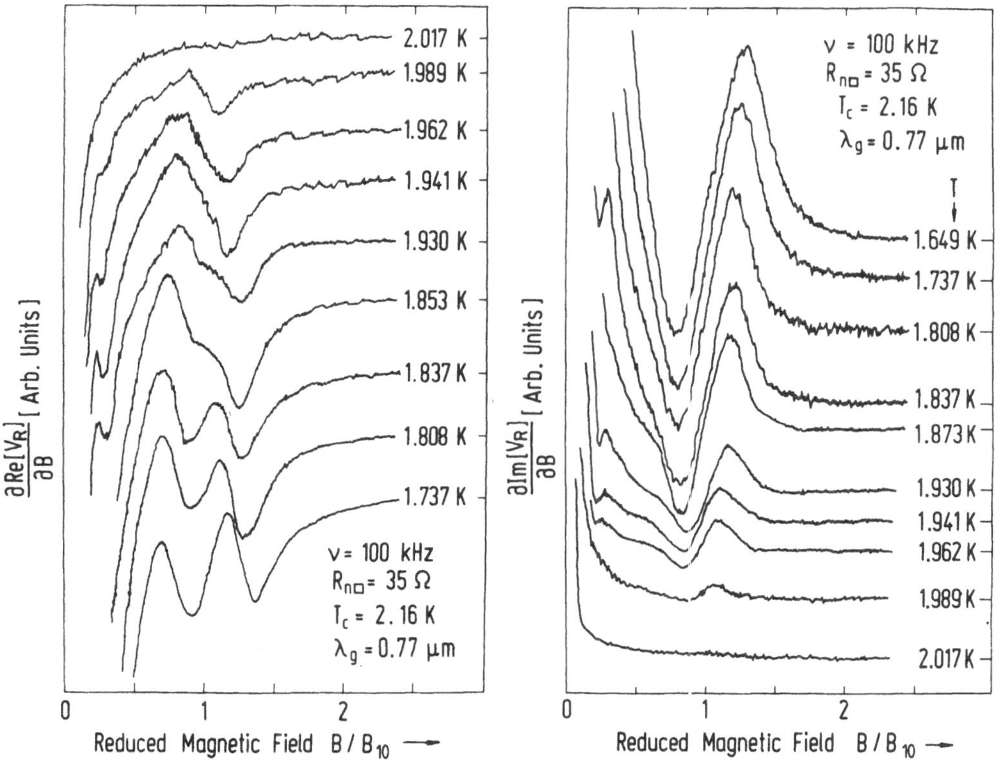


*Fig. 8. Derivative curves of the in-phase and out-of phase components of the rf-response of Al2 as a function of B. Marks on the vertical axis denote the zero-level of the signal.*

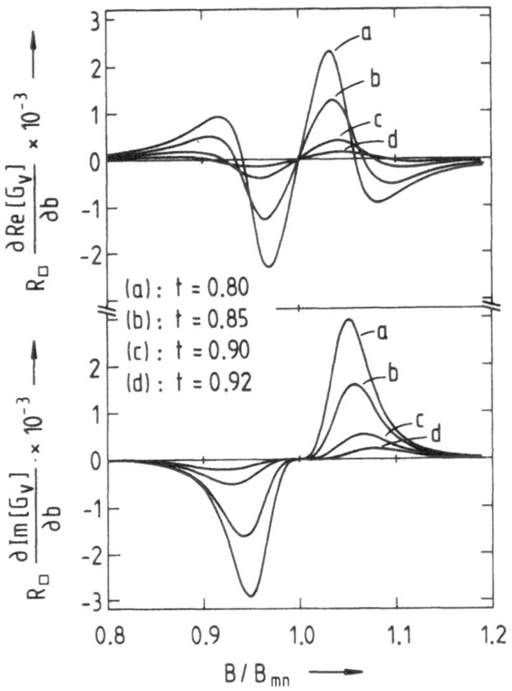


*Fig. 9. Theoretical derivative curves of the complex vortex impedance for Al2 as deduced from Eqs. (29) and (14) using $\Delta/\mu = 0.9$. See text for the determination of $\Omega\tau(T)$.*

($\delta B \approx 0.1$ Gauss) at a low frequency ($\sim 2$ Hz), the signal one actually detects at the receive coil will be essentially proportional, for $B \approx B_{mn}$, to $\partial G_{\square yy}/\partial B$. In Fig. 8 sets of curves at different temperatures for the in-phase and out-of-phase components of the 100 kHz-signal for Al2, measured with a conventional phase-sensitive technique, are shown as a function of B. There are pronounced structures in both the real and imaginary parts in the vicinity of $B_{10} \approx 30$ Gauss (weaker structures are visible also in correspondence of the unresolved $C_{11}$- and $C_{20}$-phases). The shape of these structures should now be compared with that predicted by Eq. (29). To this purpose, in Fig. 9 we show theoretical curves calculated from Eq. (29) where, in an attempt to partially include the effect of thermal fluctuations, $\Delta/\mu$ was replaced by $\Delta_R/\mu$ and $\tau = \eta/\Delta_R' q^2 = \eta/q\phi_0 d j_{cM}$ [Eq. (20)] was estimated using the $j_{cM}$-data for Al2 of Fig. 5. While there is good qualitative agreement for both components of the signal at low temperatures, the evolution of the structures about $B_{10}$ with rising temperature is quite different from that predicted by Eq. (29). In particular, as one expects from the temperature dependence of the critical mismatch $\delta_c$ tolerated by a C-phase, their width becomes narrower and narrower as T approaches $T_{LU}$, a feature which does not emerge from the theoretical curves of Fig. 9. Clearly, a more elaborated model which takes into account, in particular, diffusion phenomena of "corrugated" domain walls should be worked out in order to describe the dynamic response of the periodically pinned 2D vortex lattice at high temperatures ($T \lesssim T_{LU}$). This, as well as the study of other features of the rf-response (in particular of the origin of the drastic crossover in response at low magnetic fields in Fig. 8) will be the object of future research in our laboratory.

ACKNOWLEDGEMENTS

We would like to thank J.R. Clem, V.L. Pokrovsky and M. Puga for stimulating discussions. This work has been supported by the Swiss National Science Foundation.

REFERENCES

1. O. Daldini, P. Martinoli, J. L. Olsen, and G. Berner, Vortex-line pinning by thickness modulation of superconducting films, Phys. Rev. Lett. 32:218 (1974).
2. P. Martinoli, O. Daldini, C. Leemann, and B. Van den Brandt, Josephson oscillation of a moving vortex lattice, Phys. Rev. Lett. 36:382 (1976).
3. P. Martinoli, Static and dynamic interaction of superconducting vortices with a periodic pinning potential, Phys. Rev. B 17:1175 (1978).
4. P. Bak, Commensurate phases, incommensurate phases and the devil's staircase, Rep. Prog. Phys. 45:587 (1982).
5. J. Villain, Theories of commensurate-incommensurate transitions on surfaces, in: "Ordering in two dimensions", S. K. Sinha, ed., North Holland, Inc., New York (1980).
6. J. M. Kosterlitz, and D. J. Thouless, Ordering, metastability and phase transitions in two-dimensional systems, J. Phys. C 6:1181 (1973).
7. D. R. Nelson, and B. I. Halperin, Dislocation-mediated melting in two dimensions, Phys. Rev. B 19:2457 (1979).
8. V. L. Pokrovsky, and A. L. Talapov, The theory of two-dimensional incommensurate crystals, Zh. Eksp. Teor. Fiz. 78:269 (1980) [Sov. Phys. JETP 51:134 (1980)].

9. P. Martinoli, H. Beck, M. Nsabimana, and G.-A. Racine, Locking-unlocking transition of a two-dimensional lattice of superconducting vortices, Physica 107B:455 (1981).

10. P. Martinoli, M. Nsabimana, G.-A. Racine, H. Beck, and J. R. Clem, Locked and unlocked phases of a two-dimensional lattice of superconducting vortices, Helv. Phys. Acta 55:655 (1982).

11. A. T. Fiory, Measurements of the shear modulus of the superconducting mixed state of thin films, Phys. Rev. B 8:5039 (1973).

12. F. C. Frank, and J. H. van der Merwe, One-dimensional dislocations, Proc. Roy. Soc. (London) A198:205 (1949).

13. J. Bardeen, and M. J. Stephen, Theory of the motion of vortices in superconductors, Phys. Rev. 140A:1197 (1965).

14. M. Tinkham, in: "Introduction to superconductivity", McGraw-Hill, Inc., New York (1975).

15. M. W. Puga, E. Simanek, and H. Beck, Renormalization-group approach to commensurate-incommensurate transitions in two dimensions, Phys. Rev. B 26:2673 (1982).

16. M. W. Puga, E. Simanek, and H. Beck, Commensurate-incommensurate transitions, to appear in Helv. Phys. Acta (1983).

17. B. A. Huberman, and S. Doniach, Melting of two-dimensional vortex lattices, Phys. Rev. Lett. 43:950 (1979).

18. D. S. Fisher, Flux-lattice melting in thin-film superconductors, Phys. Rev. B 22:1190 (1980).

19. J. V. José, L. P. Kadanoff, S. Kirkpatrick, and D. R. Nelson, Renormalization, vortices, and symmetry-breaking perturbations in the two-dimensional planar model, Phys. Rev. B 16:1217 (1977) [Erratum, Phys. Rev. B 17:1477 (1978)].

20. S. Ostlund, Relation between lattice and continuum theories of two-dimensional solids, Phys. Rev. B 23:2235 (1981).

21. S. Ostlund, Incommensurate and commensurate phases in asymmetric clock models, Phys. Rev. B 24:398 (1981).

22. J. Villain, and P. Bak, Two-dimensional Ising model with competing interactions : floating phase, walls and dislocations, J. Physique 42:657 (1981).

23. S. N. Coppersmith, D. S. Fisher, B. I. Halperin, P. A. Lee, and W. F. Brinkman, Dislocations and the commensurate-incommensurate transition in two dimensions, Phys. Rev. Lett. 46:549 (1981).

24. T. Bohr, V. L. Pokrovsky, and A. L. Talapov, Commensurate-incommensurate phase transition in continuous media containing dislocations, Pis'ma Zh. Eksp. Teor. Fiz. 35:165 (1982) [Sov. Phys. JETP Lett. 35:203 (1982)].

25. T. Bohr, Dislocations in the commensurate-incommensurate transition, Phys. Rev. B 25:6981 (1982).

26. F. D. M. Haldane, P. Bak, and T. Bohr, Phase diagramms of surface structures from Bethe-Ansatz solutions of the quantum sine-Gordon model, preprint (1982).
27. S. E. Burkov, and V. L. Pokrovsky, Critical currents and electric fields of two-dimensional systems, J. Low Temp. Phys. 44:423 (1981).
28. P. Martinoli, J. L. Olsen, and J. R. Clem, Superconducting vortices in periodic pinning structures, J. Less-Common Metals 62:315 (1978).
29. P. Martinoli, and J. R. Clem, Pinning in periodic superconducting structures, in: "Inhomogeneous superconductors-1979", D. U. Gubser, T. L. Francavilla, S. A. Wolf, and J. R. Leibowitz, ed., American Institute of Physics, New York (1980).
30. A. Schmid, and W. Hauger, On the theory of vortex motion in an inhomogeneous superconducting film, J. Low Temp. Phys. 11:667 (1973).
31. J. R. Clem, Simple model for the vortex core in a type II superconductor, J. Low Temp. Phys. 18:427 (1975).
32. P. Martinoli, O. Daldini, C. Leemann, and E. Stocker, A. C. quantum interference in superconducting films with periodically modulated thickness, Solid State Comm. 17:205 (1975).
33. V. L. Pokrovsky, private communication.
34. A. T. Fiory, and A. F. Hebard, Radio-frequency complex-impedance measurements on thin film two-dimensional superconductors, in: "Inhomogeneous superconductors-1979", D. V. Gubser, T. L. Francavilla, S. A. Wolf, and J. R. Leibowitz, ed., American Institute of Physics, New York (1980).
35. A. F. Hebard, and A. T. Fiory, Recent experimental results on vortex processes in thin-film superconductors, in: "Ordering in two dimensions", S. K. Sinha, ed., North Holland, Inc., New York (1980).
36. A. F. Hebard, and A. T. Fiory, Vortex dynamics in two-dimensional superconductors, Physica 109 & 110 B:1637 (1982).

# PINNING BY ORDINARY DEFECTS IN ORDINARY SUPERCONDUCTORS

Dierk Rainer

Physikalisches Institut
Universität Bayreuth
D 8580 Bayreuth, FRG

## INTRODUCTION

Most textbooks present the BCS-Gorkov theory of superconductivity as a selfconsistent field theory which is obtained by generalizing the normal state Hartree-Fock theory to the superconducting state. The use of this familiar concept allows a fast derivation of the central equations of superconductivity but does not explain the evidently high accuracy of the BCS-Gorkov theory. Its error is for most materials of the order of a few percent. In order to understand the high precision one better views the BCS-Gorkov theory as a generalization of Landau's Fermi liquid theory to the superconducting state. One can show /1,2/ that such a theory keeps correctly the leading terms in several good expansion parameters like $kT_c/\hbar\omega_D$, $1/k_F\xi_0$, $1/k_F\ell$, etc. We borrow the notation of ref. /3/ and denote the magnitude of the expansion parameters collectively by 'small'. For a typical metal 'small' is of the order $10^{-2}$ which explains the good accuracy of a theory based on this expansion parameter. The BCS-Gorkov theory (also called 'weak-coupling theory') shall be understood in the following as the asymptotically correct theory in leading order in 'small'. The main building blocks of this theory are the electronic quasiparticles. This means, e.g., that Cooper pairs have to be formed out of two quasiparticles and not out of two electrons. A quasiparticle differs from a moving electron by some electronic backflow and an accompanying virtual phonon cloud. This 'dressing' strongly affects the quasiparticle properties. E.g., the dressing with phonons slows down the Fermi velocity by a factor $(1+\lambda) \simeq 1.5 - 2.5$. An accurate calculation of quasiparticle data for real metals is still an unsolved problem. However, given the quasiparticle data of a metal in its normal state one can calculate reliably and accurately its superconducting properties.

The message of the lengthy overture is simply that we have at hand an accurate microscopic theory of superconducting materials which we should utilize whenever experiments alone can not settle a problem of interest. We seem to be facing this situation in the problem of flux-line pinning in type-II superconductors. The observable effect, the critical current, is composed of various more fundamental partial effects such as:

a) the elementary pinning force of a single defect
b) the elastic properties of the flux-line lattice
c) the summation of pinning forces from statistically distributed pinning centers to a total force.

A critical current experiment is not able to disentangle the underlying effects, and needs theoretical support. The hardest theoretical problem is the summation theory. It is not yet settled despite of very promising recent conceptual /4/ and computational /5/ progress. Hence, it is still impossible to extract the elementary pinning forces from the observed total force. We suggest a different strategy /6,7/. One should evaluate the elementary pinning force for simple enough defects from the microscopic BCS-Gorkov theory. If this can be done accurately and reliably, one can feed the result into existing summation theories, and check the summation theory by comparison with experiments. Some feedback and active cooperation between experimental groups, summation theorists, and BCS-Gorkov theorists can probably settle the pinning problem in the near future.

We have presented our methods for calculating the elementary pinning potential in refs. /6,7/. These papers contain essentially three messages: a) conventional theories underestimate appreciably the pinning strength of small defects, b) a full calculation of the pinning potential of small defects within the framework of the BCS-Gorkov theory can be done with minor computational effort, and c) the main theoretical progress which makes our and further computations feasible comes from a very efficient formulation of the BCS-Gorkov theory, the so called 'quasiclassical method'.

## THE QUASICLASSICAL METHOD

It was realized 20 years ago /8,9/ that the theory of superconductivity is formulated most efficiently in terms of a transport equation, obtained by eliminating all unnecessary quantum-mechanical ballast from the original equations of the BCS-Gorkov theory. These ideas were extended by Eilenberger /10/, Larkin, and Ovchinnikov /11/, who developed a complete microscopic theory of superconductivity in terms of 'quasiclassical propagators' whose equation of motion resembles very much a classical transport equation. All left-over traces of quantum mechanics are contained in two internal degrees of freedom, the spin and the particle-hole degree of freedom. The latter is essential for understanding superconductivity. The internal degrees of freedom form a four-dimensional Hilbert space. Hence, the quasiclassical propagators are 4x4 matrices $\hat{g}(\hat{k},\vec{R};\varepsilon_n)$. The equation of motion for $\hat{g}$ is a first order ordinary differential equation along classical trajectories:

$$[(i\varepsilon_n+v_F e\hat{k}\cdot\vec{A}(\vec{R}))\hat{\tau}_3-\hat{\Delta}(\vec{R})-\hat{\sigma}_i(\hat{k},\vec{R};\varepsilon_n),\hat{g}(\hat{k},\vec{R};\varepsilon_n)] + \\ + iv_F\hat{k}\cdot\vec{\nabla}_R\hat{g}(\hat{k},\vec{R};\varepsilon_n) = [\hat{t}(\hat{k},\hat{k};\varepsilon_n),\hat{g}_{imt}(\hat{k},\vec{R}_0;\varepsilon_n)]\delta(\vec{R}-\vec{R}_0). \tag{1}$$

The right hand side describes the effect of a small defect located at $\vec{R}_0$. $\hat{g}_{imt}$ is the propagator without the defect, and $\hat{t}$ the quasiparticle-defect scattering matrix. The order parameter $\hat{\Delta}(\vec{R})$ and the magnetic field (vector potential) $\vec{A}(\vec{R})$ must be determined selfconsistently. The effects of random impurities (defects) are described by the impurity self-energy $\hat{\sigma}_i(\hat{k},\vec{R};\varepsilon_n)$. $\hat{\sigma}_i$ depends implicitely on the full $\hat{g}$. $\hat{g}_{imt}$ is normalized,

$$\hat{g}_{imt}(\hat{k},\vec{R};\varepsilon_n)^2 = -(\pi\not{1})^2, \tag{2}$$

and $\hat{g}$ approaches $\hat{g}_{imt}$ far away from the defect.

The defect (pinning center) is characterized by an effective potential $\hat{v}(\hat{k},\hat{k}')$, or equivalently, by a set of normal state phase shifts. The scattering matrix $\hat{t}$ and the potential $\hat{v}$ are related by the quasiclassical T-matrix equation

$$\hat{t}(\hat{k},\hat{k}';\varepsilon_n) = \hat{v}(\hat{k},\hat{k}') + \qquad (3)$$

$$+ \frac{N(0)}{\hbar}\int\frac{d\Omega''_k}{4\pi}\hat{v}(\hat{k},\hat{k}'')\hat{g}_{imt}(\hat{k}'',\vec{R}_0;\varepsilon_n)\hat{t}(\hat{k}'',\hat{k}';\varepsilon_n).$$

The impurity self-energy is given by

$$\hat{\sigma}_i(\hat{k},\vec{R};\varepsilon_n) = c\cdot\hat{t}_i(\hat{k},\hat{k},\vec{R};\varepsilon_n), \qquad (4)$$

with c the concentration of impurities, and $\hat{t}_i$ the impurity T-matrix. $\hat{t}_i$ is determined by an equation identical to (3) but with $\hat{v}$ replaced by the impurity potential $\hat{v}_i$, and $\hat{g}_{imt}$ by the full propagator $\hat{g}(\hat{k},\vec{R};\varepsilon_n)$. The final equation needed in the theory is the difference in free energy with and without the defect

$$\delta\Omega(\vec{R}_0) = \int_0^1 d\lambda\,\frac{kT}{2\hbar}\sum_n N(0)\int\frac{d\Omega_k}{4\pi}\int d^3R\,\mathrm{Tr}_4(\delta\hat{g}(\hat{k},\vec{R};\varepsilon_n;\lambda)\hat{\Delta}_b(\vec{R})) \qquad (5)$$

where $\delta\hat{g}(\hat{k},\vec{R};\varepsilon_n;\lambda) = \hat{g}-\hat{g}_{imt}$ calculated for the order parameter $\hat{\Delta}(\vec{R}) = \lambda\hat{\Delta}_b(\vec{R})$, $\hat{\Delta}_b(\vec{R})$ being the actual order parameter without the defect. $\delta\Omega(\vec{R}_0)$ is a function of the defect position. Its gradient gives us the pinning forces.

Eqs. (1) - (5) are correct for isotropic superconductors in the limit of small defects compared to the coherence length $\xi_0$ (more precisely, a small defect means $\sigma \ll \xi_0^2$, where $\sigma$ is the defect cross section). For such small defects we can let the defect shrink to a $\delta$-function scatterer, and ignore the changes in the order parameter and the magnetic field induced by the defect when calculating $\delta\Omega$*. With this simplifications a full solution of (1) - (5) takes a few minutes on a moderately fast computer. At present, we have computed the full pinning potential $\delta\Omega$ for an isolated vortex in a clean superconductor. We have studied the temperature dependence of $\delta\Omega$, its sensitivity to different types of defects (= different sets of phase shifts), and its dependence on the size of the defect. Some representative results are displayed in figs. 1 and 2.

*This is the consequence of the stationarity principle for the free energy. The first order change in $\Omega$ can be calculated with the unperturbed order parameter.

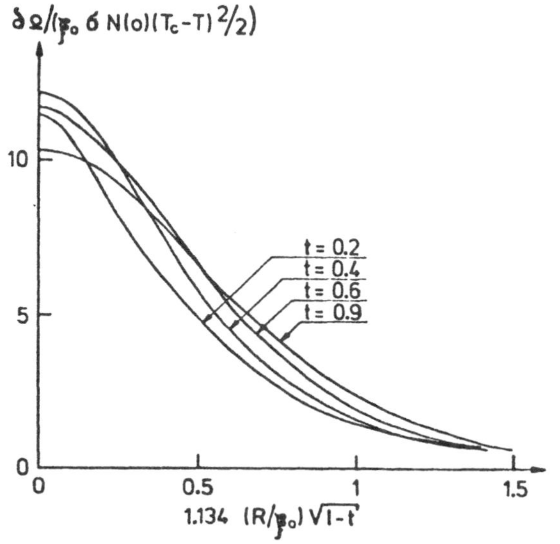


Fig. 1: Pinning potential as a function of the distance R between vortex center and defect.

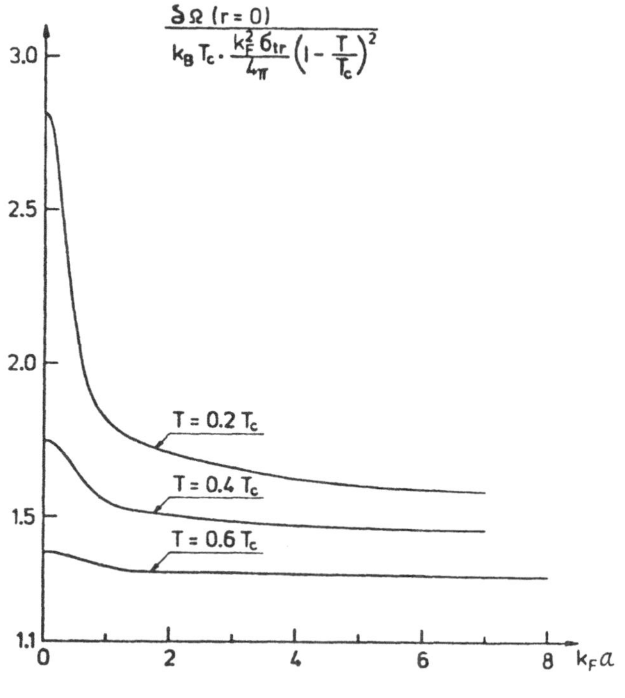


Fig. 2: Dependence of the pinning potential at the vortex center on the hard sphere radius a at different temperatures.

An extensive report on our calculations will be published shortly; preliminary results can be found in refs. /6,7/. In general, one can say that we have developed our computational techniques for solving the quasiclassical equations to a stage that further calculations, e.g., pinning in dirty materials, pinning of a vortex lattice, or other extensions could be done routinely - if needed.

## REFERENCES

/1/ A.I. Larkin, A.B. Migdal, Sov. Phys. JETP 17, 1146 (1963)
/2/ P. Morel, P. Nozières, Phys. Rev. 126, 1909 (1962)
/3/ J.W. Serene and D. Rainer, to be published in Physics Reports
/4/ A.I. Larkin and Yu. N. Ovchinnikov, J. Low Temp. Phys. 34, 409 (1979), and references cited herein
/5/ E.H. Brandt, Phys. Rev. Lett. 50, 1599 (1983)
/6/ E.V. Thuneberg, J. Kurkijärvi and D. Rainer, Phys. Rev. Lett. 48, 1853 (1982)
/7/ J. Kurkijärvi, D. Rainer and E.V. Thuneberg, proceedings of Superconductivity in d- and f-Band Metals 1982, Kernforschungszentrum Karlsruhe, 1982, W. Buckel, W. Weber, eds.
/8/ P.G. deGennes, Rev. Mod. Phys. 36, 225 (1964)
/9/ A.F. Andreev, Soviet Phys. JETP 19, 1228 (1964)
/10/ G. Eilenberger, Z. Phys. 214, 195 (1968)
/11/ A.J. Larkin and Yu. Ovchinnikov, Sov. Phys. JETP 28, 1200 (1969)

# GRANULAR SUPERCONDUCTORS AND JOSEPHSON JUNCTION ARRAYS

S. Doniach

Department of Applied Physics
Stanford University
Stanford, California 94305

## 1. INTRODUCTION

In these lectures we focus on those properties of granular superconductors which pertain to a regime of microscopic conditions such that they behave as assemblies of coupled Josephson junctions. Essentially, this should be a good picture if the Ginzburg-Landau coherence length of the bulk metal comprising the superconducting grains is of order of the grain size, while the Josephson coupling energy between grains (whether in SIS or SNS composites) is small compared to the BCS condensation energy of the individual grains. These conditions, and the fact that they lead to strongly type II superconductivity are spelled out in Section 2. In Section 3, the consequences of the resulting classical XY model behavior for the thermodynamic properties of 3-dimensional and 2-dimensional (3D and 2D) granular superconductors are summarized. The consequences of randomness of the intergrain coupling are not dealt with, apart from a brief discussion of the probable absence of localization effects in 2D systems.

In the second part of the lectures, the effect on the thermodynamic properties of including charging fluctuations, or equivalently macroscopic quantum tunneling (MQT) between grains, are discussed. In Section 4, we show how these zero-point fluctuations renormalize the vortex-antivortex coupling in 2D granular superconductors or arrays of Josephson junctions. For sufficiently small junction coupling capacitance, these effects lead to a superconductor-insulator phase boundary extending down to zero temperature. In Section 5, we discuss quantum effects in 1D junction arrays and show that in the limit that intergrain charging energies dominate self-charging effects, the system will be an insulator even in the limit of very weak tunneling. These discussions leave out, however, considerations of dissipative coupling (see lectures by Leggett and Ambegaokar, this volume).

Finally, in the third part (Section 6) we discuss briefly the nonlinear dynamics of junction arrays. Recent results on the instabilities of the simplest array - that of three junctions as a function of dc current drive are reported. In addition to three limit-cycle modes, we also find a period doubling regime leading to chaotic instability. Although this kind of behavior is undoubtedly modified in larger arrays, we believe it may give some ideas about how to analyze the instabilities of larger systems.

## 2. BASIC CONDITIONS FOR XY BEHAVIOR AND THE LANDAU SCREENING LENGTH IN GRANULAR SYSTEMS

We summarize briefly the discussion given by Deutscher, Imry and Gunther (1974). Consider an assembly of superconducting grains of size $s_0$ separated by spacing s: for SIS systems (e.g., NbN gains with oxide junctions) the Josephson coupling results from Cooper pair tunneling between grains, while for SNS systems, it may be thought of as proximity effect tunneling. Provided the Ginzburg-Landau coherence length $\xi_{GL}$ of the bulk material comprising the grains is of order of the grain size, so that the order parameter $\psi_i$ of the i grain does not vary too much across the grain, the total Hamiltonian of the system may be written

$$H = \sum_i V_0 \left\{ a|\psi_i|^2 + \frac{b}{2}|\psi_i|^4 \right\} + \tfrac{1}{2}\sum_{nn} J_{ij}|\psi_i - \psi_j|^2 \tag{2.1}$$

where $V_0$ is the volume of each grain, a and b are the Ginzburg-Landau parameters of the bulk material, and $J_{ij}$ is the intergrain coupling energy (nn denotes nearest-neighbor coupling only).

Under these conditions, Deutscher et al. show by expanding about $T_{BCS}$ (the superconducting transition temperature of the bulk material comprising the grains) that the effective coherence length for the <u>granular composite</u> is

$$\xi^2_{eff} \propto \frac{J(T)s^2}{a(T)s^3} \quad . \tag{2.2}$$

So provided $\xi_{eff} \ll s$, the intergrain spacing, the grains will fluctuate independently in the vicinity of $T \gtrsim T_{BCS}$, and there will not be a true thermodynamic transition until the phases of the grains lock together at $T_c \approx Jz$ where z is the number of nearest neighbors of a given grain. Since $a \approx T_{BCS}$ (we set Boltzmann's constant equal to unity), $\xi_{eff} \ll s$ implies that $T_c \ll T_{BCS}$ so that this class of granular materials will be characterized by two "transition" regions: as T is cooled below $T_{BCS}$, the grains will undergo a 0D transition to a paracoherent regime in which

$$\psi_i = |\psi_i| \, e^{i\phi_i} \tag{2.3}$$

with $|\psi_i| = \sqrt{|a|/b}$ a well developed order parameter, but where the $\phi_i$ are randomly disordered by thermal fluctuations. In this regime the resistivity of the composite drops to an intermediate value possibly dominated by random pinning (Trugman and Doniach 1982) or percolation effects (Imry, these lectures). Finally, at $T_c$, the phases lock together and the system is described by a classical XY model which develops true long-range order (3D) or algebraic long-range order (2D), leading to zero resistivity for $T < T_c$.

So far, we have discussed the properties of the system in the absence of an external magnetic field. The effects of a field on an XY model are not trivial to discuss, and, in order to make some estimates, we adopt a 'coarse graining' approach in which the partition function of the XY model Hamiltonian on the lattice of grains is re-expressed in terms of a Ginzburg-Landau-Wilson (GLW) continuum-type Hamiltonian. This coarse graining, which is done via a Hubbard-Stratanovich transformation, preserves the universal features of the XY model but only provides an approximate expression of the resulting parameters of the GLW model in terms of the parameters of the original XY model. Nevertheless, it gives us a useful guide to the large-scale properties of the granular system, and, in particular, allows us to estimate the Landau screening length (equivalently, penetration depth) for the granular composite. We find that it is very long compared to that of the original bulk metal in the grains, so that granular materials are very strong type-II superconductors. [This result was first shown by Rosenblatt in 1972; see Rabatou et al. (1980)].

The derivation goes as follows: (We give it in some detail, since an extension will be used in Section 4 to estimate the effects of charging fluctuations.) The partition function of the XY model may be written

$$Z \simeq \int \Pi_i \, d\phi_i e^{-H_{XY}/T} \quad (2.4)$$

where

$$H_{XY} = -2 \sum J_{ij} \psi_i^* \psi_j \equiv -\sum_q J_q \psi_q^* \psi_q \quad (2.5)$$

where $J_{ij} = J$ for $i=j$, $= -J$ for $i=nn(j)$ and

$$\psi_q = \frac{1}{\sqrt{N}} \sum_i e^{i\vec{q}.\vec{R}_i} \psi_i \quad (2.6)$$

with N the number of sites in the lattice.

Now define a set of coarse graining variables $\Phi_q$ for which the functional integral (2.4), which involves only integration over the phases of the order parameter, ($|\psi_i|$ is set to a constant value, $\psi$ ) is now transformed to a GLW model in which fluctuations of both amplitude and phase of $\Phi$ are included. This is done using the Gaussian identity

$$\int \mathcal{D}\Phi_q \, e^{-\sum_q |\Phi_{-q} - \sqrt{\beta J_q}\psi_q|^2} \equiv \int \mathcal{D}\Phi_q \, e^{-\sum_q |\Phi_q|^2} = Z_0 \quad (2.7)$$

where $\beta = 1/T$ and $J_q = \sum_{nn} e^{iq.R_{ij}} J_{ij}$. Then using (2.7), we have

$$Z = \frac{1}{Z_0} \int \Pi_i \, d\phi_i \int \mathcal{D}\Phi \, e^{-\sum_q |\Phi_q^2|} \, e^{+\sum_q \sqrt{\beta J_q} (\Phi_q^* \psi_q + h.c.)} \quad (2.8)$$

Here $\mathcal{D}\Phi$ denotes a functional integral over both phase and amplitude of $\Phi_q$ .

The course graining is now done by means of an approximate evaluation of (2.8) using a cumulant expansion:

$$Z = \int \mathcal{D}\Phi \, e^{-\mathcal{F}[\Phi]/T} \quad (2.9)$$

where $\mathcal{F}$ is expanded in powers of $\Phi_q$ as a GLW model, and the coefficients in this expansion are estimated using the cumulants of (2.8). For the quadratic term, one finds

$$\langle\psi_q\psi^*_{q'}\rangle = \int \Pi d\phi_i \frac{1}{N}\sum_{ij} e^{i\vec{q}.\vec{R}_i} e^{-i\vec{q}'.\vec{R}_j} e^{i\phi_i} e^{-i\phi_j}$$

$$x\left(\int \Pi d\phi_i\right)^{-1} = \delta_{qq'} \qquad (2.10)$$

leading to

$$\mathcal{F}^{(2)} = \sum_q |\Phi_q|^2(\beta J_q - 1) \qquad (2.11)$$

On expanding the q-dependence to second order in q this gives

$$\mathcal{F}^{(2)} \approx \sum_q |\Phi^2_q|(\beta zJ - 1) - \beta Jq^2s^2 \ . \qquad (2.12)$$

The fourth order term is of the form

$$\mathcal{F}^{(4)} = \sum_{q_1-q_4} \Phi_{q_1}\Phi^*_{q_2}\Phi_{q_3}\Phi^*_{q_4} \langle\psi^*_{q_1}\psi_{q_2}\psi^*_{q_3}\psi_{q_4}\rangle \qquad (2.13)$$

where the < > brackets denote an average over the original phase variables $\phi_i$, as in (2.10), leading to

$$\mathcal{F}^{(4)} \cong C\, \beta^2J^2 \int d^dx\, |\Phi(x)|^4 \qquad (2.14)$$

on neglecting the q-dependence of the four-$\psi$ correlation function. (C is a numerical constant). Hence, in sum, we get a GLW form

$$\mathcal{F} = \int d^dx\, \{c|\nabla\Phi(x)|^2 + a|\Phi(x)|^2 + \frac{b}{2}\,|\Phi^4(x)|\} \qquad (2.15)$$

where $c = \beta J^2s^2$ , $a = (\beta Jz-1)$ , and $b \propto \beta^2J^2$.

The bare GLW transition temperature occurs for a=0, giving the mean-field transition temperature

$$T_{MF} = zJ \ . \qquad (2.16)$$

The effect of a magnetic field is now easily included by setting $\nabla \to (\nabla - 2ieA/C)$ giving rise to a Landau screening length $\lambda_G$ for the granular composite given by (see, for instance, DeGennes 1966)

$$1/\lambda_G^2 = \frac{8\pi}{\beta} \frac{C}{4\pi^2} \frac{|\Phi|^2}{\phi_0^2} \tag{2.17}$$

where $\phi_0$ is the quantum of flux hc/2e, to be compared with

$$1/\lambda_B^2 = \xi_{GL}^2 a \ |\psi|^2/\phi_0^2 \tag{2.18}$$

for the bulk superconductor, with $\xi_{GL}$ the bulk Ginzburg-Landau coherence length. Hence

$$\lambda_G/\lambda_B^2 \sim \ \xi_{GL}/S\sqrt{\Delta/J} \quad ,$$

where $\Delta$ is the BCS gap energy. By the assumptions made for the granular system, $\lambda_G/\lambda_B$ is considerably less than 1, hence the granular material will be strongly type II even in cases (e.g., lead grains) when the bulk would be a type-I superconductor.

## 3. DISCUSSION OF THE CLASSICAL XY MODEL FOR 3D AND 2D GRANULAR MATERIALS; EFFECTS OF RANDOMNESS

In terms of thermodynamic fluctuations near the paracoherent-coherent phase transition, granular superconductors show a remarkable distinction from homogeneous superconductors: the critical region for non-linear fluctuation of the phase variables is that of a classical XY model, i.e., of order 30% of $T_c$. This is in contrast to BCS superconductors, where the unbinding of Cooper pairs above $T_{BCS}$ very rapidly washes out the inherent XY character of the phase transition: in the extreme case of pure superconductors (Thouless 1960) the critical region shrinks to something like $10^{-8}$ of $T_c$. This means that granular materials are a good system with which to study critical phenomena, both in 3D, where they should be analogous (subject to modifications due to London screening) to those occurring at the $\lambda$-point of $^4$He, and in 2D where Kosterlitz-Thouless vortex unbinding effects dominate. In this section we review briefly some recent developments in understanding of the 3D XY-model, then discuss some speculations on the effects of randomness for the 2D XY model.

## 3.1 Critical Fluctuations in 3D Granular Superconductors

In 3D, the coherent-paracoherent transition of a charged superfluid is intrinsically distinct from that of a neutral superfluid. For a type-I superconductor, Halperin, Lubensky and Ma (1974) showed that the phase transition actually becomes first order. This may be analogous to the usual phenomenon of the first order melting of classical liquids, viewed as a dislocation-loop "explosion" (Edwards and Warner 1979). For type-II superconductors, since the London length $\lambda_G$ is in principle finite, vortex loops in a 3D superconductor interact via screened current-loop-type interactions

$$U(\vec{r}_1, \vec{r}_2) = \frac{d\vec{r}_1 \cdot d\vec{r}_2}{r_{12}^3} e^{-\lambda_G r_{12}} \tag{3.1}$$

for the two line elements $d\vec{r}_1$, $d\vec{r}_2$ of a vortex a distance $r_{12}$ apart. Hence the phase transition is expected to be of 'spaghetti' type (Savit 1980) in which one can, at least at a qualitative level, neglect interactions, and simply balance the energy per unit length to grow a vortex loop, $\sim \varepsilon_{loop} L$ where L is the length of the loop, against the free energy gain due to random wandering of the loop $\approx - Ts_{eff}(T)L$ where $s_{eff}(T)$ is the entropy per unit length. By this argument, there exists a temperature $T_c = \varepsilon_{loop}/s_{eff}$ at which the loop length becomes 'infinite' and the system becomes filled with a 'spaghetti' of vortices. This situation is to be contrasted with $^4$He or dislocations loops in solids, where the vortex-vortex interactions are long ranged. Here one finds that the statistical mechanics of 'disorder fields' has the character of loops interacting via screened interactions (Kleinert 1982). These fields may be viewed as the 'dual' of the XY order parameter. For disordered fields, one has finite loops at high temperatures $T > T_c$ with a spaghetti transition as T is reduced through $T_c$, representing the 'symmetry breaking' of the disorder as a condensed phase is reached. Thus the second-order phase transition of a type-II superconductor may be viewed as a 'reverse XY model' (Das Gupta and Halperin 1981). In practice, $\lambda_G$ is large, of order of a millimeter, so finite size effects will probably wipe out the above subtleties for real samples. However, is it noteworthy that very large critical fluctuations (e.g., leading to a logarthmic divergence in the specific heat) are likely to occur in granular composites. One may also expect large critical effects in the transport properties (e.g. critical current near $T_c$) by analogy with 2D systems: vortex loops will be inherently unstable against supercurrents, and, provided pinning effects do not dominate, should lead to strongly nonlinear dependence of resistivity on current close to $T_c$.

For homogeneous 2D superconductors, the critical fluctuations are of Kosterlitz-Thouless type, leading to phase transitions dominated by vortex-antivortex unbinding (i.e., long-range forces between topological excitations) rather than by the spaghetti-type entropy as in the 3D case discussed above. Because of this difference, the response to external magnetic fields is quite different in 2D, where the injected vortices essentially wipe out the Kosterlitz-Thouless transition (Doniach and Huberman 1979, and Huberman and Doniach 1980), than in 3D, where the system remains superconducting in the presence of the Abrikosov vortex lattice. In the particular case of a periodic Josephson junction array (Voss and Webb 1982) the effects of commensurability of the vortex lattice with the underlying junction lattice modify these conclusions (Teitel and Jayaprakash 1983). For the discussion of the classical XY behavior of 2D superconductors see Mooij and Minnhagen (this volume).

### 3.2 Effects of Randomness

In 3D, the XY model with random bonds $J_{ij} \geq 0$, corresponds to a random ferromagnet if we neglect diamagnetic shielding (extreme type II limit). Note that the $J_{ij}$ all have the same sign as a result of their origin in Josephson tunneling, so that there are no frustration effects of the spin-glass type. The thermodynamics of random ferromagnets have been discussed for a number of years (Stanley et al. 1976; Harris and Kirkpatrick 1977) and, in the context of granular superconductivity, have been looked at using Monte Carlo calculations by Ebner and Stroud (1981). The basic picture is of a critical percolation probability $p_c$ below which the system cannot exhibit long-range order, owing to the absence of infinite connected clusters (see Figure 1 for the case in which $J_{ij}$ is either zero or J). For $p > p_c$ the phase transition is expected to be XY-like. At $p_c$ itself the problem becomes more complicated (and more interesting), since the coherence is occurring in a fractal-like medium (see DeGennes, Deutscher, this volume).

In 2D, the question of the effects of randomness is still open. In an interesting unpublished paper, Shastry and Bruno (1981) speculated that randomness might lead to 'localization' of the vortex-antivortex coupling, in analogy with the effects of randomness on the electrical resistance of 2D networks (Anderson et al. 1979). However, their argument may not have fully taken into account the mixed 'bond-site' character of randomness (J. Jose, private communication). We review these ideas here as an open question.

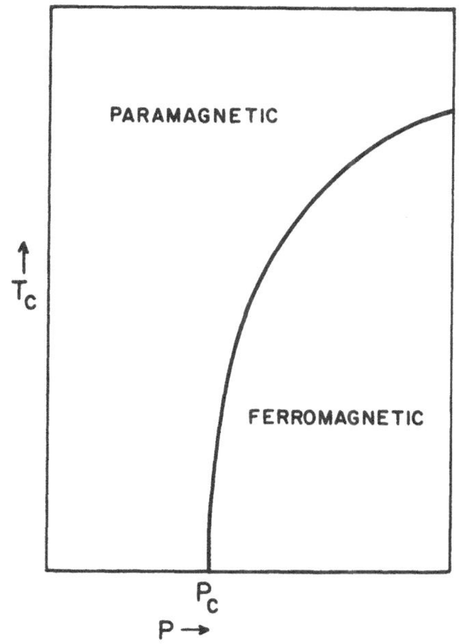


FIGURE 1. Phase diagram for the random 3D XY model. The p axis represents probability for a bond of fixed strength J.

We start with the Villain (1975) form of the XY model

$$Z = \sum_{m_{ij}} \int_{-\infty}^{\infty} \Pi d\theta_i \; e^{+\beta\Sigma_{ij}\tilde{J}_{ij}\,(\theta_{ij}+2\pi m_{ij})^2} \tag{3.2}$$

where

$$\tilde{J}_{ij} = 2[\;\delta_{ij}\sum_{\ell} J_{i\ell} - (1-\delta_{ij})J_{ij}\;] \tag{3.3}$$

and $m_{ij}$ are a set of integer 'vortex variables' which satisfy

$$\vec{\nabla}\times\vec{m} = \delta(r - r_{vortex}) \tag{3.4}$$

in a continuum approximation (Jose et al. 1977). By integrating out the angle variables in (3.2), one finds that the vortex-vortex coupling may be expressed in terms of the 'phason' propagator

$$g(R_i - R_j) \approx (\tilde{J}^{-1})_{ij} \tag{3.5}$$

where in the large $R_{ij}$ limit, g satisfies

$$\left[\left(\sum_i J_{ij}\vec{R}_{ij}\right)\cdot\vec{\nabla} + \tfrac{1}{2}\sum_{ij} J_{ij}(\vec{R}_{ij}\cdot\vec{\nabla})^2\right]g(R_i,0) = \delta(R_i) \tag{3.6}$$

on expanding (3.2) in a Taylor series.

For uniform $J_{ij}$, (3.6) just becomes the log(R) law of the 2D Coulomb gas version of the XY model. For random $J_{ij}$, g measures the propagation of phasons in a random medium. This is not the same as the propagation of electrons in a random potential because of the $\nabla$-dependence of the first term in (3.6). It seems plausible that g will still have long-range log(R) behavior, although it should be strongly modified at shorter R. However, this question has not been resolved (as far as the author is aware) up to the present time. Thus, one may expect that Kosterlitz-Thouless behavior will become modified away from $T_{2D}$, as vortex-antivortex pairs get unbound over short distances, leading to deviations from the $R(T) \propto \exp(b/\sqrt{T-T_{2D}})$ law valid for the homogeneous XY model. Possibly some of the power-law-type behavior seen in granular materials by Wolf et al. (1979) above $T_{2D}$ may be a result of randomness effects.

## 4. QUANTUM TUNNELING IN 2D GRANULAR FILMS AND JOSEPHSON JUNCTION ARRAYS (BOTH 2D AND 1D)

Anderson (1964) showed how Josephson's equation (setting $\hbar = 1$)

$$\frac{d\phi}{dt} = 2eV \tag{4.1}$$

could be re-expressed in terms of a Hamiltonian for a single

Josephson junction including the effects of Coulomb charging due to zero-point motion of Cooper pairs across the junction.

Using (4.1), and including the capacitance of the junction, the Lagrangian becomes

$$\begin{aligned} L &= \frac{CV^2}{2} - V(\phi) \\ &= \frac{C}{8e^2}\dot{\phi}^2 - V(\phi) \end{aligned} \tag{4.2}$$

where $V = -J\cos\phi$ is the Josephson energy.

Using Hamilton's equation, we then have

$$H = \frac{C}{8e^2}\dot{\phi}^2 + V(\phi) = \frac{2e^2}{C}p^2 + V(\phi) \tag{4.3}$$

where $p = (C/4e^2)\dot{\phi}$ is the conjugate momentum. This equation may be quantized by setting

$$[p,\phi] = i \ , \tag{4.4}$$

or equivalently

$$p = i\frac{d}{d\phi} \tag{4.5}$$

from which one sees that p, as the conjugate variable to the phase, measures the number of Cooper pairs transferred across the junction. The important hypothesis made here is that it is a valid approximation to quantize the macroscopic variable $\phi$, which, after all, may represent the coherent motion of $10^6$ electrons in a given grain. Although macroscopic quantum behavior is quite familiar from bulk superconductivity and liquid helium, its range of validity is still an open question for Josephson junctions, as discussed elsewhere in this volume. In this section we take the above hypothesis for granted and explore its consequences for assemblies of grains.

For a single junction, neglecting dissipation, we see that the eigenstates of (4.3) are Mathieu functions for which there is an energy gap between the ground state and first excited state. This is most easily understood by thinking of a two-well model for which the ground state is an even function

$$\psi_0 = \frac{1}{\sqrt{2}} \left[\psi(\phi-\phi^0_{left}) + \psi(\phi-\phi^0_{right})\right] \tag{4.6}$$

where $\phi^0_{left}$ and $\phi^0_{right}$ are the minima of the left and right wells, and the first excited state is an odd function

$$\psi_1 = \frac{1}{\sqrt{2}} \left[\psi(\phi-\phi^0_{left}) - \psi(\phi-\phi^0_{right})\right] \tag{4.7}$$

Here we approximate $\psi(\phi)$ as an eigenstate of a single well problem. This solution leads to a gap

$$\Delta E \cong 2 \langle \psi_{left}|H|\ \psi_{right}\rangle \tag{4.8}$$

which may be estimated by a WKB calculation. The Mathieu functions generalize this to the periodic many-well case. So, in contrast to the classical junction, the quantum junction has to be thought of as a kind of insulator, in which the phase is being averaged by zero-point motion and the average current cannot be varied continuously by varying the phase. Another way to say this is that there are constantly occuring zero point 'phase slips' preventing the coherence of a supercurrent through the junction.

## 4.1. Effects of Quantum Tunneling on Junction Assemblies

A striking feature of the charging energy in (4.2) is its strength: if C is measured by the linear dimension of the superconducting regions, then a 100 Å grain has an energy of about 1/100 Rydberg ≈ 0.1 eV. For a single junction, neglecting dissipative coupling, this clearly dominates the Josephson energy, $J \lesssim K_B T_{BCS}$ of a few degrees Kelvin. For junction assemblies, the cooperative effects to be discussed in this section increase the stability of the phase coherence, but the charging effects should still be significant (McLean and Stephen 1979). This importance may be viewed as extreme version of the enhancement of Coulomb effects due to randomness (Fukuyama, this volume), although in the granular limit we neglect the effects of Coulomb interaction on the BCS pairing itself.

The possibility that charging effects would suppress superconductivity in granular materials was raised by Abeles (1977) and discussed in more detail by Simanek (1979) and Efetov (1980). We discuss it here following Doniach (1981) using a coarse-graining approach.

The extension of Eq.(4.3) to assemblies of junctions takes the form

$$H = 2e^2 \sum_{ij} C^{-1}_{ij}\, p_i p_j - J \sum_{nn} \cos(\phi_i - \phi_j) \qquad (4.9)$$

where $p_i = i(\partial/\partial\phi_i)$ and $C_{ij}$ is a capacitance matrix. As we will show, the results will depend on the form of $C_{ij}$ and we will distinguish two special cases :

$$\text{self charging model: } C^{-1}_{ij} = \frac{1}{C_0}\,\delta_{ij} \qquad (4.10)$$

nearest neighbor charging model:

$$C^{-1}_{ij} = \begin{cases} zC^{-1}_1 & \text{for } i = j \\ -\,zC^{-1}_1 & \text{for } (i,j) = \text{nearest neighbors} \end{cases} \qquad (4.11)$$

The second case should be a reasonable model for cases where the metal grains occupy a large fraction of the sample volume, since screening lengths within a grain are extremely short (Å's) so that leakage of the electrostatic fields beyond the nearest neighbors may be strongly suppressed. Efetov has also discussed the case of longer ranged interactions. Note that if $C^{-1}_{ij}$ is short ranged, then $C_{ij}$ itself is long ranged, and vice versa.

To get a feeling for the thermodynamics of (4.9), we first discuss the limit of zero Josephson coupling (J = 0). In this limit, the Hamiltonian commutes with $p_i$ and we can characterize the eigenstates in terms of the quantum numbers $m_i = 0, \pm 1. \ldots$ of an assembly of quantum rotors. For the self-charging model, the energy is then

$$E = \frac{2e^2}{C_0} \sum_{m_i} m_i^2 \qquad (4.12)$$

which clearly has a gap of $2e^2/C$ . So the system is an "insulator" (although the model is too primitive to describe real conduction states). For the nearest neighbor charging model the partition function reads

$$Z = \sum_{\{m_i\}} e^{-2\beta e^2/C_1 \sum_{nn} (m_i - m_j)^2} \qquad (4.13)$$

which is a Gaussian solid-on-solid model. In 2D this has a Kosterlitz-Thouless transition from a low temperature phase with exponential correlations in time (i.e., an energy gap) to a high temperature phase with algebraic correlations (metallic-like)(Swendsen 1977).

So we can think of the phase diagram as exhibiting an 'insulator-metal transition'. When we add in Josephson coupling, mean field calculations show there is a critical value of the quantum tunneling parameter

$$\alpha = 2e^2/C \tag{4.14}$$

above which the system is no longer superconducting even at zero temperature. The above argument suggests this transition is from a superconducting to an insulating state. So the general phase diagram for dimension $D > 2$ may be expected to look as shown in Figure 2.

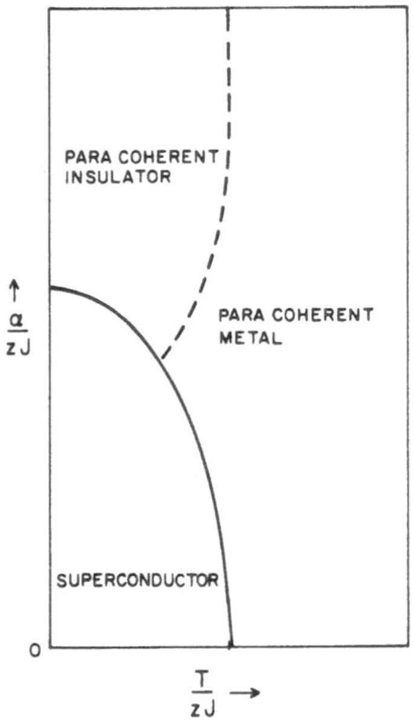


FIGURE 2. Phase diagram of granular superconductors for $D > 2$. Vertical axis: quantum coupling $\alpha = 2e^2/C$.

4.2. Quantum Renormalization of the Kosterlitz-Thouless Transition for 2D Junction Arrays

In order to discuss the thermodynamic properties of (4.9) in more detail, we extend the coarse graining treatment of Section II to include quantum fluctuations. The way to do this is to define coarse-graining field variables $\Phi(x,\tau)$ which now depend both on position and on an imaginary time variable $\tau$ lying in the range $0 \leq \tau \leq \beta$ where $\beta = 1/T$. Eq. (2.8) may now be generalized (Doniach 1981) to

$$Z = Z_0 \left< \int \mathcal{D}\Phi \; e^{-\int_0^\beta d\tau \sum_q |\Phi_q(\tau)|^2} \, e^{-\int_0^\beta J_q^{1/2} \{\Phi_q(\tau)\psi_q^+(\tau)+hc\}} \right>_0 \quad (4.15)$$

where $\langle A \rangle_0$ denotes $\mathrm{Tr}\, e^{-\beta H_0} A / \mathrm{Tr}\, e^{-\beta H_0}$, $\psi(\tau)$ is a Heisenberg operator and $H_0$ is the kinetic energy part of (4.9).

Expanding

$$\Phi_q(\tau) = \sum_m e^{i\omega_m \tau} \Phi_{q,\omega_m} \quad \text{with } \omega_m = 2\pi m T \quad (4.16)$$

we find

$$Z = Z_0 \int \mathcal{D}\Phi e^{-\mathcal{F}[\Phi]} \quad (4.17)$$

where

$$\mathcal{F} \simeq \sum_{q,\omega_m} \Omega(q,\omega_m) |\Phi_{q,\omega_m}|^2 + U \sum_i \int_0^\beta d\tau |\Phi(x_i.\tau)|^4 \quad (4.18)$$

on performing a cumulant expansion. In fact, the quadratic term in (4.18) should be nonlocal in $\omega_m$ at high temperatures, but may be shown to be local at low temperatures.

On expanding $\Omega$ to second order in $\omega_m$, we can rewrite (4.17) in terms of a Ginzburg-Landau-Wilson functional which takes into account fluctuations of $\Phi$ in time/temperature in addition to position:

$$\mathcal{F} \simeq \int d^D x \int_0^\beta d\tau \; \{c_x |\nabla\Phi|^2 + c_\tau |\tfrac{d\Phi}{d\tau}|^2 + r|\Phi^2| + \tfrac{u}{2}|\Phi^4|\} \quad (4.19)$$

where

$$r(T,\alpha) = \left( 1 - 2zJ \frac{1}{\beta} \int_0^\beta d\tau \; g_0(\tau) \right) \quad (4.20)$$

$$g_0(\tau) = \langle \psi_i(\tau)\psi_i^+(0) \rangle_0 \qquad (4.21)$$

At low temperatures this leads to a criterion for a mean field transition, renormalized by quantum fluctuations:

$$r(T,\alpha) = 0 = 1 - \frac{2zJ}{\alpha}\left(1 - \left( \frac{1-e^{-\beta\alpha}}{\beta\alpha} \right)\right) \text{ for } \beta\alpha > 1 \text{ .} \qquad (4.22)$$

As $T \to 0$ , this leads to a phase transition at a critical value of the quantum coupling $\alpha_c = 2zJ$. For finite T, we find the phase boundary

$$r\left(T,\alpha(T)\right) = 0 \qquad (4.23)$$

shown in Figure 2. Simanek and, more recently, Fazekas (1983) have found that for certain values of the self-charging term (4.10) a very slight 're-entrance' of the mean field transition temperature is possible. However, Fazekas finds this to be at most ½ % of $\alpha_c$, so the chance that this will be seen in real samples (where effects of randomness, dissipation, etc., may be expected to modify the model) seems remote.

In 2D, the form (4.19) shows us that critical fluctuations both as a function of temperature and of the quantum parameter $\alpha$, will be non-mean-field-like. At T = 0, the $\tau$ integration extends to $\beta \to \infty$ and the GLW model will correspond to a 3D XY model, two space dimensions and one time dimension. Hence the critical properties at the (correctly renormalized) critical quantum coupling $\alpha_c$ for T = 0 will correspond to those of the $\lambda$-point of $^4$He with, however, $\alpha$ playing the role of temperature.

At any finite temperature, the $\tau$ integration will become a 'slab' in time of finite thickness. Hence the critical properties will be those of the 2D XY model. However, the transition will be renormalized by quantum fluctuations. One can think of a vortex-antivortex pair in which the vortex 'core' is averaged over a finite volume by zero-point motion (Fig. 3). There will also be vortex loops, etc., renormalizing the superfluid density of the intervening superfluid.

We can obtain an estimate of the renormalization effects by using the renormalized local superfluid densities $\rho_{local}$ to calculate the vortex charge:

$$q^2_{renorm} = \frac{\hbar^2 \rho_{local}}{M} \propto q^2_{class}\left( 1 - \frac{\alpha}{\alpha_c} \right)^{2/3} \qquad (4.24)$$

where we have used the Josephson relation for the superfluid density of bulk helium. This will be valid, provided the temperature of the transition, using Kosterlitz-Thouless,

$$T_{2D} \propto q^2_{renorm} \tag{4.25}$$

is not so high that 'finite thickness effects' will invalidate the use of the bulk-helium scaling property (4.24). As shown by Doniach (1981), this assumption seems to be a reasonable one.

## 4.3 Precursor Fluctuations in the 'Superconducting Insulator'

The behavior of a paracoherent 2D granular superconductor should show some remarkable properties in the vicinity of the superconductor-insulator phase boundary. Although the sample will be an insulator for dc measurements, there should be very large fluctuating coherent domains which will show up in diamagnetic measurements of the type performed by Hebard and Fiory (1980). We can estimate these using the GLW model (4.19) within a Gaussian approximation. This calculation is simply an extension to quantum fluctuations of the classical calculations of Schmid (1969).

Extending (4.19) to include an external magnetic field:

$$\mathcal{F}(\psi) = \int_0^\beta \int d^2x \; \{ \; c_x |(\nabla - 2eiA) \; \Phi|^2 + c_\tau |\frac{\partial \Phi}{\partial \tau}|^2$$

$$+ \; r|\Phi|^2 + \frac{U}{2} \; |\Phi|^4 \; \} \tag{4.26}$$

we have for the diamagnetic susceptibility

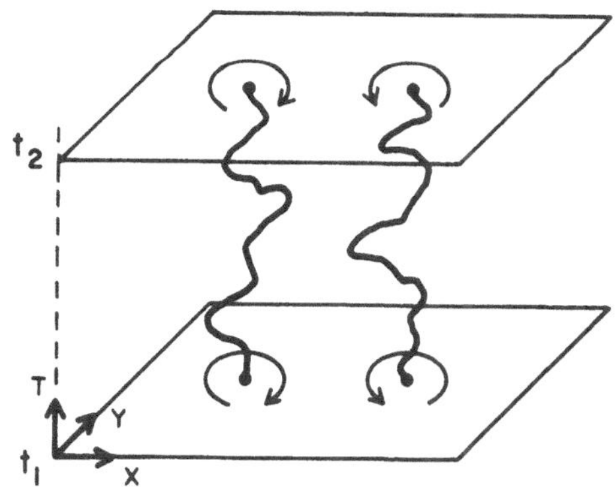


FIGURE 3. A pair of vortices execute zero-point motion in the time direction.

$$\chi(T,\alpha) = -\frac{\partial^2 F}{\partial H^2} \quad \text{with } H = \vec{\nabla} \times \vec{A} \tag{4.27}$$

$$= \chi_0 \lim_{q\to 0} \frac{\partial}{\partial q^2} \sum_{k,\omega_m} \frac{\vec{k}.\vec{q}}{\xi^{-2}+(k+q)^2+(\omega_m/v)^2}$$

$$\times \frac{1/s^2}{\xi^{-2}+ k^2+ (\omega_m/v)^2}$$

where

$$\xi^{-2} = \frac{r(\alpha,T)}{c_x} \cong s^{-2}\left( 1 - \frac{2zJ}{\alpha} + \frac{4zJT}{\alpha^2} \right) \tag{4.28}$$

and $\quad v^2 = c_x/c_\tau = (\alpha\ s)^2$ .

v is the Josephson plasma mode velocity and s is the grain spacing.

$\chi_0 = \frac{4}{3}\ \pi T\ s/\phi_0^2 =$ susceptibility per unit volume with

$\phi_0 = hc/2e$ .

Evaluating (4.28) we find

$$\frac{\chi(\alpha,T)}{\chi_0} = \sum_m \frac{2s^{-2}}{\left[\xi^{-2}+(\omega_m/v)^2\right]} = \frac{\xi v}{2Ts^2} \coth\left\{ \frac{v/\xi}{2T} \right\} . \tag{4.29}$$

For $\alpha > \alpha_c$, as $T \to 0$ this gives $\chi(\alpha,T=0) = 2/3\ \pi(\alpha\xi/\phi_0^2)$ which diverges as $\alpha \to \alpha_c^+$ . ( $\xi \sim 1/\sqrt{\alpha-\alpha_c}$ in Gaussian approximation). For $\alpha < \alpha_c$, $\chi$ diverges as $T \to T_{2D}(\alpha)$. When the argument of the coth in (4.29) becomes $\lesssim 1$, this divergence becomes classical. Hence we expect a quantum $\to$ classical crossover in the precursor fluctuations (essentially from $\sqrt{1/T-T_{2D}}$ to $1/T-T_{2D}$) as the 2D phase boundary is approached. This is shown in Figure 4. Note the remarkable size of the quantum fluctuation region.

Finally, at finite frequencies $\mathrm{Im}\chi(\omega)$ will be expected to display a Josephson plasmon gap for $T > T_{2D}(\alpha)$ (or $\alpha > \alpha_c$ at $T = 0$) , which softens as the critical line is approached. The absorption above this gap is due to the creation of propagating Josephson plasmon modes. In general, this will be damped by dissipative effects (see Doniach 1981 for a discussion).

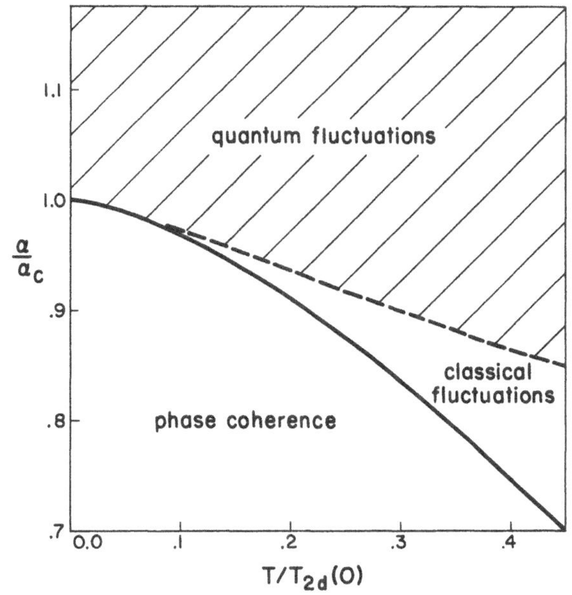


FIGURE 4. Quantum to classical crossover. For diamagnetic fluctuations of a granular film (from Doniach 1981).

## 5. QUANTUM FLUCTUATIONS IN 1D JOSEPHSON JUNCTION ARRAYS

As in other problems of 1D physics, the effects of quantum fluctuations in disrupting the phase coherence of a 1D junction array are more pronounced than in 2D. In this section, we discuss the mechanism for this disruption in two limiting cases: that of the self charging model and that of the nearest neighbor model. Unlike the discussion in 2D and 3D, we can treat the 1D problem in a much more exact way than by the coarse graining approximation.

### 5.1 Self Charging Model

In this case, which might be reached by fabricating a chain of junctions over a ground plane so that $C_0 > C_1$, the Hamiltonian becomes that of a quantum XY chain:

$$H = \sum_i \left( \alpha p_i^2 - J \cos(\phi_i - \phi_{i-1}) \right) \tag{5.1}$$

The partition function for this model may be mapped onto that of a 2D classical XY model (Fradkin and Susskind 1978) by converting the quantum formula

$$Z = \mathrm{Tr}\, e^{-\beta(H_0+H_1)} = \mathrm{Tr}\left\{ e^{-\beta H_0} e^{-\int_0^\beta d\tau H_1(\tau)} \right\} \tag{5.2}$$

to a classical Feynman path integral on dividing the $\tau$ -axis into $N_\tau$ steps of length $1/\Delta$. The BCS gap at $T = 0$, $\Delta$, is used here as a natural cut-off for the time fluctuations.

We start with the partition function for a single junction:

$$Z_s = \mathrm{Tr}\{e^{-\tau_1 H_0} e^{-H_1/\Delta} e^{-(\tau_2-\tau_1)H_0} e^{-H_1/\Delta} e^{-(\tau_3-\tau_2)H_0} \ldots\} \tag{5.3}$$

Inserting complete sets of quantum rotor states $e^{im_i\phi_i}$, this becomes

$$\lim_{\beta\to\infty} Z_s = \sum_{\{m_j\}} \int_0^{2\pi} \prod_{j=1}^{N_\tau} d\phi_j e^{-(\alpha/\Delta)m_j^2} e^{im_j(\phi_j-\phi_{j-1})} e^{H_1(\phi_j)/\Delta} \tag{5.4}$$

This can be recognized as a Villain form of a cos interaction along the $\tau$ axis:

$$Z_s = \int \prod_{j=1}^{N_\tau} d\phi_j\, e^{(\Delta/2\alpha)\cos(\phi_j-\phi_{j-1})} e^{H_1(\phi_j)/\Delta} \tag{5.5}$$

Extending to a chain of junctions, we can label each variable $\phi$ by two indices i and j to represent its position in the $x-\tau$ lattice.

$$Z_{chain} = \int\prod d\phi_{ij} \exp\{\frac{\Delta}{2\alpha} \cos(\phi_{i,j+1}-\phi_{ij}) + \frac{J}{\Delta} \cos(\phi_{i+1,j}-\phi_{ij})\} \tag{5.6}$$

which is the partition function of a classical anisotropic 2D XY model. To make this isotropic, we rescale the x and $\tau$ axes to give

$$\frac{H}{T_{eff}} = \sum_{rr'} \sqrt{\frac{J}{\alpha}} \cos(\phi_{\vec{r}} - \phi_{\vec{r}'}) \tag{5.7}$$

where r and r' are points on a rescaled 2D lattice. So we see that J/T for the classical XY model has become replaced by $\sqrt{J/\alpha}$ for the quantum 1D chain at zero physical temperature. We therefore can read off the properties from those of the usual 2D XY model with, however, the 'Boltzmann temperature' being replaced by an effective 'Planck temperature' due to the quantum fluctuations (see Figure 5). The system undergoes a phase transition at T=0 from superconducting to insulating by unbinding of vortex pairs in the time-space plane (i.e., sort of 'instantons' or correlated phase slips if one looks at a given time).

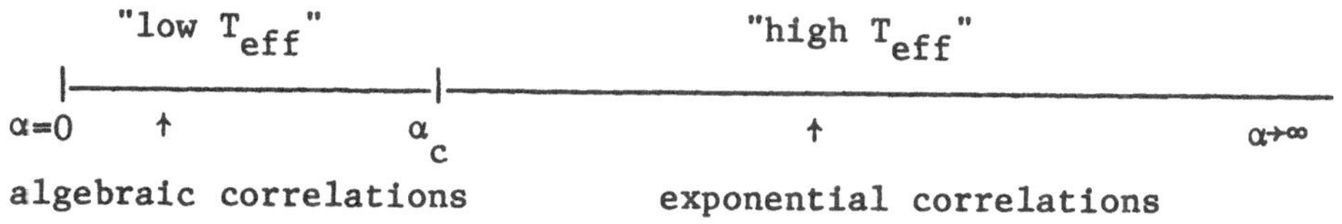


FIGURE 5. Phase diagram of the 1D self-charging chain at T = 0.

In order to explore the conduction properties of this model, R.M. Bradley (Bradley and Doniach 1983b) has looked at the current correlations

$$F_{ij}(\tau) = \langle j_i(\tau)\, j_j(0) \rangle \tag{5.8}$$

where $j_i = J \sin(\phi_i - \phi_{i-1})$. He finds (after much algebra!) that the q=0 component of the Fourier transform of F behaves like

$$\mathrm{Re}\, \sigma(\omega) = A(\alpha)\delta(\omega) \tag{5.9}$$

plus continuum absorption with a threshold at higher frequencies. Thus the response of the system is that of a superconductor if $A > 0$. He finds that $A > 0$ for $\alpha < \alpha_c$, has a 'Nelson Kosterlitz jump' at $\alpha_c$, and that $A=0$ for $\alpha > \alpha_c$. Thus the chain of junctions will be insulating (albeit with large superconducting fluctuations) when $\alpha$ exceeds its critical value.

## 5.2. The Nearest Neighbor Model

The limit $C_0 = 0$ and $C_1$ finite has been discussed by Bradley and Doniach (1983a). In this case the starting Hamiltonian

$$H = \sum_i \alpha(p_{i+1} - p_i)^2 - 2J \cos(\phi_{i+1} - \phi_i) \tag{5.10}$$

is transformed by a canonical transformation

$$\theta_i = \sum_{j>i} \phi_i \qquad P_i = p_i - p_{i-1} \tag{5.11}$$

to the form

$$H = \alpha \sum_i P_i^2 - 2J \sum_i \cos(\theta_{i+2} - 2\theta_i + \theta_{i-1}) \tag{5.12}$$

where $[P_i, \theta_i] = i$. On transforming to a 2D classical model (one space, one time coordinate) as in Section 5.1 the spin wave part of the partition function turns out to be

$$Z_{\text{spin wave}} = \int \mathcal{D}\theta e^{\int dx d\tau [K_\tau |\partial\theta/\partial\tau|^2 + K_x^2(\partial^2\theta/\partial x^2)^2]} \tag{5.13}$$

corresponding to a Lifshitz point for the phason Hamiltonian at which the coefficient of $(\partial\theta/\partial x)^2$ is zero. By means of a Villain transformation, it is found that the vortex-antivortex interaction maybe written

$$V(r) \simeq \sum_q \frac{e^{iq \cdot r}}{(\alpha/\Delta)q_x^4 a^4 + \log(\Delta/J)q_\tau^2 a^2} \tag{5.14}$$

where a is the lattice spacing. This form of interaction always increases with distance: $V = |x|$ as $x \to \infty$, $\tau = 0$ and $V = \sqrt{\tau}$ as $\tau \to \infty$, $x = 0$. Hence vortex-antivortex pairs always remain bound. However, we find that the form of the vortex coupling to the original phases, $\phi_i$, of the junctions (the above vortex pairs are in the $\theta$-variables) is such that vortex-antivortex dipole pairs oriented along the $\tau$ axis have the property that they will always disrupt the order-parameter correlations in the chain. Hence for this model, at $T = 0$, the quantum tunneling fluctuations will always lead to an insulating state no matter how small $\alpha/J$, though the coherent fluctuations become exponentially strong as $\alpha \to 0$. Therefore for this special case $\alpha_c = 0$.

For general Coulomb coupling we may expect some kind of intermediate behavior with a finite $\alpha_c$, though reduced by the nearest neighbor coupling-induced fluctuations.

## 6. NONLINEAR DYNAMICS OF JOSEPHSON JUNCTION ARRAYS

The dynamical behavior of Josephson junctions driven by an external current source, or by a radio-frequency 'pump' (as in an r.f. SQUID) are of considerable interest both as analog models to study basic questions of nonlinear dynamics in complex systems (Crutchfield, Farmer and Huberman 1982), and for their possible applications as rf devices, low noise amplifiers, and the like (Jain et al. 1982). So far, most work has been done on systems with only one or two junctions (d'Humieres et al. 1982, Miracky et al. 1983). However, one set of experiments on an NbN granular film by Wolf et al. (1982) is of particular interest in showing the richness of phenomena in a complex system - in their case, of order $10^7$ junctions randomly coupled in a 2D array.

In this section we focus on recent work by Strenski (Strenski and Doniach 1983), who has investigated one of the simplest non-trivial arrays, consisting of just three junctions. To introduce this, we remind the reader of the properties of a single dc-driven Josephson junction. These are studied using the resistively shunted junction (RSJ) model in terms of the classical equation of motion derived from Eq. (4.2) by the addition of a resistance term:

$$\frac{C}{4e^2}\frac{d^2\phi}{dt^2} + G\frac{d\phi}{dt} + J\sin\phi = I \tag{6.1}$$

where I is the external driving current and G = 1/R is the passive conductivity. In this approach we neglect the effect of thermal or quantum fluctuations on the dynamics. In the limit of strong damping, the system is superconducting ($\phi$ = constant) for $I < I_c = J$, and has an instability at $I = I_c$, leading asymptotically to a voltage $V = \theta = IR$ for $I \gg I_c$. (See Figure 6).

As G is reduced, a threshold is reached at which the junction becomes underdamped and exhibits hysteretic behavior: for increasing I, the behavior is qualitatively, as in Figure 6, for decreasing I, the effect of the $d^2\phi/dt^2$, or 'inertial term' in Eq. (6.1) is to maintain the 'ball rolling down the washboard' even for $I < I_c$.

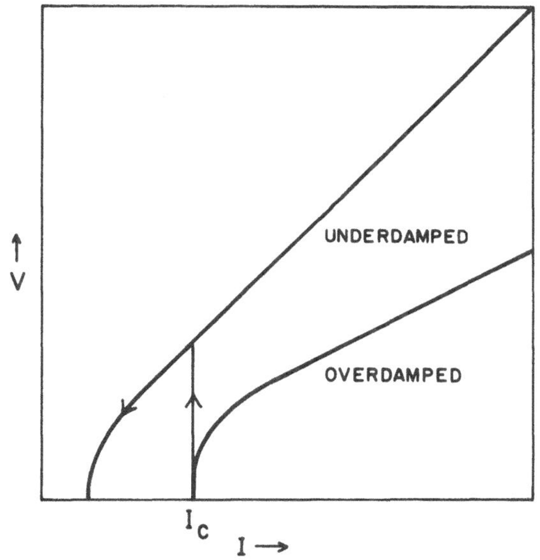


FIGURE 6: V-I characteristics of a single driven Josephson junction (schematic).

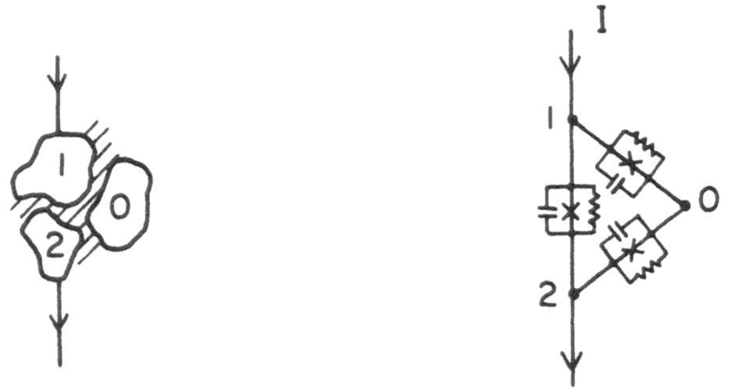


FIGURE 7. Primitive Three-junction Cell with Equivalent Circuit.

## 6.1 Nonlinear Dynamics of Three Coupled Junctions

We consider the smallest 'unit cell' of a junction lattice, which might be formed by three superconducting grains joined by an oxide junction (Figure 7). To investigate the dynamics of this system, we change the time scale to a dimensionless scale $\underset{\sim}{t} \leftarrow t/RC$ and rescale J and I to $\tilde{J} = J(CR^2)$ and similarly for $\tilde{I}$, so that the single junction equation (6.1) becomes

$$\ddot{\phi} + \dot{\phi} + \tilde{J} \sin \phi = \tilde{I} \quad . \tag{6.2}$$

Using these units (R,C the same for all three junctions) and fixing the phase of $\phi_0$ in Figure 6.2 without loss of generality, the equations of motion of the three junction system become

$$\ddot{\phi}_1 + \dot{\phi}_1 + \tfrac{2}{3}J_{10}\sin \phi_1 + \tfrac{1}{3}J_{20}\sin \phi_2 + \tfrac{1}{3}J_{12}\sin(\phi_1 - \phi_2) = I/3 \tag{6.3}$$
$$\ddot{\phi}_2 + \dot{\phi}_2 + \tfrac{1}{3}J_{10}\sin \phi_1 + \tfrac{2}{3}J_{20}\sin \phi_2 - \tfrac{1}{3}J_{12}\sin(\phi_1 - \phi_2) = - I/3$$

where we have dropped the tilde on I and J for simplicity.

The nonlinear dynamics of the four-dimensional (4D) vector $\underset{\sim}{u}(t) = \{ \phi_1, \dot{\phi}_1, \phi_2, \dot{\phi}_2 \}$ has been investigated both numerically and analytically by Strenski in the strongly underdamped regime $J > 1$ . He finds the system has one fixed point: $\phi_1, \phi_2$ = constant, $\dot{\phi}_1 = \dot{\phi}_2 = 0$, i.e., the superconducting case,

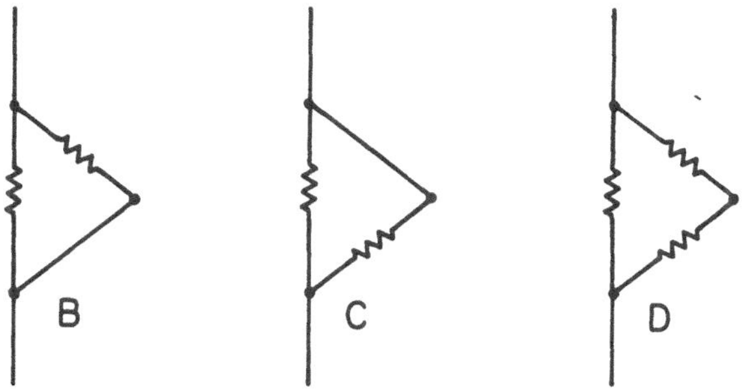


FIGURE 8: Three limit cycles of the array.

denoted by A, and three distinct limit cycles, denoted by B, C and D. Away from threshold the limit cycle behavior asymptotes the dc behavior represented in Figure 8.

In cycles B and C, one of the three junctions remains superconducting while the other two run down the washboard. In cycle D, all three are washboarding. The phase diagram is indicated schematically in Figure 9.

Cycles B and C are stable only in a limited range of I. For $J \gg 1$, the upper limit $I_1'$ is found by a strong coupling expansion to be of order 2J, while the lower limit $I_1$ is of $O(J^{1/2})$. Cycle D is unbounded for large I. At its lower end, Strenski finds a series of period doubling instabilities followed by a chaotic regime including period three and period five intervals. The dimension of this strange attractor has been investigated using the method of Grassberger and Procaccia (1983). Results will be reported elsewhere.

## 6.2 Discussion - Many Junction Arrays

As seen from Figure 9, the three-junction array has two instability thresholds with the effective resistance increasing as one goes from the limit cycles B and C (degenerate for equal resistance in each arm of the circut) to limit cycle D. This, in miniature, mimics the behavior seen by Wolf and collaborators (unpublished) for granular films (Figure 10).

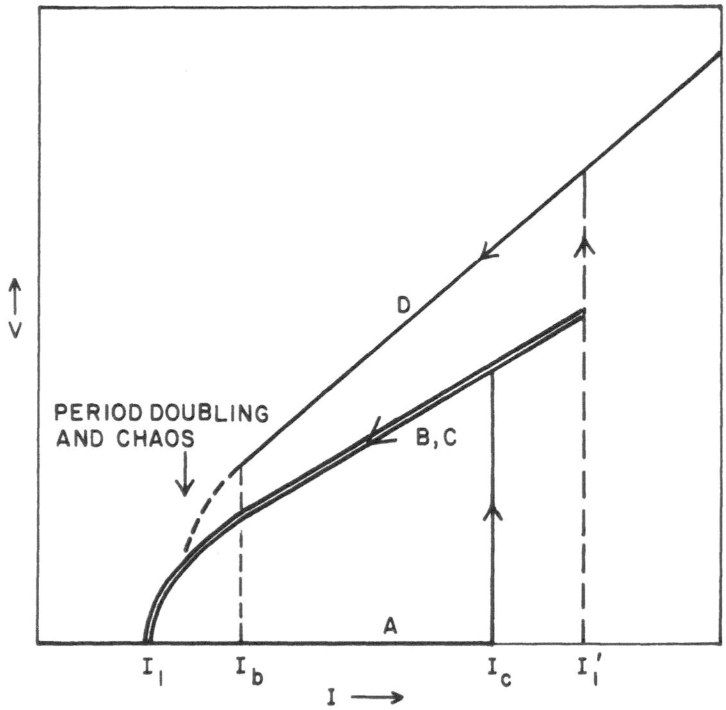


FIGURE 9: Phase diagram of the three junction array (schematic).

Recent numerical studies for a 1D array of junctions by Bishop and coworkers (1983) shows that coupling between junctions tends to inhibit chaotic behavior, giving rise instead to collective instabilities. We may speculate that the peaks seen by Wolf et al. (with resistivities corresponding to 100's of junctions running in series) correspond to some kind of collective behavior of the 2D array; perhaps the generation of a series of driven vortex lattices as I is increased. Undoubtedly, the random nature of the coupling will tend to nucleate certain instabilities at 'weak link' regions of the array. However, the systematic washing out of the instability threshold observed as T is increased above the Kosterlitz-Thouless temperature of the film, does suggest that we are dealing with some kind of collective behavior rather than with isolated flux jump nucleation centers. Clearly this is a very fruitful field for future study, both theoretically, and experimentally using artifical junction lattices.

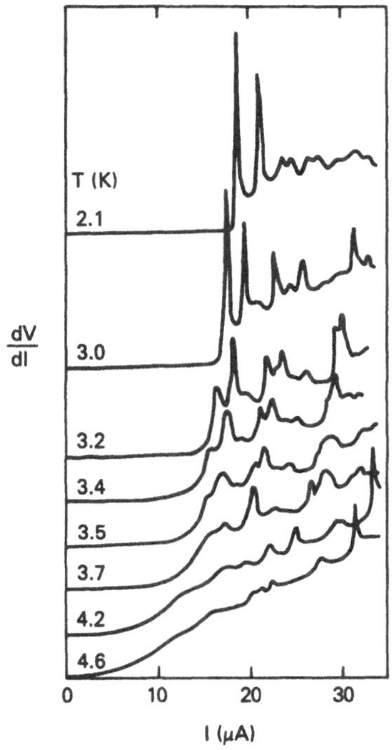


FIGURE 10: Dynamical Resistance of an NbN granular film, Wolf and collaborators (unpublished).

Acknowledgments: This work was partly supported by NSF grant DMR 80-07934. I thank Mark Bradley and Phil Strenski for their contribution to these notes and for many stimulating discussions. I am grateful to Stuart Wolf for letting me use his unpublished data (Figure 10).

## References

Abeles, B., 1977 Phys. Rev. B, 15:2828.

Abrahams, E., Anderson, P.W., Liciardello, D.C., and Ramakrishnan, T.V. 1979, Phys. Rev. Lett., 42:673.

Anderson, P.W., 1964, in "Lectures on the Many Body Problem," E.R. Caianiello, ed., Academic Press, New York. 2:127.

Bishop, A.R., Fesser, K., Lomdahl, P.S., Kerr, W.C., Williams, M.B., and Trullinger, S.E., 1983, Phys. Rev. Lett., 50:1095.

Bradley, R.M., and Doniach, S., 1983a, J. de Physique (Colloques) (in press).

Bradley, R.M., and Doniach, S., 1983b, (manuscript in preparation).

Crutchfield, J.P., Farmer, D., and Huberman, B.A., 1982, Phys. Reports, 92:45.

d'Humieres, D., Beasley, M.R., Huberman, B.A., and Libchaber, A., 1982, Phys. Rev. A, 26:3483.

Das Gupta, C., and Halperin, B.I., 1981, Phys. Rev. Lett., 47:1556.

DeGennes, P.G., 1966, "Superconductivity of Metals and Alloys," W.A. Benjamin Inc., New York.

Deutscher, G., Imry, Y., and Gunther, L., 1974, Phys. Rev. B, 10:4598.

Doniach, S. 1981, Phys. Rev. B., 24:5063.

Doniach, S., and Huberman, B., 1979, Phys. Rev. Lett., 42:1169.

Ebner, C., and Stroud, D. 1981, Phys. Rev. B, 23:6164; also Phys. Rev. B 25:5711 (1982).

Edwards, S.F., and Warner, M., 1979, Philos. Mag., 40:257.

Efetov, K.B., 1980, Zh. Eksp. Teor. Fiz., 78:17 [Sov. Phys. JETP, 51:1015].

Fazekas, P., 1982, Z. fur Physik B 45:215.

Fradkin, E., and Susskind, L., 1978, Phys. Rev. D, 17:2637.

Grassberger, P., and Procaccia, I., 1983, Phys. Rev. Lett., 50:346.

Halperin, B.I., Lubensky T.C., and Ma, S.-K., 1974, Phys. Rev. Lett., 32:292.

Harris, A.B., and Kirkpatrick, S., 1977, Phys. Rev. B, 16:542.

Hebard, A.F., and Fiory, A.T., 1980, Phys. Rev. Lett., 44:291.

Huberman, B.A., and Doniach, S., 1979, Phys. Rev. Lett., 43:950.

Jain, A.K., Likharev, K.K., Lukens, J.E., and Sauvageau, J.E., 1982, App. Phys. Lett., 41:566.

Jose, J.V., Kadanoff, L.P., Kirkpatrick, S., and Nelson, D.R., 1977, Phys. Rev. B, 16:1217.

Kleinert, H., 1982, Physics Letts., 93A:86.

McLean, W.L., and Stephen, M.J., 1979, Phys. Rev. B, 19:5925.

Miracky, R.F., Clarke, J., and Koch, R.H., 1983, Phys. Rev. Lett., 50:856.

Rabatou, A., Rosenblatt, J., and Peyral, P., 1980, Phys. Rev. Lett. 45:1035.

Savit, R., 1980, Rev. Mod. Phys. 52:453.

Schmid, A., 1969, Phys. Rev., 180:527.

Shastry, R., and Bruno, J., 1981, preprint, University of Utah.

Simanek, E., 1979, Phys. Rev. B, 22:459.

Stanley, H.E., Birgeneau, R.J., Reynolds, P.J., and Nicoll, J.F., 1976, J. Phys. C., 9:L553 .

Strenski, P., and Doniach, S., 1983, manuscript in preparation.

Swendsen, R.H., 1977, Phys. Rev. B., 15:689.
Teitel, S., and Jayaprakash, C., 1983, Phys. Rev. B, 27:598.
Thouless, D., 1960, Annals of Phys. (New York), 10:553.
Trugman, S., and Doniach, S., 1983, Phys. Rev. B. 26:3682.
Villain, J., 1975, J. de Physique, 36:581.
Voss, R.F., and Webb, R.A., 1982, Phys. Rev. B., 25:3446.
Wolf, S.A., Gubser, D.U., and Imry, Y., 1979. Phys. Rev. Lett., 42:324.
Wolf, S.A., Gubser, D.U., Fuller, W.W., Garland, J.C., and Newrock, R.S., 1982, Phys. Rev. Lett., 47:1071.

DISORDER IN WEAKLY COUPLED GRANULAR SUPERCONDUCTORS

J. Rosenblatt

Laboratoire de Physique des Solides, Equipe de Recherche Associée N°980, Institut National des Sciences Appliquées B.P. 14A 35043 Rennes Cedex, France

## INTRODUCTION

Granular superconductors are usually prepared as thin films of metal-insulator mixtures. The grain size depends on the nature of both the metal and the insulator and the details of the codeposition process, giving diameters typically in the range 30 Å to 1000 Å. We report here on the properties of systems which differ significantly from this description from the point of view of grain size and fabrication process.[1] Our samples are made of commercially available slightly oxidezed metallic powder (usually Nb of Ra), with grain diameters typically in the range 10 to 50 μm, mixed with epoxy resin to keep them together[2]. Samples can be moulded into any desired shape. The metal filling factor is well above the percolation threshold in all cases, and close to that of a simple cubic arrangement of hard spheres (∿0.5). Therefore the grains are expected to be in electrical and mechanical contact with their nearest neighbors. By adjusting the pressure exerted during hardening of the resin we can control the resistivity of our samples, with typical values going from 0.1 to 1 Ω-cm.

The motivation to study this system is that it is a particularly simple one. Since the grains are bulk superconductors with linear sizes $a \gg \lambda_s(T)$, $\xi_s(T)$, where $\lambda_s$ and $\xi_s$ are the superconducting penetration depth and coherence length, respectively, their individual behavior with temperature is in principle well known. Furthermore, the electromagnetic properties of the system will depend on the Josephson coupling between grains, and therefore will be sensitive to currents and fields too small to affect the amplitude $\Delta(T)$ of the order parameter in the grains. We have then three-dimensional arrays of Josephson junctions which can be modelled[3,4] as assemblies of vectors $\Delta(T)\exp(i\phi_\alpha)$ in the complex plane coupled through the energy

$$H_{\alpha\alpha'} = - J_{\alpha\alpha'} \cos\phi_{\alpha\alpha'} , \tag{I.1}$$

where $\alpha$ and $\alpha'$ designate neighboring grains, $\phi_{\alpha\alpha'} = \phi_\alpha - \phi_{\alpha'}$ indicates a phase difference and

$$J_{\alpha\alpha'}(T) = \frac{\Phi_o}{4} \frac{\Delta(T)}{eR_{\alpha\alpha'}} \tanh \frac{\Delta(T)}{2k_BT} \tag{I.2}$$

is the temperature-dependent Josephson coupling energy[5], with $R_{\alpha\alpha'}$ the normal state resistance of the junction between grains $\alpha$ and $\alpha'$ and $\Phi_o$ is the flux quantum.

Similar results and additional physical insight can be obtained from a microscopic description[2,6] where the BCS[7] and pair tunneling[8] Hamiltonians are expressed in terms of pseudospin operators $\vec{s}_{k\alpha}$ :

$$H_{BCS} = \sum_\alpha H_\alpha = \sum_\alpha \left(-2\Sigma\varepsilon_k s_{3k\alpha} - \sum_{kk'} V_{kk'}(s_{1k\alpha}s_{1k'\alpha} + s_{2k\alpha}s_{2k'\alpha})\right) \tag{I.3}$$

$$H_T = \sum_{<\alpha\alpha'>} \left(- \sum_{kk'} M_{k\alpha k'\alpha'}(s_{1k\alpha}s_{1k'\alpha'} + s_{2k\alpha}s_{2k'\alpha'})\right) \tag{I.4}$$

Here $\varepsilon_k$ is a quasiparticle energy measured from the Fermi level, $V_{kk'}$ (=0 for $|\varepsilon_k| > \hbar\omega_D$), is the electron-electron interaction and $M_{k\alpha k'\alpha'}$ is a pair tunneling matrix element. The pseudospin vectors obey spin commutation relations :

$$\vec{s}_{k\alpha} \wedge \vec{s}_{k'\alpha'} = i \, \vec{s}_{k\alpha} \, \delta_{kk'} \, \delta_{\alpha\alpha'} \tag{I.5}$$

The z-component $s_{3k\alpha} = \frac{1}{2}(1 - n_{k\alpha\uparrow} - n_{-k\alpha\downarrow})$ is a pair number operator (the n's being single particle number operators) : $s_{3k\alpha} = \pm\frac{1}{2}$ implies absence or presence, respectively, of a pair in state k in grain $\alpha$. The angle of rotation around z of the average pseudospin $\phi_\alpha$, is shown[8] to be the same for all k and k' in each grain and to correspond to the phase of the superconducting order parameter. Annihilation and creation operators $s^{\pm}_{k\alpha} = s_{1k\alpha} \pm i\, s_{2k\alpha}$ for pairs in state k in grain $\alpha$ can be defined ; they are related to the fermion creation and destruction operators, $c^+_{k\sigma}$ and $c_{k\sigma}$, $s^+_{k\alpha} = c^+_{k\uparrow}c^+_{-k\downarrow}$ , $s^-_{k\alpha} = c_{-k\downarrow}c_{k\uparrow}$.

When performing thermal averages, we can, as long as $M_{k\alpha k'\alpha'} << V_{kk'}$ use unperturbed BCS states for the sums over k and k'. Furthermore, if the grains are large enough so that their condensation energy $\frac{1}{2} N(0)\Delta^2 a^3 >> k_BT$ (for 1K this corresponds to $a \simeq 50$ Å), the amplitude of the superconducting order parameter in each grain is a non-fluctuating, temperature-dependent quantity. Once the k-sums are thus performed at finite temperature, one finds[6]

$$H_T = -\frac{1}{2}\sum_{<\alpha\alpha'>} J_{\alpha\alpha'}\, S^+_\alpha S^-_{\alpha'} + HC \tag{I.6}$$

where $J_{\alpha\alpha'}$ is given by Eq.(I.2) and results from the thermal average

$$J_{\alpha\alpha'}\, S^+_\alpha S^-_{\alpha'} = < \sum_{kk'} M_{k\alpha k'\alpha'}\, s^+_{k\alpha}\, s^-_{k'\alpha'} >_{BCS}.$$

In the classical limit $S^{\pm}_\alpha \to \exp(\pm\, i\phi_\alpha)$, thus confirming the intuitive result (I.1). Eq.(I.6) gives the only phase-dependent part of the total Hamiltonian and therefore governs the statistics of all phase-dependent quantities. On the other hand, it is the exact analog of the Hamiltonian of a system of two-dimensional spins $\vec{S}_\alpha$ coupled ferromagnetically through exchange integrals $J_{\alpha\alpha'}$, known as the X-Y model. From this point of view, disordered weakly coupled granular superconductors (WCGS) are a particular example of a ferromagnetic glass.

Numerical studies of translationally invariant three-dimensional (3D) X-Y systems show that they present a second order phase transition at a critical temperature $T_c$ with order parameter $\Psi = <\exp(i\phi)> = |\Psi| \exp(i\hat{\phi})$ and critical exponents given by[9] $(t = (T-T_c)/T_c)$ :

$$\Psi \sim |t|^\beta \quad , T < T_c \; , \quad \beta \simeq 0.33, \tag{I.7}$$

susceptibility :

$$\chi \sim |t|^{-\gamma} \; , T \neq T_c \; , \gamma \simeq 1.33, \tag{I.8}$$

correlation length :

$$\xi \sim |t|^{-\nu} \; , T \neq T_c \quad \nu \simeq 0.66 \tag{I.9}$$

correlation function :

$$G(\vec{r}) = \frac{1}{r^{d-2+\eta}}\; D\left(\frac{r}{\xi}\right) \; , \quad T \neq T_c \; , \quad \eta \simeq 0 \tag{I.10}$$

with d the dimensionality of space and $D(r/\xi) \sim \exp(-r/\xi)$ for $r > \xi$. Finally the critical magnetization is :

$$\Psi \sim H^{1/\delta} \quad , \quad T = T_c \quad , \quad \delta \simeq 5. \tag{I.11}$$

In granular superconductors, the existence of an order parameter $\Psi \neq 0$ below $T_c$ corresponds to the establishment of long range phase coherence in the whole system, and this is a necessary condition for the appearance of bulk superconducting properties. We expect $T_c \sim J_{\alpha\alpha'}/k_B < T_{cG}$ the superconducting critical temperature of the grains, because $J_{\alpha\alpha'} \to 0$ as $T \to T_{cG}$. We refer to the region $T_c < T < T_{cG}$ as the paracoherent region, as opposed to the coherent domain $T \leq T_c$.

Effects described by the Hamiltonians (I.3) and (I.4) or (I.6) are of course present in all granular superconductors, irrespective of the grain size. But as the grains become smaller, other effects have to be taken into account and make the situation more complex : an electrostatic energy term proportional to $a^{-1}$, due to charge transfer between grains, can no longer be neglected, and for $a \lesssim 10Å$ the electronic level separation becomes important. Weakly coupled large grains $a > \lambda_s, \xi_s$ , on the contrary, should be described by the Hamiltonian (I.6) alone. There is however, an important reason why the numerical values in Eqs.(I.7) to (I.11) may not hold : disorder. Indeed, our samples lack translational invariance and present many types of structural disorder : (i) a distribution of grain sizes and shapes ; (ii) randomly oriented junctions ; (iii) distributions of the number of nearest neighbours $z_\alpha$ and of coupling energies $J_{\alpha\alpha'}$. We may expect, however, that a simple property, statistical self-similarity[10], will replace translational invariance. In our case this implies that in regions of linear size $L > a_{max}$ (the maximum grain size), the average number of grains is

$$N(L) \sim L^d \qquad (I.12)$$

and therefore a scaling transformation $L \to L/b > a_{max}$ gives

$$N(L/b) = b^{-d}N(L), \qquad (I.13)$$

with d = 3, the dimensionality of our samples.

Our previous work on WCGS[1-4] was aimed at demonstrating the existence of the paracoherence-coherence transition. Indeed, reasonable fits to the paracoherent resistance data were obtained with a mean-field description of the transition.[3] In this lecture we review recent work done at Rennes studying in more detail the critical properties of disordered WCGS near the cohenrence terperature.

In the following section II we discuss experiments giving the values of $\beta$, $\gamma$ and $\nu$ for the coherence transition of WCGS. We shall see that in spite of disorder, they provide evidence for the existence of a well-defined coherence temperature $T_c$. However, the exponent values obtained are significantly greater than expected from results on translationally invariant systems, Eqs. (I.7) to (I.9), while obeying known scaling relations. Section III presents a description of phase transitions in ferromagnetic glasses, which may explain the above results. Concluding remarks appear in section IV.

## II - CRITICAL EXPONENTS OF THE COHERENCE TRANSITION

When the critical current and the penetration depth of WCGS are measured[11], they are found to obey power laws as functions of t, below $T_c$. On can then expect that the corresponding exponents are related to those defined in Eqs. (I.7) to (I.11).

II.1 - Critical current exponent

The current density operator $\vec{j}$ is easily obtained from the continuity equation for the supercurrent, $d\vec{P}/dt = \partial\vec{P}/\partial t - \vec{j} = 0$, where $\vec{P} = -2e\sum_\alpha S_3 \vec{r}_\alpha/L^d$ is the pair polarisation operator with $\vec{r}_\alpha$ the position vector of grain $\alpha$. The operators $S_{3\alpha}$ and $S^\pm_\alpha$ obey spin commutation relations

$$\left(S_{3\alpha}, S^\pm_{\alpha'}\right) = \pm S^\pm_\alpha \delta_{\alpha\alpha'} \quad , \quad \left(S^+_\alpha , S^-_{\alpha'}\right) = 2S_{3\alpha}\delta_{\alpha\alpha'} \qquad \text{(II.1)}$$

which together with Eq.(I.6) give

$$\vec{j} = \frac{\partial\vec{P}}{\partial t} = \frac{i}{\hbar}\left(\vec{P}, H_T\right) = \frac{i}{\hbar}\,\frac{e}{L^d}\sum_{<\alpha\alpha'>} J_{\alpha\alpha'} S^+_\alpha S^-_{\alpha'} \vec{r}_{\alpha\alpha'} + \text{H.C.} \qquad \text{(II.2)}$$

with $\vec{r}_{\alpha\alpha'} = \vec{r}_\alpha - \vec{r}_{\alpha'}$.

We now look for coherent current-carrying steady states characterised by a phase gradient $\vec{q}$ and therefore by the Fourier components :

$$S^\pm_q = N^{-1/2}\sum_\alpha \exp\left(\mp i\vec{q}.\vec{r}_\alpha\right) S^\pm_\alpha \qquad \text{(II.3)}$$

$$J_q = N^{-1}\sum_{<\alpha\alpha'>} \exp\left(-i\vec{q}.\vec{r}_{\alpha\alpha'}\right) J_{\alpha\alpha'} \qquad \text{(II.4)}$$

Replacing the inverse transforms of Eq. (II.3) and (II.4) in (II.2) one obtains

$$<\vec{j}(\vec{q})> = \frac{2\pi c}{\Phi_o} L^{-d} \frac{\partial J_q}{\partial\vec{q}} \,\text{Im}\,\chi_q \qquad \text{(II.5)}$$

where

$$\chi_q = -i\sum_\alpha \exp\left(i\vec{q}.\vec{r}_\alpha\right) <S^+_o S^-_\alpha>_{XY} \qquad \text{(II.6)}$$

is similar to the static susceptibility of wave vector $\vec{q}$. The subscript 0 indicates the origin, and $<\ldots.>_{XY}$ is a thermal average with the X-Y model Hamiltonian (I.6). Transforming the sum into an integral and taking into account the isotropic form of the spin-spin correlation function given in Eq. (I.10),

$$\chi_q \sim \xi^{2-\eta}\int_0^1 D(x)\, x^{-\eta}\,\frac{\sin q\xi x}{q\xi}\,dx = \xi^{2-\eta} f(q\xi) \quad (q\neq 0) \qquad \text{(II.7)}$$

with $L \simeq \xi(T)$. A similar result obtains when $L \to \infty$ . In the same way, if the distribution of $J_{\alpha\alpha'}$ is isotropic, and neglecting for the time being the dispersion in grain sizes, i.e. $|\vec{r}_{\alpha\alpha'}| = a$, angular integration in Eq. (II.4) gives

$$J_q = \frac{1}{2}\,\overline{zJ}\,\frac{\sin qa}{qa} \qquad \text{(II.8)}$$

where the bar indicates spatial average of the quantity $\sum_{\alpha'} J_{\alpha\alpha'}$ , that is, the coupling energy of a grain.

Let us now make the almost trivial assumption that there exists a maximum (critical) supercurrent $j_c = j(q_c)$ and the less trivial one that it depends only on the static equilibrium properties of the system. Then, when $\xi(T)$ is large, as $T \to T_c$, we expect $q_c \simeq \xi^{-1} \ll a^{-1}$. Indeed, $\xi(T)$ is of the order of the linear size of the smallest region displaying the same behavior as the bulk. If $q_c \ll \xi^{-1}$, say, the order parameter $\langle S_q^+ \rangle$ will be modulated, according to (II.3), with a period $2\pi/q_c \gg \xi$ and bulk properties, including critical current density, would require a region of at least this size to show up. The inverse argument applies to $q_c \gg \xi^{-1}$. Taking $L \simeq \xi(T)$, using (II.7) and differentiating (II.8) in the limit $q_c a \ll 1$, we obtain from (II.5):

$$j_c \sim \xi^{1-\eta-d} \sim t^{(d-1+\eta)\nu} \sim t^{2\beta+\nu} \quad , \qquad \text{(II.9)}$$

where, in the last step, we have used the hyperscaling relation[12]

$$(d - 2 + \eta)\nu = 2\beta. \qquad \text{(II.10)}$$

We studied the critical current[13] $I_c$ as a function of temperature of cylindrical samples, 13 mm long and 3mm in diameter. Since the V(I) characteristic is rounded near $I_c$, we rather measured the steeper $R_d(I)$ characteristic with a lock-in detector. Its output is fed into a microcomputer which commands a current generator so as to obtain a predetermined small value of $R_d$, typically $\simeq 10^{-3} R_n$, where $R_n$ is the normal resistance of the sample. The corresponding current gives the experimental $I_c$. The experimental results for five Nb granular samples of widely different coherence critical temperatures are shown in Fig.1 in a log-log plot. The slope obtained is

$$2\beta + \nu = 2.5 \pm 0.3 \qquad \text{(II.11)}$$

which is well above the value predicted by the X-Y model from Eqs.(I.7) and (I.9) or by a mean field description, $\beta = \nu = 0.5$, $2\beta + \nu = 1.5$.

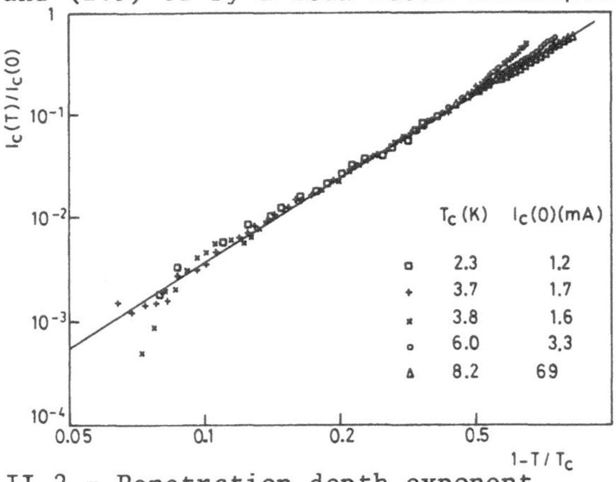

Fig. 1.

Critical data of granular niobium samples. Coherence temperatures and extrapolated critical currents are shown. The straight line has a slope 2.5. From Ref. 13.

|   | $T_c$ (K) | $I_c(0)$(mA) |
|---|---|---|
| □ | 2.3 | 1.2 |
| + | 3.7 | 1.7 |
| × | 3.8 | 1.6 |
| ○ | 6.0 | 3.3 |
| △ | 8.2 | 69 |

II.2 - Penetration depth exponent

When an external field $\vec{H}$ is applied to the sample, a "microscopic" field $\vec{h}$ meanders around and into the grains down to a superconducting penetration depth $\lambda_s$. A spatial average over a volume element containing many grains defines $\vec{B}(\vec{r}) = \overline{\vec{h}(\vec{r})} = \nabla \wedge \vec{A}$. It is important to distinguish here between effects due to the almost perfect ($a >> \lambda_s$) diamagnetism of the grains giving a permeability $\mu \simeq 1 - f$, where f is the metal filling factor of the sample, and those due to the Josephson coupling between grains. The latter will develop screening currents in a region of the order of $\lambda >> a$ as a response to the applied field. We then consider the behavior of such currents in a medium of permeability $\mu$[13]. Recall that the coherence order parameter $\Psi = |\Psi| \exp(i\vartheta) = \langle S^+ \rangle$ may change appreciably only over distances greater than the correlation length $\xi(T)$ and assume $\lambda(T) >> \xi(T)$. We can then define a slowly varying gauge invariant phase gradient $\vec{q}(\vec{r}) = \nabla\vartheta - 2\pi\vec{A}/\Phi_0$. Neglecting time derivatives and dissipative terms in Maxwell's equations one obtains, in the limit $H \to 0$ :

$$\nabla^2\vec{q} = \frac{8\pi^2\mu}{c\Phi_0}\,\vec{j}_q(T) \simeq \frac{\overline{zJ}}{3}\,\frac{8\pi^3\mu}{\Phi_0^2}\,\frac{|\Psi|^2}{a}\,\vec{q} \qquad \text{(II.12)}$$

where the last expression results from the $q \to 0$ limit of Eqs.(II.5), (II.6) and (II.8). This defines

$$\lambda(T) = \frac{\Phi_0}{2\pi}\left(\frac{3a}{2\pi\mu\,\overline{zJ}\,|\Psi|^2}\right)^{1/2} \qquad \text{(II.13)}$$

as the low-field penetration depth of WCGS. At temperatures below $T_{cG}/2$, $\overline{J}(T)$ given by (I.2) is practically a constant, and the main temperature dependence of $\lambda(T)$ is just $\lambda \sim |\Psi|^{-1} \sim t^{-\beta}$. At $T = 0$ $|\Psi| = 1$ and one can estimate $\lambda(0)$ from $\overline{zJ} \simeq 2k_BT_c$, $\mu \simeq 0.5$, which gives $\lambda(0) \simeq 1$ mm.

A solution of Eq.(II.12) for the cylindrical geometry of our samples gives

$$\Phi_c = 2\pi R\lambda\,\left(I_1(R/\lambda)/I_0(R/\lambda)\right)\,\mu\,H, \qquad H \to 0 \qquad \text{(II.14)}$$

where R is the radius of the sample and $I_0$ and $I_1$ are modified Bessel functions. The solution (II.14) assumes that the field is low enough to produce no vortices in the sample. In the opposite limit, namely, that of a strong field, but still too low to modify the superconducting properties of the grains, we expect to induce a sufficiently high vortex density to give $B \simeq \mu H$ and a flux

$$\Phi_s = \pi R^2\mu\,H \qquad H \to \infty \qquad \text{(II.15)}$$

When one comes to actual numbers, the difficulties of an experimental determination of $\lambda$ become clear. The electrodynamics described by Eq.(II.12) is the same as for type II superconductors, and this analogy predicts a field of first vortex penetration in an infinite sample $H_{c1} = \Phi_0/4\pi\lambda^2 \simeq 10^{-6} - 10^{-5}$ Oe. Fortunately enough, it also predicts a surface barrier field $H_s$ opposing the entrance of new vortices in a finite sample, of the order of the thermodynamic critical field. The latter can be estimated at T = OK by equating the corresponding magnetic energy density and the coupling energy density $k_B T_c/a^3$, which gives $H_s \simeq 10^{-3} - 10^{-2}$ Oe.

A classical method for measuring $\lambda$ in superconductors uses the voltage induced by a small ac field on a coil containing the sample to infer the induced flux and gives $\lambda$ from a solution of Eq.(II.14) with $\mu = 1$. The superposition of a dc field eventually allows us to reach the normal state described by Eq.(II.15). This differential measurement proved to be unreliable in our case, the samples showing small but significant hysteresis due to vortex pinning. Taking into account the above estimates, we applied instead a purely ac field of varying amplitude Hm. The idea is that for $H_m < H_s$ Eq.(II.14) should apply, while for $H_m \gg H_s$ vortices would be swept in and out of the sample, obeying Eq.(II.15) over most of the ac cycle. Let V be the voltage output of the pick-up coil, proportional to $H_m$. Then the quantity $v = V/H_m$ is proportional to the average permeability and should be independent of the applied field in the regions where Eqs.(II.14) and (II.15) apply, with values $v_c$ and $v_s$ respectively. This behavior is clearly shown in Fig.2(a). Since $v_s$ measures the diamagnetic response of the grains alone, it should be practically constant in the range of temperatures studied, which is confirmed by experiment. The penetration depth can then be obtained by solving the trascendental equation

$$\frac{\lambda}{R}\,\frac{I_1(R/\lambda)}{I_0(R/\lambda)} = \frac{1}{2}\,\frac{v_c}{v_s} \qquad \text{(II.16)}$$

Results for two different samples are shown in Fig.2(b), together with fits to the law $\lambda(T) \simeq \lambda(0)\ t^{-0.7}$. The values obtained for $\lambda(0)$ are in good order-of-magnitude agreement with the estimate $\lambda(0) \simeq$ 1mm. The exponent $\beta = 0.7 \pm 0.1$ is about twice as large as the *ordered* X-Y model prediction (I.1), thus confirming the trend observed in section II.1.

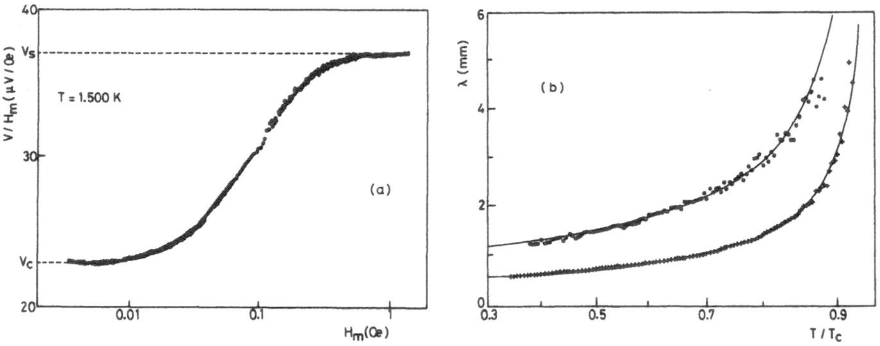


Fig.2 (a) Response of a coil wound around a cylindrical sample to an ac field of amplitude $H_m$. (b) Measured penetration depths of two samples as a function of temperature. Solid lines are fits by the power law $\lambda(T) = \lambda(0)t^{-0.7}$. From Ref. 13.

## II.3 - Susceptibility exponent

In principle, scaling relations allow to obtain all the exponents when only two of them are known. Measuring more than two provides an interesting check on experimental consistency and increased accuracy.

An interesting quantity to measure is the order parameter susceptibility (*not* the magnetic susceptibility). However, a major difficulty here is that the order parameter of the coherence transition, like the superconducting order parameter, does not couple to any external field. A way out of this difficulty has been suggested by Scalapino[14] who showed that the supercurrent across a junction between two superconducting thin films with critical temperatures $T_c$ and $T_c'$, at temperature T, $T_c \lesssim T << T_c'$, is proportional to the superconducting order parameter susceptibility of the film. More precisely, linear response theory gives the supercurrent with junction voltage V and applied field H

$$I_1(V,H) = A|C'|^2 \,\mathrm{Im}\,\chi(\omega,q), \qquad \text{(II.17)}$$

where A is a geometry-dependent constant, C' is a coupling constant proportional to $\Delta'(T)$ the order parameter in the high critical temperature (primed) side and inversely proportional to $R_{JN}$, the normal resistance of the junction and $\omega = (2\pi/\Phi_0)V$, $q = (2\pi/\Phi_0)H(\lambda_s'+l/2)$, with $\lambda_s'$ the penetration depth of the primed side and $l$ the thickness of the unprimed side. Finally $\chi(\omega,q)$ is the Fourier transform of the response function

$$\chi(\vec{r},t) = -\, i\langle[\Delta(\vec{r},t),\, \Delta^+(0,0)]\rangle\, \theta(t)/\hbar\ , \qquad \text{(II.18)}$$

where $\theta(t)$ is the unit step function (of time).

Eqs.(II.17) and (II.18) have been established for a superconductor-normal junction above $T_c$. Will they still be valid for a junction between a superconductor and a WCGS above the coherence temperature ? The physical phenomena are the same[15], but in the paracoherent region the grains are fully superconducting, so the quantum operators $\Delta^+$, $\Delta$ in Eq.(II.18) become classical vectors with fluctuating phases and non-fluctuating amplitudes. The preceding derivation should still be valid using classical linear response theory and thermal averages with the tunneling Hamiltonian (I.6) in the classical limit. The latter imply Boltzmann factors of the type $\exp(J_{\alpha\alpha'} \cos\phi_{\alpha\alpha'}/k_BT)$. But since the $J_{\alpha\alpha'}$'s are themselves funtions of temperature according to (I.2), the behavior of the system is that of one with temperature-independent coupling energies $J_{\alpha\alpha'}(T=0)$ and inverse temperature

$$\theta = \frac{1}{T}\frac{\Delta(T)}{\Delta(0)} \tanh \frac{\Delta(T)}{2k_BT} \qquad \text{(II.19)}$$

This is important in analysing data in the paracoherent region : a few degrees between $T_c$ and $T_{cG}$ correspond to the interval $(\theta_c^{-1},\infty)$, with $\theta_c = \theta(T_c)$.

We measured[16] the susceptibility exponent $\gamma$ with a rather unusual Josephson junction configuration, shown in Fig.3(a). Two 25 μm thick niobium disks are located in recesses on both ends of cylinders made of the grain-epoxy mixture. Concentric with the disks, copper washers allow voltage measurements either along the cylinder, or with respect to the closest Nb foil. Actually, the junction voltage we need is that across the oxide barrier between the Nb electrode and the nearest layer of grains. The Cu washers measure this through granular material which is itself resistive. The recess geometry chosen insures that when current flows along the cylinder from one disk to the other, relatively few current lines will exist in the region between adjoining Nb and Cu electrodes and therefore the latter will be very nearly at the same potential as the grains in contact with the disk.

In the limit of low voltage $\hbar\omega \ll k_BT$ and H = 0 the supplementary conductance due to fluctuations in the X-Y order parameter is, from (II.17) and (II.18),

$$\lim_{V\to 0} \frac{I_1}{V} = \frac{1}{R_J} - \frac{1}{R_{Jq}} = \frac{e}{\hbar}\frac{|C'|^2}{k_BT} A \; G_\Delta(0,0) \qquad \text{(II.20)}$$

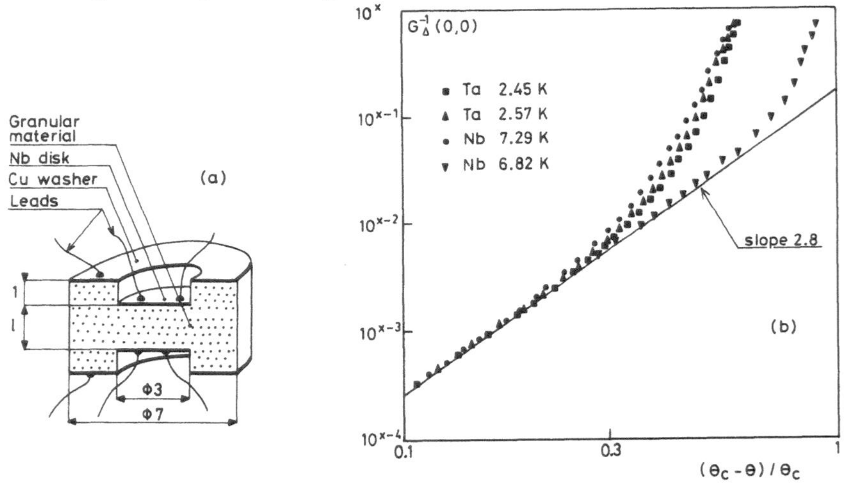


Fig.3 (a)Sample geometry to measure the susceptibility exponent. Josephson junctions are formed between the Nb disks and the granular material. Dimensions are in millimeters; the thickness $l$ can be adjusted to different values. (b) Log-log plot of the inverse of the Fourier transform of the order parameter-order parameter correlation function vs reduced termperature. Characteristic parameters of the samples are given in Table 1. The exponent $x$ in the ordinates means that the curves have been displaced vertically to make them superpose. From Ref.16.

where $G_\Delta(\omega,q)$ is the Fourier transform of the order parameter-order parameter correlation function, $R_J(T)$ the dynamic resistance of the junction at zero bias and $R_{Jq}$ its quasiparticle resistance. Thus $G_\Delta(0,0)$ can be obtained from experimentally accessible quantities and from the values of $\Delta'(T)$.

Experimental results are shown in Fig.3(b) for four different samples, whose main properties are given in Table 1.

Table 1 : Characteristic parameters of samples in Fig.3

| Symbol | Metal | Normal Resistivity (mΩ.cm) | $R_{JN}$(Ω) | $T_c$(K) | $l$(mm) | Nominal Grain Size (μm) |
|---|---|---|---|---|---|---|
| ■ | Ta | 146.0 | 5.05 | 2.45 | 6.9 | 45 |
| ▲ | Ta | 17.5 | 1.50 | 2.57 | 2.5 | 45 |
| ● | Nb | 5.0 | 0.098 | 7.29 | 2.0 | 10 |
| ▼ | Nb | 243 | 0.078 | 6.82 | 3.2 | 45 |

The data of Fig.3 is the first *direct* evidence of the transition to coherence of an array of superconducting grains at a temperature well below $T_{cG}$. Indeed, the junction is resistive above $T_c$ in spite of the grains being individually superconducting. On the other hand, the zero voltage currents observed to flow below $T_c$ cannot exist unless phase coherence is established on both sides of the junction. The granular side must then have undergone a transition to coherence which is distinct from the grains' superconductive transition at $T_{cG}$.

Linear response theory assumes that the junction Hamiltonian is a perturbation, i.e. that the Josephson coupling between the Nb foil and a neighboring grain is much smaller than the grain-grain coupling. Experimentally, this implies that the junction critical current at low temperatures must be much smaller than the critical current of the granular material itself, which is the case of all our samples. On the other hand, too weak a coupling in the junction may result in a finite voltage at $T \lesssim T_c$ due to quasiparticle noise[17]. The consequence is that zero resistance appears in the junction at a temperature slightly below the $T_c$ inferred from the resistive transition of the WCGS. We find experimentally, however, that

$$G_\Delta^{-1}(0,0) \sim \frac{\Delta'^2(T) R_{Jq} R_J}{T(R_{Jq} - R_J)} \sim \left(\frac{\theta_c - \theta}{\theta_c}\right)^\gamma \qquad \text{(II.21)}$$

is fitted by the same value of $\gamma$ for samples where the difference between these two temperatures took values going from 0.05K to 0.5K.

Inspection of Fig.3 and Table 1 shows that samples of different granular metals, and differing widely in normal resistivity, length $l$, grain size and coherence temperature give responses compatible with a single susceptibility exponent $\gamma = 2.8 \pm 0.3$ near $T_c$. Again, this is much larger than the X-Y model result $\gamma \simeq 1.33$ (Eq.(I.8)).

II.4 - Discussion of exponent values

From the critical current exponent $2\beta + \nu = 2.5 \pm 0.3$ and $\beta = 0.7 \pm 0.1$ we can obtain $\nu = 1.1 \pm 0.3$. Hyperscaling gives us a more accurate value : $2\beta + \gamma = d\nu$, $2\beta + \gamma + \nu = (d + 1)\nu = 5.3 \pm 0.4$ which results in $\nu = 1.3 \pm 0.1$ for $d = 3$, within experimental error of the previous value. The specific heat exponent $\alpha = -2.2 \pm 0.3$ comes from the scaling relation $2 - \alpha = 2\beta + \gamma$. Furthermore $\gamma/\nu = 2 - \eta$ and $\delta = 1 + \gamma/\beta$ give $\eta = 0.1 \pm 0.3$ and $\delta = 5.0 \pm 0.3$. These last two values are in good agreement with the X-Y model predictions, Eqs.(I.10) and (I.11), and the ratios $\tilde{\beta} = \beta/\nu$ and $\tilde{\gamma} = \gamma/\nu$ are also those of the X-Y model. Actually, this is the content of the hypothesis of weak universality[18], according to which the above ratios, together with the exponents $\delta$ and $\eta$, belong to a stronger class of universality than the other exponents. We shall see in the next section that the assumption of an inhomogeneous transition in disordered WCGS makes weak universality appear as a consequence of the usual hypothesis of universality.

## III - POSSIBLE NONUNIFORM TRANSITIONS IN FERROMAGNETIC GLASSES

We already mentioned in the introduction that disordered WCGS were a particular example of ferromagnetic glasses, and suggested that disorder could provide a clue to the change in the values of the exponents compared to translationally invariant systems. Among the various types of disorder touched upon in section I, we shall dwell particularly on effects resulting from the distribution of coupling energies between grains, or, in the general case of ferromagnetic glasses, between sites.

Our starting point is the well-known fact that in all phase transitions above the critical temperature, thermodynamic fluctuations bring about the formation of clusters with an out-of-equilibrium value (different from zero) of the order parameter. In the usual homogeneous transition of an ordered system such clusters have no preferred nucleation sites. On the contrary, we may expect that in an extremely disordered medium the clusters will be preferentially (inhomogeneously) nucleated in regions of higher-than-average coupling energies. Let us go a step further and *assume* that critical properties at $T > T_c$ are determined by "typical" clusters of linear size $L \simeq \xi(T)$ and coupling energies $J > J^*(T)$. The precise meaning of $J$ is unessential here. We may for example take the coupling energy of a single bond, $J = J_{\alpha\alpha'}$. Or $J = J_\alpha = \sum_{\alpha'(\neq\alpha)} J_{\alpha\alpha'}$, the coupling energy of grain $\alpha$ to its neighbors $\alpha'$. And then restrict the sum to grains $\alpha'$ such that $J_{\alpha'} > J^*(T)$, etc. The main point, as we shall see below, is that all these definitions determine a percolation problem with parameter $p = \text{Prob}(J > J^*)$ and *percolation* correlation length $\xi_p \sim \rho^{-\nu_P}$ where $\rho = (p_c-p)/p_c$ with $p_c$ the percolation threshold. In certain respects, this approach is similar to that of G. Deutscher et al[19], who considered the physically appealing assumption $J^*(T) \simeq k_B T$. We discuss instead the scaling properties of clusters and of their numbers, which make the actual determination of $J^*(T)$ superfluous. So let s be the number of elements (sites or bonds, depending on the definition of J) in a cluster and $G_s(T)$ its *total* free energy. Then s is a function of L and the condition for a cluster to be "typical" with $L \simeq \xi(T)$ is that it should have maximum probability ($\propto \exp(-G_s/k_B T)$) of being formed when $s = s_\xi = s(L \simeq \xi)$ as compared with other values of s. We assume at the same time that there is a $J^*(T)$ (and therefore a $p(J^*)$) such that clusters with $J > J^*$ are thermodynamically favored irrespective of their size, at least for a range of L's.

III.1 - Percolation clusters and the thermal transition

Consider a system with a distribution $m(J)$ of coupling energies. Then

$$p(J^*) = \text{Prob}\,(J > J^*) = \int_{J^*(T)}^{\infty} m(J)dJ \qquad \text{(III.1)}$$

defines a percolation problem with threshold $p_c = p(J_c)$. Now, large percolation clusters[20,21] have certain well-known properties at $p \simeq p_c$ of which two will be used here : (i) $L = s^{1/D} R(z)$, where $z = s^{1/D\nu_P} (p - p_c)/p_c \simeq \pm (L/\xi_p)^{1/\nu_P}$ is a scaling variable, $R(z)$ is regular around $z = 0$ and[22] $D = d - \beta_p/\nu_p$ is a fractal dimensionality[10]. The exponent $\beta_p$ characterises the probability that a site or bond belongs to the infinite cluster

$$P(p) \propto (p - p_c)^{\beta_P} \qquad \text{(III.2)}$$

The name of fractal dimensionality is justified because the relation $s \sim L^D$ at $p = p_c$ for a cluster is similar to that between number of sites and linear size in d dimensions, as in Eq. (I.12), with D replacing d. (ii) The number of clusters $n_s$ of s elements (per lattice point) is

$$n_s \sim s^{-1-d/D} f(z), \qquad \sum_s n_s\, s = p \qquad \text{(III.3)}$$

where $f(0) = 1$ and $f(z)$ has a maximum at $z = z_m \simeq -1$ ($z_{\bar{m}} \simeq -0.8$ for $d = 3$ (Ref.20)). The second equation (III.3) gives the normalization of $n_s$ per lattice point at $p < p_c$.

These properties[23] enable us to obtain the temperature dependence of $\rho = (p_c - p)/p_c$ for $T > Tc$. Consider equilibrium fluctuations giving thermally excited clusters of given s and p or equivalently, of given L and $\xi_p$. Their contribution to the free energy per lattice point is, from Eq.(III.3)

$$F_s = - k_B T\, n_s < \ln Z_s >_c \sim s^{-1-d/D} f(z) G_s \tag{III.4}$$

where $Z_s$ is the partition function of s sites forming the cluster and $<\dots>_c$ indicates an average over all cluster configurations. The total free energy of the cluster, $G_s$, depends on three lengths : L, $\xi_p$ and $\xi(T)$. The free energy per unit volume g of the cluster should have the scaling property

$$g(L,\xi_p,\xi) = L^{-d} G_s = b^{-d} g\left(\frac{L}{b}\, , \frac{b}{\xi_p}\, , \frac{b}{\xi}\right) = L^{-d} G_s\left(z\, , \frac{L}{\xi}\right)\, , \tag{III.5}$$

where the last equality is obtained by making b = L. Then $F_s$ has a minimum as a function of p (or $\xi_p$), which is given by the condition, obtainable from (III.4) and (III.5),

$$\frac{\partial \ln G_s}{\partial z} = - \frac{d \ln f}{dz} \tag{III.6}$$

If this is to be true for a range of L's and therefore of z's with a single value of $\xi_p$, we must have

$$G_s = f^{-1}(z)\; g_s(L/\xi) \tag{III.7}$$

where $g_s(L/\xi)$ is an unknown function. But $G_s$ must be a minimum for $L \simeq \xi(T)$ while from property (ii) above $f^{-1}(z)$ is a minimum when $L \simeq \xi_p$ . Then $G_s$ is a minimum in all its variables when $L \simeq \xi_p \simeq \xi(T)$, where the last equality implies

$$\frac{p_c - p}{p_c} = \rho \sim t^{\nu/\nu_p} = \left(\frac{T - T_c}{T_c}\right)^{\nu/\nu_p} \tag{III.8}$$

for typical or dominant clusters. Equation (III.8) has the important consequence that in spite of the fact that two transitions, thermal and percolative, proceed simultaneously, there is always a single characteristic length in the system. Another consequence of (III.4) and (III.7) is that the singular part of the free energy of the whole system of clusters per lattice point

$$F = \sum_s n_s G_s \sim \sum_s s^{-1-d/D} g_s \sim \xi^{-d} \tag{III.9}$$

as expected for a phase transition. Nonanalytic parts of sums like (III.9) can be obtained by transforming them into integrals, or, more easily by the following rule of thumb[20] : replace s by the typical value $s_\xi \sim \xi^D$ (or L by $\xi$) everywhere and multiply by a factor $s_\xi$ corresponding to the number of such terms.

## III.2 - Fractal hyperscaling and weak universality

Roughly speaking, universality asserts that systems interacting through a short range Hamiltonian will all have critical exponents which depend only on the space dimensionality d and on the number of components n of the order parameter (n = 2 for the X-Y model) irrespective of the details of the interaction. However, we have already seen that WCGS fail to satisfy universality. The fact that percolation clusters have, for $p \lesssim p_c$, a fractal dimensionality $D < d$, raises the question of how the values of the exponents of an inhomogeneous transition will be affected by this new dimensionality.

Consider an extensive quantity Q, measured *per lattice point* in an infinite system, and having a singularity of the form $Q \sim t^{-\phi}$ for $T \to T_c^+$. The same quantity in a finite system of linear size L, $Q_L$, will satisfy finite size scaling[24] (FSS)

$$Q_L = L^{\tilde{\phi}} F_Q (L^{1/\nu} t) \tag{III.10}$$

where $\tilde{\phi} = \phi/\nu$ and $F_Q$ is a scaling function, regular about $T_c$. Actually, Eqs.(I.12) and (III.5) are examples of FSS. If the finite system is a cluster (or more precisely $Q_L$ is the thermal and configurational average of Q in clusters of size L), it is interesting to study the behavior of $Q_s = L^d Q_L/s$, i.e., the same physical quantity measured per element belonging to the cluster. From property (i) in the preceding section,

$$Q_s = L^{d+\tilde{\phi}-D} F'_Q (L^{1/\nu} t) = L^{\tilde{\phi}_c} F'_Q (L^{1/\nu} t) \tag{III.11}$$

with $F'_Q(x)$ a new scaling function and

$$\tilde{\phi}_c = \frac{\phi_c}{\nu_c} = \frac{\phi}{\nu} + \frac{\beta_P}{\nu_P} \tag{III.12}$$

We call the exponents $\beta_c$, $\gamma_c$, $\nu_c$ ... defined by Eq.(III.12) "cluster exponents" or c-exponents, because they describe the critical behavior of clusters leading to the critical properties of the whole system. In fact Q is recovered by applying the above mentioned rule of thumb to the sum

$$Q = \sum_s n_s \, s \, Q_s = \sum_s n_s \, s^{1+\tilde{\phi}_c/D} F'_Q \sim \xi^{-d+D+\tilde{\phi}_c} \sim \xi^{\tilde{\phi}} \tag{III.13}$$

Cluster exponents have a very important property : they satisfy hyperscaling in D-dimensional space. Taking into account Eq.(III.12) and using the hyperscaling relations in d dimensions of $\beta$, $\gamma$, $\nu$ ... and $\beta_P$, $\gamma_P$, $\nu_P$ ... one obtains

$$D\nu_c = 2 - \alpha_c = \beta_c(\delta_c + 1) = 2\beta_c + \gamma_c \; ;$$

$$(D - 2 + \eta_c)\nu_c = 2\beta_c \; ; \quad \delta_c = (D + 2 - \eta_c)/(D - 2 + \eta_c) \tag{III.14}$$

In other words, $\beta_c$, $\gamma_c$, $\nu_c$ ... are just the exponents of a phase transition in *D dimensions*. It is interesting to point out that crossover exponents for dilute Ising models in random fields near the percolation threshold[25], also satisfy fractal hyperscaling of the type (III.14).

Another consequence[26] of FSS is that a characteristic fractal dimensionality can be ascribed to every extensive quantity, reflecting the nonlinear coupling of its fluctuations near the critical point. In effect, the total value of Q in a region of size $L^d Q_L \sim L^{d+\phi} \sim L^{D+\phi_c}$ defines

$$D_Q = d + \phi/\nu = D + \phi_c/\nu_c \qquad \text{(III.15)}$$

In particular, the order parameter fractal dimensionality $D_\psi = d - \beta = D - \beta_c < D$. In other words, percolation clusters can contain thermal clusters even if, according to weak universality, $\beta/\nu$ has the same value as in the homogeneous transition.

What about intensive quantities ? The most conspicuous one is the correlation length. It would seem at first sight that in addition to $\xi \sim t^{-\nu}$ whe have $\xi_c \sim t^{-\nu_c}$, the correlation length which would be found in a very large D-dimensional cluster. However, the correlation length in a region with $L < \xi$, $\xi_L$, should be the same when derived from one or the other expression by application of FSS. This gives

$$\xi_L(T) = L\, F_\xi\, (L^{1/\nu}t) = L^{\nu_c/\nu}\, F_{\xi_c}\, (L^{1/\nu}t) \qquad \text{(III.16)}$$

which can be satisfied if $\nu = \nu_c$ and the scaling functions $F_\xi$ and $F_{\xi_c}$ are the same. This is equivalent to saying trivially that the dimensionality of the correlation length is always unity, irrespective of the imbedding space.

A number of important conclusions can be drawn from the preceding results :

a) The exponent $\nu$ for the inhomogeneous transition is, from Eqs. (III.14) and (III.16), that of a phase transition in D dimensions, while the homogeneous transition has $\nu_h = \nu(d) < \nu(D)$ because[26] $D < d$. Disorder can only make the exponents $\beta$, $\gamma$, $\nu$, to *increase* compared to the ordered case.

b) In six dimensions and above $\beta_p = 1$, $\nu_p = 1/2$, while in four dimensions and above $\beta_h = 1/2$, $\nu_h = 1/2$, the mean-field exponents. We remark that D (d = 6) = 4. This implies that the critical dimensionality above which mean-field exponents apply in a ferromagnetic glass is $d_c = 6$ rather than $d_c = 4$.

c) Universality does *not* require that the ferromagnetic glass exponents be those characteristic of d dimensions, but rather that the *c-exponents* of Eqs.(III.14) be equal to the exponents of a *homogeneous* transition in D dimensions. In this way weak universality appears as a consequence of, rather than in contradiction with universality.

d) The above suggests an immediate numerical check of our calculations: c-exponents should be obtainable from well-known renormalisation group formulas[27] in D dimensions and then thermal exponents should follow from Eq.(III.13). Unfortunately, both the $\varepsilon$-expansion with $\beta_p/\nu_p = 0.48$ (Ref.21) and $\varepsilon = 4 - D = 1.48$ and the $1/n$ expansion with $n = 2$ are hardly expected to work, the former because $\varepsilon$ is too large and the latter because n is too small. However, it is rewarding to see that the $1/n$ expansion gives $\nu = \nu_c = 1.3 \pm 0.3$, where the error is estimated from the spread in numerical values of $\beta_p/\nu_p$ found in Ref.21. This strikingly good (and perhaps coincidental) agreement with the experimental value results in equally good predictions for the other exponents discussed in section II.4 : one obtains from scaling $\beta = 0.65$, $\alpha = -2.0$, $\gamma = 2.7$, etc.

## III.3 - Temperature range of inhomogeneous transitions and stability

Using FSS we showed in section III.1 that clusters had maximum probability of being formed in a fraction p of the system. Clusters formed elsewhere correspond to fluctuations in p away from the equilibrium value given by Eq. (III.8). A further question is whether the equilibrium value of p is stable against such fluctuations. This point is discussed in a recent publication[23], and we shall not repeat the arguments here. Suffice it to say that the percolation transition is known to be stable[25] against fluctuations in p, but that these fluctuations introduce correlations[28] in local critical temperatures $T_c(\vec{r})$ which become effectively long range at the critical point. This is understandable, because $T_c(\vec{r})$ depends on the coupling energy at point $\vec{r}$ and the inhomogeneous transition correlates nucleation sites with the highest possible coupling energy.

The constraining dimensionality D is the same at $T \lesssim T_c$ as at $T \gtrsim T_c$. In fact, the dominant cluster below $T_c$ ($p \gtrsim p_c$) is the infinite percolation cluster. The number of sites belonging to this cluster, among the $\xi^d$ lattice points in a region of linear size $\xi$, goes as $\xi^d p$, or, from Eq.(III.2), as $\xi^{d - \tilde{\beta}_p}$. Therefore critical exponents are the same above and below $T_c$. But away from the critical percolation region clusters are known[20] to become more compact, reaching the dimensionality d. A crossover to a different behavior, probably to a homogeneous transition with short range correlations in local critical temperatures[28], should be expected.

At temperatures well above $T_c$ we also expect a crossover to a homogeneous transition. This is simply because any real system will have a maximum coupling energy, $J_m$, and will start sensing its own inhomogeneities when $k_B T \simeq J_m$. The temperature interval over which a nonniform transition will be observable is then $\Delta T \simeq (J_m - J_c)/k_B$. In three dimensions $p_c$ is about[20] 0.2 - 0.3 depending on the lattice and only the tail of the distribution m(J) contributes to the transition when $p < p_c$. A straight-line approximation to this tail gives

$$p_c = \frac{1}{2} m_c (J_m - J_c), \tag{III.17}$$

where $m_c = m(J_c)$. Therefore

$$\Delta T \simeq 2 \, p_c / m_c k_B \tag{III.18}$$

depends on the details of the distribution, particularly above $J_c$.

The above sheds some light on a rather puzzling phenomenon, appearing in Table 1 : large variations of normal resistivity $\rho_n$ may result in comparatively small changes in critical temperature. Actually, the normal resistivity depends on the backbone of the same percolating cluster that defines $T_c$ *and* on the rest of the distribution, branched in parallel with this cluster. Loosely speaking, $\rho_n$ is affected by the peak of the distribution m(J) while $T_c$ is determined *only* by the tail of this destribution.

Finally let us discuss briefly possible effects of a grain size distribution between a minimum and a maximum value, $a_{min} < a < a_{max}$. Figure 3(b) shows that in the high temperature region the slope of $G^{-1}(0,0)$ is higher than when $T \to T_c^+$. Assuming a crossover between two transitions, one sees that the high temperature one has $\gamma' > \gamma$. This precludes the possibility of mean-field behavior as is usually expected at high temperatures, because this would give $\gamma' = 1 < \gamma$. The probability that samples behave effectively as two-dimensional when $\xi(T) > l$ (see Fig.3(a)) and as three-dimensional otherwise must also be dismissed because[26] in such a case, $\gamma' = \gamma(d = 3)$ would be smaller than $\gamma(d = 2)$.

In a granular system small grains will be mainly clustered in regions left empty by larger gains. These regions will have on the average the highest coupling energy *density* and therefore will nucleate thermal clusters at high temperatures. Consider the situation where $a_{min} < \xi(T) < a_{max}$. Then an equation like (I.12) with L replaced by $\xi$ is no longer valid. *If* the number of grains in a volume $\xi^d$ can still be represented by a power law

$$N(\xi) \sim \xi^{d'} , \tag{III.19}$$

the exponent $d' < d$ because as $\xi$ increases grains of increasing size have to be included and their number cannot increase as fast as $\xi^d$. Furthermore, if we *assume* that (III.19) is the right definition of dimensionality for the thermal transition, $\gamma' = \gamma(d') > \gamma(d)$. These assumptions allow a definite prediction to be made : the exponent $\gamma(d)$ will be recovered when

$$\xi(T) = \xi_o \left(\frac{\theta_c - \theta}{\theta_c}\right)^{-\nu} \simeq a_{max} \tag{III.20}$$

We expect $\xi_o \simeq a_{min}$. The kink in the curves of Fig.3(b) appears at

$(\theta_c - \theta)/\theta_c \simeq 0.3$ for three samples and at 0.5 for the fourth (which happened to be supplied by a different firm),which gives $a_{max}/a_{min} \simeq 5$ and 2.5, respectively, with $\nu = 1.3$. These are reasonable values for the spread in grain size of our samples.

## IV - CONCLUDING REMARKS

Our experimental results show that the transition to coherence of WCGS is described by abnormally large critical exponents. The hypothesis of a nonuniform transition provides a possible explanation for this behavior, but needs further support from microscopic theory. A confirmation of this idea would mean that disordered WCGS provide the first example of a real transition of fractal dimensionality. It would raise at the same time a number of questions. For example, it is customary to divide disorder into two types : quenched and annealed. In our case the dividing line is somewhat blurred : although bond randomness is temperature independent, the strength of bonds nucleating thermally excited clusters is an equilibrium temperature-dependent quantity.

Further investigation is also needed on the effects of correlations in local critical temperatures[28]. As we mentioned briefly, a nonuniform transition is characterised by intermediate-range correlations in local critical temperatures directly related to fluctuations of the percolation parameter about its equilibrium value. On the other hand these correlations are short-ranged in the homogeneous transition of a disordered system. Possible crossovers between these two transitions well above and well below the critical temperature and their eventual relation with a change of dimensionality of percolation clusters are still open questions.

Finally, a description of V(I) characteristics and more generally dynamic critical phenomena, which have been excluded from the present study, require further development of the model of inhomogeneous transitions.

Acknowledgments

The use of "we" in this paper goes well beyond a clause of literary style. It is also a recognition of the fact that it describes results of collective work done with Dr. Paulette Peyral and Dr. Alain Raboutou, in an atmosphere of pleasant comradeship.

## References

1. J. Rosenblatt, H. Cortès and P. Pellan, Evidence of a Phase Transition in an Assembly of Point Contacts, Phys. Lett. A33:143(1970).

2. J. Rosenblatt, A. Raboutou and P. Pellan, X-Y Model Description of a Granular Superconductor, in "Low Temperature Physics LT14, M. Krusius and M. Vuorio, ed., New York (1975).
3. P. Pellan, G. Dousselin, H. Cortès and J. Rosenblatt, Phase Coherence and Noise Resistivity in Weakly Connected Granular Superconductors , Sol. St. Comm. 11:427(1972).
4. J. Rosenblatt, Coherent and Paracoherent States in Josephson Coupled Granular Superconductors, Rev. Phys. Appl. 9:217(1974).
5. V. Ambegaokar and A. Baratoff, Tunneling Between Superconductors, Phys. Rev. Lett. 10:486(1963).
6. R.H. Parmenter, Characteristic Parameters of a Granular Superconductor, Phys. Rev. 167:387(1968).
7. P.N. Anderson, The Random Phase Approximation in the Theory of Superconductivity, Phys. Rev. 112:1900(1958).
8. P.R. Wallace and M.J. Stavn, Quasi-Spin Treatment of Josephson Tunneling Between Superconductors, Can. J. Phys. 43:411(1965).
9. D.D. Betts, in "Phase Transitions and Critical Phenomena", C. Domb and M.S. Green, ed., Academic Press, New York (1974).
10. B.B. Mandelbrot, "The Fractal Geomety of Nature", Freeman ed., San Francisco (1982).
11. A. Raboutou, P. Peyral and J. Rosenblatt, Critical Currents and Penetration Depth of a Randomly Oriented Three-Dimensional Assembly of Point Contacts , in "Inhomogeneous Superconductors", D.U. Gubser, T.L. Francavilla, J.R. Leibowitz and S.A. Wolf, ed., American Institute of Physics, New York (1980).
12. H.E. Stanley, "Introduction to Phase Transitions and Critical Phenomena", Clarendon Press, Oxford (1971).
13. A. Raboutou, J. Rosenblatt and P. Peyral, Coherence and Disorder in Arrays of Point Contacts, Phys. Rev. Lett. 45:1035(1980).
14. D.J. Scalapino, Pair Tunneling as a Probe of Fluctuations in Superconductors, Phys. Rev. Lett. 24:1052(1970).
15. D.R. Tilley, Private Communication.
16. J. Rosenblatt, P. Peyral and A.Raboutou, Susceptibility Exponent of Disordered Bulk Granular Superconductors, Phys. Lett. 98A:463(1983).
17. V. Ambegaokar and B.I. Halperin, Voltage due to Thermal Noise in the dc Josephson Effect, Phys. Rev. Lett. 22:1364(1969).
18. M. Suzuki, New Universality of Critical Exponents, Prog. Theor. Phys. 51:1992(1974).
19. G. Deutscher, O. Entin-Wohlman, S. Fishman and Y. Shapira, Percolation Description of Granular Superconductors, Phys. Rev.B 21:5041(1980) and G. Deutscher's Lecture in this A.S.I..
20. D. Stauffer, Scaling Theory of Percolation Clusters, Phys. Rep. 54:1(1974). The exponent $\tau$ in Stauffer's paper is equivalently written here as $1 + d/D$.

21. J.W. Essam, Percolation Theory, Rep. Prog. Phys. 43:833(1980).
22. S. Kirkpatrick in "Ill Condensed Matter", R. Balian, R. Maynard and G. Toulouse, ed. North-Holland, Amsterdam (1979).
23. J. Rosenblatt, Inhomogeneous Transitions in Extremely Disordered Media, Phys. Rev. B, 28:5316(1983).
24. M.E. Fisher and M.N. Barber, Scaling Theory for Finite-Size Effects in the Critical Region, Phys. Rev. Lett. 28:1516(1972).
25. Y. Gefen, A. Aharony, Y. Shapiro and A. Nihat Berker, Hyper-scaling and Crossover Exponents near the Percolation Threshold, J. Phys. C 15:L801(1982).
26. M. Suzuki, Phase Transition and Fractals, Prog. Theor. Phys. 69:65(1983).
27. S. Ka Ma, in "Modern Theory of Critical Phenomena", Benjamin,ed., Reading, MA (1976).
28. A. Weinrib and B.I. Halperin, Critical Phenomena in Systems with Long-Range-Correlated Quenched Disorder, Phys. Rev.B 27:413(1983).

PARTICIPANTS OF THE NATO ADVANCED STUDY SERIES
ON PERCOLATION, LOCALIZATION, AND SUPERCONDUCTIVITY,
held June 19 - July 1, 1983,
at Les Arcs, Savoie, France

PARTICIPANTS

D. ABRAHAM, Gordon McKay Lab, Harvard University, Cambridge, MA 02138, USA

D. ABRAHAM, Department of Physics, Tel Aviv University, Tel Aviv, 69978, Israel

V. AMBEGAOKAR, Lab of Atomic and Solid State Physics, Cornell University, Ithaca, NY 14853, USA

M. R. BEASLEY, Department of Applied Physics, Stanford University, Stanford, CA 94305, USA

E. BEN-JACOB, Institute for Theoretical Physics, University of California, Santa Barbara, CA 93106, USA

H. BERNAS, C.S.N.S.M. B$^{r}$108, B.P.n°1, 91406 Orsay, France

D. BOL, Kamerlingh Onnes Lab, Nieusteeg 18 2311SB Leiden, The Netherlands

R. M. BRADLEY, Department of Physics, Stanford University, Stanford,CA 94305, USA

A. BRAGINSKI, Westinghouse R&D Center, 401-3A15, 1310 Beaulah Rd., Pittsburgh, PA 15235, USA

Y. BRUYNSERAEDE, Physics Department, University of Leuven, Celestijnenlaan 200D B-3030 Leuven, Belgium.

M. BUETTIKER, T. J. Watson Research Center, P.O. Box 218,
Yorktown Heights, NY 10598, USA

M. BURNS, Department of Physics, University of California,
Los Angeles, CA 90024, USA

C. CAMERLINGO, Instituto di Cibernetica, CNR, Via Tioano 6,
80072 Arco Felice, Mapoli, Italy

J. CLAASSEN, Code 6800, Naval Research Laboratory, Washington,
D.C., 20375, USA

L. COFFEY, James Franck Institute, University of Chicago,
Chicago, IL 60637, USA

M. A. A. COX, Department of Engr. Math., University of
Newcastle Upon Tyne, United Kingdom

C. CRISTIANO, Instituto di Cibernetica, CNR, Via Tioano 6,
80072 Arco Felice, Napoli, Italy

R. DEBRUYN OUBOTER, Kamerlingh Onnes Lab, University of Leiden,
Nieuwsteeyid, 2311 S.B. Keudeb, The Netherlands

P. G. DE GENNES, College de France, Physique de La Matiere
Condensee, 11 Place Marcelin-Berthelot, 75221 Paris Cedex 05,
France

A. DE LOZANNE, Department of Physics, Stanford University,
Stanford, CA 94305, USA

G. DEUTSCHER, Department of Physics and Astronomy, Tel Aviv
University, Ramat Tel Aviv, Israel

S. DONIACH, Department of Applied Physics, Stanford University,
Palo Alto, CA 94305, USA

C. FALCO, Department of Physics, University of Arizona, Tucson,
AZ 85721, USA

H. J. FINK, Department of ECE, University of California, Davis,
CA 95616,USA

H. FUKUYAMA, Institute for Solid State Physics, University of
Tokyo, Roppongi, Minato-ku, Tokyo 106, Japan

J. C. GARLAND, Department of Physics, Smith Laboratory, Ohio
State Univesity, Columbus, OH 43210, USA

Y. GEFEN, Institute for Theoretical Physics, University of California, Santa Barbara, CA 93106

M. GIJS, Department of Physics, University of Leuven, Celestijnenlaan 200D, B03030 Leuven, Belgium

B. GIOVANNINI, Section de Physique, 32 bd d'Yvoy, 1211 Geneve 4, Switzerland

F. GOLDIE, Solid State Group, Blackett Lab, Imperial College of Science and Technology, London, SW7 2BZ, United Kingdom

A. M. GOLDMAN, University of Minnesota, School of Physics and Astronomy, 116 Church St. S.E., Minneapolis, MN 55455, USA

J. GORDON, Gordon McKay Lab, Harvard University, 9 Oxford St., Cambridge, MA 02138

J. GRAYBEAL, Ginzton Lab, Stanford University, Stanford, CA 94305, USA

D. GUBSER, Code 6630, Naval Research Lab, Washington, D.C. 20375, USA

C. Y. HUANG, Los Alamos Lab, Los Alamos, NM 87545, USA

Y. IMRY, Physics Department, Tel-Aviv University, Tel Aviv 69978, Israel

D. INGLIS, Solid State Chemistry, National Research Council of Canada, Montreal Road, Ottawa K1A OR9, Canada

A. M. KADIN, Energy Conversion Devices, 1675 W. Maple Rd., Troy, MI 48084, USA

R. LAIBOWITZ, IBM Research Center, P.O. Box 218, Yorktown Heights, NY 10598, USA

C, LEEMAN, Institut de Physique, Universite de Neuchatel, 2000 Neuchatel, Switzerland

A. LEGGETT, Department of Physics, University of Illinois, Urbana, IL 61801, USA

P. E. LINDELHOF, H. C. Ørsted Institute, Universitets-parken University of Copenhagen DK21000, Copenhagen, Denmark

P. LINDENFELD, Department of Physics, Rutgers University, New Brunswick, NJ 08903, USA

C. LOBB, Gordon McKay Lab, Harvard University, 9 Oxford St., Cambridge, MA 02138, USA

J. MALBOUISSON, Department of Physics, Imperial College, London SW7 AZ, England

P. MARTINOLI, Institut de Physique, Universite de Neuchatel CH-2000 Neuchatel, Switzerland

A. MASON, University of Minnesota, School of Physics and Astronomy, 116 Church St. S.E., Minneapolis, MN 55455, USA

S. MCALISTER, National Research Council of Canada, Ottawa K1A OR9, Canada

D. S. MCLACHLAN, Physics Department, University of Witwatersrands, Johannesburg, South Africa

W. L. MCLEAN, Serin Physics Lab, Rutgers University, New Brunswick, NJ 08903, USA

P. MINNHAGEN, Department of Theoretical Physics, Lund University, Solvegatan 14 A, Lund, Sweden

J. E. MOOIJ, Department of Applied Physics, Delft University of Technology, Delft, The Netherlands

B. MÜHLSCHLEGEL, Institut for Theoretical Physics, Cologne University, Zulpicher Str. 77, 5000 Köln 41, West Germany

J-P. NADAL, Grove de Physique du Solide, Ecole Nordal Superieure 24 rue Lhomond, 75231 Paris Cedex 05, France

R. S. NEWROCK, Department of Physics, ML-11, University of Cincinnati, Cincinnati, OH 45221, USA

R. OPPERMANN, Inst. of Theoretical Physics, University of Heidelberg, 19 Philosophenweg, 69 Heidelberg, F. R. Germany

B. ORR, School of Physics and Astronomy, University of Minnesota, 116 Church St. S.E., Minneapolis, MN 55455, USA

A. PALEVSKI, Department of Physics and Astronomy, Tel Aviv University, Ramat Aviv, Tel Aviv, Israel

B. PANNETIER, C.R.T.B.T., PB 166X, 25 Avenue des Markys, 38042 Grenoble, France

S. PERKOWITZ, Physics Department, Emory University, Atlanta, GA 30322, USA

P. PEYRAL, INSA, 20 Ae des Buttes de Coesmes BP 14A, 35043 Rennes Cedex, France

J-L. PICHARD, Comissariat A L'Energi DPH/G/SPSRM, Orme des Merisiers, 91191 Gif sur Yvette Cedex, France

B. PLACAIS, Grove de Physique du Solide, Ecole Nordal Superieure, 24 rue Lhomond 75231 Paris Cedex 05, France

M. POMERANTZ, IBM Research Center, P.O. Box 218, Yorktown Heights NY 10598, USA

D. PROBER, Department of Applied Physics, Yale University, P.O. Box 2157, New Haven, CT 06520, USA

G-A. RACINE, Institut de Physique, University de Neuchatel rue A.L., Brequet 1, CH-2000 Neuchatel, Switzerland

H. RAFFY, Laboratoire de Physique de Solide Bat 510, Universite Paris-Sud 91405, France

D. RAINER, Physikalisches Inst., Universitat D 8580 Bayreuth, Germany

S. RAMMER, Phys. Lab I, M.C. Orsted Institute, Universitetsparken 5, 2100 KBM.O, Copenhagen, Denmark

H. ROMIJN, Department of Applied Physics, Delft University of Technology, Lountsweg 1, Delft, The Netherlands

J. ROSENBLATT, INSA, 20 Av des Buttes de Coesmes BP 14A, 35043 Rennes, France

D. A. RUDMAN, Smith Lab, Ohio State University, Columbus, OH 43210, USA

P. SANTHANAM, Section of Applied Physics, Becton Center, Yale University, New Haven, CT 06520, USA

M. SCHACK, C.S.N.S.M. P.B.N°L, University of Paris-Sud Bat 108 91406 Orsay, France

A. SCHMID, Institut fur Theorie der Kondensierten Materie Physikhochhaus, Universitat Karlsruhe, Postfach 6380, 9500 Karlsruhe, 1, West Germany

G. SCHON, Institute for Theoretical Physics, University of California, Santa Barbara, CA 93106, USA

I. SCHULLER, BLDG. 223, Material Science and Technology Div., Argonne National Labs, Argonne, IL 60439, USA

R. SIMON, Code 6634, Naval Research Lab, Washington, D.C. 20375, USA

J. STANTON, Department of Physics, Cornell University, Ithaca, NY 14850, USA

T. STREIT, Dammpfad 1, 6904 Eppelheim, F.R. Germany

D. STROUD, Department of Physics, Ohio State University, Columbus, OH 43210, USA

S. TEITEL, Department of Physics, Ohio State University, Columbus, OH 43210, USA

M. TINKHAM, Physics Department, Harvard University, Cambridge, MA 02138, USA

A. TRAVERSE, C.S.N.S.M. B[r] 108, P.P. n°1 91406 Orsay, France

C. UHER, Physics Department, University of Michigan, Ann Arbor, MI 68109, USA

C. VAN HAESENDONCK, Department of Physics, University of Leuven, Celestijnenlaan 200D, B-3030 Leuven, Belgium

D. VAN HARLINGEN, Department of Physics, University of Illinois, Urbana, IL 61801, USA

C. VANNESTE, Laboratorie d'Electrophique Parc Valrose, 06034 Nice France

D. VAN VECHTEN, Code 6634, Naval Research Labs, Washington, D.C. 20375, USA

M. WHITEMORE, Memorial University of Newfoundland, St.John's, Newfoundland, Canada

E. L. WOLF, Department of Physics, Ames Lab, USDoE, Iowa State University, Ames, Iowa 50011, USA

S. A. WOLF, Code 6634, Naval Research Lab, Washington, D.C. 20375, USA

W. ZWERGER, I. Inst. fur Theor. Physik, Universitat Hamburg, Jungiusch. 3, 2000 Hamburg 36 FRG

INDEX